The Periglacial Environment

The Periglacial Environment

Third Edition

Hugh M. French
The University of Ottawa

John Wiley & Sons, Ltd

Copyright © 2007   John Wiley & Sons Ltd, The Atrium, Southern Gate, Chichester,
West Sussex PO19 8SQ, England
Telephone (+44) 1243 779777

First Editon published in 1976 by Longman Group Limited © Longman Group Limited 1976
Second Edition published by Addison Wesley Longman Limited 1996 © Addison Wesley Longman
Limited 1996

Email (for orders and customer service enquiries): cs-books@wiley.co.uk
Visit our Home Page on www.wiley.co.uk or www.wiley.com

Other Wiley Editorial Offices

John Wiley & Sons Inc., 111 River Street, Hoboken, NJ 07030, USA

Jossey-Bass, 989 Market Street, San Francisco, CA 94103-1741, USA

Wiley-VCH Verlag GmbH, Boschstr. 12, D-69469 Weinheim, Germany

John Wiley & Sons Australia Ltd, 33 Park Road, Milton, Queensland 4064, Australia

John Wiley & Sons (Asia) Pte Ltd, 2 Clementi Loop #02-01, Jin Xing Distripark, Singapore 129809

John Wiley & Sons Canada Ltd, 6045 Freemont Blvd, Mississauga, Ontario, L5R 4J3, Cananda

Wiley also publishes its books in a variety of electronic formats. Some content that appears in print
may not be available in electronic books.

Library of Congress Cataloging-in-Publication Data

French, Hugh M.
   The periglacial environment / Hugh French. – 3rd ed.
      p. cm.
   Includes bibliographical references.
   ISBN-13: 978-0-470-86588-0
   ISBN-13: 978-0-470-86589-7
   1. Frozen ground. 2. Glacial landforms. 3. Cold regions. I. Title.
   GB641.F73 2007
   551.3′84–dc22
                                                        2006022730

Anniversary Logo Design: Richard J. Pacifico

British Library Cataloguing in Publication Data

A catalogue record for this book is available from the British Library

ISBN 978-0-470-86588-0 (HB)
ISBN 978-0-470-86589-7 (PB)

Typeset in 9.5/11.5 pt Times by SNP Best-set Typesetter Ltd., Hong Kong

This book is printed on acid-free paper responsibly manufactured from sustainable forestry in which at
least two trees are planted for each one used for paper production.

Contents

Preface to First Edition

This book is intended for use by second- and third-year level geography students in universities or colleges of higher education in the United Kingdom. It is also suitable as a text for an undergraduate course on periglacial geomorphology at the honors level in Canada and the United States. On a more general level, the book may prove useful to high school teachers and other individuals interested or specializing in the physical geography of cold regions. I have assumed, however, that the reader will already possess some understanding of the physical environment, such as might be provided by a first-year physical geography or elementary geomorphology course.

In writing this book I had two aims in mind. The first was to give a realistic appraisal of the nature of the geomorphic processes and landforms in high-latitude periglacial environments. The second was to provide some guide to the recognition and interpretation of periglacial features in the now temperate regions of North America and Europe. The regional emphasis is oriented towards areas of which I have personal field experience, notably the western Canadian Arctic, central Siberia, southern England, and central Poland. Thus, the overall focus is more towards lowland, rather than alpine, periglacial conditions. Notwithstanding this comment, I have attempted to give a balanced world picture; important literature pertaining to other areas has been incorporated.

The reasons for writing this book are also twofold. First, the majority of students will never have the opportunity to experience, at first hand, high-latitude periglacial environments. However, since cold conditions prevailed over large areas of middle latitudes at several times during the last one million years, the appreciation of such conditions is essential for a balanced interpretation of these landscapes. Second, the vast northern regions of North America and Siberia are assuming an ever-increasing importance in man's quest for natural resources. Their development will be possible only if we understand the terrain and climatic conditions of these regions. For both these reasons, I hope this book will serve a useful purpose.

I have divided the book into three parts. Part 1 is a general introduction to periglacial conditions in which the extent of the periglacial domain and the variety of periglacial climates are briefly considered. Part 2 presents a systematic treatment of the various geomorphic processes operating in present-day periglacial environments. Wherever possible, I have attempted to show the relationship between process and form and to stress the multivariate nature of many landforms. The sequence of chapters is important since they are planned to be read successively. Part 3 serves only as an introduction to Pleistocene periglacial phenomena. Emphasis in this part is upon forms rather than processes and their interpretation in the light of our understanding of similar phenomena in present-day periglacial environments.

I have not attempted to be comprehensive in my treatment of the literature. By selecting information, I have attempted to give a viewpoint. Inevitably, this viewpoint is biased to reflect my own prejudices and field experience. For example, if I had worked extensively in alpine rather than high-latitude lowland environments, probably I would not have given the same emphasis to permafrost, ground ice and thermokarst as I do. However, I believe a viewpoint is necessary since my experience with students is that they require some guidance in coping with the increasing volume of literature which appears each year.

I would like to acknowledge the help and encouragement given me by a number of individuals and organizations, without which this book would not have been written. The late Professor Jan Dylik of the University of Lódź, Poland, provided me with much inspiration and encouragement in the early stages, as well as friendship and hospitality. He was instrumental in planning the organization of many of the chapters and it is to be regretted that his untimely death in 1973 did not permit him to see the final product. Professors Ron Waters and Stan Gregory of the University of Sheffield, England, were also extremely helpful in encouraging me to write this book and identifying its basic thrust. In Canada, the opportunity to work in the Arctic since 1968 has been made possible by the active support of the Geological Survey of Canada and the Polar Continental Shelf Project. Numerous individuals both in Canada and the United Kingdom have helped in many ways, by discussion, providing material, and reading some of the early draft chapters; they include R. J. E. Brown, M. J. Clark, J. G. Fyles, P. G. Johnson, D. Mottershead, A. Pissart, D. A. St-Onge, R. J. Small, and P. Worsley. To all, I extend my thanks.

Last, and most important of all, the unfailing encouragement and support of my wife, Sharon, is acknowledged with deep gratitude and affection.

Hugh M. French
Ottawa, 1976

Preface to Second Edition

When I drafted the first edition of *The Periglacial Environment* over twenty years ago, I had worked in the cold, non-glacial regions of the world for only six years, mostly in the Canadian Arctic. Having previously completed my graduate studies upon the Pleistocene periglacial phenomena found on the Chalklands of Southern England, I looked to the polar region of North America as a natural analogue for the cold-climate conditions which had largely fashioned the Chalk landscape. Now, after nearly twenty more years of field work in many of the so-called periglacial regions of the world, I am not sure that a simple analogue exists. Instead, I am impressed by the complexity and diversity of periglacial environments, both today and in the past.

The last twenty years has also seen a dramatic expansion in our understanding of the geomorphic conditions, especially permafrost, which typify these environments. A distinct process-oriented geomorphology, termed *geocryology* or *permafrost science*, has developed in North America and elsewhere, building largely but not exclusively upon Russian concepts and principles. There has also been an integration into periglacial geomorphology of modern instrumentation and technology. At the same time, rapid advances in Quaternary dating techniques mean that our understanding of Pleistocene events is now more precise.

Underlying much of our interest in the cold non-glacial regions of the world is an appreciation of the natural resources known to occur in such regions. For example, the oil and gas resources of the Western Siberian Plain, some of the largest in the world, necessitate an understanding of the tundra and taiga environments. Modern environmental protection attitudes dictate that the exploitation of these resources is undertaken in a manner which minimizes harmful impacts upon the terrain, flora, fauna, and indigenous peoples of these regions. The same is true in North America, and in many of the alpine periglacial regions of the world. Finally, the significance of the cryosphere, of which the periglacial domain is an important component, is now being examined in the context of ongoing and predicted global changes. For various reasons, it is thought that global climate changes will be first apparent and most magnified in the high latitudes. Hence, there is an urgent need for the monitoring of change in the boreal forest, tundra, and polar desert environments.

For all these reasons, a second edition of my earlier work is justified. My aim has been to incorporate the results of these new developments while at the same time not altering the overall level, scope and organization of the book. I have tried to maintain the original flavor and style; however, many chapters are new and others have been entirely rewritten. The volume also looks different since I have deliberately tried to incorporate a large number of new or different diagrams and photos. As with the first edition, my selection of material is deliberately subjective; I have not attempted a comprehensive coverage of the literature, and the book is certainly not meant to be reference text. Rather, the second edition continues to be my own personalized view of the cold non-glacial environments of the world. There is a heavy emphasis upon those areas with which I have familiarity, such as the North American and Eurasian polar and mid-latitude lowlands. The alpine, high-altitude periglacial environments of middle and low latitude are not neglected but, because there is a relative abundance of accurate up-to-date information available else-where, I feel justified in my lack of emphasis in this area. A second omission in the cover-age of this book, also apparent in the first edition, is the relative lack of examples from the southern hemisphere, especially the ice-free areas of Antarctica. Here, I plead my partial ignorance. Finally, the extensive periglacial region of the Qinghai-Xizang (Tibet) Plateau has so far not been adequately described in the western literature, yet it is the same size as the North American or Eurasian periglacial regions. Thus, I have attempted to incorporate, wherever appropriate, data from this unusual environment.

Throughout my academic career I have taught in the Departments of Geography and Geology at the University of Ottawa. This has been a fruitful and productive milieu for pursuing my periglacial interests. I have been fortunate in supervising, over the years, a number of talented and energetic graduate students and employing undergraduate field assistants, all without whom my visits to the Arctic would have been much lonelier and certainly less productive. In this regard, special mention and warm appreciation must be given to the stimulus provided by Paul Egginton, Toni Lewkowicz, David Harry, Wayne Pollard, Dana Naldrett, Lorne Bennett, Julian Murton, and Baolai Wang. They have contributed much to my understanding of the geomorphology and Quaternary geology of the cold non-glacial regions of the world. At the same time numerous colleagues, both in Canada and elsewhere, have encouraged me, or collaborated with me; these include Mike Clark, Jan Gozdzik, Cheng Guodong, Charlie Harris, Stuart Harris, Alan Heginbottom, Alfred Jahn, Johannes Karte, Vyacheslav Konishchev, Eduard Koster, J. Ross Mackay, Derek Mottershead, Troy Péwé, Albert Pissart, Anders Rapp, Nikolai Romanovskii, Mike Smith, and Link Washburn. Several colleagues, sadly no longer alive, have also influenced me: the late Roger J. E. Brown was instrumental in encouraging me to become more involved in the geotechnical aspects of permafrost, and in the administration of perma-frost science and engineering in Canada; the late Brian Rust, my longtime friend and colleague in Geology at the University of Ottawa, always provided critical support and fostered in me an appreciation of Pleistocene and Recent sedimentation. Finally, NSERC and PCSP have generously provided operating grants and Arctic logistics respectively over a twenty-five year period, and I have received close support and cooperation from many officers of the Geological Survey of Canada and from the Department of Indian and Northern Affairs, both agencies located in Ottawa.

My secretary, Pierrette Gouin, has provided outstanding service in the preparation of the text.

To all mentioned above, I owe a debt of gratitude since this second edition is as much their work as it is mine.

Hugh M. French
Ottawa, 1995

Preface to Third Edition

It is more than 30 years since I wrote the first edition of this text. In hindsight, that was a bold, possibly premature, venture that justified the *Second Edition* in 1996. Now, the last 10 years has witnessed continued advances in periglacial geomorphology, geocryology and Quaternary science. These have been combined with a growing awareness of the importance of high latitudes in the context of global climate warming. It is now widely accepted that high latitudes will be significantly impacted. Several positive feedback mechanisms will accentuate climate warming. These include progressive shrinkage of the snow and sea-ice covers and the thaw-degradation of permafrost. The latter will release increasing amounts of greenhouse gases (CO_2, CH_4) into the atmosphere. A different group of recent concerns relate to the nine million or more people, mostly in Russia and northern Eurasia, who live in the periglacial environments. Their future health and economic well being are issues that need to be addressed. Closely linked is the fact that high latitudes contain significant natural resources, notably hydrocarbons and minerals, the future exploitation and development of which is almost inevitable. For all these reasons, there is a continued need to understand the periglacial environment, and a third edition is required.

Although the overall plan of the book remains the same, all chapters have been rewritten or reorganized. However, the balance of the text remains unchanged. There is a deliberate emphasis upon cold-climate weathering, permafrost, ground ice, and thermokarst (Chapters 4–8). This is because I believe these topics lie at the heart of periglacial geomorphology and supply the solid scientific base upon which the discipline rests.

The third edition continues to be my personalized view of the landscapes of the cold, non-glacial regions of the world. As before, it does not attempt a comprehensive coverage of the literature. At the same time, I have attempted to give a balanced interpretation. The professional reader may find fault in my treatment of certain controversial topics, such as cold-climate weathering, the relative importance of azonal versus permafrost-related processes, and global climate change. The geotechnical reader will find my treatment of ground freezing, certain geomorphic processes, and the engineering aspects, to be descriptive and somewhat superficial. Likewise, techniques of investigation, either in the field, laboratory, remotely-sensed, or involving numerical simulation, are brief. On the other hand, I hope that my treatment is sufficiently provocative and in-depth as to stimulate further research.

There are some areas where I have deliberately not ventured far. As in earlier editions, I have continued to emphasize high latitude, at the expense of alpine, periglacial environments. This is because other colleagues have more in-depth experience of the latter, especially in the mid-latitudes of the world, and of mountains. Glaciers, and surface snow and ice in general, I also leave to others, yet an understanding of these topics is clearly relevant to modern periglacial geomorphology. I am also conscious of the vast amount of Quaternary knowledge now available and the sophistication of dating and other techniques. Therefore, I view my treatment of Pleistocene periglacial environments in Part III as merely an introduction to a highly specialized field that deserves separate treatment. The same must be said for Part IV. I have always been aware of the need to apply periglacial knowledge and believe that periglacial geomorphology should not be a narrowly focused academic discipline. Part IV serves as an introduction to a field that deserve separate and much fuller treatment elsewhere.

For nearly forty years, I have conducted field work in the cold, non-glacial environments of the high latitudes. Since publication of the *Second Edition*, I have undertaken additional fieldwork in Antarctica and Svalbard. I am grateful to the Italian Antarctic Program (PNRA) for providing the opportunity to work in Southern Victoria Land in the late 1990s, and to GAP Adventures for allowing me to visit the Falkland Islands, South Georgia, and many localities in the Antarctic Peninsula in 2005–2006. As regards Svalbard, I thank UNIS and Professor Ole Humlum for inviting me each spring between 1999 and 2004. In recent years, I also commenced Pleistocene periglacial investigations in the mid-latitudes of eastern North America, notably the Gaspésie Mountains of Québec, Canada, and the Pine Barrens of Southern New Jersey, USA. I have attempted to incorporate all these recent experiences into the third edition.

The *Second Edition* owed much to the research of my Arctic graduate students at the University of Ottawa. A number continue to provide me with inspiration. Equally important, in my role as Editor-in-Chief of *Permafrost and Periglacial Processes* (PPP) between 1990 and 2005, my interaction with numerous authors, reviewers and Editorial Board members has broadened significantly my understanding of periglacial geomorphology, geocryology, and Quaternary science. Finally, I have received invaluable support, guidance and encouragement from numerous good friends and colleagues, several of more than thirty years standing. The latter include the late Alfred Jahn, J. Ross Mackay, the late Troy Péwé, Albert Pissart, and Link Washburn. To all the above, I express my sincere thanks because you have all contributed, in numerous ways that you cannot recognize, to the third edition.

Finally, I thank Jill, who patiently allowed me the time to read, write, and travel in the periglacial world.

Hugh French
Ottawa, 2006

Acknowledgments

A number of individuals have generously allowed use of their photographs and other material in this edition. These are: Professor L. Eissmann (Figure 12.2), Dr J. Gozdzik (Figure 13.6), Professor O. Humlum (Figures 6.9B, 6.12), Professor V. N. Konishchev (Figure 4.12), Professor A. G. Lewkowicz (Figure 9.13), Professor J. R. Mackay (Figures 5.4, 7.3, 7.17), Professor D. N. Mottershead (Figure 13.3B), D. Nasagaloak (Figure 10.2C), Professor T. L. Péwé (dec.) (Figure 6.2), Professor M. Seppälä (Figure 6.8), Professor H. Svensson (Figure 12.5), Dr S. C. Zoltai (dec.) (Figure 6.20B), Dr R. Zurawek (Figure 13.3A).

PART I
The Periglacial Domain

1 Introduction

This chapter outlines the periglacial concept and the historical growth of periglacial geomorphology, and its interaction with geocryology, geomorphology, and Quaternary science. The core of periglacial geomorphology concerns the study of freezing processes, the associated ground ice, and related landforms. Permafrost occupies a central, but not defining, position. Periglacial geomorphology is a component of cold-region geomorphology and incorporates a mix of glacial, periglacial, and azonal processes that assume distinct characteristics in the cold non-glacial regions of the world.

Approximately one quarter of the earth's land surface currently experiences periglacial conditions. During the Pleistocene an additional one fifth was affected to greater or lesser extent.

1.1. THE PERIGLACIAL CONCEPT

The term "periglacial" was first used by the Polish geologist Walery von Lozinski in the context of the mechanical disintegration of sandstones in the Gorgany Range of the southern Carpathian Mountains, now part of central Romania (Lozinski, 1909, 1912). Subsequently, at the XI Geological Congress in Stockholm in 1910, he introduced the concept of a "periglacial zone" to describe the climatic and geomorphic conditions of areas peripheral to the Pleistocene ice sheets and glaciers. Theoretically, this was a tundra zone that extended as far south as the treeline. In the mountains, it was a zone between timberline and snowline (Figure 1.1).

Almost certainly, Lozinski was influenced by the Swedish geologist J. G. Andersson, who, a few years earlier (1906, pp. 94–97), had summarized his observations on mass-wasting on Bear Island (latitude 74°N), a cold, wet, and windswept island in the North Atlantic. It was Andersson who introduced the term "solifluction" to the scientific literature. He also described the "stone runs," or quartzite blockfields, that characterize the valley-side slopes of the equally cold and damp Falkland Islands, located in the South Atlantic (Andersson, 1906, pp. 97–101). On hearsay alone, these phenomena had already been compared to the "rubble-drift" and "head" deposits of southern England by James Geikie (1894, pp. 722–723), who attributed the latter to a cold climate "more severe than the present."

Lozinski referred to his rock-rubble accumulations as *periglacial facies*. He observed that similar deposits occurred on many of the upland massifs of central Europe. In subsequent years, coarse angular rock-rubble accumulations on upland slopes and summits were widely reported in the scientific literature. Today, they are usually referred to as blockstreams, blockfields, felsenmeer, or kurums.

Two criteria are regarded as diagnostic of periglacial environments. First, there is ground freezing and thawing. According to J. Tricart, "the periglacial morphogenetic

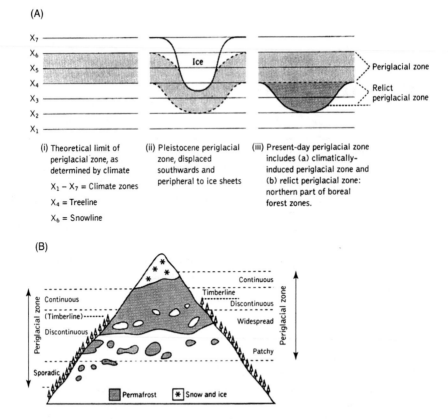

Figure 1.1. Schematic diagram illustrating limits of the periglacial zone: (A) high latitudes; (B) alpine areas.

milieu is that where the influence of freeze–thaw oscillations is dominant" (Tricart, 1968, p. 830). Second, there is the presence of perennially-frozen ground, or permafrost. According to T. L. Péwé, "permafrost is the common denominator of the periglacial environment, and is practically ubiquitous in the active periglacial zone" (Péwé, 1969, p. 4).

Periglacial geomorphology developed rapidly in the 1950–1970 period as a sub-discipline of a European-dominated climatic geomorphology (Büdel, 1963, 1977; Tricart, 1950, 1963; Tricart and Cailleux, 1967). During this period, the International Geographical Union (IGU) supported a Periglacial Commission under the leadership of Professor J. Dylik between 1952 and 1972. An international journal, *Biuletyn Peryglacjalny*, was started in Łódź, Poland, in 1954.

In hindsight, there were several weaknesses associated with the growth of periglacial geomorphology. First, there was a largely uncritical acceptance of mechanical (frost) weathering and of rapid cold-climate landscape modification. Second, the variability, duration, and severity of cold-climate conditions, both today and during the Quaternary, were not fully appreciated. Another concern was that insufficient consideration was given to the influence of lithology in controlling landforms. A final concern related to terminology. Although "cryo" terminology had been proposed earlier to reflect the cold-climate (cryogenic) processes involved (Bryan, 1946, 1949; Capello, 1959), the Pleistocene orientation of periglacial geomorphology led to acceptance of the term "periglacial" (Dylik, 1964a; Dylikowa, 1962; Hamelin, 1964). This raised criticism because the term was used to refer to both processes and areas. The apparent misuse of the term led to the suggestion that it be replaced by more specific terms such as permafrost, ground-ice, or soil-ice environment (Linton, 1969). More recently, cryological terminology has been reintroduced (ACGR, 1988) in order to accommodate a major semantic problem posed by the thermal and physical states of water. The problem appears simple. Unfrozen water can exist in soil or rock at temperatures below 0 °C due to the presence of mineral salts, pressure, or other causes. Because of this, "perennially-frozen ground" (i.e. ground at a temperature >0 °C, or "cryotic" in nature) may not, in fact, be "frozen". Thus, cryotic ground can be either "frozen" or "unfrozen" depending upon the amount of unfrozen water present. It follows that not all permafrost is necessarily frozen!

These terminological problems persist today. An obvious example is in the definition of the active layer proposed by the ACGR (1988). Traditionally, the active layer was defined as the near-surface layer above permafrost which thaws during summer (see Chapter 5). The ACGR definition would include the uppermost part of permafrost in those situations where either salinity or clay content allows permafrost to thaw and refreeze annually, even though the material remains cryotic (below 0 °C). Under this definition, it is argued that the active layer cannot be determined consistently in time and space (Burn, 1998b) and the term is impossible to use effectively in the field. A partial solution has been to recognize a "transient layer" at the top of permafrost and at the base of the active layer (Shur et al., 2005). The active layer and the transient layer are discussed in detail in Chapter 5.

Today, the utility of the periglacial concept requires careful assessment. Lozinski referred to a specific and limiting environment that is not typical of the vast majority of periglacial environments today. In fact, few, if any, modern-day analogues can be identified (French, 2000). Lozinski also used the term to refer primarily to areas or regions. However, one may argue that typical "periglacial regions" do not exist, and, if they do, lack well-defined boundaries. Thus, André (1999) explicitly questions the effectiveness of freeze–thaw in fashioning the periglacial landscape and refers to the "smokescreen of the periglacial scenery." In a similar vein, French and Thorn (2006) suggest that periglacial areas are cold-climate "zones" in which seasonal and perennial frost, snow, and normal azonal processes are present to greater or lesser degree. The reality is that many periglacial landscapes inherit the imprint, in varying degrees, of either glacial or non-cold-climate conditions.

A further complication is that so-called "periglacial" conditions often extend south of the latitudinal treeline and below the altitudinal timberline (Figure 1.2). This is because many areas of northern boreal forest are underlain by relict permafrost and, in alpine regions, glaciers may extend below timberline and into the forest zone. Finally, the treeline is a zone rather than a line and may extend over a latitudinal distance of ~100–150 km.

To summarize, therefore, modern usage of the term "periglacial" refers to a range of cold, non-glacial processes. We can define periglacial geomorphology as the sub-discipline of geomorphology concerned with cold non-glacial landforms. Because permafrost is a central, but not defining, element, much of the core of periglacial geomorphology is also a component of geocryology. However, periglacial geomorphology also includes the impact of seasonal freezing and the roles of seasonal snow, and of ice of a fluvial, lacustrine, and marine nature. Finally, periglacial geomorphology must embrace the azonal processes such as running water, wind, and waves that exhibit distinct characteristics in cold-climate environments.

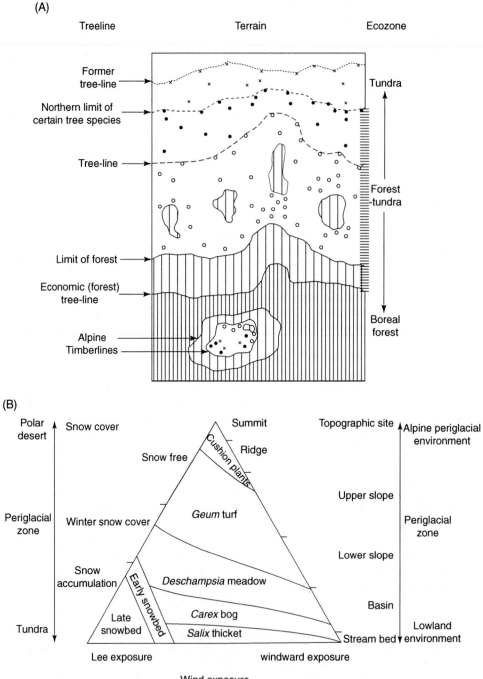

Figure 1.2. Relationship between the periglacial zone, the treeline and the timberline. (A) Forest limits and tree-lines according to Hustich (1966). (B) The plant type distribution, from tundra to polar desert, and from lowland to alpine (upland) environments, according to Billings and Mooney (1968).

1.2. DISCIPLINARY CONSIDERATIONS

Periglacial geomorphology currently maintains a bridging position between geomorphology, geocryology, and Quaternary science. Here, the disciplinary overlaps are considered. They are illustrated schematically in Figure 1.3.

1.2.1. The Growth of Geocryology

Geocryology, or permafrost science, is one of the cryospheric sciences. Clearly, the cryolithosphere (i.e. perennially and seasonally cryotic – that is, below 0°C, – ground) is central, and the cryohydrosphere (i.e. snow cover, glaciers, and river, lake, and sea ice) less central, to periglacial geomorphology.

For several reasons, the relations between geocryology and periglacial geomorphology are complex. First, for many years, permafrost studies were conducted in North America and the former Soviet Union (Russia) not only in relative isolation to each other but also in isolation from mainstream (geographical) geomorphology. Second, Russian and Chinese

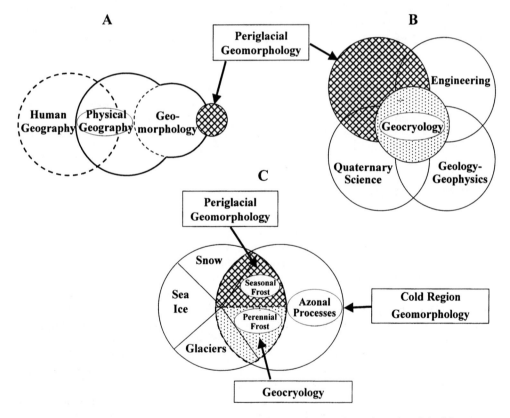

Figure 1.3. Diagram illustrating the disciplinary interactions and overlap of periglacial geomorphology. (A) The relations between physical geography, geomorphology, and periglacial geomorphology. (B) The relation between periglacial geomorphology and geocryology and the interactions of these disciplines with Quaternary science and other natural sciences. (C) Periglacial geomorphology and its overlap with the cryospheric sciences. Note: all the disciplinary boundaries are porous and those marked by broken lines are particularly so.

geocryology adopt holistic all-encompassing approaches whereas North American permafrost studies are characterized as being either "science" or "engineering" in nature. Thus,
there is no North American text that equals the breadth and depth presented by the most
recent Russian and Chinese texts, *General Geocryology* (Yershov, 1990) and *Geocryology
in China* (Zhou Youwu et al., 2000). Third, permafrost studies sit awkwardly between the
disciplines of geography and geology. For example, in North America and much of Europe,
periglacial geomorphology is taught usually in geography departments while permafrost
is within geology, geophysics, or earth science departments. Similar fractionation occurs
in Russia and China.

The most obvious areas of overlap between periglacial geomorphology and geocryology
lie in the problems associated with ground freezing and the occurrence of ground ice.

1.2.2. The Changing Nature of Quaternary Science

Advances in Quaternary science, and in particular the expansion and proliferation of
sophisticated dating techniques, mean that studies involving paleo-environmental reconstruction no longer rely solely upon the morphological and stratigraphical evidence (Büdel,
1951, 1953; Poser 1948; Smith, 1949) that was typically used in traditional Pleistocene
periglacial studies. Instead, a broader range of evidence now includes biological phenomena such as fauna and flora, arboreal and non-arboreal pollen, temperature-sensitive
insects such as beetles (coleoptera sp.), and geochemical indicators such as isotopes.
Stratigraphy is increasingly being supplemented by the study of ground ice, or cryostratigraphy. Thaw unconformities, truncated ice bodies, and cryostructures are now used to
infer previous freezing and thawing events or early Holocene climate change (Burn, 1997;
French, 1999; Melnikov and Spesivtsev, 2000). One must conclude that much of traditional
Pleistocene periglacial geomorphology has been largely replaced by cryostratigraphy and
cryolithology.

1.2.3. Modern Periglacial Geomorphology

As noted by M. Church (2005), geomorphology is no longer the preserve of geographers
and geologists. However, periglacial geomorphology continues to be identified as a process
sub-discipline of geomorphology that is distinct from both geocryology and Quaternary
science. The key processes are those associated with seasonal and perennial frost.

While geocryologists concentrate upon the thermal implications of terrain and the
presence of ice within the ground, periglacial geomorphologists emphasize the associated
landforms, and their growth and modification through time. Obviously, there is considerable overlap between the two. For example, the recognition of anti-syngenetic wedges on
hillslopes (Mackay, 1990a, 1995b) is an illustration of the overlap between landscape
evolution (geomorphology) and permafrost-related processes (geocryology). Likewise, as
discussed above, there is a complex overlap between periglacial geomorphology and
Quaternary science via cryostratigraphy and cryolithology.

Permafrost cannot be the only diagnostic criterion for periglacial geomorphology. This
is because permafrost is a thermal concept while geomorphology is concerned with landforms that are not controlled by ground temperature alone. On the other hand, permafrost
and ground ice must be central to periglacial geomorphology in the same way that hillslopes and running water are central to geomorphology at large.

It is easy to forget that the broad features of cold-climate terrain are largely influenced
by lithological variability, the nature and distribution of ice contained within bedrock or

surficial materials, and the enhanced action of azonal processes. The earlier editions of this text, like that of A. L. Washburn (1979), gave insufficient attention to the geological control over periglacial landscapes.

1.3. THE GROWTH OF PERIGLACIAL KNOWLEDGE

Even before Lozinski proposed his periglacial concept, a scattered body of geomorphic knowledge was available concerning the cold non-glacial regions of the world.

As might be expected, many of the earliest observations were by the European explorers of the vast sub-arctic regions of North America and Eurasia. These were casual, opportunistic, and non-scientific. For example, in Russian Alaska, the peculiarities of frozen ground were observed in 1816 by members of the Otto von Kotzebue expedition (von Kotzebue, 1821) as they traveled through the Bering Strait region (Figure 1.4). The presence of massive bodies of ground ice, portrayed in Figure 1.4, was to subsequently become a major component of periglacial study in the latter part of the twentieth century. Elsewhere in Russia, Karl Ernst von Baer, an Estonian–German naturalist who had traveled to Novaya Zemblya and Lapland in 1837, was the first to report (Baer, 1838) upon the excavation of a well in perennially-frozen ground at Yakutsk, central Siberia. Subsequently, Alexander von Middendorf, Karl Baer's younger traveling companion from his expedition to Lapland, descended the shaft, known today as Shergin's Well. The temperatures that he measured (Middendorf, 1862) are the earliest published information on the thermal regime of what is now termed permafrost. Middendorf correctly interpreted the ground temperature variations with depth and recognized what is now referred to as the "depth of zero-annual amplitude."

In North America, the eighteenth-century employees of the Hudson Bay Company occasionally made observations related to the terrain over which they traveled. Then, in 1839, Dr John Richardson, the physician who accompanied the explorer John Franklin on his expeditions of 1819–22 and 1825–27, presented observations upon frozen ground in North America (Richardson, 1839, 1841). Later, he sketched one of the distinctive pingos of the Mackenzie Delta region, known locally today as Aklisuktuk ("the little one that is

Figure 1.4. Members of the privately-financed Russian expedition led by Otto von Kotzebue examine exposed ground ice on Kotzebue Sound in 1816. 'Vue des Glaces dans le Paris', 1822, plate IX. Painting from the Rasmuson Library Collection, University of Fairbanks-Alaska, donated by the National Bank of Alaska.

growing") (Richardson, 1851, p. 234; see Mackay, 1981c). Following upon the disappearance of John Franklin's 1848–49 expedition to the Arctic and the numerous Franklin searches and other expeditions in the subsequent decades, data on the depth of frost penetration at various latitudes on the North American continent were published in a series of reports by the Royal Geographical Society in Great Britain (Lefroy, 1887, 1889a, b).

The beginning of the twentieth century saw a sharp increase in knowledge concerning the cold non-glacial regions of the world. This was the time of the 1898 Klondike gold rush in northwestern Canada and the subsequent migration of many of its miners to Alaska in 1901–1903. It was also the time of heroic exploration in Antarctica, culminating in the race to reach the South Pole between Scott and Amundsen in 1910–11. Many of the individuals involved in these historic activities made observations upon the frozen ground, and the harsh, cold-climate conditions that they experienced.

In the Klondike, miners had to remove a frozen overburden ("muck"), often many meters thick, in order to reach the placer gold that rested upon bedrock. Typically, a fire was built on the surface and the thawed ground beneath was progressively removed, thereby creating vertical pits through to the underlying bedrock. Alternatively, streams were damned and water diverted across the claim, thereby causing "natural thawing" of the ground beneath. The early Canadian government geologists who were assigned to the area reported typical rates of thaw of 5–10 cm per day (McConnell and Tyrrell, 1898). These methods, known as "frost prospecting," were attributed to earlier Russian mining practices in the Ural Mountains (Perrett, 1912). Later, steam thawing was used. Ultimately, the most efficient method of thawing frozen ground was found to be through the application of cold running water (Weeks, 1920). The relevance of these early mining experiences to our understanding of thermokarst and related processes is now obvious but at the time, seemed obscure.

In Antarctica, the scientific reports prepared by members of the heroic expeditions are also of great interest. For example, there is considerable anecdotal evidence concerning the exceptional strength of the katabatic winds blowing off the Antarctic ice sheet. C. E. Borchgrevink (1901, pp. 128, 140) first commented on the ability of strong and persistent wind to transport sediment particles, small boulders, and even objects such as heavy boots, over considerable distances. Observations by members of Scott's Northern Party, who spent two winters of incredible hardship in Northern Victoria Land in 1910–11, confirm this: "pebbles were flying about the beach like small bullets . . ." and "the sea ice was strewn with pebbles up to half an inch in diameter" (Priestley, 1914, p. 139). Almost certainly, comments like these contributed to a general acceptance of the importance of wind in periglacial environments. R. E. Priestley was also the first to record, in popular writing (Priestley, 1914, p. 290), the audible sound of thermal-contraction cracking, a process that, a few years later, was to be corrected inferred as the cause of ice wedges in northern Alaska (Leffingwell, 1915, 1919). Griffith Taylor, another member of the 1910–13 British Antarctic Expedition, was the first to describe the large polygons ("tesselations") of the McMurdo Sound region (Taylor, 1916, 1922) that were subsequently identified as sand wedges by T. L. Péwé (1959).

Given this context, it is not surprising that the periglacial concept was enthusiastically embraced by European geologists in the years following Lozinski's presentations at Stockholm in 1910. Several influential benchmark papers soon followed. Cold-climate patterned ground was described by W. Meinardus (1912) and the importance of frost-shattering of rocks was highlighted by B. Högbom (1914).

Because of the inaccessibility of most northern regions at that time, it was perhaps inevitable that periglacial geomorphology subsequently developed as a branch of a European-dominated climatic geomorphology. The primary aims were Pleistocene paleo-geographic reconstruction and global regionalization (Büdel, 1944, 1953; Cailleux, 1942; Dylik 1953, 1956; Edelman et al., 1936; Poser, 1948; Troll, 1944). From the privileged viewpoint of history, it is now easy to see how the concept of a "periglacial environment" or a "morpho-climatic zone" (Büdel, 1951, 1977; Peltier, 1950) became popular. In later years, a trend towards study of the northern polar region can be discerned (Büdel 1963; Jahn 1975; Tricart and Cailleux 1967).

The early 1970s witnessed a dramatic increase in awareness of the high latitudes in North America and the USSR. This was partly for geopolitical reasons but also the result of the search for natural resources, notably oil and gas. An increase in geotechnical engineering prompted an upsurge in the study of permafrost-related processes, and permafrost science, or geocryology, became a priority research discipline in the United States, Canada, Scandinavia, and the USSR. Often, there was substantial government involvement. As a result, traditional Quaternary-oriented periglacial studies became overshadowed. Texts by A. Jahn (1975), H. M. French (1976a), and A. L. Washburn (1979) document the changes in periglacial geomorphology in this period. Others, by V. A. Kudryavtsev (1978) and the Desert Research Institute of the Chinese Academy of Science (Academica Sinica, 1975), summarize Soviet and Chinese advances.

Over the last 40 years, a series of international permafrost conferences, held first in 1963 and then at five-year intervals since 1973, progressively record increasing international collaboration in periglacial geomorphology. Of special significance was the formation of the International Permafrost Association (IPA) in 1983. Several summaries of periglacial geomorphology during this period are available (Pissart, 1990, Thorn, 1992, Barsch, 1993).

The last 10 years have seen further growth. An international peer-reviewed journal, *Permafrost and Periglacial Processes*, was launched in 1990. A Chinese journal, *Bingchuan Dongtu (Journal of Glaciology and Geocryology)*, first published in 1978 by the newly-formed Lanzhou Institute of Glaciology and Geocryology of the Chinese Academy of Sciences, now publishes four issues a year as part of a reorganized Cold and Arid Regions Environmental and Engineering Research Institute (CAREERI). In Russia, an international journal, *Earth Cryosphere*, was launched in 1997 by the Institute of Earth's Cryosphere (Tyumen), Siberian Branch, Russian Academy of Sciences. Since 1988, the IPA has published an annual newsletter, *Frozen Ground*, and in 1993 it created a Periglacial Working Group that coordinated with an IGU Periglacial Commission that continued to be active from 1980 until final dissolution in 2004.

Other international journals that record advances in periglacial geomorphology include *Earth Surface Processes and Landforms, Journal of Quaternary Science, Geografiska Annaler, Geomorphology, Progress in Physical Geography, Polar Geography, Arctic,* and *Arctic, Antarctic and Alpine Research.*

1.4. THE PERIGLACIAL DOMAIN

The periglacial domain refers to the global extent of the so-called periglacial zone. Based upon the spatial association of certain microforms and their climatic threshold values, several different periglacial zones can be recognized (Figure 1.5). They occur not only as tundra zones in the high latitudes, as defined by Lozinski's concept, but also as forested zones south of treeline and in the high-altitude (i.e. alpine) regions of temperature latitudes. They include (a) polar desert and semi-desert (frost-debris zones) of the High Arctic, (b) tundra, (c) boreal forest, and (d) sub-arctic areas of either maritime or continental nature. Not included in Figure 1.5 is the vast high-elevation Qinghai-Xizang (Tibet)

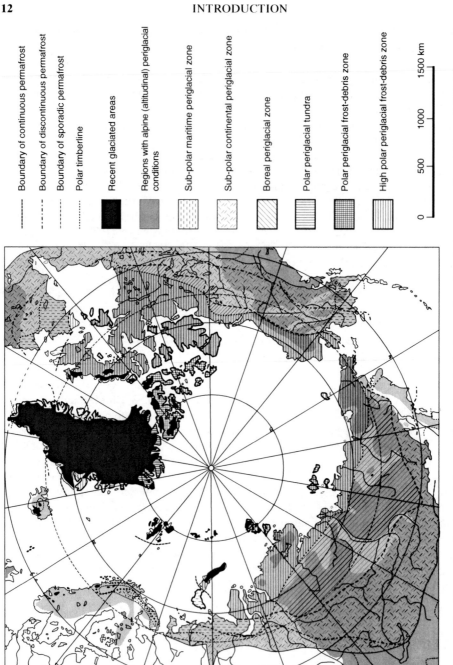

Figure 1.5. The regional extent of the periglacial domain in the northern hemisphere according to **Karte** (1979; **Karte** and **Liedtke**, 1981). Reproduced by permission of Bochumer Geographische Arbeiten.

plateau and its surroundings. Also not included are the alpine periglacial zones that occur in the various mid- and low-latitude mountain ranges. A similar map of the southern hemisphere would include the higher elevations and southern tip of South America, the sub-antarctic islands, the Antarctic Peninsula and the various ice-free areas on the Antarctic continent.

It should be emphasized that there is no perfect spatial correlation between areas of intense frost and areas underlain by permafrost. For example, a number of sub-arctic, maritime, and alpine locations experience frequent freeze–thaw oscillations but lack permafrost. Furthermore, the fact that relict permafrost underlies extensive areas of the boreal forest in Siberia and North America makes any simple delimitation of periglacial environments difficult. In practice, the relict permafrost of Siberia and North America extends the periglacial domain beyond its normal (i.e. frost action) limits.

Using the diagnostic criteria presented earlier, a conservative estimate is that approximately 25% of the Earth's land surface currently experiences periglacial conditions. There are all gradations between environments in which frost processes dominate, and where a whole or a major part of the landscape is the result of such processes, and those in which frost processes are subservient to others. Having said this, there are two complicating factors. First, certain lithologies are more prone to frost action than others, and hence more susceptible to periglacial landscape modification. Second, many periglacial landscapes show the imprint of previous glacial conditions or, less frequently, of non-glacial (i.e. temperate or tropical) conditions.

During the cold periods of the Pleistocene, large areas of the now-temperate middle latitudes experienced reduced temperatures because of their proximity to the ice sheets. Permafrost would have formed, only to have degraded during later climatic ameliorations. In all probability, an additional 20–25% of the Earth's land surface experienced frost action and permafrost conditions at some time in the past.

In summary, the present-day periglacial domain extends over two major vegetation types: (1) sub-arctic and boreal forests and (2) arctic tundra and polar deserts. The periglacial domain includes high-latitude areas in both northern and southern hemispheres. It includes the high-altitude, or alpine, zones which exist in many mountain ranges of the world, the largest of which is the Qinghai-Xizang (Tibet) plateau. The periglacial domain must also include the ice-marginal areas adjacent to modern glaciers and ice sheets.

1.5. THE SCOPE OF PERIGLACIAL GEOMORPHOLOGY

The components of modern periglacial geomorphology can be summarized under four headings, as outlined in the following subsections.

1.5.1. Permafrost-Related Processes and Landforms

Processes that are unique to periglacial environments relate to ground freezing. These include the growth of segregated ice and associated frost heaving, formation of permafrost, development of cryostructures and cryotextures, thermal-contraction cracking, and growth of frost mounds. Although the cryostratigraphic study of frozen earth material, especially the amount, distribution, and origin of the ice that is contained within it, is not strictly geomorphological in nature, it is uniquely relevant to understanding permafrost history, the interpretation of permafrost-related landforms, and to making inferences as to past climate.

A number of frost-action processes operate in the near-surface layer subject to seasonal thaw (the active layer), the near-surface permafrost located above the depth of zero-annual amplitude, and the zone of seasonal freezing and thawing in non-permafrost regions. These processes include moisture migration within frozen ground and those associated with repeated freezing and thawing (soil churning or cryoturbation, frost creep, solifluction and gelifluction, the upfreezing of stones, and particle-size sorting).

1.5.2. Azonal Processes and Landforms

A number of important periglacial processes center upon the seasonal freezing of soil and bedrock. These include the weathering of rock by either mechanical (frost) wedging or the complex of physical, biochemical or physico-chemical processes. Better known are those associated with running water, wind, snow, and waves, all of which assume distinctive characteristics under cold-climate conditions.

Some azonal processes are of especial interest to periglacial geomorphologists. For example, snow is an important source of moisture and an abrasive agent at low temperatures. Furthermore, it can act as a local source of soil moisture for ground heaving and frost action. Wind is also of special interest to periglacial geomorphology because the paucity of vegetation in cold regions provides wind with important erosional and transportational abilities. Finally, sea ice and river ice (commonly referred to as "ice-infested waters"), by restricting the time duration of wave action and/or open-channel flow, and through ice-pushing, ice-jams, and other impacts, can produce relatively distinct coastal, river channel, and lake conditions.

One aim of geomorphology is to create models of landscape evolution. These embody assumptions as to the processes involved, their speed of operation, and the manner in which surface morphology changes. In the case of periglacial geomorphology, the peculiarities of frozen ground and intense frost action impart a unique landform response. For example, slopes that are frozen, or are thawing, experience relatively unusual conditions associated with pore-water expulsion and thaw consolidation. These may promote mass failures that are distinct from the more well-known failures that occur on slopes that evolve under non-frozen conditions.

1.5.3. Paleo-Environmental Reconstruction

The growth of cryostratigraphy, when combined with the increased sophistication of Quaternary science, is largely replacing traditional Pleistocene periglacial geomorphology as a component of modern periglacial geomorphology. Morphological and stratigraphic evidence is now interpreted within the context of a more realistic appreciation of permafrost terrain and its climatic significance combined with, and constrained by, the use of isotopic and other absolute dating techniques and by proxy data sets.

1.5.4. Applied Periglacial Geomorphology

Many components of periglacial geomorphology are of applied significance and have societal relevance. For example, the periglacial environments of the world are home to over nine million people, mostly in northern Russia and Eurasia, and their health and economic well-being are of concern. The provision of water, municipal services, housing, roads, and other forms of infrastructure must take into account the nature of the cold-climate environment. Also, natural resource development in the northern polar region requires an understanding of the peculiarities of permafrost terrain. Human-induced thermokarst and other disturbances need to be minimized through appropriate geo-technical engineering, and by sound management and regulatory practice. The fact that climate change may first become apparent at high latitudes, and that permafrost can be regarded as an archive of past temperatures, has promoted long-term monitoring of the thickness of the active layer and of permafrost temperatures. In the sub-arctic, the potential thaw of permafrost is a concern to human and economic activity in general. In alpine regions, the increased utilization of upper slopes for recreation activities, and the potential for slope instability consequent upon permafrost thaw, has promoted studies of mountain permafrost. Likewise, the thinning of sea ice and the potential expansion of arctic shipping lanes has prompted international attention towards cold-climate coasts.

ADVANCED READING

André, M.-F. (2003). Do periglacial landscapes evolve under periglacial conditions? *Geomorphology*, **52**, 149–164.

French, H. M. (2000). Does Lozinski's periglacial realm exist today? A discussion relevant to modern usage of the term "periglacial". *Permafrost and Periglacial Processes*, **11**, 35–42.

French, H. M. (2003). The development of periglacial geomorphology: 1 – up to 1965. *Permafrost and Periglacial Processes*, **14**, 29–60.

Thorn, C. E. (1992). Periglacial geomorphology. What? Where? When? In: Dixon, J. C., Abrahams, A. D., eds., *Periglacial Geomorphology*, John Wiley & Sons, Chichester, pp. 1–30.

DISCUSSION QUESTIONS

1. How useful is Lozinski's periglacial concept?

2. What constitutes the periglacial domain?

3. Has periglacial geomorphology become a branch of geocryology?

4. What does cryostratigraphy contribute to periglacial geomorphology and to Quaternary science?

2 Periglacial Landscapes?

This chapter critically examines the concept of a "periglacial landscape." Many such landscapes do not contain typical periglacial landforms because they have only recently emerged from beneath Quaternary ice sheets. As such, periglacial processes are merely modifying already-existing glacial landscapes.

Two cold-climate landscapes are described from the western Canadian Arctic. Both were never covered by moving continental ice sheets during the Quaternary and can be regarded, therefore, as models to which landscapes might evolve under prolonged periglacial conditions. Two landscapes are also described from now-temperate mid-latitude regions where there is also no glacial legacy to consider. These are regarded as largely relict periglacial landscapes because, for special reasons, each has experienced relatively little subsequent modification.

2.1. INTRODUCTION

It is commonly assumed that periglacial landscapes are distinct. This is because most periglacial landforms are usually associated with permafrost. The most widespread are tundra polygons. Ice-cored hills, or pingos, are a less widespread but equally "classic" periglacial landform. Thermokarst features, such as ground-ice slumps, thaw lakes, and irregular depressions, are also permafrost-related. Other phenomena result from frost wedging and/or cryogenic weathering of exposed bedrock. For example, coarse, angular rock debris occurs widely over large areas of the polar deserts and semi-deserts, and angular frost-shattered bedrock protuberances (tors) reflect more resistant bedrock. Superimposed upon these "typical" landscape features is the occurrence of patterned ground. A final assumption relates to the overall smoothing of slopes thought typical of periglacial regions. This is generally attributed to mass wasting and "cryoplanation."

It is important that we establish the accuracy of these widely held impressions. Unfortunately, the manner and speed of landscape evolution under periglacial conditions is still unclear. The assumption of rapid change, implicit in many early periglacial investigations, may merely reflect either the relatively rapid adjustment from glacial to non-glacial conditions or the susceptibility of certain lithologies to intense frost action.

2.2. PROGLACIAL, PARAGLACIAL OR PERIGLACIAL?

Given the fluctuating climates of the Pleistocene, and the fact that many areas of current periglacial conditions have only recently emerged from beneath continental ice sheets, relatively few areas of the periglacial domain can be regarded as being in geomorphic equilibrium. In mid-latitudes, which currently experience temperate climates, the major problem is to identify the degree to which the effects of the Pleistocene cold climates have been eliminated. Often lithology is the important variable controlling the degree of preservation of periglacial landforms. Careful field investigation is required before the nature and magnitude of the periglacial legacy can be established.

It is useful at this point to distinguish between the terms "proglacial" and "paraglacial," and to relate the conditions implied by these terms to landscape modification. Paraglacial refers to the disequilibrium that occurs as one geomorphic environment moves from one equilibrium condition to another (Church and Ryder, 1972). In the case of the periglacial environment, the transition is usually to, or from, a glacial or temperate environment, the so-called glacial and interglacial periods of the Pleistocene. The proglacial environment, which refers specifically to ice-marginal conditions, is a periglacial environment in the original sense of Lozinski.

If we bear these two concepts in mind, one can illustrate the temporal relationships that exist between the glacial, periglacial, and temperate landscapes (Figure 2.1). Whether or not it is correct to assume that glacial landscapes are relatively high energy and temperate latitude landscape are relatively low energy is open to question. However, this is not of fundamental concern at this point. What is more relevant to our discussion is the assumption that periglacial landscapes reflect a constant energy condition, or equilibrium. Equally debatable is whether or not such equilibrium characterizes periglacial environments today. For example, the classic periglacial terrain that was examined by participants on the XI International Geological Congress field trip to Svalbard in 1910 constitutes the recently ice-free margin of a heavily-glaciated and mountainous island that is currently undergoing isostatic rebound. In short, they were visiting either a paraglacial or proglacial environment, rather than the periglacial environment as defined by Lozinski.

One perspective to this dilemma may be found in the cold regions of the world that have remained unglaciated throughout much, if not all, of the Quaternary. Such environments may reasonably be regarded as being in geomorphic near-equilibrium, although the rate and efficacy of the processes involved may not have remained constant. The possibility exists that certain landforms may be inherited from earlier, warmer times.

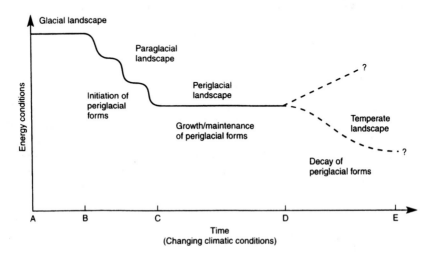

Figure 2.1. Temporal relationship between glacial, periglacial, and temperate landscapes under changing energy conditions. A to B: glaciation; B to C: deglaciation; C to D: periglacial equilibrium; D to E: climatic amelioration. Modified from Thorn and Loewenhertz (1987) with permission from Cambridge University Press.

The most extensive areas of such terrain occur in Eurasia, central and eastern Siberia, and northwestern Arctic North America (Alaska and interior Yukon Territory). The Tibet Plateau, thought by most Chinese scientists to have been unglaciated during the Quaternary, cannot be regarded as an equilibrium periglacial landscape because it is tectonically young, the result of exceptionally rapid uplift during the Late Pleistocene (~10mm/year for the last 130k/years). Thus, the permafrost and frost action which currently characterize the plateau reflect the ever-increasing elevation of the plateau. Equally, most alpine periglacial environments are, by definition, proglacial in nature. As for the boreal forest, tundra, and much of the High Arctic, the majority have only recently emerged from beneath continental ice sheets; these regions have moved temporally and spatially from being proglacial to paraglacial. The degree to which they are in balance with current cold-climate conditions varies and depends largely upon the susceptibility of local bedrock to cryogenic weathering and landscape modification.

A second perspective is to examine landscape where it is clear that no glacial legacy exists. The Holocene and Recent deltaic environments of the large northern rivers, such as the Ob, Lena, Yenisei, Yukon, and Mackenzie, or the emergent coastal relief of many of the High Arctic islands, are obvious examples. But it is also clear that these are special cases which do not easily permit historic insight of the time scale required. Accordingly, our attention must focus first upon the cold and never-glaciated regions of the world. There are surprisingly few.

2.3. UNGLACIATED PERIGLACIAL TERRAIN

In this section, we examine two unglaciated cold regions as being possible type-examples of periglacial terrain. One is upland, the other is lowland. Both are located in the western Canadian Arctic. The prevailing climatic conditions are summarized in Table 2.1. Figure 2.2 outlines the limits of Quaternary glaciation in this part of Arctic North America.

2.3.1. Beaufort Plain, Northwest Banks Island, Arctic Canada

The northwest corner of Banks Island (latitude 73°30′N, longitude 124°00′W) consists of a gentle surface developed upon Late-Tertiary age gravel and sand of the Beaufort Formation. The area appears to fulfill the essential requirements for periglacial designation.

First, a near-complete Quaternary record is available from a sequence of sediments on southwestern Banks Island that indicates the climate was cool in this part of the Arctic throughout Quaternary times (Vincent, 1982, 1989). The oldest sediments, know as Worth Point Formation, contain macrofossils of larch, shrubs, and herbaceous plants that suggest an open sub-arctic forest-tundra existed in Middle Pleistocene times (<790ka/years). Glacial sediments veneer the central part of the island, a spatially-complex association of glacigenic sediments and much-modified glacial landforms occurs in the east-central area, and fresh glacial landforms are found only in the east, south, and along the north coast. No glacial sediments are found in the northwest corner of the island. It seems reasonable to conclude that northwest Banks Island has experienced uninterrupted cold-climate conditions throughout at least the last one million years.

Second, during the most recent glaciation (the Late Wisconsinan), a major ice lobe extended west through McClure Strait, impinged on the north coast, and left a well-developed lateral moraine system. Part of this northern coastal area is illustrated in Figure 2.3. A series of sinuous, flat-floored ice-marginal channels parallel the former ice lobe (French, 1972b). Some appear to have been plugged by flow-till that mass-wasted

Table 2.1.   Climatic data: (A) Western Canadian Arctic stations; (B) Stations closest to the Barn Mountains, interior Yukon Territory.

(A)

Physiographic Region	Mackenzie Delta		Low Arctic	High Arctic
Location:	Inuvik	Tuktoyaktuk	Sachs Harbour (Ikaahuk)	Resolute Bay
Latitude:	69°N	69°30'N	72°N	74°N
Ecological region:	Northern Sub-arctic	Low Arctic	Low Arctic	High Arctic
Elevation (m):	60	18	84	64
Mean temperature (°C):				
Annual	−9.6	−10.7	−14.5	−16.4
January	−29.0	−27.2	−29.0	−32.6
July	+13.2	+10.3	+5.0	+4.3
Average frost-free period:	45	55	15	9
Average precipitation:				
Annual (mm)	260	130	102	136
June-August (mm)	93	64	49	69
Snowfall (cm)	174	56	47	78

(B)

	Temperature (°C)			Precipitation (mm)			No. of days with precipitation	No. of days with frost
	Mean Daily Year	Extreme Max.	Min.	Rain	Snow	Total		
Shingle Point: 55 m a.s.l. 68°57'N 137°13'W	−10.6°C	31.0°C	−52.2°C	110	86	196	63	286
Eagle Plain: 116 m a.s.l. 66°26'N 136°45'W	−6.9°C	27.5°C	−48.0°C	277	166	441	105	n/a
Old Crow: 253 m a.s.l. 67°34'N 139°43'W	−10.5°C	31.6°C	−59.4°C	108	104	212	71	302

Source: Wahl et al. (1987).

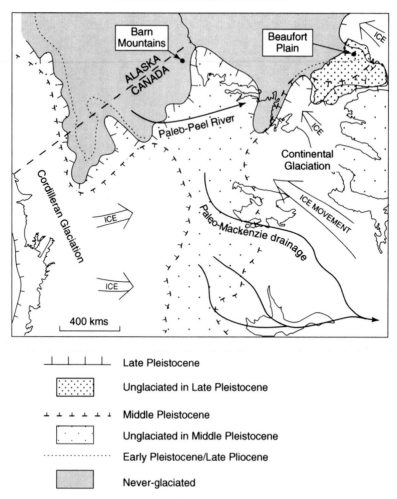

Figure 2.2. Maximum Limits of Quaternary glaciation in the western Canadian Arctic (after Vincent, 1982, 1989; Duk-Rodkin et al., 2004). Also indicated are the main paleo – drainage systems and locations of Beaufort Plain and Barn Mountains. See Figure 11.4 for approximate ages and correlations between Cordilleran and Continental glaciations.

from the flanks of the moraine, probably associated with melt-out of buried glacier-ice bodies. At some point, the northward drainage of the Beaufort Plain became impounded against the McClure Strait ice lobe and formed a series of proglacial water bodies that subsequently overflowed westward. A shallow, box-shaped, and progressively widening spillway channel was eroded. This is also visible on Figure 2.3. Clearly, the extreme north-western part of Banks Island experienced a classic Lozinski-type, ice-marginal location during the Late Pleistocene.

Third, although climatic data are unavailable for this remote and uninhabited area, one can assume that conditions are similar to those at Sachs Harbour (Ikaahuk) (72° N) and Resolute Bay (74° N) (Table 2.1A). The climate is cold, with mean annual air temperatures approximately –12 to –15 °C. Winter extremes may be as low as –50 °C. The region has less than 200 mm of rain or snowfall-equivalent a year. The frost-free period is limited to between 10 and 30 days per year. Permafrost typically exceeds 250 m in thickness and the maximum active-layer thickness probably never exceeds 60–75 cm (French, 1970).

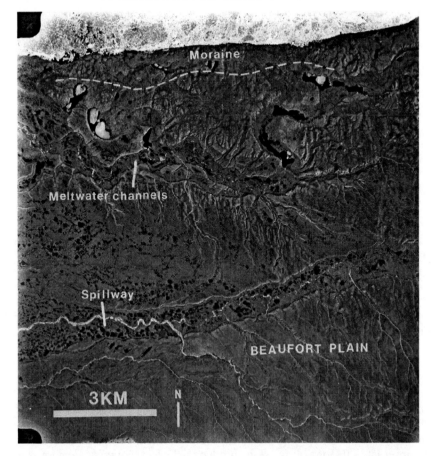

Figure 2.3. Vertical air photo of part of northwestern Banks Island showing the morainal contact between Late-Wisconsinan ice lobe in McClure Strait and never-glaciated Beaufort Plain. A series of meandering meltwater channels, some plugged by flow till, parallel the coastal moraine. A spillway channel, formed by overflow from a glacially-dammed lake to the east, is shown incised within the fluvially-dissected surface of the Beaufort Plain. Air Photograph A 17381-137, National Air Photo Library © 1996. Produced under licence from Her Majesty the Queen in Right of Canada, with permisssion from Natural Resources Canada.

In this regional context, it is informative to describe the landscape. J. G. Fyles, the Canadian Quaternary geologist, was the first to study the geomorphology. He descerided the Beaufort Plain in these terms:

Northwestern Banks Island is a fluvial plain, . . . underlain by thick gravels and sands. The Beaufort Plain rises gradually from sea level at the west coast to an altitude of about 800 feet at its eastern boundary. It is drained by more or less parallel, west-flowing, consequent major rivers and by dendritically arranged tributary streams . . . River valleys are cut a few feet to 500 feet below the plain surface. The larger valleys . . . have flat floors and steep walls. Some are asymmetrical in cross section, with a steep undercut north wall. Many of the rivers are strikingly braided . . . Large abandoned valleys just south of the morainal belt appear to be ancient, west-draining courses of the rivers that now flow to the north coast . . . The valleys are separated by broad areas of flat plain, and dissection has only locally progressed beyond early youth. The plain surface is remarkably uniform and completely free of lakes . . . The plain has not yielded clear evidence of having been overridden by glacial ice. (Fyles, 1962, in Thorsteinsson and Tozer, 1962, pp. 15–17).

Figure 2.4. Oblique air view of the Beaufort Plain, northwest Banks Island, showing fluvial dissection of the plain, winter snow remnants, mass-wasting on lower slopes, and large-scale polygonal patterns on the upland surfaces. Photo taken in early July, 1969.

The annotated aerial photograph (Figure 2.3) suggests this never-glaciated periglacial terrain is not all that different from terrain existing in many non-periglacial regions of the world. However, it is unusual. First, extensive areas of the plain are covered by large polygons, the borders of which form shallow troughs. These are similar to the "tessellations" first described by Griffith Taylor (1916, 1922) from Antarctica (see p. 10).

The valleys that dissect the Beaufort Plain are the most conspicuous landscape feature (Figure 2.4). Many are asymmetrical with steeper slopes facing west or southwest (French, 1970, 1971b). During summer, the gradual melt of snow banks on east-facing slopes provides moisture for mass wasting (solifluction) to occur. Thus, the stream in the valley bottom moves laterally towards the slope producing the least colluvium and an asymmetry is produced in which the two valley-side slopes are constantly adjusting to each other and to the basal stream channel. Slope asymmetry is interpreted, therefore, as a quasi-equilibrium form. Despite its fluvial dissection, the upland surface of the Beaufort Plain is relatively featureless. Evidence of ground ice is conspicuously lacking; pingos and frost mounds are absent, and thaw lakes and standing water bodies are non-existent. Patterned ground is restricted to the occasional non-sorted circle. A wind-veneer of small pebbles covers the surface and vegetation is sparse.

In summary, the Beaufort Plain suggests that a periglacial landscape is one of low relief and fluvial dissection. Wind, snow, and running water are the crucial geomorphic processes. Permafrost is indicated by the occurrence of thermal-contraction crack polygons but evidence of high ice content in near-surface sediments is lacking. Mass wasting is confined to snowbank localities. The braided nature of the stream channels reflects the unconsolidated nature of the bedrock, the availability of abundant bedload, and the highly variable and seasonal (nival) discharge regime.

2.3.2. Barn Mountains, Northern Interior Yukon Territory, Canada

The Barn Mountains (latitude 68°30′N: longitude 138°30′W) are part of the British Mountains and rise to elevations in excess of 1200 m a.s.l. In the foothills, sequences of gently folded Mesozoic-age shale and sandstone produce anticlinal escarpments and synclinal ridges. The main range of the Barn Mountains consists of faulted Paleozoic shale and limestone bedrock interbedded with quartzite, chert, and conglomerate.

The Barn Mountains, like the Beaufort Plain, appear to fulfill the main requirements for periglacial designation. First, the mountains lie beyond the maximum limit of Pleistocene glaciation. The last major ice advance, termed the Buckland Glaciation (Duk-Rodkin et al., 2004; Hughes, 1972; Rampton, 1982), was from the Mackenzie Delta along the coastal plain to the north. Evidence of glaciation, in the form of either till deposits or glacially-sculptured landforms, is absent from the Barn Mountains. Second, the present climate is cold and dry. Data recorded at Shingle Point, on the Beaufort Sea coast, and at Eagle Plain, some 150 km inland, indicate local conditions (Table 2.1B). In winter, because of inversion effects, the Eagle Plain data are more representative. In summer, because of elevation, the Barn Mountains are slightly cooler than both Shingle Point and Eagle Plain. Precipitation, in the form of both rain and snow, is probably similar to that at Eagle Plain, with more than half falling as rain, between June and September.

The landscape of the Barn Mountains is also instructive as far as periglacial terrain is concerned. The mountains consist of structurally-controlled upland massifs surrounded by extensive pediments. The latter are dissected by streams that drain towards the Arctic coast (Figure 2.5). The pediment surfaces, first reported upon by O. L. Hughes (1972), are best developed where less resistant Mesozoic-age shale outcrops adjacent to more resistant sandstone. In places, sandstone outliers are completely surrounded by gentle pediment surfaces, forming isolated inselberg-like hills (Figure 2.6A). The pediment

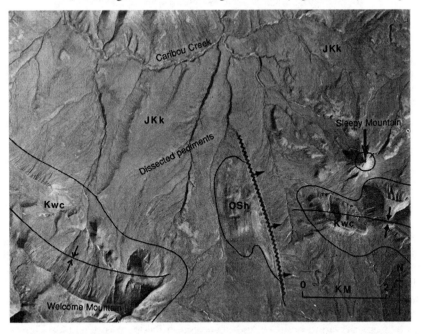

Figure 2.5. Vertical air photo of part of the Barn Mountains is annotated to show main structural elements. The geology is after Norris (1977). The area has been subject to folding and faulting. The uplands are synclinal in structure and formed by relatively resistant Lower Cretaceous sandstone and quartzite (Kwc). The pediments are developed upon softer shale and siltstone (JKk). Faulting has resulted in an inlier of Ordovician-age shale (Osh) that also results in upland terrain. The pediments are dissected. Air Photograph A 24502-160, National Air Photo Library © 1996. Produced under licence from Her Majesty the Queen in Right of Canada, with permisssion from Natural Resources Canada.

(A)

(B)

Figure 2.6. Landscape elements of the Barn Mountains, interior Yukon Territory. (A) Synclinal ridges and outliers form upland massifs and inselberg-like hills near Sleepy Mountain.
(B) Pediment-like surfaces surround upland massifs. See Figure 2.5.

surfaces truncate dip of the underlying rock (Figure 2.6B), are dissected, occasionally
cross fault lines (see Figure 2.5), and sometimes encroach upon the adjacent sandstone.
For the most part, they are "covered" pediments (Twidale, 1987); that is, they are cut in
weak bedrock and carry a protective veneer of detritus derived from upslope. In the
European literature, such features are commonly termed "glacis" (Dresch, 1982, p. 90) or
"erosion glacis" (Joly, in Demek, 1972b, pp. 189–192). Closely-related features are large
terraces or steps carved in bedrock and occupying hillslope positions (see Figure 9.6).

In all probability, the Barn Mountain landscape is old, dating from the Middle to
Late Tertiary. The pediments, probably in existence prior to the onset of the Quaternary,
would have been effective slopes of transportation of waste material under cold-climate
conditions. At the meso-scale, the bedrock terraces (see Figure 9.6) show no obvious
relationship with geological structures. At the micro-scale, angular rock rubble veneers
many of the upland surfaces. A striking observation is that all these features show little
sign that they are forming under today's climate.

   In summary, the Barn Mountains landscape contains a different suite of landforms to that of the Beaufort Plain. To a large extent, the difference reflects geology. Certain elements resemble those of hot, semi-arid regions.

2.4. RELICT PERIGLACIAL LANDSCAPES

While certain lithologies may be highly susceptible to frost action and to distinct landscape modification, it is equally true that other lithologies are better suited to the preservation of periglacial features long after the disappearance of periglacial conditions. These ideas are explored in this section, where two lowland landscapes from the never-glaciated and now-temperate mid-latitudes are described. Both may be viewed, to varying degrees, as relict periglacial landscapes.

2.4.1. Chalk Uplands, Southern England and Northern France

The chalklands of Southern England and Northern France (~latitudes 48–52°N) are underlain by thick sequences of relatively pure, soft calcareous sedimentary rock of Cretaceous age. It is highly porous and permeable. The Chalk forms broad plateaus, striking in-facing escarpments, and gentle dip-slope surfaces. The extent of the Chalk outcrop along the English Channel coastlands is illustrated in Figure 2.7.

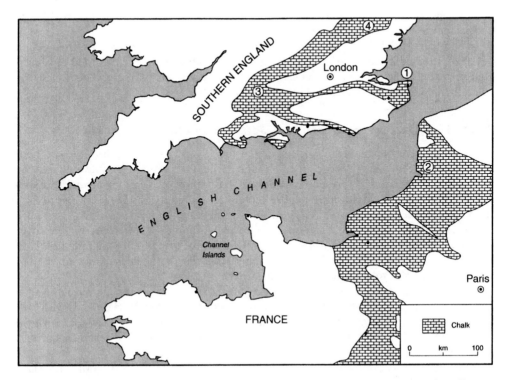

Figure 2.7. Map showing extent of the Chalk outcrops in southern England and northern France. Legend: 1-Isle of Thanet (see Figure 13.2); 2-Pays de Caux (see Figure 13.5B); 3-Clatford Bottom (see Figure 13.10); 4-Walton Common (see Figure 12.6).

Near the surface, the Chalk is frequently weathered into thin, platy fragments and a calcareous "paste," or colluvium, often mantles lower slopes. For over a century, these and other phenomena were attributed to the action of frost (Bull, 1940; Reid, 1887; Williams, 1987). Recent experimental studies confirm that the chalk is highly frost-susceptible (Murton et al., 2001). The Chalk outcrops are also veneered by loess and coversand ("brickearth") that locally reach a thickness in excess of 3–4 m. Most loess deposits date from the last glacial, the Late Devensian, but more extensive records in the Somme basin, northern France, indicate that accumulation began as early as 900 ka (Antoine et al., 1998, 2003).

A further characteristic is that the near-surface 5–10 m of Chalk bedrock is frequently brecciated (Murton, 1996a; Murton et al., 2003) (see Figure 13.2). This is attributed to frost heave within and below an ice-rich layer in chalk that formed when permafrost was present during the cold periods of the Pleistocene (Murton and Lautridou, 2003). The brecciated chalk has been reworked as a diamicton that is known locally as "coombe rock" (Reid, 1887), or more generally as "head" (Bates et al., 2003; Geikie, 1894).

The Chalk landscapes are also characterized by a network of valleys, many of which are either dry or occupied by misfit streams. Many are asymmetrical with steeper slopes facing west or southwest. The apparent freshness of the valleys, the density of the network, and the absence of large headwater catchments favor the explanation that they were incised when the surface of the chalk was perennially-frozen and when semi-perennial snow banks would have nourished flow (Bull, 1940). Relatively rapid valley incision would have been facilitated by brecciated bedrock. Wind, snow, and differential mass wasting at the time of incision probably produced the asymmetry (French, 1972a; Ollier and Thomasson, 1957). When permafrost degraded, the Chalk regained its permeable and porous nature to leave the valleys mostly dry.

In summary, the lithology of the chalk favored not only modification by frost action but also subsequent preservation of landscape. The modifications are recorded by brecciated bedrock, surficial ("head") materials, and dry valleys. As permafrost degraded, the cold-climate landscape elements became largely fossilized.

2.4.2. Pine Barrens, Southern New Jersey, Eastern USA

The Pine Barrens of Southern New Jersey (latitude 39–40° N) lie south of the southern limit of Late-Pleistocene (Wisconsinan) ice, and earlier glaciations did not extend into the region. Today, the area is a preserved wilderness tract of over 1.4 million acres of heavily-wooded lowland that extends through the center of the Outer Coastal Plain (Figure 2.8). It consists of flat, sandy, and gravely terrain with maximum elevations of 25–30 m a.s.l. The area is underlain by sand and gravel of Cretaceous and Tertiary age. A surface cover of fluvial sand and gravel, known locally as Bridgeton Formation, is regarded as Late Miocene to Early Pleistocene in age. In many respects, the geology is not unlike the Beaufort Plain of Banks Island, described earlier.

Denudation rates since the Late Miocene on the Atlantic coastal plain have averaged 10 mm/year (Stanford et al., 2002). Slope retreat has replaced low-level uplands with low-angled pediments, thereby preserving relict topography at higher elevations. Surficial mapping (Newell et al., 2000) indicates the lower slopes are mantled by several sequences of cold-climate colluvium.

Broad shallow valleys contain terraces that grade into wetlands in the valley bottoms. Although the dense forest cover makes landforms difficult to identify, air photographs taken in the early 1930s, when the region had been largely deforested by nineteenth-century lumbering activity, reveals the current drainage to be misfit and located within large

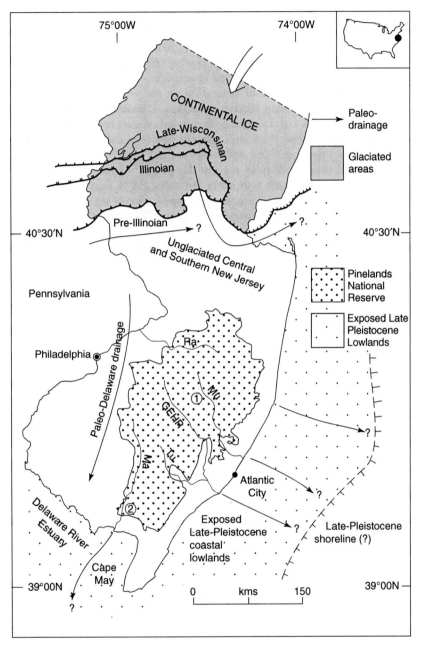

Figure 2.8. Map of New Jersey, eastern USA, showing location of the Pine Barrens and limits of Pleistocene glaciations. Legend: 1-Newtonville, Atlantic County (see Figure 12.7, 12.8A); 2-Port Elizabeth, Cumberland County (see Figure 12.4A). Rivers: GEHR – Great Egg Harbor River, Tu – Tuckerhoe River, Ma – Maurice River, Mu – Mullica River, Ra – Rancocas River.

braided paleo-channels. Ancient sand wedges are regularly exposed in aggregate extraction pits (French et al., 2003) and indicate the previous existence of permafrost. Deformed near-surface sediments and dislocated bog ironstone beds are interpreted in the context of the thaw of icy beds that formed within the permafrost (French et al., 2005).

The reason why the upland surfaces of the Pine Barrens preserve periglacial structures relates to hydrogeology. Although the current regional groundwater table is within 1–2 m of the surface, this has not always been the case. Similar to chalk, the underlying sand and gravel is highly porous and permeable. During the cold periods of the Pleistocene, the regional water table would have dropped in accordance with the fall in Atlantic Ocean sea level. At the same time, a lack of vegetation on the well-drained sandy and gravely terrain would have allowed deep frost penetration. Under these conditions, moisture would have migrated upward from the depressed water table towards downward-advancing frost or permafrost. Although modern geotechnical wisdom (see Chapters 4 and 14) suggests that coarse-grained sediments are not usually associated with high ice content, there is field experience from present-day permafrost regions that demonstrates ice volumes and porosities in such sediments can exceed 30% (see Chapter 7).

Following thaw-degradation and eventual disappearance of permafrost, the terrain of the Pine Barrens regained its highly permeable nature. This has resulted in the low-level upland surfaces being largely preserved from Holocene fluvial landscape modification. The lack of post-colonial agriculture on the infertile sandy soils has been a further factor that has allowed sand-wedge structures to be preserved in the near-surface sediments and for palimpsests of the paleo-drainage to be visible on aerial photographs.

2.5. CONCLUSIONS

Three general conclusions can be drawn from these four regional examples.

First, a basic advance in our understanding of periglacial environments is to recognize that the geomorphic footprint of periglacial conditions is not always achieved, and most periglacial environments possess some degree of inherited paraglacial or proglacial characteristics.

Second, the importance of lithology should not be underestimated in any consideration of periglacial landscapes. Lithology determines not only the susceptibility of the landscape to cold-climate modification but also the speed of modification and the degree to which it can be preserved. Ground ice dynamics appear crucial. For example, the growth and thaw of segregated ice near the top of permafrost and in seasonally-frozen ground is largely a function of (i) frost susceptibility of bedrock and surficial deposits, and (ii) hydrogeology. It is the eventual thaw of ground ice that largely controls periglacial landscape modification because this influences the supply of sediment to hillslopes and streams. If ground ice were not a factor, the similarities in landscape evolution between the hot and cold deserts of the world would be striking.

Third, it appears that the preservation of relict periglacial features in now-temperate regions relies largely upon their current desiccation and removal from runoff and fluvial activity.

ADVANCED READING

(a) General

Büdel, J. (1977). *Klima-Geomorphologie*. Gebruder-Bortraeger, Berlin. English translation, 1982, by Lenore Fischer and Detlef Busche, *Climatic Geomorphology*, Princeton, NJ, Princeton University Press, 443 pp. (especially sections 2.2 "The polar zone of excessive valley-cutting (the active periglacial zone)" and 2.3 "The peritropical zone of excessive planation").

Dylik, J. (1957). Tentative comparison of planation surfaces occurring under warm and cold semi-arid conditions. *Biuletyn Peryglacjalny*, **5**, 175–186.

Peltier, L. C. (1950). The geographic cycle in periglacial regions as it is related to climatic geomorphology. *Annals, Association of American Geographers*, **40**, 214–236.

(b) Northwest Banks Island, Canada

French, H. M. (1971). Slope asymmetry of the Beaufort Plain, northwest Banks Island, NWT, Canada. *Canadian Journal of Earth Sciences*, **8**, 717–731.

French, H. M. (1972). Proglacial drainage of northwest Banks Island, District of Franklin, NWT. *Musk-Ox*, **10**, 26–31.

Fyles, J. G. (1962). Physiography. In: Thorsteinsson, R., Tozer, E. T., *Banks, Victoria and Stefansson Islands, Arctic Archipelago*. Geological Survey of Canada, memoir **330**, 8–17.

Vincent, J.-S. (1982). The Quaternary history of Banks Island, NWT, Canada. *Géographie physique et Quaternaire*, **36**, 209–232.

(c) Barn Mountains, Northern Yukon Territory, Canada

Duk-Rodkin, A., Barendregt, R. W., Froese, D. G., Weber, F., Enkin, R., Smith, I. R., Zazula, G. D., Waters, P., Klassen, R. (2004). Timing and extent of Plio-Pleistocene glaciations in north-western Canada and east-central Alaska. In: Ehlers, J., Gibbard, P. L., eds., *Quaternary Glaciations – Extent and Chronology*, part II. Elsevier, Amsterdam, pp. 313–345.

Rampton, V. N. (1982). Quaternary geology of the Yukon coastal plain. *Geological Survey of Canada, Bulletin*, **317**, 49 pp.

French, H. M., Harry, D. G. (1992). Pediments and cold-climate conditions, Barn Mountains, unglaciated northern Yukon, Canada. *Geografiska Annaler*, **74A**, 145–157.

(d) Chalk Uplands, Southern England and Northern France

Antoine, P., Catt, J., Lautridou, J.-P., Somme, J. (2003). The loess and coversands of northern France and southern England. *Journal of Quaternary Science*, **18**, 309–318.

Murton, J. B., Lautridou, J.-P. (2003). Recent advances in the understanding of Quaternary periglacial features of the English Channel coastlands. *Journal of Quaternary Science*, **18**, 301–307.

Murton, J. B., Bateman, M. D., Baker, C. A., Knox, R., Whiteman, C. A. (2003). The Devensian periglacial record on Thanet, Kent, UK. *Permafrost and Periglacial Processes*, **14**, 217–246.

Ollier, C. D., Thomasson, A. J. (1957). Asymmetrical valleys in the Chiltern Hills. *Geographical Journal*, **123**, 71–80.

(e) Pine Barrens, Southern New Jersey, Eastern USA

French, H. M., Demitroff, M., Forman, S. L. (2003). Evidence for Late-Pleistocene permafrost in the New Jersey Pine Barrens (latitude 39°N), Eastern USA. *Permafrost and Periglacial Processes*, **14**, 259–274.

French, H. M., Demitroff, M., Forman, S. L. (2005). Evidence for Late-Pleistocene thermokarst in the New Jersey Pine Barrens (latitude 39°N), Eastern USA. *Permafrost and Periglacial Processes*, **16**, 173–186.

Newell, W. L., Powars, D. S., Owens, J. P., Stanford, S. D., Stone, B. D. (2000). Surficial geologic map of Central and Southern New Jersey. United States Geological Survey, Miscellaneous Investigations Series, Map 1-2540-D.

Stanford, S. D., Ashley, G. M., Russell, E. W. B., Brenner, G. J. (2002). Rates and patterns of late-Cenozoic denudation in the northernmost Atlantic coastal plain and piedmont. *Bulletin, Geological Society of America*, **114**, 1422–1437.

DISCUSSION TOPICS

1. Distinguish between periglacial, proglacial, and paraglacial environments.

2. What are the main landscape elements of (a) the Beaufort Plain and (b) the Barn Mountains?

3. What is the role played by lithology in (a) influencing frost-action processes and (b) preserving relict periglacial landscapes?

4. In what ways does the landscape of the Barn Mountains differ from that of a hot arid region?

3 Periglacial Climates

A number of cold-climate environments exist in which frost action is important. Their combined spatial extent constitutes the periglacial domain. Their extent is difficult to establish because all gradations exist, from environments in which frost-action processes dominate and where all, or a major part, of the landscape depends upon such processes, to those in which frost-action processes are subservient to others.

Ground temperature is as important as air temperature in influencing the periglacial landscape. Snow, wind, vegetation, soil type, and ground-thermal properties all influence ground temperature.

The periglacial domain is a component of the cryosphere. Global climate models (GCMs) predict warming will be greatest in high latitudes and that cold-climate environments will be significantly affected.

3.1. BOUNDARY CONDITIONS

The boundary between periglacial and non-periglacial conditions is arbitrary and, to a large extent, varies according to the criteria used.

The presence or absence of perennially-frozen ground would appear to be a reasonably easy boundary definition. Unfortunately, the outer limits of the discontinuous permafrost zone are often hard to delineate. The location of the treeline is a second relatively unambiguous boundary definition, since most investigators equate severe periglacial conditions with tundra or polar desert ecozones. However, the treeline is not static but either advancing or retreating depending upon environmental and regional climatic conditions. Moreover, the treeline is not a line but rather a zone, sometimes over 50–100 km wide in places, lying between the biological limit of continuous forest and the absolute limit of tree species (see Figure 1.2).

For our purposes, and in keeping with a simple frost-action definition, we adopt an empirical definition and define the periglacial domain as including all areas where the mean annual air temperature is less than +3°C. This closely follows the limits proposed by Williams (1961) for solifluction and patterned ground. Because it includes not only areas where frost-action conditions dominate but also areas that are marginally periglacial in character, it gives some idea of the maximum extent of the periglacial domain. We may further subdivide the periglacial domain by the −2°C mean annual air temperature into environments in which frost action dominates (mean annual air temperature less than −2°C) and those in which frost-action occurs but does not necessarily dominate (mean annual air temperature between −2°C and +3°C).

The Periglacial Environment, Third Edition   Hugh M French
Copyright © 2007 John Wiley & Sons Ltd

3.2. PERIGLACIAL CLIMATES

The concept of a periglacial climate was first proposed by C. Troll (1944) in his global survey of frost-action conditions. It was subsequently incorporated into a scheme of morphogenetic regions by L. C. Peltier (1950), who characterized the periglacial climate as having a mean annual air temperature of between −15 °C and −1 °C, precipitation of between 120 mm and 1400 mm per annum, and "intense frost action, strong mass movement, and the weak importance of running water." This definition masks a wide range of climatic conditions and implies a relative uniformity to the periglacial climate. This is misleading because it is far more realistic to stress the variety of cold climates that exist in which frost-action processes are important.

Some of the first to appreciate this were the French geomorphologists J. Tricart and A. Cailleux (Tricart, 1963; Tricart and Cailleux, 1967, pp. 45–67). They distinguished between three types of periglacial climatic environments. The first (Type I: Dry climates with severe winters) experiences seasonal and deep freezing while the third (Type III: Climates with small annual temperature range) experiences shallow and predominantly diurnal freezing. The second (Type II: Humid climates with severe winter) is intermediate. Permafrost is characteristic of the first, is irregular in occurrence and distribution in the second, and absent in the third. A problem with this three-fold division is that the first two types are identified primarily in terms of humidity while the third is identified in terms of temperature. As a result, Type I includes the rather different climatic environments of, for example, the Canadian High Arctic and sub-arctic central Siberia. Moreover, an arctic subtype of category II, as typified by Spitsbergen, is more similar to the environments of other high latitudes than to a mountain variety, which is the other subtype of that category.

In a more pragmatic fashion, D. Barsch (1993) suggests that periglacial environments simply exist in areas with polar climates. In other words, they exist in those areas dominated by E (Ef and ET) climates of Köppen (1923). As regards alpine periglacial climates, the boundaries are similar but problems arise in arid to semi-arid mountains, where the upper timberline (i.e. the lower limit of periglacial conditions) does not exist. As regards the upper limit of periglacial conditions, the existence of glaciers and permanent snow and ice for both the polar and alpine periglacial environments is relatively unambiguous. Unfortunately this approach fails to recognize the important boreal forest component of the periglacial domain, and the fact that relict permafrost can be just as important as contemporary frost action in determining the character of the periglacial landscape.

In the second edition of this book five broad categories of periglacial climates were proposed using the criteria of insolation, temperature, and elevation. This was to incorporate the relatively unique climatic conditions of the Qinghai-Xizang (Tibet) plateau. Basic to the classification is the fact that radiant energy from the sun influences both air and ground temperatures. For example, Figure 3.1(A) shows potential insolation, expressed as a percentage of the amount received at the equator, as a function of latitude. If one assumes potential insolation at the equator to be 3.12×10^5 ly/year (Dingman and Koutz, 1974), then the potential insolation of a horizontal surface on the Qinghai-Xizang plateau at latitude 29–38° N is approximately 2.68×10^5 ly/year. This can be compared to the comparable value for the Mackenzie Delta, Canada (latitude 68–69° N) of 1.58×10^5 ly/year.

Elevation also influences the amount of solar radiation received. For example, Fenghuo Shan, on the Tibet plateau, is at the same latitude as, but 3500 m higher than, Xi'an in central China. However, Figure 3.1(B) indicates the actual solar radiation received at Fenghuo Shan is much higher than at Xi'an. This is because the water vapor and aerosol content in the atmosphere decrease with increasing elevation, which, in turn, cause less

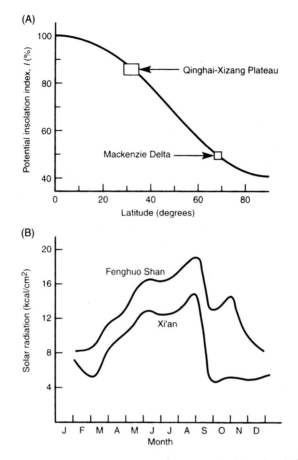

Figure 3.1. Diagrams showing the variability of solar radiation with latitude and altitude. (A) Potential insolation expressed as a percentage of the potential insolation at the equator (according to Dingman and Koutz, 1974). (B) Incoming solar radiation received at two localities in China at the same latitude (34° N) but at different elevations: Fenghuo Shan at 4800 m a.s.l. and Xi'an at 1300 m a.s.l. From Wang and French (1994). Reproduced by permission of John Wiley & Sons Ltd.

cloud cover and a greater transparency in the atmosphere at higher elevations. Therefore, the combination of low latitude and high elevation makes the Tibet plateau a strong solar receiver. This effect is still present, but less marked, in the alpine environments of mid-latitudes.

Based upon elevation, insolation, and temperature, the periglacial climatic environments can be summarized as follows:

1. *High Arctic climates* – in polar latitudes; extremely weak diurnal pattern, strong seasonal pattern. Small daily and large annual temperature range. Examples: Spitsbergen (Green Harbour, 78° N); Canadian Arctic (Sachs Harbour (Ikaahuk), 72° N).

2. *Continental climates* – in sub-arctic latitudes; weak diurnal pattern, strong seasonal pattern. Extreme annual temperature range. Examples: Central Siberia (Yakutsk, 62° N); Interior Alaska and Yukon Territory (Fairbanks, 65° N; Dawson City, 64° N).

3. *Alpine climates* – in middle latitudes in mountain environments; well-developed diurnal and seasonal patterns. Examples: Colorado Front Range, USA (Niwot Ridge, 40°N); European Alps (Sonnblick, 47°N).

4. *Qinghai-Xizang (Tibet) plateau* – a high-elevation, low-latitude mountain environment. Well-developed diurnal and seasonal patterns. Above-normal insolation due to elevations of 4200–4800 m a.s.l. Example: Fenghuo Shan (34°N).

5. *Climates of low annual temperature range* – two types of azonal locations: (a) island climates in sub-arctic latitudes. Examples: Jan Mayen (71°N), South Georgia (54°S). (b) Mountain climates in low latitudes. Examples: Andean summits, Mont Blanc Station, and El Misti, Peru (16°S); Mauna Kau, Hawaii (20°N).

In addition, a sixth category must be added to accommodate the intense cold, windiness, and aridity of the ice-free areas of the Antarctic continent.

Climatic data for some of the locations mentioned above are given in Table 3.1 and the nature of their typical frost-action conditions are graphically illustrated in Figure 3.2. The number of frost-free days, days of alternating freezing and thawing, and days in which temperatures remain below 0°C are plotted for eight localities. The graphs clearly show that the least number of freezing and thawing days occur in the High Arctic and continental climates, which are dominated by seasonal patterns, while the greatest number occur in the climates of low annual temperature range, which are dominated by diurnal or cyclonic (i.e. short-term) fluctuations. The characteristics of each climatic type are described below in more detail.

3.2.1. High Arctic Climates

Several characteristics make these very distinct periglacial climates. First, extremely low winter temperatures occur for periods of several months, when there is perpetual darkness and temperatures may fall to between −20°C and −30°C. The ground freezes to form permafrost and the ground surface contracts under the intense cold. Second, temperatures rise above freezing for only 2–3 months of the year, when the surface thaws to depths varying between 0.3m and 1.5m and average air temperature rises 4–6°C above 0°C. Third, precipitation amounts are low. In parts of the Canadian Arctic, precipitation is less than 100mm per annum, of which approximately one-half falls as rain during the summer period. Because of low evaporation rates, and the inability of the water to percolate the soil on account of permafrost, the effectiveness of this precipitation is high. As a consequence, although arid in terms of total precipitation, these regions are surprisingly wet during summer. In more maritime locations, such as Spitsbergen, precipitation amounts increase to as much as 250–400mm per annum. In these regions, snow assumes importance in protecting the ground surface from the extremes of cold and increases the magnitude of the spring runoff.

In all High Arctic climates, winter snow cover is thin and often discontinuous, even in the more humid regions. Upland surfaces and exposed areas are swept bare by wind, allowing frost to penetrate deeply, and snow banks accumulate only in hollows and lee-slope positions.

In the northern hemisphere, the approximate spatial limits of these climatic conditions may be arbitrarily defined by the glacial limit to the north and by either the treeline or the 8–10°C annual July air isotherm to the south. For the most part, we refer to these areas as either the tundra or polar regions of the world.

Table 3.1. Climatic data for selected localities in different types of periglacial climatic environment.

(a) Temperature (°C) and precipitation (mm) data	J	F	M	A	M	J	J	A	S	O	N	D
High Arctic Climates:												
Spitsbergen, Green Harbour T (°C)	−16	−19	−19	−13	−5	+2	+6	+5	0	−6	−11	−15
P (mm)	35	33	28	23	13	10	15	23	25	30	25	38
Canadian Arctic, Sachs Harbour (Ikaahuk) T (°C)	−29	−31	−27	−20	−8	+2	+5	+4	−1	−10	−24	−27
P (mm)	2	2	3	2	5	4	24	16	16	12	4	3
Continental climates:												
Central Siberia, Yakutsk T (°C)	−43	−36	−22	−7	+6	+15	+19	+15	+6	−8	−28	−40
P (mm)	11	9	6	11	18	33	43	42	26	20	16	12
Yukon, Canada, Dawson City T (°C)	−31	−23	−16	−3	+8	+13	+14	+12	+6	−4	−17	−25
P (mm)	20	20	12	18	23	33	40	40	43	33	33	28
Alpine climates:												
Alps, Sonnblick, 3060 m a.s.l. T (°C)	−13	−14	−11	−9	−4	−1	+1	+1	−1	−5	−9	−11
P (mm)	124	124	160	167	157	140	142	129	116	129	117	134
Rockies, Niwot Ridge, 3750 m a.s.l. T (°C)	−10	−10	−8	−4	+3	+9	+12	+11	+7	+2	−5	−9
P (mm)	137	91	105	102	68	70	80	57	72	39	112	88
Qinghai-Xizang Plateau (Tibet):												
Fenghuo Shan, 4800 m a.s.l. T (°C)	−18.2	−14.7	−11.6	−6.5	−2.2	1.6	4.2	4.8	1.5	−5.1	−13.5	−18.5
P (mm)	1.4	2.2	4.4	6.2	35.1	109.6	86.3	47.2	47.0	5.4	0.0	0.5
Climates of low annual temperature range:												
Jan Mayen T (°C)	−3	−3	−4	−2	−1	+3	+5	+5	+3	−1	−2	−4
P (mm)	40	38	30	29	13	15	23	23	48	43	33	30
South Georgia, Gruytviken T (°C)	+5	+5	+4	+2	0	−2	−2	−2	+1	+2	+3	+4
P (mm)	84	104	129	134	139	127	139	129	86	66	86	86

Table 3.1. *Continued*

(a) Temperature (°C) and precipitation (mm) data

Antarctica:
Inexpressible Island, Northern Victoria Land
80m. a.s.l.

	J	F	M	A	M	J	J	A	S	O	N	D
T(°C)	-4	-12	-20	-24	-26	-25	-25	-26	-25	-19	-10	-4

Wind speed (V):
(knots)

<1 –	2.3%
1–6 –	12.7%
7–16 –	18.3%
17–27 –	23.2%
28–40 –	24.9%
>41 –	18.6%

Direction: 320°
Constancy: 0.80

(b) Summary data	Latitude	Mean annual temperature (°C)	Annual temperature range	Total precipitation (mm)
High Arctic Climates				
Spitsbergen	78° N	-8	25	298
Sachs Harbour (Ikaahuk)	72° N	-14	36	93
Continental climates				
Yakutsk	62° N	-10	62	247
Dawson City	64° N	-5	45	343
Alpine climates				
Sonnblick	47° N	-7	15	1638
Niwot Ridge	39° N	-3	22	1021
Qinghai-Xizang Plateau (Tibet)				
Fenghuo Shan	34° N	-6	23	345
Low temperature range				
Jan Mayen	71° N	0	8	365
Gruytviken	54° S	+2	7	1309
Antarctica				
Inexpressible Island	74° S	-18	19	N/A

Sources: Troll (1944) and local records (AES, Canada; INSTAAR, USA; PNRA, Italy; Northwest Railway Institute, China).

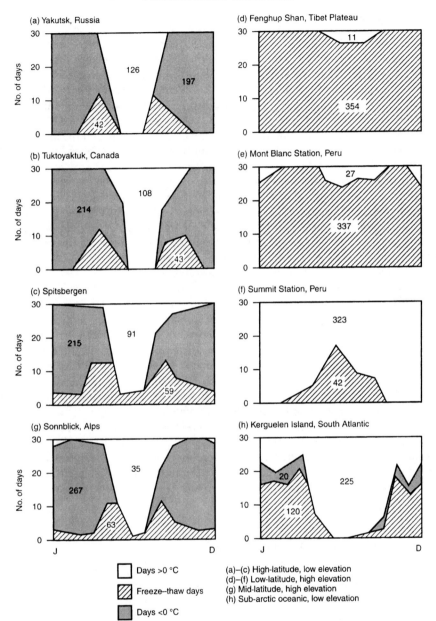

Figure 3.2. Freezing and thawing conditions in various periglacial environments of the world: (a) Yakutsk (lat. 62°01′N, long. 129°43′E, altitude 108 m), Siberia, Russia; (b) Tuktoyaktuk (lat. 69°27′N, long. 133°02′W, altitude >10 m), Mackenzie Delta, Canada; (c) Green Harbor (lat. 78°02′, long. 14°14′E, altitude 7 m), Spitsbergen; (d) Fenghuo Shan (lat. 34°20′N, long. 92°52′E, altitude 4800 m), Qinghai-Xizang (Tibet) plateau, China; (e) Mont Blanc Station, El Misti (lat. 16°16′S, long. 71°25′W, altitude 4760 m), South Peru; (f) Summit Station, El Misti (lat.16°16′S, long. 71°25′W, altitude 5850 m), South Peru; (g) Sonnblick (lat 47°03′N, long. 12°57′E, altitude 3060 m), Austria; (h) Kerguelen Island (lat 49°30′S, long. 69°30′E, altitude at sea level), Southern (Indian) Ocean. Sources: (a), (c), (e), (g), and (h) from Troll (1944); (b) from AES records, Canada; (d) constructed from monthly means of air temperature recorded at Fenghuo Shan by Northwest Railway Institute personnel.

3.2.2. Continental Climates

Large areas south of the treeline in sub-arctic and continental locations experience a much wider range of temperatures than High Arctic areas. Here, although mean annual air temperature is approximately the same or slightly higher than in areas north of the treeline, the mean value disguises excessively low temperatures in winter and remarkably high temperatures in summer. Yakutsk, in central Siberia, and Dawson City, in Yukon Territory, illustrate this phenomenon (Table 3.1). For example, at Yakutsk, January temperatures of $-40\,°C$ and July temperatures of $+20\,°C$ are common, and the annual average temperature range is $62\,°C$. The latter value can be compared to Sachs Harbour, where the annual range is only $36\,°C$.

Besides being hotter, summers are longer than in High Arctic climates, and above-freezing temperatures occur for 5–6 months of the year. Typically, seasonal thaw may penetrate to depths in excess of 2.0–3.0 m. Precipitation amounts are also greater than in High Arctic regions because disturbances associated with the Arctic and polar fronts are more frequent at these latitudes. Between 250 mm and 600 mm per annum are typical, with the majority falling during the summer months. However, evaporation rates are also high during the summer and there is usually a soil moisture deficit.

The continental climatic regime supports a northern or boreal forest. This is termed taiga in Siberia. In North America, the boreal forest is composed largely of spruce (*Picea glauca* and *Picea mariana*) whereas in Siberia it is composed predominantly of pine (*Pinus silvestris*) and tamarack (*Larix dahurica*). In northern Eurasia, birch (*Betula pubescens*) is more common. The trees, the high summer air temperatures, and the dryness of the summer months constitute a favorable environment for forest fire. According to Viereck (1973b), wildfires burn an average area of 400,000 ha/year in Alaska alone. Some ecologists regard the northern boreal forests as fire-climax communities. Fires constitute, therefore, a distinct feature of the continental climatic environment. They are rare, but not unknown, in the southern tundra but have no counterpart in the treeless and cooler High Arctic environments. Winter snowfall is another differentiating characteristic of continental climates since amounts are considerably greater than in High Arctic climates. Together, snowfall and tree cover interact to determine whether permafrost forms or not. Beneath trees, where snow depth is less, there is deeper frost penetration and a greater probability of permafrost occurrence (Viereck, 1965).

3.2.3. Qinghai-Xizang (Tibet) Plateau

Because the Tibet plateau has low latitude and high elevation, it experiences both diurnal and seasonal rhythms. The mean annual air temperature ranges between $-2.0\,°C$ and $-6.0\,°C$ and the annual range of monthly air temperature is relatively low when compared to continental and High Arctic climates. In contrast to alpine environments, however, precipitation is low, being more similar to high latitudes. Because evaporation rates of between 1200 mm/year and 1800 mm/year are typical and annual total precipitation ranges between 200 mm and 600 mm, most of it concentrated in the June–September period, the plateau has a marked water-budget deficit. As a consequence, vegetation is steppe-tundra in nature and more barren then the Low Arctic tundra. Snow cover is also generally thin, <6 cm on average. It often disappears two to three days after deposition when air temperatures rise above $0\,°C$ in mid-afternoon, or when evaporation and/or sublimation occurs. In general, the mean annual air temperature on the plateau changes with both latitude and altitude: lapse rates range between $0.44\,°C$ and $0.55\,°C$ per 100 m.

The annual fluctuations in solar radiation and air temperature create a distinct freeze–thaw regime. Data for Fenghuo Shan are illustrated in Table 3.1 and in Figure 3.2; these are typical for most stations on the plateau. Although the plateau experiences approximately the same number of freeze–thaw cycles as other localities of similar latitude and altitude (e.g. Mont Blanc Station, Peru; see Figure 3.2), the number of frost-free days is approximately one order of magnitude less on the Plateau than at high latitudes. For example, the average number of frost-free days at Fenghuo Shan is only 11, while at Tuktoyaktuk, in the Canada Arctic, it is 108, despite the fact that the mean annual air temperature at Tuktoyaktuk is about 4–5 °C lower than at Fenghuo Shan. This apparently anomalous situation reflects the fact that the Tibet plateau has a much smaller temperature range than the high latitudes, where seasonal rhythms dominate over diurnal ones.

In summary, the unusual climate of the vast Tibet plateau justifies its recognition as a distinct periglacial climate.

3.2.4. Alpine Climates

Alpine periglacial climates are characteristic of tundra regions lying above timberline (treeline) in mid-latitude locations. They are not as extensive as any of the three previous climatic environments so far described. In the European Alps and the North American Rockies, the timberline occurs at elevations varying between 2000 m and 4000 m, but in sub-arctic localities, such as northern Scandinavia, Iceland, and northern Labrador-Ungava, the timberline approaches sea level. Irrespective of elevation, however, all these climatic environments experience a seasonal and diurnal rhythm of both temperature and precipitation.

None of the alpine climates experiences the severe winter cold of either High Arctic or continental climates. On the other hand, the diurnal and seasonal rhythm imposed by their mid-latitude locations results in a higher frequency of temperature oscillations around the freezing point. Precipitation is also heavy, the result of either orographic or maritime effects. Total amounts often exceed 750–1000 mm per annum, much of which occurs as snow. Permafrost is often lacking or discontinuous because of the high mean annual temperature and the protection given to the ground surface by winter snow cover.

3.2.5. Climates of Low Annual Temperature Range

There are certain areas of the world which experience not only a mean annual air temperature below +3 °C but also a remarkably small range of temperatures. These rather unique climatic conditions occur in two types of localities.

The first is a sub-arctic, oceanic location. Here, the surrounding water exerts a moderating influence upon temperature. In the northern hemisphere, Jan Mayen and Bear Island experience this climate, while in the southern hemisphere, the islands surrounding Antarctica, such as the Falkland Islands, Kerguelen, South Georgia, and the South Orkneys, are similar. In these climatic environments, the mean annual amplitude of temperature is in the order of only 10 °C (Table 3.1). Not surprisingly, these areas experience a high frequency of freeze–thaw cycles of short duration and shallow ground penetration. Because of their maritime locations, these islands have considerable amounts of precipitation, often varying between 1000 mm/year and 2000 mm/year, and unstable cyclonic weather, with much low cloud and fog.

The second is an alpine location in low latitude where diurnal temperature variations dominate over weak seasonal influences. Numerous shallow freeze–thaw cycles occur throughout the year and precipitation may vary from near-arid to humid, depending upon location. These climatic conditions exist near the summits of the various mountain ranges in South America, East Africa, and near the summit of Mauna Loa, Hawaii. Clearly, they are not extensive in area.

3.2.6. Antarctica: A Special Case

The climate experienced by the ice-free areas of Antarctica finds no easy place in any climatic classification. This is because Antarctica has a range of climates that reflect the vast size and unique location of the continent. The mildest are those experienced by the Antarctic Peninsula that extends northwards into the South Atlantic Ocean. The harshest are those experienced by the ice-free summits of the Trans-Antarctic Mountains and the other small nunataks that protrude through the vast Antarctic ice sheet. In between in character are a number of relatively small ice-free locations that surround the continent. These are close to open water during 1–3 months of the year but may be over 100–200 km away from the edge of the maximum expansion of the surrounding sea ice in winter.

The ice-free areas in southern and northern Victoria Land constitute a special case. They result from strong katabatic winds that promote sublimation to such an extent that perennial snow and ice is unable to form. These environments are best described simply in terms of their air temperature and wind characteristics (see Table 3.1). For example, Table 10.7 presents data from several automatic weather stations maintained by the Italian Antarctic Program (PNRA) within a 50 km radius of their coastal station at Terra Nova Bay, northern Victoria Land. The mean annual temperature ranges between –15 °C at the coast to –24 °C inland. Only the coastal station records average air temperatures that rise above –5 °C. Equally important is the exceptional strength and duration of the wind. For example, the station located on Inexpressible Island records near-continuous wind flowing outwards from the Priestley Glacier at speeds in excess of 50 km/hour for over 51% of the time. Clearly, the ice-free areas that surround the Antarctic continent are some of the coldest, windiest, and driest areas on Earth.

3.3. GROUND CLIMATES

Air temperatures are a useful way of characterizing periglacial climates. However, temperature at the ground surface and the immediate near-surface are probably more relevant when considering cold-climate weathering processes and permafrost formation.

Typically, there are significant differences between air and ground temperatures. These reflect either the effects of vegetation or the ameliorating effect of snow cover (Burn and Smith, 1988a; Goodrich, 1982; Mackay and MacKay, 1974). However, the situation is complex because wind also influences snow cover and its redistribution. The latter can protect the ground from the negative effects of wind or, conversely, can promote desiccation and abrasion by wind-driven snow particles. These and other wind-induced effects become especially important near the northern treeline (Scott et al., 1993) and in the tundra regions. Water bodies are a further complication.

Some of these interactions can be illustrated by data that lists annual mean air, ground (surface), and lake-bottom temperature at a number of sites in the Mackenzie Delta region

Table 3.2. Annual mean air, ground, and lake-bottom temperatures (2000–2004) at sites in the Mackenzie Delta region, Canada, that span the tundra–taiga (forest) transition zone.

Ecotone	Locality	Temperature (°C)			
		Annual mean air temperature	Summer air temperature	Winter air temperature	Ground temperature
Southern Arctic	Illisarvik	−12.5			−5.6
(tundra)	Tuktoyaktuk	−9.0	9.9	−27.0	
	Todd Lake				
	(10 m depth)	3.4			
	(littoral bench)	−2.0			
Taiga Plain	Inuvik	−6.7	13.4	−25.5	
(forest)	Inuvik peat lands				−1.2
	Dempster Lake	5.7			
	(3 m depth)				

Source: Burn (2005). Reproduced by permission of John Wiley & Sons Ltd.

of Canada (Table 3.2). Collectively, the sites span the tundra–taiga (forest) boundary. This is a zone 50–150 km in extent that extends southwards from the arctic coast to beyond the treeline. Over this area the mean annual air temperature varies by 6 °C, there is a difference between air and 3 m lake-water depth temperature of ~18 °C, and the range between maximum and minimum air temperature is ~40 °C. Clearly, the use of average annual or monthly air temperatures values to describe periglacial climatic conditions is a gross oversimplification.

At a more fundamental level, the annual change of state of near-surface material from its frozen to thawed state, and vice-versa, is of specific interest to periglacial geomorphology. This is now discussed briefly in the context of n-factors.

3.3.1. The n-Factor

V. J. Lunardini (1978) first described n-factors as transfer functions between air and ground surface temperatures. Seasonal values can then be multiplied with air temperature indices to obtain suitable approximations for ground temperature. For example, some typical n-factor values assigned to different vegetation types in central Alaska are presented in Table 3.3. Similar values have been successfully applied to the variability of air and soil-surface temperatures in a number of localities (Hinkel et al., 1997; Karunaratne and Burn, 2003, 2004; Klene et al., 2001) but calculation of n-factors for the thaw season is difficult (Taylor, 1995, 2000). This is illustrated by data from five sites in the boreal forest in central Yukon Territory, Canada (Table 3.4). As might be expected, variation in n-factor over the winter seasons was largely the result of differences in snow depth. However, in summer, as the thaw front penetrated below the depth of diurnal temperature fluctuation, the air and ground surface temperatures converged. Because the rate of thaw penetration is governed by soil-thermal diffusivity, this suggests that summer n-factors vary directly with soil-thermal diffusivity.

Table 3.3. Thermal parameters (n-factors) and soil moisture values assigned to certain vegetation types in central Alaska.

Vegetation type	n-factor		Soil Moisture (% wt.)
	thawed (^N_t)	frozen (^N_f)	
Closed aspen forest	1.00	0.30	15
Closed birch forest	0.90	0.35	20
Mixed birch-spruce forest	0.85	0.35	20
Closed white-spruce forest	0.80	0.35	50
Open black-spruce forest	0.60	0.30	50
Closed black-spruce forest	0.50	0.30	45
Low shrub-scrub	0.85	0.30	40
Tussock bog	0.90	0.30	55

Source: Jorgenson and Kreig (1988). Reproduced by permission of Tapir Academic Press.

Table 3.4. Summary of snow characteristic at five sites in the Takhini valley, central Yukon Territory, for the winters 1997/98, 1998/99, and 1999/2000. The relation between air and near-surface (ground) temperatures is expressed by values of the coefficient of determination (r^2). Freezing n-factors were computed using air and near-surface freezing degree-days for the five sites.

Snow characteristic	Forest	Burned	Meadow	Snow fence	Cleared
Max snow depth (cm)	19	28	24	25	0
Mean snow depth (Dec-Feb) (cm)	10	23	15	19	0
Mean duration of snow cover (weeks)	21	20	21	21	0
Coefficient of determination (r^2) (Air: near-surface temperatures)	0.31	0.14	0.21	0.09	0.27
Freezing n-factors	0.54	0.32	0.47	0.32	0.55

Source: Karunaratne and Burn (2003).

3.3.2. The Thermal Offset

If the ground surface is conceived as a thermal rather than a physical boundary, the mean annual temperature immediately above, at, and below the ground surface can be represented schematically (Figure 3.3). Three levels can be recognized (Smith and Riseborough, 2002): (1) air temperature, measured at standard height above seasonal snow cover (MAAT), (2) ground surface temperature (MAGST), and (3) temperature at the top of permafrost (TTOP).

The mean annual air temperature (MAAT) can then be expressed as the thawing index for air temperature (thawing degree-days, TDD) minus the freezing index for air temperature (freezing degree-days, FDD, expressed as a positive number) divided by the number of days in a year. In contrast, the mean annual ground surface temperature (MAGST) reflects the effect of plant vegetation in summer and the effect of snow cover in winter. Because of the insulating effect of the winter snow cover (known as the "nival" offset), the MAGST exceeds the MAAT in winter. However, due to vegetation effects (known as the "vegetation" offset), the situation is reversed in summer, when the MAGST is less than the MAAT. The exact magnitude of these offsets varies depending upon the

Mean Annual Temperature Profile

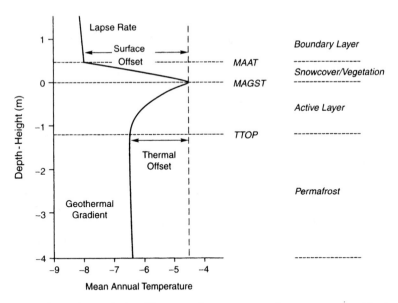

Figure 3.3. A schematic graph that illustrates the mean annual temperature profile through the surface boundary layer in a cold region. It shows the relation between air temperature and ground temperature. From Smith and Riseborough (2002). Reproduced by permission of John Wiley & Sons Ltd.

magnitude of the freezing and thawing indices at the locality concerned, the nature of the vegetation, and the amount, nature, and duration of snow cover. Because the effects of snow cover in winter are generally greater than those of vegetation cover in summer, the MAGST is usually higher than the MAAT.

Therefore, the thermal offset is defined as the difference between the MAGST and the temperature at the top of permafrost (TTOP). While the magnitude of the offset depends directly upon the air-thawing index, it is also determined by the thermal conductivity ratio that must exist between thawed and frozen earth materials. In the case of solid bedrock, where moisture amounts are minimal, this ratio approaches unity and there is little or no thermal offset. However, in most mineral soils the thermal conductivity ratio, or typical n-factor, ranges between 0.6 and 0.9. The greatest range occurs in organic materials (peaty soils), where values range from about 0.3 or less under saturated conditions to close to 1.0 for dry conditions (Riseborough and Smith, 1998).

The thermal offset concept can be applied to tundra lakes. The seasonal maximum ice thickness in the lake environment can be regarded as analogous to the active layer or the zone of seasonal freezing in the terrestrial environment (Riseborough, 2006). The lake-offset value largely reflects the different thermal regime that must exist under ice-covered and ice-free conditions. Offsets in excess of 10 °C are not uncommon in deep lakes and are often the cause, in permafrost terrain, of unfrozen zones existing beneath lakes. This topic is discussed more fully in Chapter 5.

In summary, while air temperature is an important defining characteristic of periglacial climates, ground and surface temperatures are probably more important in the geomorphological context. Seasonal snow cover, soil moisture (thermal conductivity), and ground-thermal properties are crucial site-specific factors that need to be considered.

3.4. PERIGLACIAL CLIMATES AND THE CRYOSPHERE

The periglacial domain is part of the cryosphere. Terrestrial components include seasonal-snow cover, mountain glaciers, ice sheets, perennially-frozen ground (permafrost), and seasonally-frozen ground. The other important component of the cryosphere is Arctic sea ice. Table 3.5 lists the extent of the terrestrial components of the cryosphere.

The cryosphere is particularly sensitive to global warming. Most global climate models (GCMs) predict warming to be greatest is high latitudes. In particular, winter warming is expected to exceed the global annual average. Projected changes in Arctic climates will dramatically reduce sea-ice cover and significantly degrade permafrost. In the Antarctic, satellite sensors show changes in ice extent, including accelerated flow and thinning of glaciers, and the break-up of parts of the Nansen and Ross Sea ice shelves (Shepherd et al., 2001).

To appreciate the magnitude of some of these global changes, Table 3.6 lists what might happen in Fennoscandia if there were a doubling of the CO_2 content in the atmosphere, as modeled by the GCM developed by the Goddard Institute for Space Studies (GISS). The model predicts mean winter temperatures will increase by 5–6 °C and mean annual temperature will increase by 4–5 °C. Also, the growing season (daily mean temperature

Table 3.5. Relative extent of terrestrial areas of seasonal snow cover, ice, and permafrost.

	Area (million km²)	Volume (million km³)
Land Ice:		
Antarctica	13.9	30.1
Greenland	1.8	2.7
Small ice caps	0.35	0.2
Mountain glaciers	0.2	0.03
Major permafrost regions:		Percentage underlain by permafrost
Russia	16.84	50
Canada	9.38	50
China	9.38	22
Greenland	2.18	100
USA (Alaska)	1.52	82
World land area:	140.7	20–25
Northern Hemisphere		
Continuous	7.6[1]	
Discontinuous	17.3[1]	
Alpine permafrost	2.33[1]	
Land ice and seasonal snow:	Area	Volume (million km³)
Northern Hemisphere		
Early February	46.3	0.002
Late August	8.7	
Southern Hemisphere		
Late July	0.85	
Early May	0.07	

[1] Approximate.
Sources: Washburn (1979), Rott (1983).

Table 3.6. Simulated climate change in the Fennoscandia region due to a doubling of the atmospheric CO^2 content, according to a scenario derived from the GISS model.

Climate parameter	Climate change Increase (+) or decrease (−)		Magnitude
Mean annual temperature	+		4–5 °C
Mean winter temperature	+		5–6 °C
Mean summer temperature		+	2–3.5 °C
Annual temperature amplitude		+	2–4 °C
Length growing season (daily mean temp. ≥5 °C)		+	70–150 days
Effective temperature sum (threshold ≥5 °C)	+		500–1000 degrees
Length thermal winter (daily mean temp. ≤0 °C)		−	2–4 months
Length thermal summer (daily mean temp. ≥10 °C)	+		2–3 months
Number heating degree-days (daily mean temp. ≤17 °C)		−	30–40%
Mean annual precipitation	+		150–300 mm
Precipitation excess		+	10–50%

Source: Boer et al., (1990).

≥5 °C) would increase by 70–150 days, and precipitation would increase by 150–300 mm. Clearly, if the changes predicted for Fennoscandia are typical of those predicted for other high-latitude regions, then the extent of periglacial climates, and by extension the periglacial domain, will change radically. Even in low latitudes, the effects of global warming will be important. For example, a recent rise in the mean annual ground temperature on the Tibet plateau between 1962 and 1989 suggests that permafrost may disappear from the Tibet plateau in approximately 150 years.

Many periglacial environments have high negative radiation budgets. Changes in their budgets will prompt major changes in atmospheric and oceanic circulation. For example, the melt of sea ice and ice shelves promotes the formation of cold bottom waters that drain equator-ward and influence oceanic circulation patterns. The polar seas are also large sinks for CO_2 and play an important role in the exchange of CO_2 between ocean and atmosphere.

Positive feedback mechanisms will accelerate global warming in cold regions. For example, the melt of sea ice and seasonal snow cover will reduce albedo and increase solar radiation inputs at the ground surface. Large terrestrial CO_2 and CH_4 (methane) reservoirs will be released as temperatures rise and permafrost thaws in boreal forest and tundra regions.

In summary, the periglacial domain is susceptible to significant change, both in character and extent, if present and predicted global warming trends continue. The cryosphere, of which the periglacial domain is a major component, will play an important role. The implications of climate warming are discussed further in Chapter 15.

ADVANCED READING

IPCC (Intergovernmental Panel on Climate Change) (2001). Third Assessment Report. WGI. *Climate Change 2001: The Scientific Basis* (Houghton, J. T., et al., eds.), 881 pp., and WGII, *Climate Change 2001: Impacts, Adaptations and Vulnerability* (McCarthy, J. J., et al., eds.), 1032 pp., Cambridge University Press, Cambridge.

Karunaratne, K. C., Burn, C. R. (2003). Freezing n-factors in discontinuous permafrost terrain, Takhini River, Yukon Territory, Canada. In: Phillips, M, Springman, S. M., Arenson, L. U., eds., *Permafrost, Proceedings of the Eighth International Conference on Permafrost*, 21–25 July, Zurich, Switzerland. Balkema, Lisse, vol. 1, pp. 519–524.

Smith, M. W., Riseborough, D. W. (2002) Climate and the limits of permafrost: a zonal analysis. *Permafrost and Periglacial Processes*, **13**, 1–15.

Troll, C. (1944). Structure soils, solifluction and frost climate of the Earth. (English translation, 1958: Translation 43, US Army Snow Ice and Permafrost Research Establishment, Corps of Engineers, Wilmette, Illinois, 121 pp.)

DISCUSSION TOPICS

1. How can one measure the variability of periglacial climates?

2. What controls the difference between air and ground temperature in periglacial regions?

3. What is the importance of the cryosphere in global climate change scenarios?

PART II

Present-Day
Periglacial Environments

4 Cold-Climate Weathering

Frost action in soil refers to two processes: (1) frost heave that occurs as water changes state during the freezing period, and (2) thaw weakening and settlement that occurs as seasonally-frozen ground thaws. Frost action in bedrock is caused by hydro-fracturing, ice segregation, thermal shock, or a combination of these processes, and leads to mechanical disintegration of material. Frost action is one component of cryo-genic weathering, a poorly-understood group of physico-chemical processes that operate in cold climates. Specific chemical weathering processes include salt weather-ing, solution, and biological weathering. Cryogenic processes result in distinctive soil morphology patterns.

4.1. INTRODUCTION

Since publication of the *Second Edition*, significant advances have been made in our understanding of periglacial weathering processes. This statement does not necessarily imply that new and important processes have been discovered. Rather, it reflects a cautious and considered re-evaluation of known processes combined with a growing appreciation that much is still unknown about the nature of cold-climate, or cryogenic, weathering. To reflect these changes, the title of this chapter has been modified from "Frost action and cryogenic weathering" to the more general title "Cold-climate weathering."

This change may appear semantic but it reflects a growing acceptance that frost action has been overemphasized in the past and that temperature alone is not necessarily the controlling factor in cold-climate weathering. A related assumption that is now question-able is that cold-climate landscape modification is relatively rapid. While this might be the case in certain areas, it may also be interpreted as a reflection of the paraglacial adjust-ment that many landscapes are currently experiencing rather than the implied efficacy of frost action. In addition, it is increasingly realized that physico-chemical processes, espe-cially those that operate at sub-zero temperatures, are far more complex, and certainly less well understood, than traditional weathering processes.

Nevertheless, frost action remains the defining characteristic of periglacial environ-ments and demands detailed consideration.

4.2. GROUND FREEZING

Frost action is a collective term used to describe a number of distinct processes which result from alternate freezing and thawing in soil, rock, and other materials. Frost action also refers to the effects of frost on material and structures placed on, or in, the ground.

The latter are dealt with in Part IV; here we deal with the geomorphic processes associated with repeated freezing of ground and the manner by which destruction of rock and soil proceeds under cold-climate conditions.

4.2.1. The Freezing Process

The freezing process has attracted much attention from geotechnical engineers in Canada, Norway, the USA, China, and Russia, and from government agencies such as the US Army Cold Regions Research and Engineering Laboratory (CRREL), the US Highway Research Board, the Norwegian Geotechnical Institute, and the National Research Council of Canada. The detailed physics of the freezing process need not concern us; the text by P. J. Williams and M. W. Smith (1989, pp. 1–8, 174–201) includes a good discussion. However, a few comments are necessary to preface this section.

First, different soils cool at different rates depending upon heat conductivity and moisture content, and thus freeze at different rates. Second, soils do not necessarily freeze when their temperatures fall to $0\,°C$ since they are known to exist in a supercooled state. For example, a common condition is for saline groundwater to lower the temperature at which soil freezes. Third, the duration and intensity of a temperature drop below $0\,°C$ will affect the rate and amount of soil freezing.

Bearing these comments in mind, it is also necessary to describe a few basic relationships that relate to the freezing of soil and bedrock.

First, pure water freezes at $0\,°C$ and in doing so expands by approximately 9% of its volume. This phase transition, between liquid and solid, is fundamental to our understanding of frozen and freezing soils. The most obvious result is the volume increase in soil; this is commonly known as frost heave and has considerable geomorphic and practical significance, the latter by displacement of buildings, foundations, and road surfaces (see Part IV), the former by heaving of bedrock, uplifting of stones and objects, and frost sorting (see Chapter 8). However, it must be emphasized that the 9% expansion associated with the change from water to ice is not what permafrost scientists normally regard as frost heave. For soil to heave, ice must first overcome the resistance to its expansion caused by the strength of the overlying frozen soil. This usually occurs only when segregated ice lenses form.

Second, not all water freezes at $0\,°C$. This is because soil moisture, or water in the ground, commonly contains dissolved salts. These result in depression of the freezing point. This concept is illustrated in Figure 4.1. Usually, because the concentration of dissolved salts is weak, the freezing-point depression is only $0.1\,°C$ or so below $0\,°C$.

A third physical process relates to the molecular forces that exist between phases when the interface is confined. This is known as capillarity. In soils, the capillarity between soil particles increases as the soil particles become smaller. Moreover, during soil freezing, the formation of ice results in water being confined progressively in a smaller space. Accordingly, the free energy of the water falls as freezing takes place and this is most apparent in fine-grained sediments. This effect is termed cryosuction and is the cause of water migration to the freezing zone.

Fourth, the amount of water which freezes at any one particular temperature will also depend on the nature of the soil in question. This is because different mineral particle surfaces have different adsorption properties. Adsorption refers to forces emanating from the particle surface which lower the free energy of water in a thin layer near the surface of the particle. For freezing to continue, this moisture needs to be converted to ice, and consequently lower temperatures are required. Usually, a film of water separates soil ice

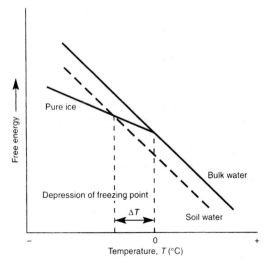

Figure 4.1. Schematic diagram illustrating how the lower free energy of soil water, a weak solution,leads to a depression of the freezing point. From Williams and Smith (1989). Reproduced by permission of Cambridge University Press.

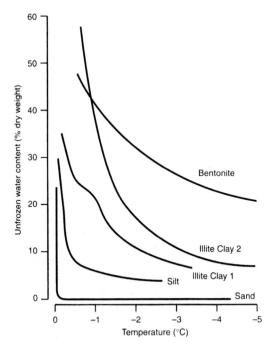

Figure 4.2. Diagram illustrating the different amounts of unfrozen water existing in different soils at temperatures below 0 °C. From Williams and Smith (1989). Reproduced by permission of Cambridge University Press.

from solid particles. Thus, "frozen" soil still contains unfrozen water. However, as temperatures fall, these films become thinner. This explains how, in a series of laboratory experiments, it can be demonstrated that different soils retain varying amounts of unfrozen water at temperatures below 0°C (Figure 4.2).

Several other soil-moisture freezing relationships need brief mention. First, latent heat is involved in the phase change of H_2O. For example, when ground freezing occurs, ground temperatures initially remain approximately constant due to latent heat flux. This is the so-called "zero-curtain" effect (see Figure 4.6). It should be emphasized that heat flow always takes place in response to a thermal gradient. The latent heat released at the freezing front flows upwards along the thermal gradient, so temperatures do not rise; rather, they remain constant for as long as sufficient latent heat is released to satisfy the thermal gradient and thermal conductivity. When ice formation decreases, due to reduced supply of water, less latent heat is released and the soil cools down, allowing the freezing front to advance once again. Second, the progressive freezing of soil water helps determine the thermal properties of the soil. For example, the apparent heat capacity will rise and the thermal conductivity of frozen soil will change as the ice and water contents change. Third, pressure lowers the freezing point of water and this can result in unfrozen conditions because of high overburden pressures (as beneath advancing glaciers) at depth.

4.2.2. Ice Segregation

As soil freezes the water present within the soil either freezes in-situ, to form pore ice, or moves towards the freezing plane to form lenses of segregated ice (Smith, 1985a). Early work carried out upon the mechanism of segregated-ice formation by S. Taber (1929, 1930) led to two important conclusions: (a) ice segregation is favored in materials having a grain-size composition of 0.01 mm diameter or less, and (b) ice crystals grow within the material in the direction in which heat is being most rapidly conducted away, i.e. normal to the surface. These conclusions have been verified by a number of later studies.

The cryosuction that results from capillarity increases by approximately 1.2 MPa per °C below 0 °C (Williams and Smith, 1989, p. 190). Cryosuction is expressed using the terms P_i (pressure of ice) and P_w (pressure of water):

$$P_i - P_w = \text{soil constant } (2\sigma/r)$$

The soil constant is a function of σ, the surface tension and soil particle volume, and r, the radius of curvature of the soil particle surface (i.e. grain size). This relationship determines whether segregated ice or pore ice will form. For example, Figure 4.3 considers two soil particles as representing a section of soil located approximately at the freezing level. When the ground freezes, the freezing plane may either remain stationary above the soil particles, or tongues of ice may descend through the pores. In the case of the latter, pore ice will result. However, if the freezing plane remains above the soil particles, water will move upwards through the soil pores from the unfrozen ground beneath and towards the freezing plane. Ice crystals will develop and, as long as the supply of water is maintained, a lens of segregated ice will develop. This promotes upward heave of overlying sediment.

The main control over whether the freezing level will remain stationary or descend is cryosuction because this tension controls the supply of water to the freezing plane. If the tension is low, a tongue of ice moves down through the soil by progressively freezing the pore water in place. If the tension is high, it will maintain the freezing plane at its original level. In this case, a lens of segregated ice will form. When it is appreciated that fine-grained soils possess small interstices, and cryosuction can be more easily maintained in such a situation, it is clear that segregated ice formation is favored in fine-grained materials. Such sediments are termed "frost-susceptible" by engineers. Pore ice, by contrast,

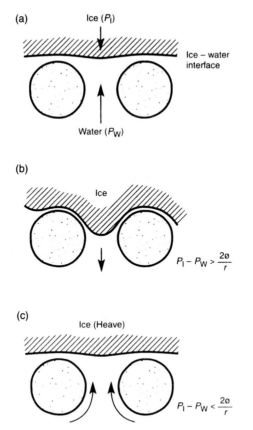

Figure 4.3. Schematic illustration of two mineral soil particles with ice and water: (a) general conditions for segregated or pore ice formation in freezing soil; (b) growth of pore ice; (c) growth of segregated ice + frost heave.

commonly develops in coarser-grained material, often sand and gravel, where it is harder for the tension of cryosuction to be maintained in the correspondingly larger soil interstices. These sediments are termed "non-frost-susceptible" and are used as aggregate in building and for construction purposes, or wherever ground heaving needs to be minimized.

Under certain conditions, apparently non-frost-susceptible materials may contain large segregated ice lenses. This is commonly explained by saturated conditions associated with pore-water expulsion ahead of the freezing plane.

4.2.3. The Frozen Fringe

The frozen fringe refers to the transition zone in a freezing, frost-susceptible soil that lies between the warmest isotherm in which ice exists in pores and the isotherm at which the warmest ice lens is growing (Miller, 1972). The concept was introduced to accommodate the fact that freezing of soil moisture does not occur instantaneously at $0\,^{\circ}\mathrm{C}$. It was subsequently developed by geotechnical engineers interested in predicting frost heave (Konrad and Morgenstern, 1983). The freezing front refers to the downward-advancing boundary between frozen (or partially-frozen) ground and unfrozen ground. As such, it does not necessarily coincide with the 0 °C isotherm (the cryofront). Under these conditions, water is drawn upwards to the advancing freezing plane by cryosuction, first to form pore ice when it reaches the freezing plane, and subsequently to form segregated ice lenses. Thus, the "frozen fringe" refers to the unstable conditions in the cryogenic zone immediately adjacent to the frost line.

4.2.4. Frost Heave

Frost heave refers to the raising of the ground surface following formation of segregated ice. The latter develops when soil moisture migrates through the frozen fringe to form discrete layers or lenses. To understand this process, we need to consider the phase transition of substances. To recap, two phases of a single substance can coexist only when the free energies of the phases are equal. In the case of water (H_2O), this exists when water is at the freezing (or melting) point. However, at temperatures below the freezing point, the free energy of the liquid state exceeds that of the solid state. The result is an unstable condition in which the phase of lower free energy is increased at the expense of the higher energy phase. In other words, as temperature drops below 0 °C, water progressively changes to ice (i.e. the solid phase).

It is also useful to distinguish between primary and secondary frost heave. Primary heave usually refers to the heave that occurs near to the frost line while secondary heave refers to heave that occurs within frozen layers at various temperatures. Primary heave occurs predominantly in the autumn freeze-back period and is related to the volumetric expansion of water as it freezes combined with the growth of segregational ice lenses. Secondary heave is not so clearly understood, but occurs at temperatures below 0 °C and at some distance behind the freezing front. Several experiments have demonstrated that moisture migrates through frozen soil (Burt and Williams, 1976; Ershov et al., 1980; Hoekstra, 1969; Vtyurina, 1974). The importance of secondary heave lies in the large heaving pressures that may develop over time.

Two quantitative parameters are used to describe frost heave. The first, the frost-heave coefficient (η), is defined as the ratio of total heave (Δh) to frost penetration (h):

$$\eta = \Delta h/h \tag{4.1}$$

A second parameter, the frost-heave strain (ε), is often used in preference to the frost-heave coefficient. This is defined as the ratio of total heave (Δh) to frost penetration (h) minus the total heave (Δh):

$$\varepsilon = \Delta h/h - \Delta h \tag{4.2}$$

The reason why frost-heave strain (ε) is usually preferred to the frost-heave coefficient (η) relates to the fact that it emphasizes relative heave and better reflects frost-heave susceptibility and ice content. For example, consider a hypothetical frost-susceptible regolith that is 35 cm deep before freezing and 47 cm deep after freezing. The magnitude of frost heave (Δh) is $47 - 35 = 12$ cm, frost penetration (h) is 35 cm, and ($h - \Delta h$) is $35 - 12 = 23$ cm. These give values of $\eta = 34.3\%$ and $\varepsilon = 52.2\%$. Real data are provided by the following two examples. First, typical field values of ε for saturated, fine-grained, and highly frost-susceptible lake-bottom sediments in the Mackenzie Delta, Canada, range between 20% and 60% (Burn, 1990b). Second, and by contrast, frost heave measurements on well-drained slopes on the Tibet Plateau, China, indicate values of ε that are only between 3%

and 6%. These reflect the moisture-poor nature of the Tibet Plateau (Wang and French, 1994).

In spite of the fact that frost-heave pressures are relatively easily explained, prediction of their amounts appears difficult and controversial, as in the case of chilled gas pipelines (Konrad and Morgenstern, 1984; Williams, 1986). Frost-heave pressures as high as 100–300 kPa may develop relatively quickly (Williams and Smith, 1989, p. 197) and there is still no widespread agreement upon the basis for quantitative predictions of heave.

4.3. FREEZING AND THAWING

The frequency of freezing and thawing is of particular significance with respect to frost wedging and rock shattering. While the calculation of thawing and freezing degree-days, as described in Chapter 3, provides an index of the severity of the climate and the magnitude of the thaw and freezing periods, these data give no indication of the frequency at which temperatures oscillate above and below the freezing point. Yet the frequency of freezing and thawing ("freeze–thaw cycles"), and the formation of pore and segregated ice, is significant with respect to frost wedging (Matsuoka, 2001a, b) and rock shattering (Douglas et al., 1983; Mackay, 1999).

Unfortunately, several problems limit the usefulness of freeze–thaw cycles as a measure of frost-action effectiveness. First, there is the basic difficulty of defining the exact point of freezing across which the oscillations should be measured. Second, the use of air temperatures to define cycles is not satisfactory because, as demonstrated in Chapter 3, significant differences exist between air and ground temperatures. Third, even when direct ground temperature measurements are available, just what constitutes a freeze–thaw cycle is debatable. Each occasion when water either freezes or melts requires a different degree of heating and cooling dependent upon such factors as the ground temperature, the moisture and/or unfrozen water content, and the nature of the soil or bedrock. Surface conditions, such as the intensity of solar radiation and the character and depth of any snow cover, are also relevant. Fourth, cycles can be of different intensities (i.e. different temperature ranges) and this makes any comparison of cycle frequency difficult. Finally, the duration of a cycle can range from seconds to several days (Hall and André, 2001), and it is probably unwise to assign equal significance to the various cycles.

Numerous studies conducted in many different cold-climate environments now seriously question the assumption that numerous freeze–thaw cycles occur. Direct ground surface measurements indicate the number is surprisingly few (Table 4.1). The greatest occur at the ground surface and these are twice as numerous as air cycles. With depth, however, there is a rapid drop in frequency such that beneath 5.0–10.0 cm, only the annual cycle takes place.

Cold oceanic climates of low annual temperature range (see Chapter 3) are traditionally assumed to be the most suited for freeze–thaw processes. Even here, where there is a relative abundance of moisture for ice segregation, the number of freeze–thaw cycles is low. For example, on Signy Island, only 19 and 14 freeze–thaw cycles were recorded at the 1.0 and 5.0 cm depths, respectively (Chambers, 1966), and at depths in excess of 10.0 cm only the annual cycle occurred (Table 4.1). Mid-latitude alpine climates are also regarded as being particularly suited to numerous freeze–thaw cycles since they experience marked diurnal temperature fluctuations. In the Japanese Alps, an annual average of 88 cycles occurred at a depth of 1.5 cm over a 5-year period, while in the Colorado Front Range, USA, over 50 fluctuations across the 0 °C threshold were monitored within 24 hours at two sites over one year.

Table 4.1. The frequency of some typical freeze–thaw cycles, variously defined, as recorded in different types of periglacial environments.

Location	Climatic Type	Ground Cycles	Cycles at 1.0–2.0 cm	Cycles at 5.0 cm	Cycles at 20 cm	Definition of Cycle
Resolute Bay, Canada, 74°N	High Arctic	23	n.d.	0	n.d.	−2 to 0°C (28–32°F) (Cook and Raiche, 1962)
Mesters Vig, Greenland, 72°N	High Arctic	23	n.d.	18	n.d.	Amplitude ≥ 0°C (Washburn, 1967)
Siberia						
(a) Kolyma, 67°N	Continental – tundra	50	n.d.	n.d.	2	Temperature transitions across 0°C (Ushakova, 1986)
(b) Vilyusk, 63°N	Continental – taiga	94	n.d.	n.d.	2	
(c) Krasnoyarsk, 56°N	Continental – taiga/steppe	59	18	14	4	
Signy Island, 61°S	Low temperature range		19	4		+0.5 to −0.5°C (Chambers, 1966)
Front Range, Colorado, 39°N	Alpine (3750 m)	n.d.	9	n.d.	n.d.	>0°C and <0°C within 24 hours (Fahey, 1973)
	Subalpine (3000 m)	n.d.	50	n.d.	n.d.	
Japanese Alps, 35°N	Alpine (2800 m)	n.d.	88	n.d.	n.d.	−2 to +2°C (Matsuoka, 1990)
Terra Nova Bay 74°S	Antarctica	42	n.d.	n.d.	n.d.	−4.0°C (French and Guglielmin, 1999)

n.d., not determined.

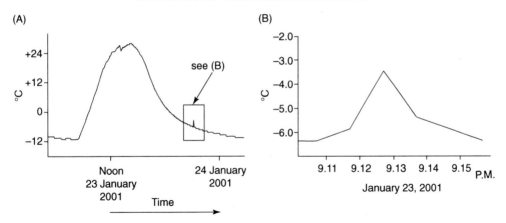

Figure 4.4. High-frequency temperature measurements observed on a south-facing vertical brick surface, January 23–24, 2001, Prince George, northern British Columbia, Canada. Modified from Hall (2004). Adapted with permission from John Wiley & Sons Ltd.

Collectively, the type of data presented in Table 4.1. led K. Hall to explicitly question the effectiveness of freeze–thaw weathering (Hall, 1995, 1997a, 1998). More recently, in a series of studies that involved high-frequency surface and near-surface rock temperature measurements, attempts have been made to further quantify the magnitude and nature of freeze–thaw (Hall, 2004; Hall and André, 2001, 2003). One study involved temperature measurements made at 1-minute intervals upon a set of paving bricks exposed to freezing and thawing in northern British Columbia, Canada (Hall, 2004). The high frequency of measurement enabled the recognition of so-called "exotherms" (cf. exothermic (adj.), chemical change that generates heat) (Figure 4.4). The temperature at which freezing occurred varied through the year, with some inferred exotherms indicating that freezing occurred when air temperatures were approaching –30 °C. On one occasion, the 1 cm depth froze before the surface of the brick. These changes in freeze temperature may be related to the chemical composition of the paving brick coupled with salt from the melt of snowfall. More details concerning the possible interactions between chemical and mechanical weathering at cryogenic temperatures are discussed later in this chapter.

4.4. THE GROUND-TEMPERATURE REGIME

The majority of periglacial environments experience some form of seasonal change through the year. It is possible, therefore, to recognize a summer thaw period of limited duration and a longer winter period during which temperatures remain below 0 °C. The exception is the alpine periglacial environment of low latitudes, which experiences a diurnal rather than a seasonal cycle.

4.4.1. The Seasonal Regime

Numerous data sets record seasonal ground temperature changes. The typical regime is illustrated by data from Fenghuo Shan on the Tibet Plateau, China (Figure 4.5). The most obvious characteristic is that, at a depth of 1.2 m, the ground remains frozen throughout

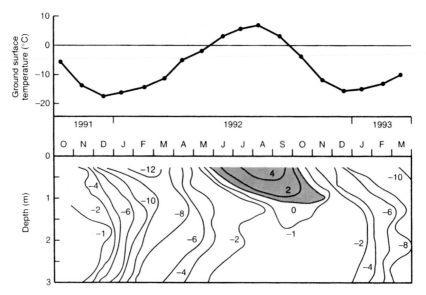

Figure 4.5. Annual ground temperature regime recorded at Fenghuo Shan, Tibet Plateau, 1991–1993. The thawed active layer is indicated. The elevation of the station is 4850 m a.s.l. From Wang and French (1995c). Reprinted with permission from Elsevier.

the year. Second, the deeper sections of the seasonally-thawed layer are in a transitional freezing zone, between +2 °C and −4 °C, for over seven months of the year. Third, the length of time during which ground temperature remains within the transitional freezing boundary progressively decreases towards the ground surface. Fourth, the near-surface experiences an increasing number of short, shallow temperature fluctuations related to diurnal, cyclonic, or other influences.

From a geomorphic viewpoint, the spring thaw is of special interest because it influences the nature of spring runoff. Thaw occurs quickly, and approximately 75% of the soil thaws within the first five weeks of air temperature rising above 0 °C. The rapidity of thaw is the result of meltwater percolating through the soil, thereby transferring heat to frozen material beneath. Coarse sediments are particularly suited to rapid thaw since percolation is easy and thermal conductivity of such material is high.

The fall freeze-back is more complex and may extend over 8–10 weeks. This is illustrated with data from Adventdalen, Svalbard (Figure 4.6). Freezing occurs mostly from the surface downwards but also, to a lesser extent, from the permafrost surface upwards. This so-called "two-sided" freezing is an important differentiating criterion between areas experiencing perennial frost (permafrost) and those experiencing seasonal frost. In Adventdalen, freezing begins in early September and ends in early November. During the majority of this time, ground temperatures remain remarkably constant in the range of −2 °C to 0 °C. This transitional temperature zone is the "zero curtain" (see earlier) and occurs because the onset of freezing during autumn releases latent heat to the soil, which temporarily compensates for the upward heat loss associated with the drop in air temperatures. It is analogous to the inferred "exotherm" phenomenon reported by Hall (2004) (see above) but is a large-scale effect.

A second feature of interest in the freeze-up period is the initially slow rate of freezing from the surface downwards and then the dramatic speed-up in freezing at depth. This

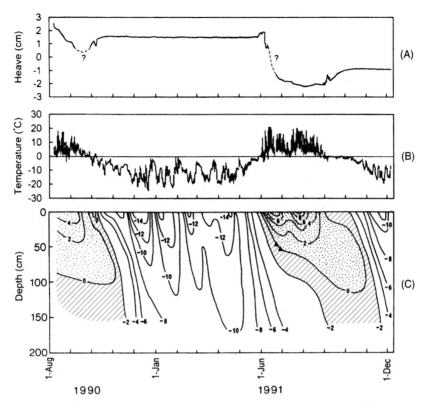

Figure 4.6. Frost heave and ground temperature data recorded at a site in Adventdalen, Svalbard, between August 1990 and December 1991. (A) Frost heave; (B) surface temperature; (C) subsurface isotherms. The zero-curtain lies between 0 °C and −2 °C and is indicted by diagonal shade. The thawed active layer is indicated by stipple. From Matsuoka and Hirakawa (2000). Reproduced by permission of the National Institute of Polar Research, Japan.

increase in the rate of freezing at depth is the result of two factors. First, soil moisture decreases with depth because it has already been drawn upwards towards the freezing plane to form segregated ice lenses. Therefore, the amount of latent heat released as a result of the water–ice phase change at depth is reduced and no longer offsets the temperature fall. Second, up-freezing from the perennially-frozen ground combines with down-freezing from the surface to quicken the overall movement of the freezing plane at depth.

The implications of these observations with respect to weathering processes are uncertain. It is clear that much of the soil profile remains within a transitional freezing boundary (−2 °C to 0 °C) for a long and continuous period. The restriction of short-term temperature fluctuations to the surface layers implies that they are incapable of inducing mechanical weathering at depth. It follows that freezing boundary conditions, which are related to the annual cycle, are probably more relevant when considering weathering at depth. However, it is an open question whether mechanical or chemical weathering is more important within a zone of near-zero constant temperatures combined with high moisture content.

There are certain areas for which the annual cycle, as described above, is not entirely representative. Alpine environments of low latitudes are the obvious example since they experience diurnal rather than seasonal conditions. Sub-arctic locations which experience seasonal freezing and thawing but which are not underlain by permafrost are a second. In both environments, the absence of an impermeable frozen layer means that meltwater percolates to the groundwater table. There is also an absence of up-freezing in the autumn freeze-up period. The net effect of both is to reduce the importance of the zero curtain.

4.4.2. Short-Term Fluctuations

Short-term temperature fluctuations are frequently imposed upon the long-term temperature cycle just described. These fluctuations may be classified as being either diurnal or cyclonic in nature, although it is often difficult to distinguish between the two.

Diurnal variations usually relate to changes in solar insolation and surface heating brought about by variations in angle and azimuth of the sun. In high latitudes, these diurnal influences are relatively weak because the sun is above the horizon for much of the Arctic day. In middle latitudes, diurnal ranges are better developed. For example, the range between mean daily maximum and minimum air temperatures, at 3050 m in the Colorado Front Range, can reach 15 °C in July (Fahey, 1973). It is also in middle latitudes that the effects of orientation become well developed, with south- and west-facing slopes in the northern hemisphere experiencing significantly higher near-surface soil temperatures.

Local climatic events of a cyclonic nature may also produce marked fluctuations in soil temperature. This is especially the case in High Arctic regions, where periods of direct solar radiation may heat the ground surface to give a 20–30 °C temperature difference between air and ground surface. In Antarctica, for example, thawing of snow when air temperature is −20 °C may be caused by the localized heating of dark rock surfaces. The passing of a cloud, or the onset of precipitation, will also affect soil temperatures. For example, on Banks Island, the author documented the rapid changes in soil temperatures which followed a summer snowfall of 2.5 cm (French, 1970); within 3–4 hours, the entire active layer had attained a near-isothermal state due to downward percolation of the snowmelt. Even on the Tibet Plateau, at altitudes in excess of 4800 m a.s.l., air temperatures fluctuate significantly during the day in accordance with periods of direct sunshine (Wang and French, 1994, 1995a). A final example is provided by data obtained over a 31-day period in early 1992 on Livingston Island, in the South Shetlands, maritime Antarctic (Serrano et al., 1996); a total of 14 daily fluctuations above and below freezing were recorded.

4.5. ROCK (FROST?) SHATTERING

The disintegration, or mechanical breakdown, of bedrock in periglacial regions is well known. The most dramatic features that result are the extensive surfaces of angular rock fragments, commonly referred to as "blockfields" in the English-language literature (Washburn, 1969), or "kurums" in the Russian literature (Romanovskii and Tyurin, 1983, 1986; Romanovskii et al., 1989). Other features include tors and near-vertical rockwalls with extensive debris slopes beneath (see Chapter 9), and brecciated bedrock (see Chapters 7 and 13). Traditionally, this mechanical disintegration has been attributed to intense frost action, usually combined with the presence of moisture and ice segregation. In recent

(A)

(B)

Figure 4.7. Examples of bedrock disintegration under cold-climate conditions. (A) Fractured, dark brown, fine-grained diorite boulder lying on ablation till surface ("Younger Drift") in vicinity of Simpson Crags, Northern Victoria Land, Antarctica. The disintegration was probably caused by thermal stress. (B) In-situ bedrock disintegration of fissile sandstone of Jurassic age, Prince Patrick Island, NWT, Canada. Disintegration is presumably by subaerial frost action.

years, there has been renewed emphasis placed upon thermal stress as a possible mechanism. Here, we discuss both mechanisms and conclude that each may be effective under different circumstances and for different rock types. For example, a cracked granodiorite boulder lying on an ablation till surface is almost certainly the result of thermal stress (Figure 4.7A) while an exposure of shattered fissile sandstone is almost certainly the result of frost action and ice segregation (Figure 4.7B). More problematic are the angular rock-rubble accumulations that veneer many upland surfaces (Dahl, 1966; Dredge, 1992; Ives, 1966). These are the present-day equivalents of the so-called "periglacial facies" of Lozinski (see Chapter 1).

4.5.1. Frost Action and Ice Segregation

Several investigators stress the role of micro-fractures in bedrock as the method by which water penetrates bedrock and subsequently freezes to cause rock disintegration. However, when this cracking takes place, and what causes it, are still not clear. For example, using sandstone samples and evidence derived from acoustic emissions, B. Hallet et al. (1991) inferred that freezing-induced micro-fracture propagation occurred at temperatures between −3 °C and −6 °C. Earlier, it had been found that high frequencies of freeze–thaw cycles are not necessary for crack propagation and that the latter can occur at temperatures anywhere between −4 °C and −15 °C (Walder and Hallet, 1985, 1986). More recent studies (Matsuoka, 2001a; Prick, 2003) report upon the monitoring of rock temperatures, the opening of joints, and the frequency of rockfall in sedimentary rocks. For example, in the Japanese Alps, joint widening occurs in both autumn and spring. The autumn events are associated with freeze–thaw cycles, and the magnitude of the widening reflects freezing intensity and water availability. By contrast, widening in spring likely originates from refreezing of melt water entering the joint. In the high-latitude environment of Svalbard, there was little evidence for thermal-stress fatigue. Instead, when high moisture content was associated with freezing conditions, this favored rock disintegration. Frost shattering acted by wedging along bedrock joints and cracks. Similar conclusions were reached earlier by L. Dredge (1992) when describing the disintegration of limestone bedrock in the eastern Canadian Arctic. Joints and weaker bedding allow water infiltration and provide sites for fracturing by volumetric expansion, either in open-system or closed-system environments, or by hydraulic freeze-back processes. Joint expansion may also cause the upward buckling of slabby limestone along joint lines.

In a series of experiments, V. N. Konishchev and V. V. Rogov (1993) found that the speed of crack growth in water (ice)-saturated samples far exceeded that of dry samples when subject to numerous freeze–thaw cycles. These data (Table 4.2A), compiled for different rock types, give some indication of the approximate maximum speed of frost weathering. The average thickness of the disintegration layer for ice-saturated rocks during one freeze–thaw cycle ranged from a high of 3.5 mm in marl to a low of $30–50 \times 10^{-5}$ mm

Table 4.2. Some Russian data concerning rates of bedrock weathering by frost action: (a) average thickness of disintegration layer for one freeze–thaw cycle in various rocks of different water saturation; (b) data on fissuring of feldspar sandstones in an opencast mine, Taimyr, Siberia.

(A)	Average Thickness of Disintegration Layer (mm)			
Rock, deposits	Dry Samples		Water-saturated Samples	
	Range (mm)	Average (mm)	Range (mm)	Average (mm)
Granite	$(6.4–8.6) \times 10^{-5}$	8.0×10^{-5}	$(6.4–34.9) \times 10^{-5}$	14.5×10^{-5}
Weathered granite	$(8.7–11) \times 10^{-5}$	10×10^{-5}	$(10–32) \times 10^{-5}$	20×10^{-5}
Gneiss-granite, gneiss	$(7.0–9.2) \times 10^{-5}$	8.0×10^{-5}	$(7.0–9.9) \times 10^{-5}$	9.0×10^{-5}
Porphyry	$(8.1–25) \times 10^{-5}$	11×10^{-5}	$(3.1–92.4) \times 10^{-5}$	30×10^{-5}
Diabase	$(0.8–8.3) \times 10^{-5}$	4.5×10^{-5}	$(1.1–8.3) \times 10^{-5}$	4.5×10^{-5}
Metamorphic shale	$(2.8–27) \times 10^{-5}$	12×10^{-5}	$(8–35) \times 10^{-5}$	18×10^{-5}
Limestone	$(3.8–7.3) \times 10^{-5}$	5.5×10^{-5}	$(22.3–24.7) \times 10^{-5}$	23.5×10^{-5}
Sandstone	$(4.0–7.1) \times 10^{-5}$	6.0×10^{-5}	$(4.0–160) \times 10^{-5}$	48×10^{-5}
Limestone, dolomite	$(400–1000) \times 10^{-5}$	600×10^{-5}	1–8.7	1.0
Marl	$(400–1000) \times 10^{-5}$	600×10^{-5}	1–8.7	3.5

(B) Fissures	Year of Open-pit Stripping	Width of Fissures (mm)	
		Min.	Max.
Horizontal	1973	15	1
	1977	18	0.5
Vertical and subvertical	1973	50	3
	1970	60	1

Source: Konishchev and Rogov (1993). Reproduced by permission of John Wiley & Sons Ltd.

in sandstone and porphyry. Similar conclusions were reached following field observations in Antarctica, Svalbard, and the Japanese Alps (Matsuoka, 1990, 1991). Another study found that micro-fracture density and crack width in basalt correlated best with the overall pattern of annual rock removal (Douglas et al., 1991). Although peaks did occur in spring and autumn, presumably due to frost, tensile strength of the basalt was equally important. Collectively, these data sets indicate that rock type and water availability are important variables that control frost shattering.

Other field observations (Konishchev and Rogov, 1993), provide further insight into the nature of bedrock disintegration by frost. For example, in the Ukraine, the speed of frost weathering of water-saturated limestones on buildings in Simpheropol City has been 1–10 mm/year, while on dry, or only locally-saturated, limestone rocks, rates were lower, between 10^{-1} and 10^{-2} mm/year. Finally, in the Taimyr region of northern Siberia, arcose sandstone exposed in an opencast mine showed a marked increase in cracks and fissures in the years immediately following exposure to periodic freeze–thaw (Table 4.2B).

4.5.2. Frost Weathering Models

Frost weathering is thought to occur in a number of ways. The simplest model assumes that frost shattering results from the 9% volumetric increase which occurs during the water–ice phase transition. This process is termed "hydro-fracturing." Theoretically, the maximum pressure set up by the freezing of water is 2100 kg/cm^2 at $-22\,°C$. At temperatures below this, pressure decreases because ice actually begins to contract. In reality, however, this maximum value is almost certainly never reached, or even approached, because several factors reduce the pressures developed. First, water or ice must be contained within a closed system for high pressures to develop; this usually means conditions of extremely rapid freezing from the surface downward, which seals the pores and cracks in the rock. Second, air bubbles in ice and pore spaces within the rock reduce pressures considerably. Third, and probably most important of all, the rock itself, and certainly the soil mantle, is not strong enough to withstand such extreme pressure, especially because it is a tensile force rather than a compressive force that is being considered. As a result, the actual pressures developed by the freezing of water in rocks are probably much less than the theoretical maximum.

A more realistic model is the segregation-ice model. This treats freezing in water-saturated rock as analogous to slow freezing in fine-grained soil. Expansion is primarily the result of water migration to growing ice lenses and only secondarily the result of the 9% volumetric expansion (Hallet et al., 1991; Walder and Hallet, 1985). It follows that frost shattering results from the progressive expansion of micro-cracks and pores that are wedged open by ice growth. Circumstantial evidence which supports this model includes (1) experimental studies that document water migration towards the freezing front in freezing rocks (Fukuda, 1983) and (2) laboratory and field studies that record dilatation of bedrock during freezing and thawing (Prick, 1997; Prick et al., 1993; Wegmann and Keusen, 1998). Other laboratory experiments have involved periodic measurements of weight loss following repeated and varied freezing and thawing but these must be regarded as inconclusive because the actual time and temperature under which disintegration occurred is poorly constrained. Moreover, the spatial distribution of the weight loss and the growth of micro-cracks or fractures are not usually recorded. Table 4.3 compares the two frost-weathering models.

Table 4.3. A comparison of the volume-expansion ("hydrofacturing") and segregation-ice models of frost weathering.

Volumetric-Expansion (Hydro-fracturing) Model	Segregation-Ice Model
1. No frost weathering if pore fluid contracts upon freezing	1. Frost weathering does not depend on the volumetric expansion of water during freezing. Frost weathering results from heaving pressures that are universal in freezing porous solids, whether the pore fluid expands or contracts upon freezing
2. No frost weathering under conditions common in nature: saturation level less than about 91%, and pores not effectively sealed off (hydraulically-closed system)	2. Saturation level influences rate of water migration in hydraulically-connected pores (open system). Low saturation does not preclude water migration and crack growth.
3. Water may be expelled from freezing sites, but never drawn towards such sites	3. Water attraction to freezing sites, due to chemical potential gradients, is a key factor in frost weathering. If crack growth cannot accommodate water-to-ice expansion, water is expelled from freezing sites
4. Crack growth should occur in bursts as water freezes and expands	4. Slow, steady crack growth should occur as water migrates towards ice bodies within cracks. Predicted crack-growth rates are compatible with values inferred from experimental data

Source: Hallet et al. (1991). Reproduced by permission of John Wiley & Sons Ltd.

4.5.3. Insolation Weathering and Thermal Shock

Insolation weathering, or "spalling," refers to the cracking in bedrock that is thought to be caused by temperature-induced volume changes such as expansion and contraction (Ollier, 1963, 1984). For many years, this mechanism was not considered viable for the natural disintegration of rock in cold regions. This was because early studies by E. Black-welder (1925) and D. T. Griggs (1936) considered the process in the context of hot deserts. In laboratory experiments, it was found that no thermal shock failure occurred when rocks were heated and subsequently quenched through several hundred degrees Celsius. Subsequently, it was demonstrated that the larger the rock specimen and the less polished the surface, the larger were the thermal stresses generated. Still later, A. Rice (1976) presented a plot of a spalling-tendency index versus the number of quenches required to spall commercial bricks heated to ~500°C (Figure 4.8A) and concluded that insolation may be sufficient to break rocks.

Theoretical considerations suggest that heating is certainly capable of breaking rocks if it is combined with the availability of small amounts of moisture. For example, Figure 4.8B compares the differential expansion of granite with that of pure water. It appears that the heating of granitic rock from 10°C to 50°C can develop 250 atm as tensile strength, a force strong enough to disrupt the soundest rock with time. Accordingly, there is renewed interest in this mechanism, especially as it relates to cold environments. Today, the term "thermal-shock resistance" is used to describe the ability of a rock to withstand sudden and severe temperature change without fracturing (Marovelli et al., 1966).

In general terms, the tendency of any rock to resist disintegration under thermal cycling depends upon such factors as its thermal diffusivity, tensile strength, thermal-expansion

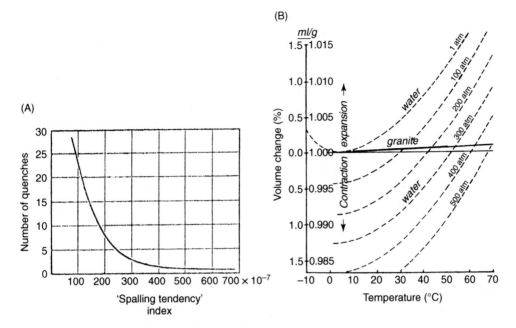

Figure 4.8. The nature of insolation (spalling) weathering. (A) Graph showing a theoretical spalling-tendency index (*x* axis) versus number of quenches (*y* axis). See Rice (1976, p. 61) for details of the spalling-tendency index. (B) Temperature–volume–pressure diagram for water (from International Critical Tables) superimposed on a plot of the average temperature-induced expansion of granite. From Winkler (1977).

coefficient, and the modulus of elasticity. Therefore, weathering by thermal shock demands two requirements. First, the rock must experience temperature contrasts. However, the magnitude of the contrast, the time over which the contrast occurs, and the lag between the time of maximum surface temperature and that of the maximum at depth within the rock, must also be considered. Second, the rock must have pores of a capillary size such that entrapped moisture cannot escape when water expands during the heating cycle. Here, micro-crack porosity appears crucial.

Several laboratory studies suggest that the threshold value for thermal shock approximates to a rate of temperature change of 2 °C/min (Richter and Simmons, 1974; Yatsu, 1988).

It is significant that most weathering phenomena attributed to thermal shock have been observed in arid environments, irrespective of whether these are hot or cold. For example, V. N. Konishchev and V. V. Rogov (1993) observe that an engineering study in Tashkent, Tadjikstan, reported numerous micro-cracks in granite of heterogeneous mineralogical composition that were the result of sharp temperature changes and large temperature amplitudes. Furthermore, it appears that there are many similarities in the rock-thermal regimes of hot and cold regions where aridity is a central characteristic. For example, in a quartz-monzonite boulder in the Mohave Desert, California, the maximum rock surface temperature over an 8-hour period ranged between 48 °C and 23 °C (Roth, 1965). In Antarctica, Russian scientists have measured daily temperature ranges of 40–42 °C in Victoria Land and monitored rates of heating and cooling of 0.8 °C/min and 15–20 °C/h on the ice-free nunataks of the Queen Maud Mountains (see Bardin et al., 1965, and

Myagkov, 1973, both cited in Hall and André, 2001). Even on the relatively cloudy and damp Antarctic Peninsula, rock temperature measurements on a granodiorite outcrop indicate rates of change in excess of the critical threshold value of $2\,^{\circ}C/min$ (Hall and André, 2001).

Despite several papers that invoke thermal stress as the cause of rock disintegration in cold regions (Hall, 1999; Hall and André, 2001; Hall et al., 2002, pp. 589–592), there are few direct measurements of the actual tensile stresses that must be involved. Furthermore, the thermal stresses that are associated with bedrock at the cold end of the temperature spectrum are often neglected by physicists and others interested in rock properties. The few field studies that are of relevance usually deal with either thermal-contraction cracking in unconsolidated sediments (Mackay, 1974c) or frost jacking (Dyke, 1984). The most useful information comes from geocryology. Here, E. Yershov (1990, pp. 139–140) states that the coefficients of linear and volumetric thermal expansion are highest in rocks and minerals that have a low energy of the crystalline lattice. It follows that the greater the SiO_2 content of a rock the larger the coefficient becomes. Figure 4.9 provides summary data that show the coefficient of linear thermal expansion of acidic igneous rocks to be several times greater than that of ultrabasic rocks. Linked to this is the observation that quartz-containing igneous rocks appear subject to more micro-cracking than quartz-free igneous rocks (Winkler, 1977). This suggests that the anomalous behavior of the mineral quartz during the process of cooling from magma leads to intra-granular and circum-granular cracking, when the rock is eventually exposed to sub-aerial conditions. It is significant that many examples of cold-climate rock shattering attributed to thermal stress are associated with quartzitic sandstones and silica-rich igneous and metamorphic rocks.

In summary, thermally-induced stress is probably an important rock weathering process in cold arid environments. However, until more experimental, field, and laboratory data become available, it is difficult to make a more definitive statement.

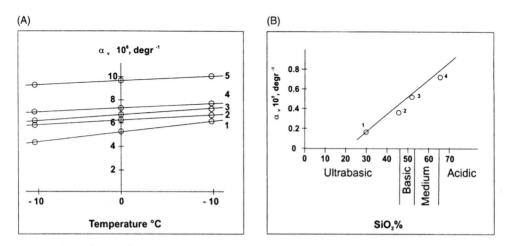

Figure 4.9. Variations in thermal expansion of different rock types. (A) Dependence of the coefficient of volumetric thermal expansion (α,v) on rocks in the temperature range $-10\,^{\circ}C$ to $+10\,^{\circ}C$. Legend: 1, granite; 2, diabase; 3, limestone; 4, tuffola; 5, sandstone. (B) Dependence of coefficient of linear thermal expansion (α) on silica content in rocks. Legend: 1, dunite; 2, gabbro; 3, diorite; 4, granite. From Yershov (1990). Reproduced by permission of Cambridge University Press.

4.5.4. Discussion and Perspective

It is important to understand why coherent rock disintegrates under cold-climate conditions, yet the preceding discussion presents at least two different mechanisms. To provide a degree of closure, let us consider some simple field observations undertaken on the cracking of rocks and the inferences that can be made.

Table 4.4 summarizes the statistics associated with 200 glacial erratic pebbles that were monitored by J. R. Mackay for signs of mechanical disintegration over a 19-year period in the western Canadian Arctic (Mackay, 1999). All were exposed on the floor of a recently-drained lake basin. The climate of the area is cold and relatively damp, being located close to the Arctic coast (see Table 2.1 for Inuvik and Tuktoyaktuk, and Table 3.1 for freezing and thawing degree-days at Tuktoyaktuk). Prior to the period of observation, many of the erratics had undergone thousands of freeze–thaw cycles when the rocks were either exposed at the ground surface or buried in the active layer. Thus, the first inference is that rock does not experience regular disintegration on an annual basis because all of the pebbles at the beginning of the observation period were unaltered. Second, during the period of observation (1973–1994), only 10 rocks (i.e., 5%) were observed to experience sub-aerial weathering. Therefore, cold-climate rock shattering is not an especially active process. Third, Mackay identifies three possible mechanisms: (i) traditional freeze–thaw volume expansion ("hydro-fracturing"), (ii) ice segregation (the Hallet model), and (iii) thermal stress. If ice segregation were the mechanism, fracture planes should be observed as sub-parallel to the rock surface, i.e. the cooling surface. When the specific rocks were examined, and using this criterion, a fine-grained granite pebble that had fractured indicated hydro-fracturing or thermal stress rather than ice segregation as the cause (see Mackay, 1999, figure 3). Likewise, a fine-grained granite pebble also indicated hydro-fracturing or thermal stress as the cause (see Mackay, 1999, figure 4a, b). By contrast, a coarse-grained granite pebble suggested ice segregation or hydro-fracturing rather than thermal stress (see Mackay, 1999, figure 4c, d). Finally, three dolomite pebbles all cracked along foliations approximately parallel to the ground surface, suggesting ice segregation as the cause. In addition, eight rocks, four of which were dolomite, showed evidence of what Mackay termed "explosive shattering," and what others have termed "frost-bursting" (Michaud et al., 1989), because rock fragments were scattered in the immediate vicinity. Whether this explosive shattering was the result of ice segregation, hydro-fracturing, or thermal shock is unknown. However, Mackay cites anecdotal evidence given by A. L. Washburn (1969, pp. 32–33) of an occasion in mid-February 1933 at Letty Harbour

Table 4.4. Summary of observations made by J. R. Mackay between 1974 and 1993 upon the shattering of rocks on the floor of a drained lake, Mackenzie Delta region, Canada.

Rocks:	Igneous	Sedimentary (Sandstones dolomites)	Metamorphic	Total
Initial number (1974)	151	28	21	200
Final number (1993)	136	24	6	166
No. that shattered:	2	1 (sandstone) 6 (dolomite)	1	10

Source: Mackay (1999). Reproduced by permission of John Wiley & Sons Ltd.

(latitude 69°N), approximately 200 km east of Tuktoyaktuk, when there was bright sunshine and the air temperature was approximately −43°C. A sound like a rifle shot was heard to emanate from a rock about 60 cm in diameter. Upon examination, the surrounding snow was seen to be covered in rock fragments. Mackay cautions against interpreting this anecdotal evidence as proof of either thermal shock or freeze–thaw activity. He merely emphasizes the sub-zero temperature at the time, the low angle of the sun (6°), and limited daylight at that time of the year. These are all factors that might affect, in various ways, the disintegration mechanism, whatever that might be.

In summary, it is difficult to differentiate between the ice segregation, hydro-fracturing, and thermal-shock mechanisms. It seems wise to conclude that all are valid, depending upon specific climatic conditions, moisture availability, and physical properties of the rock concerned.

4.6. CHEMICAL WEATHERING

4.6.1. General

There is a tendency to underestimate the role of chemical weathering in periglacial environments. This is because attention usually focuses upon the more dramatic effects of mechanical weathering, as described above. However, limited evidence suggests that chemical weathering processes in cold climates are just as effective, if not more so, than in non-periglacial environments. If one focuses attention upon soil climate (Pope et al., 1995), geographical variations in chemical weathering appear less obvious than previously thought.

There are several reasons why the importance of so-called "normal" chemical weathering processes in cold environments is often underestimated. First, the presence of permafrost in many periglacial environments imparts a distinct character to soils and soil-forming processes. Most pedologists are more interested in soils where horizon differentiation is more marked. Cold-climate soils, now recognized as cryosols, are discussed later in this chapter. Second, a number of unusual weathering phenomena, such as tafoni and honeycomb weathering, divert attention from more traditional weathering studies. These phenomena reflect an intimate, but largely unknown, interaction between saline conditions, the freezing process, and sub-zero (cryogenic) temperatures. These are also discussed later in this chapter. Third, the few detailed studies upon cold-climate weathering that are available are not necessarily of widespread applicability. For example, the classic study of A. Rapp (1960a) was undertaken in a glacially-overdeepened and recently-deglaciated valley (Kärkevagge) in northern Sweden that is currently experiencing paraglacial transition. As seen in Chapter 2, this is not truly representative of the periglacial domain. Moreover, the importance of solution processes that was so convincingly demonstrated in that study (see Table 9.3) is now seen to reflect the presence of pyrite-rich rock that, upon weathering, produces sulfuric acid, which enhances the weathering of muscovite and calcite (Darmody et al., 2000, 2001; Thorn et al., 2001). Furthermore, clay mineral evidence suggests that pedogenesis is more advanced on the upland surfaces that surround the valley than in the warmer and wetter valley-bottom itself (Allen et al., 2001). This raises the possibility that the parent materials on the ridges are not residual but are pre-weathered materials that have been transported from elsewhere.

It is not unreasonable to suggest that sub-aerial weathering and soil formation in cold environments is not very different to that in other environments. For example, when a number of machine-polished disks were inserted into the ground at Kärkevagge and then measured for loss of mass over a 5-year period, limestone and dolomite disks weathered much fastest than granite disks (Thorn et al., 2002). This is not surprising. It is also not surprising that poor drainage (increasing wetness) and decreasing PH (increasing acidity) correlate with weathering of the dolomite disks. Elsewhere in northern Sweden, post-glacial rates of mechanical weathering are not only relatively slow (André, 1995a, 1996, 2002) but also dependent on rock type. Again, as might be expected, rates of general surface lowering on carbonate rocks are an order of magnitude greater (~5 mm/ka) than on homogeneous crystalline rocks (~0.2 mm/ka) (André, 2002).

Apart from these investigations in northern Sweden, there is little systematic data available on the absolute and relative rates of chemical weathering processes in cold polar latitudes. Several data sets from mid-latitude alpine regions (Caine and Thurman, 1990; McCarroll, 1990; Williams et al., 2006) suggest that chemical weathering is of importance. The one definitive conclusion that can be made is that the limitation to chemical weathering in cold climates relates primarily to moisture rather than to cold temperature (Balke et al., 1991).

4.6.2. Solution and Karstification

A chemical weathering process that always attracts attention from geomorphologists is that of limestone solution. This reflects the solubility of calcium carbonate ($CaCO_3$) in water. Because the solubility of $CaCO_3$ in water increases with a decrease in temperature, it has been suggested that solution in cold regions is higher than in other regions. For example, Rapp (1960a) concluded that solution loss was by far the most important agent of denudation at Kärkevagge (see earlier). Yet studies in other cold environments suggest that solution activity is no greater than elsewhere. For example, in the central Canadian Arctic, solution rates are actually smaller than in low latitudes, carbonate concentrations in standing water bodies are of the same order of magnitude as those of temperate regions, and elevated concentrations of carbon dioxide in snow banks do not always exist (Smith, 1972). Some typical limestone solution values and rates of denudation are listed in Table 4.5. Other factors being equal, precipitation controls the solution rate. We might conclude, therefore, that the generally low rates of limestone solution denudation that typify many cold environments probably reflect the relative aridity of these areas rather than any "weakness" of solution activity.

Where hard and relatively pure limestone outcrops over large areas, solution processes that are concentrated along joints, bedding planes, and other discontinuities may assume special importance. The result is cold-climate "karst" terrain. Initially, it was thought that permafrost inhibited solution activity and karst terrain. Thus, D. A. St-Onge (1959) remarked upon the absence of solutional effects in terrain developed in gypsum on Ellef Ringnes Island, and J. B. Bird (1967, pp. 257–270), in a detailed account of the limestone scenery of the central Canadian Arctic, concluded that solution effects were weak. However, recent studies by D. C. Ford now suggest that a "sub-cutaneous karst" model is applicable to limestone terrain in areas of permafrost. The model asserts that groundwater circulation and solution are limited to the seasonally-active zone, favoring the development of spreads of shallow karren (i.e. solution-pitted) ground at the expense of sinkhole and cavern topography.

A number of relationships between karst terrain and glaciation have been considered (Ford, 1984, 1987, 1996) in the context of northern Canada. The first is where continuous permafrost exists. Here, bedrock has remained frozen, initially beneath Pleistocene ice and subsequently during deglaciation. Thus, karst development is postglacial in age and restricted to the active layer, and mostly along the edges of bedrock outcrops. Today, minor

Table 4.5. Some typical rates of limestone solution under periglacial and non-periglacial conditions.

Location	Mean Annual Precipitation (mm)	Net Rate of Denudation (mm/1000 years)	Total Carbonate Hardness (ppm)	
			$CaCO_3$	$MgCO_3$
Arctic/Sub-arctic:				
Hudson Bay, Canada	600	6.0		
Somerset Island, Canada	130	2	84	22
Northern Yukon, Canada	215	n.d.	80	106
Svalbard	300	27		
Tanana River, central Alaska	450	40		
Svartisen, northern Norway	740–4000	400		
Abisko, northern Sweden	800	5.3		
Mountains:				
Alps (mid-latitude)	2200	13		
Equatorial	4000	32		
Non-periglacial:				
Western Ireland	1000–1250	51–63	66	11
Southern Algeria	60	6		
Indonesia	200–300	83		

Sources: Jennings (1971), Sweeting (1972, tables X and XI), Smith (1972), Thibaudeau et al. (1988), André (2002).

solutional forms (karren) on exposed surfaces accelerate rock disintegration by other weathering (frost) processes (see Dredge, 1992). A second relationship is characterized by reticulate patterns of large "corridor" troughs. It is hypothesized that these were created by meltwater erosion beneath weakly erosive glacier ice of the ice-cap interior. A third relationship results from the hydrological conditions associated with warm-based marginal ice, or from sustained thaw during the postglacial. This terrain occurs in the sub-arctic zone of widespread discontinuous permafrost. Here, large fluvio-karst systems have formed, as in the Great Bear Lake region of the Mackenzie Valley region of northern Canada (van Everdingen, 1981), and Akpatok Island in Eastern Canada (Lauriol and Gray, 1990). The former is characterized by large closed depressions, powerful karst springs, and ephemeral lakes, the latter by an extensive limestone plateau dissected by dry valleys and large enclosed depressions or sink holes. In all probability, all these karst systems pre-date the last glaciation and were either sustained beneath the ice or re-established during the deglacial phase. The Nahanni karst, of the Mackenzie Mountains, NWT, deserves special mention. This area is close to the southern limit of widespread permafrost and lay beyond the limit of the Wisconsinan and Cordilleran ice sheets. Here, the latest glaciation is thought to have occurred more than 300000 years ago (Brook and Ford, 1978, 1982). Thus, an extensive period of cold non-glacial conditions has allowed the development of "the most accidented cold regions karst that is known" (Ford, 1987, p. 520). It possesses impressive canyons, gorges, caves, springs, and subterranean drainage.

  Two models for karstification (Figure 4.10) within the discontinuous permafrost zone of northern Canada summarize the major relief features associated with cold-climate

(A)

Permafrost extent	Continuous		Widespread		Discontinuous		Sporadic
Latitude	75°N				65°N		55°N
Mean annual temperature (°C)	−15°		−10°			−5°	0°
Surface terrain conditions	Active layer ~0.5 m		lake		lake		(i) thermokarst depression (ii) Active layer ~1.5 m

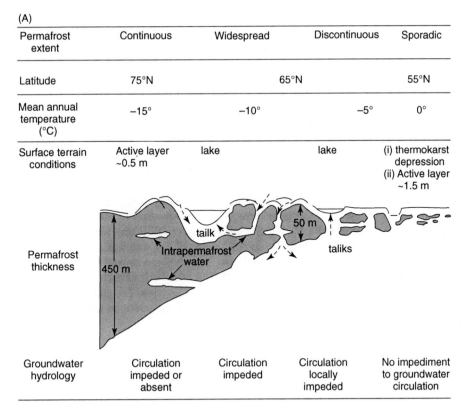

Permafrost thickness							
Groundwater hydrology	Circulation impeded or absent		Circulation impeded		Circulation locally impeded		No impediment to groundwater circulation

(B)

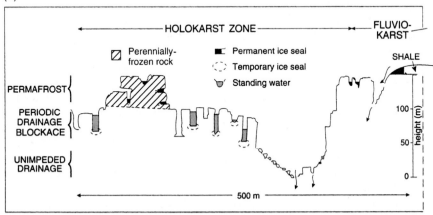

Figure 4.10. Karst morphologies associated with permafrost terrain. (A) Karst–permafrost relations in the lowlands of northern Canada. (B) Nahanni karst, NWT. Both from Ford (1987), in French and Slaymaker (1993). Reproduced by permission of McGill-Queens Press.

karst. They are especially applicable to the areas of discontinuous permafrost in Canada, but similar terrain that exists on Svalbard (Salvigsen and Elgersma, 1985) and in Siberia (Popov et al., 1972) suggests these models are of wide applicability.

4.6.3. Salt Weathering

Frost weathering, via ice segregation, relies upon the crystallization pressures generated by growing ice crystals that feed off migrating pore water. Commonly, the water is not pure. This leads to solute effects, which range from the relatively simple pressure generated by salt crystallization itself to the complexities of depressed freezing points and physico-chemical weathering.

Laboratory experiments have shown that some rock types normally resistant to frost weathering become highly susceptible once they are immersed in salt solution. A recent study indicates a range of salts are capable of this effect (Williams and Robinson, 2001). As a result, it is difficult to predict the aggressiveness of salt weathering because different combinations of salts may act in different ways. The damage caused by the combination of salt and frost to buildings, bridges, highways, and other infrastructure is well known in cold countries such as Canada, the United States, and Sweden. For example, a considerable literature exists upon the durability of cement and concrete in frost-dominated climates (Rosli and Harnik, 1980).

Geomorphological studies into salt and frost weathering usually concentrate upon their effects on limestone and dolomite terrains (Dredge, 1992; Goudie, 1974; St-Onge, 1959), on coastal rock platforms (Trenhaile and Rudakas, 1981), or on how rock type is affected by different freezing regimes and varying strengths of salt solutions (Fahey, 1985; Jerwood et al., 1990a, b). In addition, a number of descriptive studies document honeycomb weathering and granular disintegration phenomena that are partially or wholly attributed to a combination of salt and frost weathering (André and Hall, 2005; Calkin and Cailleux, 1962; Cailleux and Calkin, 1963; Czeppe, 1964; French and Guglielmin, 2000, 2002a; Selby, 1971a; Watts, 1983).

Various explanations have been suggested as to why salt accelerates frost weathering. The more important are described here. First, salt can accumulate in the outer layers of rocks as a result of surface evaporation. This tends to block the pores and seal the surface, often leading to surface ("case") hardening. As a result, water cannot escape by extrusion through freezing, thus increasing the stresses within the rock. Second, salt may intensify frost weathering because the rock is subject to both ordinary frost action (i.e. volume expansion and/or ice crystallization) and salt crystallization. In theory, however, it is only at temperatures below the depressed freezing point that weathering can be due to the combined pressures. Therefore, this simple explanation does not account for the enhanced frost weathering, in combination with salt, which is known to occur at temperatures above this. Third, frost damage may result from the expansion and contraction of adsorbed water on clay particles. This is sometimes referred to as "hydration shattering" (White, 1976). If salts are present, freezing of adsorbed water is delayed and salts continue to expand as freezing progresses, causing enhanced rock disintegration. This explanation does not account for frost and salt weathering of non-sorption-sensitive rocks such as sandstones. Fourth, the enhanced damage associated with salt and rock weathering may partly be the result of the slower rates of freezing associated with dilute salt concentrations. Thus, the resulting ice crystals will be larger than normal. In this case, experimental studies have yet to demonstrate that slower rates of freezing cause more frost disintegration than faster rates.

There is much that is not understood about the relationship between frost action and salt weathering. Equally, it must be stressed that calcareous rock surfaces the world over are frequently pitted with a range of cavernously-weathered forms of varying sizes. Larger forms are generally termed tafoni whereas smaller cavities are termed alveoles or honeycombs. Thus, salt weathering is not unique to cold climates. One must also remember that, even if salt weathering is proven to cause rock disintegration, it is still unclear how cavernous forms develop through this mechanism. For example, cavernously-weathered forms are often characterized by a case-hardened surface or "weathering rind." This suggests tafoni formation is a two-stage process, the first involving case-hardening and the second involving the breaking of this patina and the exposure of the rock interior to a saline environment. The latter might then cause enhanced swelling and contraction of any clay minerals within the rock. In support of this mechanism, it is significant that many examples of tafoni weathering occur in coastal locations where, presumably, a salt supply is provided by the ocean. On the other hand, tafoni occur on a wide range of rock types, some of which (e.g. sandstone) do not possess clay minerals in any abundance, and are particularly widespread in extremely arid areas, such as central Australia, the Sahara, SW USA, and Jordan. Case-hardening is an additional complication since this characteristic introduces considerations of wind, aspect, insolation (thermal shock), and rock strength.

In spite of all these uncertainties and ambiguities, the pervasive impression remains that salt is important in the development of cavernous weathering. This can be illustrated by the range of cavernously-weathering forms that occur in the ice-free areas of Antarctica (Table 4.6).

Two specific examples illustrate the ambiguity of any salt-weathering hypothesis.

Table 4.6. Some reports of the occurrence of tafoni, alveoles, and honeycomb weathering in Antarctica.

Location	Rock Type	Reference
(i) Antarctic Peninsula:		
Alexander Island	Sandstone	André & Hall (2005)
West coast, Antarctic Peninsula	Gabbro	Muscoe (1982)
(ii) Antarctic Continent:		
Queen Maud Land	Granite, gneiss	Muscoe (1982)
Sor Rondane Mountains	Not specified	Sekyra (1969)
Freedom Archipelago	Dolerite, granite, quartz, pegmatite	Muscoe (1982)
Bunger Oasis, Schirmacher Oasis	Granite gneiss, schist	Sekyra (1969); Muscoe (1982)
(iii) Southern Victoria Land:		
Bull Pass, Dry Valleys	Dolerite	Conca and Astor (1987)
Cape Evans	Granite, gneiss, dolerite	Muscoe (1982)
Cape Royds	Alkaline lavas	Muscoe (1982)
Taylor Valley, McMurdo Oasis	Gneiss	Selby (1971a); Muscoe (1982)
Victoria Valley	Granite, gneiss	Calkin and Cailleux (1962); Cailleux and Calkin (1963)
(iv) Northern Victoria Land:		
Terra Nova Bay	Monzogranite	French and Guglielmin (2000; 2002a)

First, in the Terra Nova Bay area of Northern Victoria Land, tafoni and cavernously-weathered forms are well developed in coarse-grained granites and metasedimentary rocks (Figure 4.11A). The area is extremely dry and cold (see Table 3.1). Thin sections of the weathered bedrock indicate that physical disintegration is associated with the preferential fracturing of quartz minerals while softer feldspar minerals remain relatively intact (Figure 4.11B). It has been hypothesized that micro-cracks, formed in the quartz minerals at the time of magma cooling, permit penetration and subsequent freezing of gas–liquid intrusions at sub-zero temperatures (French and Guglielmin, 2000). The salt is provided by snowmelt and acts to lower the freezing point. Disintegration is further helped by the increasing brittleness of quartz at low temperature, and case hardening accentuates the unusual morphology by providing a hardened outer shell. Thus, the process is best

(A)

(B)

Figure 4.11.   Cavernous weathering in coarse-grained monzogranite, Terra Nova Bay area, Northern Victoria Land, Antarctica. (A) Taffoni-weathered boulder. Backpack for scale. (B) Thin section photo of tafoni-weathered monzogranite viewed under cross-polarized light. The photo dimensions cover 8 mm. The photo shows a highly fractured quartz mineral (left) and a largely unfractured microcline mineral (right). A large void separates the two minerals and smaller voids are present within the quartz mineral.

described as being both physico-chemical and cryogenic in nature. Unfortunately, this explanation is not entirely satisfactory because it is inapplicable to the hot deserts of the world and therefore, by implication, is not necessarily the answer for cold environments either.

A second example is provided by the interpretation of tafoni that has formed upon Mesozoic-age sandstone blocks on Alexander Island, Antarctic Peninsula (André and Hall, 2005). In some ways, a similar conclusion involving both physical and chemical weathering is reached. In order to accommodate apparently conflicting evidence that suggests both current activity and inactivity of tafoni, the authors attribute its initiation to coastal spray weathering by halite during the last 6500 years. Current activity, in the form of flaking and granular disintegration, is then explained in terms of thermal shock. The latter is invoked on the basis of rock temperature measurements made on the back wall of a hollow that indicated a number of thermal events in which the rate of temperature change exceeded the assumed critical threshold of >2 °C/min for thermal-shock effectiveness (see pp. 64–66). While we may assume that thermal shock is a viable mechanism, this explanation leaves unanswered the more fundamental questions concerning salt weathering and the initiation of cavernous weathering.

4.7. CRYOGENIC WEATHERING

Cryogenic weathering refers to the combination of mechanico-chemical processes which cause the in-situ disintegration of rock under cold-climate conditions. Salt weathering is one such example (see above) but here, discussion is more wide-ranging.

A number of field and experimental studies in the former Soviet Union show that one of the main effects of cold-climate weathering is the production of silty particles with grain sizes of between 0.05 mm and 0.01 mm in diameter. Analysis of this material indicates that it consists mainly of primary minerals such as quartz, feldspar, amphibolite, and pyroxene (Konishchev, 1982; Konishchev and Rogov, 1993). This evidence suggests that frost weathering occurs within the layer of unfrozen water that is adsorbed on the surfaces of these particles. The susceptibility of these particles to weathering depends not so much on their mechanical strength but more on the thickness and properties of this unfrozen water film. According to V. N. Konishchev (1982), the protective role of this stable film of unfrozen water is highest with silicates, such as biotite and muscovite, and lowest with quartz. Cryogenic disintegration occurs when the thickness of the protective unfrozen water film becomes less than the dimensions of the various micro-fractures and defects that characterize the surface of mineral particles.

In a series of laboratory experiments in which minerals were subject to repeated freeze–thaw, it was established that 0.05–0.01 mm grain sizes are the limit for the cryogenic disintegration of quartz, amphibole, and pyroxene, 0.1–0.5 mm for feldspar, and 0.25–0.1 mm for biotite. Quartz grains, in all the size fractions investigated, proved less resistant when compared to corresponding grain sizes of unchanged feldspar (Figure 4.12). These relationships are opposite to the normal weathering behavior of these minerals under temperate or warm climates. The yield of heavy minerals is also unusual in that the 0.05–0.01 mm fraction is lower than that in the 0.05–0.1 mm fraction.

The relatively high degree of instability of quartz under cold-climate conditions has been confirmed in several other Russian studies (Minervin, 1982; Rogov, 1987; see also Table 4.7). The mechanism of cryogenic disintegration is thought to be based on the wedging effect when ice forms in micro-cracks and produces volume widening following repeated freeze–thaw. The freezing of gas–liquid inclusions, commonly containing salts, is the specific cause of the widening. Konishchev and Rogov (1993) have also been able

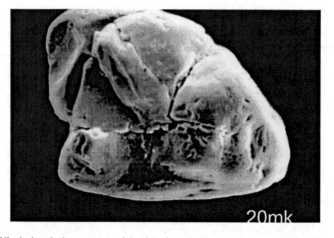

Figure 4.12.   Wind-abraded quartz particle showing cracks formed as a result of freezing, probably of saline gas–liquid inclusions at cryogenic temperature. The quartz grain is of Sartan (15–27 ka) age, and is from loess-like sediments, Cape Chukochyi, Kolyma region, Siberia. Magnification ×400; the grain size is ~20μm. The photo and caption information are supplied courtesy of Professor V. N. Konishchev.

to characterize unconsolidated surficial sediments taken from many parts of Siberia on the basis of a coefficient of cryogenic contrast (CCC), defined as: $CCF = Q_1/F_1/Q_2/F_2$ where Q_1 = quartz content (%) in fraction 0.05–0.01 mm, F_1 = feldspar content (%) in fraction 0.05–0.01 mm, Q_2 = quartz content (%) in fraction 0.1–0.05 mm, and F_2 = feldspar content (%) in fraction 0.1–0.05 mm. It was concluded that deposits affected by cryogenic weathering have CCC values in excess of 1.5.

Even more poorly understood are the physico-chemical changes which occur under negative temperatures. Three further examples illustrate the complexities of cold-climate (cryogenic) weathering.

First, a specific problem concerns the behavior of the fine-grained (clay) fraction. In early experiments, V. N. Konishchev et al. (1976) and others found that, after repeated freeze–thaw, the size of kaolin and montmorillonite particles decreased while the crystalline lattice and inter-parcel distances generally increased. In later experiments, particle disintegration dominated in the early stages of cryogenic transformation but then, after 10–100 freeze–thaw cycles, aggregation and coagulation occurred together with marked changes in pH and the ion exchange complex (Ershov, 1984). The results of Russian studies, summarized by E. D. Yershov (1990, pp. 124–126), indicate the complexity of physico-chemical changes that occur in fine-grained sediments subject to freeze–thaw. One set of data is presented in Table 4.8.

A second illustration is provided by evidence that secondary precipitates (iron oxide, calcite) may originate by the freezing process. This results in coatings that display fibrous crystalline fabrics (Vogt and Corte, 1996). Because water must both freeze and thaw to produce these precipitates, it is concluded that precipitation must take place either in the active layer (or zone of seasonal freezing and thawing) or during the formation of the permafrost table (Figure 4.13). Other studies suggest transformation and neoformation of minerals in the active layer (Vogt and Larqué, 1998). In all probability, transformations take place not only during freeze–thaw cycles, when temperature transitions cross 0 °C, but also within negative temperature fluctuations.

Table 4.7. Some experimental data that demonstrate the instability of quartz particles when subject to cryogenic weathering.

Condition of Experiment		Mineral	Content of Grain Size (%) Diameter of Grain (mm)				
			0.25–0.1	0.1–0.05	0.05–0.01	0.01–0.005	0.005–0.001
1. Before experiment		Quartz	100	–	–	–	–
		Feldspar	100	–	–	–	–
		Calcite	100	–	–	–	–
		Biotite	100	–	–	–	–
2. Freezing-thawing water-saturated	−10°C to +15°C	Quartz	11	20	68	1	–
		Feldspar	7	44	48	1	1
		Calcite	6	1	20	30	29
		Biotite	98	–	1	–	1
3. Freezing-heating dry	−10°C to + 50°C	Quartz	98.5	1	0.5	–	–
		Feldspar	98	1.5	0.5	–	–
		Calcite	93.5	5	1	0.5	–
		Biotite	100	–	–	–	–
4. Wetting-drying in laboratory condition	+18 to +20°C	Quartz	100	–	–	–	–
		Feldspar	100	–	–	–	–
		Calcite	100	–	–	–	–
		Biotite	100	–	–	–	–

Source: Konishchev and Rogov (1993). Reproduced by permission of John Wiley & Sons Ltd.

Table 4.8.   Chemical changes observed following repeated freezing and thawing of clay minerals.

Mineral	Number of freeze-thaw cycles	Content (mmol/100 g)		
		Na⁺	K⁺	Ca²⁺
Kaolinite	0	0.08	0.19	13.86
	100	0.20	–	11.65
Bentonite	0	8.26	6.00	44.22
	100	15.66	0.38	47.53
Polymineral clay	0	0.16	0.13	57.42
	100	6.00	0.64	75.25

Source: Datsko and Rogov (1988).

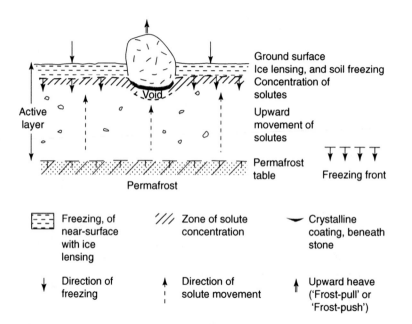

Figure 4.13.   The probable mechanism of solute precipitation by freezing. As the freezing front progresses downwards, it attracts solutes from the active layer that concentrate beneath coarser particles. Modified from Vogt and Corte (1996).

    A third example concerns anomalous micro-erosional phenomena that occur in some of the cold deserts of Antarctica. Their mechanics of formation are unclear. For example, grooves, pits, and furrows ("pseudo-rinnenkarren") have been described from several localities (French and Guglielmin, 2002a, b; Richter, 1985). In Northern Victoria Land, the grooves are developed on steeply inclined (>35°) biotite–monzogranite bedrock surfaces. They are generally straight, 10–30 cm deep, 20–80 cm in width, and up to 10 m in length. Many join and bifurcate and others meander, all with no apparent structural control. One possible explanation is snowmelt erosion at sub-zero temperatures, together with prior granular disintegration. Rock varnish on adjacent surfaces suggests a chemical or biogeochemical origin, perhaps aided by wind (Campbell and Claridge, 1987, pp. 124–129; Dorn and Oberlander, 1982; Dorn et al., 1992) but similar, and equally puzzling, pseudo-karren exists in other regions of the world (Doerr, 1999; Ford and Williams, 1989). It may be that similar weathering features result from different weathering processes.

4.8. CRYOBIOLOGICAL WEATHERING

It is well known that cold climates preserve things from decay. For example, the excavation of bodies buried in permafrost in Alaska and Svalbard of individuals who died as a result of the 1918 influenza epidemic has resulted in a better understanding of the genetic code of the 1918 virus. More relevant to periglacial geomorphology is the discovery that viable micro-organisms are present within Antarctic ice that are dated at 420 000 years and in northern hemisphere permafrost with ages in excess of 2–3 million years (Gilichinsky, 2002a). In Svalbard, frozen in-situ soil and vegetation beneath a cold-based glacier indicates microbes have survived for over 1100 years in a subglacial frozen state (Humlum et al., 2005). All these observations indicate that one must not dismiss the potential for biological weathering at cold temperatures.

Cold-adapted micro-organisms were first described in the late nineteenth century when bioluminescent bacteria were observed in dead fish preserved by low temperatures (Gilichinsky et al., 1995). Since then, the science of cryogenics has developed, with numerous applications in medicine and food technology. Surprisingly, there are relatively few geomorphological studies that examine the role of biological activity in cold climates. Most concern the growth of lichens, algae, or fungi in promoting rock weathering through flaking and exfoliation (Etienne, 2002; Fry, 1927; Hall and Otte, 1990; McCarroll and Vines, 1995). More detailed reviews of the various micro-organisms known to exist in permafrost are provided by D. Gilichinsky and colleagues (Gilichinsky, 1994; Gilichinsky and Wagener, 1995; Gilichinsky et al., 1995). Although this literature is sparse, it indicates the potential for significant biological weathering at low temperatures, and further studies are required.

Understanding the role played by bacteria and other organisms at low temperature is important not only to soil development (see Section 4.9.1) but also to planetary science, where the potential for life is under investigation (Gilichinsky, 2002b). The organisms that survive in perennially-frozen conditions on Earth may possess unique mechanisms that allow them to maintain viability for very long periods. Science needs to establish if these mechanisms exist and, if so, what they are. Organisms that survive in permafrost may be ancient bacteria and their analysis may reveal the nature of microbial life as it was several million years ago. Today, the thawing of permafrost consequent upon global climate warming (see Chapter 15) may release these organisms into the modern world, with unpredictable consequences. This increase in microbial activity in the presence of viable micro-organisms and radiative gases (CH_4, CO_2) plus the availability of organic substrates in permafrost will further stimulate the emission of greenhouse gases. On the other hand, increased microbial activity may also increase the ability of the tundra to function as a sink for greenhouse gases. In summary, the complexities and uncertainties associated with potential changes in the cryosphere indicate that cryobiological activity, in terms of both weathering and life forms, needs further investigation.

4.9. CRYOPEDOLOGY

Soil is the end product of weathering. Therefore, any discussion of cold-climate weathering must consider this topic. Cryopedology refers to the study of soils at temperatures below 0 °C, with particular reference to those subject to intense frost action, and to soils overlying permafrost. The term "cryosol," first coined by N. Federoff (1966), is now widely accepted (Kimble, 2004). Furthermore, a Cryosolic Order, initially developed in Canada (Canada Soil Survey Committee, 1978; Soil Classification Working Group, 1998), is gaining acceptance in the soil-science community.

In spite of early studies by Russian and German pedologists (Dokuchaev, 1900; Meinardus, 1912, 1930), study of the soils of cold regions was neglected for many years. This was because these soils are generally shallow, lack well-defined soil profiles, and are characterized by instability of the mineral soil, the result of frost action. Soil-horizon designation was often elementary, difficult, and unsatisfactory. Traditionally, these soils were described in the context of their relation to the major vegetation zones of the world (Hill and Tedrow, 1961; Tedrow, 1977).

4.9.1. Cryosols

A cryosol is defined as "soil formed in either mineral or organic materials having permafrost either within 1 m below the surface or, if the soil is strongly cryoturbated, within 2 m below the surface, and having a mean annual ground temperature below 0 °C" (van Everdingen, 1998). Cryosols are of three types (Table 4.9). Static cryosols develop primarily in coarse-textured mineral parent material, where cryoturbation and small-scale patterned ground phenomena are largely absent. Turbic cryosols show marked evidence of cryoturbation within the active layer (i.e. disrupted, mixed, or broken horizons and displaced material). Organic cryosols develop in organic material containing more than 17% organic carbon by weight and where permafrost is usually less than 1 m from the ground surface. Common landforms associated with organic cryosols include palsas, peat polygons, and high-centered polygons.

Table 4.9. Summary of the nature of materials, degree of cryoturbation, and depth to permafrost of the major cryosolic soils.

Parameter	Cryosolic Great Groups		
	Turbic Cryosol	Static Cryosol	Organic Cryosol
Soil material	Mineral	Mineral	Organic
Cryoturbation	Marked: usually patterned ground	None	Uncommon
Permafrost	Within 2 m of surface	Within 1 m of surface	Within 1 m of surface

N.B. A more detailed description of turbic cryosols, static cryosols, and organic cryosols is given in Tarnocai (2004a, tables 5.1–5.4, pp. 607–609).

The pedon, the basic soil unit, is the three-dimensional unit by which cryosolic soils are described. The pedon is also a geomorphological concept because it allows one to consider the numerous small-scale patterned ground phenomena that typify many periglacial landscapes. The pedon of a non-sorted circle, the most widespread of all small-scale patterned ground forms, is illustrated in Figure 4.14. In the case of low- and high-centered polygons, one pedon would characterize the polygon trench area while a second would characterize the central part of the polygon (Tarnocai, 2004a, p. 600). The various horizon designators for cryosolic soil description, as identified in the Canadian System of Soil Classification (Soil Classification Working Group, 1998), need not concern us here and are summarized in Tarnocai (2004a, pp. 605–607).

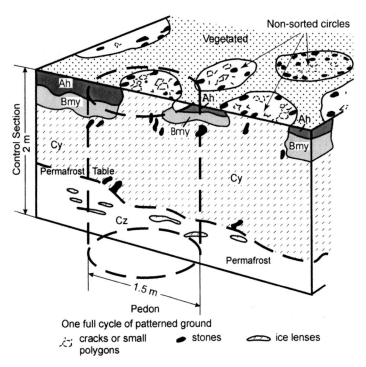

Figure 4.14. Location of the pedon of an orthic turbic cryosol in patterned ground (non-sorted circle), together with typical cryosolic soil nomenclature. Legend to soil horizon: Ah = mineral near-surface horizon (A) with zone of accumulation of organic matter (h); Bmy = mineral horizon (B), slightly altered by hydrolysis, oxidation or solution (m), with broken (cryoturbated) layers (y); Cz = mineral parent material (C) that is perennially frozen (z). See also Table 4.9. From Tarnocai (2004a). Reproduced by permission of John Wiley & Sons Ltd.

4.9.2. Soil Micromorphology

Cryogenic processes result in distinctive soil-morphology patterns. These are described by soil scientists (Brewer, 1976; Fox and Protz, 1981; Pawluk, 1988; van Vliet-Lanoë, 1982, 1998; van Vliet-Lanoë et al., 2004). The terminology associated with soil micromorphology is complex.

Essentially, when material is displaced by freezing and thawing, coarse and clay-sized minerals are realigned, aggregates and mineral particles are rotated, and coarse and fine particles are sorted (Fox, 1994). If segregated ice lenses form during the freezing process, the result is desiccation, compaction, and compression of the soil. Typically, this leads to the formation of a platey or lamellar micro-structure. Under certain situations, hardpans or indurated layers may result (Fitzpatrick, 1956; Nikiforoff, 1955). Planar voids and cracks may also appear. The main factors that contribute to the evolution of cryosols are waterlogging of the ground during spring and summer and frost desiccation during the winter when soil is frozen. Because cryosuction potential at $-20\,^{\circ}\mathrm{C}$ can extract adsorbed water from soil particles, the activity of the unfrozen water film may be high. This may explain the formation of iron coatings, rock varnish, and case hardening at the surface of rocks. Likewise, a thermally-induced water potential may explain the migration of iron, aluminum hydroxide, and silica ions to the permafrost table. Weak humification is another characteristic of the weathering of cryosols. Cryptogamic crusts are also important in certain aridic, cold, or post-fire environments. In summarizing the cryopedological literature, B. van Vliet-Lanoë et al. (2004, p. 386) conclude that the distinct features of cryosols reflect (i) the dominance of mechanical processes over chemical or biological activity, and (ii) the presence of two-sided freezing, the growth of segregated ice lenses, and the associated displacement of material.

ADVANCED READING

Kimble, J. M. (2004). Section 3. Properties and processes of cryosols. In: *Cryosols. Permafrost-Affected Soils*. Springer-Verlag, New York, pp. 341–458.

Konishchev, V. N., Rogov, V. V. (1993). Investigations of cryogenic weathering in Europe and Northern Asia. *Permafrost and Periglacial Processes*, **4**, 49–64.

Williams, P. J., Smith, M. W. (1989). *The Frozen Earth. Fundamentals of Geocryology*. Cambridge University Press, Cambridge, pp. 1–26, 83–121.

Yershov, E. D. (1990). *General Geocryology*. English translation by P. J. Williams, (1998), Cambridge University Press, Cambridge, pp. 93–144.

DISCUSSION TOPICS

1. How does ice segregation and frost heaving of the ground take place?

2. Compare the volumetric expansion and segregation ice models of frost weathering.

3. What processes constitutes cryogenic weathering?

4. How effective is freeze–thaw in promoting rock disintegration?

5. What processes make cryosols different to other soil types?

5 Permafrost

More than 20% of the world's land area is underlain by permafrost. Russia possesses the largest extent of permafrost; Canada is second, and The People's Republic of China ranks third. Elsewhere, >80% of Alaska possesses permafrost, bodies of alpine permafrost with areas >100000 km^2 occur in the contiguous western part of the United States, and permafrost also occurs in Scandinavia, Greenland, Antarctica, and the various high mountains of the world, including the European Alps.

The distribution of permafrost is controlled by a number of factors, the most important being climate. Site-specific controls include thermal conductivity and diffusivity of earth materials, vegetation and snow cover, topography, aspect, fire, and water bodies.

5.1. INTRODUCTION

In many periglacial environments, a consequence of the long period of winter cold and the relatively short period of summer thaw is the formation of a layer of ground that does not completely thaw during the summer. In the English-language literature, this perennially-frozen ground was termed "permafrost" by S. W. Muller, a Professor of Geology at Stanford University who had been seconded to the US Army Corps of Engineers during the Second World War to advise upon construction of the Alaskan Highway (Muller, 1943).

5.1.1. Definition

Permafrost is defined on the basis of temperature: it is ground (i.e. soil and/or rock) that remains at or below 0 °C for at least two consecutive years. However, permafrost is not "permanently-frozen ground," as Muller (1943) initially described it, but "perennially-frozen ground." Furthermore, permafrost may not necessarily be frozen since the freezing point of included water may be depressed several degrees below 0 °C. The typical temperature–depth relationship is illustrated in Figure 5.1A together with relevant terminology.

Moisture, in the form of either water or ice, may, or may not, be present in permafrost. Therefore, it is sometimes useful to differentiate between the temperature (i.e. thermal) and state (i.e. frozen or unfrozen) conditions of permafrost. In these circumstances, the terms "cryotic" and "non-cryotic" are useful. These terms, also indicated on Figure 5.1A, refer solely to the temperature of the material independent of its water/ice content (ACGR, 1988). Perennially-cryotic ground is, therefore, synonymous with permafrost, and permafrost may be "unfrozen," "partially-frozen," and "frozen," depending upon the state of the ice/water content. The term "dry" permafrost is sometimes used, especially in the Antarctic literature, when referring to permafrost that contains little or no ice.

The Periglacial Environment, Third Edition   Hugh M French
Copyright © 2007 John Wiley & Sons Ltd

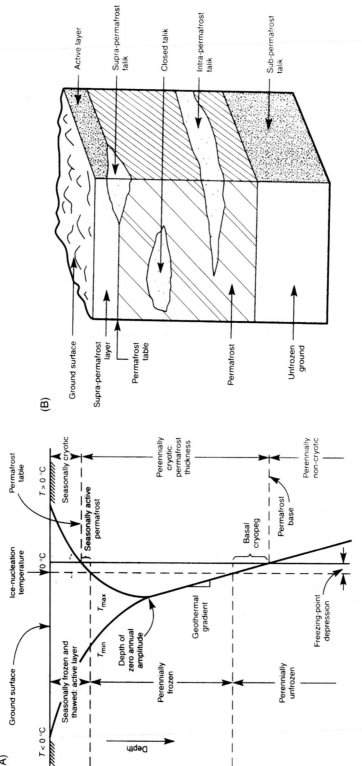

Figure 5.1. Permafrost. (A) Typical ground-thermal regime indicating maximum and minimum temperatures, the decrease in temperature with depth, the geothermal gradient, the depth of zero-annual amplitude, and the depth of seasonal thaw (the active layer). From: ACGR (1988). (B) The relationship between permafrost, the permafrost table, the active layer, and supra-, sub-, and intra-permafrost taliks. From Ferrians et al. (1969), courtesy of the United States Geological Survey.

Several other terms need immediate definition. They are illustrated in Figure 5.1B. The permafrost table is the upper surface of the permafrost, and the ground above the permafrost table is called the supra-permafrost layer. The active layer is that part of the supra-permafrost layer that freezes in winter and thaws during summer – that is, it is seasonally-frozen ground. Although seasonal frost usually penetrates to the permafrost table in most areas, in some areas where permafrost is relict (see Section 5.5), an unfrozen zone exists between the bottom of seasonal frost and the permafrost table. This unfrozen zone is called a talik. Unfrozen zones within and below the permafrost table are also termed taliks.

5.1.2. Moisture and Ice within Permafrost

The terminological subtleties described above are more than mere semantics. Many important problems posed by permafrost are related, either directly or indirectly, to the water and/or ice content of permafrost. These may be summarized under three main categories.

First, the freezing of water results in ice segregation, as explained in Chapter 4. The magnitude of frost heave varies according to the amount and availability of moisture. Poorly-drained silty soils usually possess some of the highest ice or water contents and are termed "frost-susceptible" while coarse, free-draining sediments are termed "non-frost-susceptible." The nature of primary and secondary heave is explained in Chapter 4.

Second, ground ice is a major component of permafrost, particularly in unconsolidated sediments (see Chapter 7). Frequently, the amount of ice held within the ground in a frozen state exceeds the natural water content of that sediment in its thawed state. Therefore, if permafrost thaws, ground subsidence is the result. Thaw consolidation occurs as thawed sediment compacts and settles under its own weight. High pore-water pressures generated in the process may favor soil instability and mass movement. The various processes associated with permafrost degradation are discussed more fully in Chapter 8. A related problem is that the physical properties of frozen ground, especially where soil particles are cemented together by pore ice, may be quite different to those in the same material but in an unfrozen state (Tsytovich, 1973). For example, in unconsolidated and/or soft sediments there is often a significant loss of bearing strength upon thawing. This is discussed further in this chapter.

Third, the hydrologic and groundwater characteristics of permafrost terrain are different from those of non-permafrost terrain (Hopkins et al., 1955; van Everdingen, 1990). For example, the presence of both perennially- and seasonally-frozen ground prevents the infiltration of water into the ground or, at best, confines it to the active layer. At the same time, subsurface flow is restricted to unfrozen zones or taliks. A high degree of mineralization in subsurface permafrost waters is often typical, caused by the restricted circulation imposed by the permafrost and the concentration of dissolved solids in the taliks. These characteristics are examined in Chapter 6.

5.2. THERMAL AND PHYSICAL PROPERTIES

An understanding of the properties of frozen ground is necessary to understand the surface features of permafrost terrain. While mechanical properties are important to geotechnical engineering (see Chapter 14), physical properties are more useful to periglacial geomorphology and geocryology.

Latent heat is released when ground freezes and is largely responsible for the "zero-curtain" effect discussed in Chapter 3. Frozen soil thermal conductivity is greater than that of unfrozen soil because ice conductivity is approximately 4 times that of water; therefore, other things being equal, heat penetrates the ground more quickly when frozen than when unfrozen. As a result, thermal conductivity is a major control over the permafrost temperature gradient and the active-layer thickness (see below). Porosity influences the degree of saturation of soil or rock, and moisture is described by its gravimetric water content and bulk density. The latter is useful when discussing ground ice (see Chapter 7).

Typical values for thermal conductivity and volumetric heat capacity are given in Table 5.1. Figures 5.2A and 5.2B summarize the temperature dependence of thermal conductivity of different materials of different grain size.

5.2.1. The Geothermal Regime

The growth of permafrost reflects a negative heat balance at the surface of the earth. The minimum time for the duration of permafrost is two years. Thus, if the ground freezes one winter to a depth of 60 cm and only thaws during the following summer to a depth of 55 cm, 5 cm of permafrost comes into existence in the second year. If the climatic conditions are repeated in following years, the layer of permafrost will thicken and extend downwards from the base of the seasonal-frost layer. Ultimately, permafrost several hundreds of meters in thickness can be formed.

The thickness to which permafrost develops is determined by a balance between the internal heat gain with depth and heat loss from the surface. According to A. Lachenbruch (1968), heat flow from the Earth's interior normally results in a temperature increase of approximately 1 °C per 30–60 m increase in depth. This is known as the geothermal gradient (see Figure 5.1A). The permafrost base is indicated on Figure 5.1A. Also shown is the thickness of a basal "cryopeg," i.e. that part of permafrost that is unfrozen but at a temperature <0 °C.

The amplitude of the surface seasonal temperature wave decreases with depth. This can be expressed by the following equation:

$$Az = As.e-z\sqrt{\pi/\alpha P} \tag{5.1}$$

where Az is the amplitude of the temperature wave, at depth z, As is the amplitude of the surface temperature wave (i.e. amplitude = ½ range), α is thermal diffusivity of the soil or rock, and P is the period of the wave (one year). The depth of zero-annual amplitude varies according to both air temperature and thermal diffusivity but generally occurs within 20 m of the ground surface.

If the ground-surface temperature and geothermal gradients are known, it is possible to calculate the ground temperature at any depth by using a one-dimensional heat flow equation:

$$Tz = Ts + Gg \times Z \tag{5.2}$$

where Tz is the ground temperature (°C) at depth Z (m), Ts is the mean annual ground surface temperature (°C), Gg is the geothermal gradient (°C m^{-1}) (i.e. the temperature increase over unit depth) , and Z is depth (m). For example, if $Ts = -4$ °C and $Gg = 0.02$ °C/m, the temperature at 100 m depth is calculated as follows: $T100 = -4 + 0.02 \times 100 = -4 + 2 = -2$ °C.

Table 5.1. Some typical values of thermal conductivity, thermal diffusivity, and volumetric heat capacity for various materials.

(A) Thermal conductivity of various materials

Material	Thermal Conductivity (k) (W/m K)	Material	Thermal Conductivity (k) (W/m K)
Air	0.024	Rocks	
Water	0.605	Shale	1.5
Ice (at 0°C)	2.23	Granite	1.7–4.0
Snow		Building materials	
Loose, new	0.086	Concrete	1.3–1.7
On ground	0.121	Steel	35–52
Dense	0.340	Wood	0.12–0.16
Organic material		Asbestos	0.07
Peat, dry	0.05	Polystyrene	0.033
Peat, saturated unfrozen	0.50		
Peat, saturated frozen	2.00		

(B) Heat capacities of various materials

Material	Heat Capacity, c		Volumetric Heat Capacity, C	
	Btu/lb.°F	kJ/kg-K	Btu/ft.³°F	MJ/m³-K
Air	0.24	1.000	0.0187	0.00125
Water	1.00	4.187	62.4	4.187
Ice	0.50	2.094	28.1	1.88
Soil minerals	0.17	0.710	28.0	1.875
Organic soil	0.40	1.674	37.5	2.52
Polystyrene insulation	0.24	1.000	0.65	0.0435
Concrete	0.21	0.895	30.0	2.01
Asphalt	0.4	1.674	37.5	2.52

(C) Thermal diffusivities of various materials

Material	Thermal Diffusivity, α	
	ft.²/hr. × 10^{-3}	m²/s × 10^{-7}
Water	5.61	1.45
Fresh snow	12.8	3.3
Ice	46.2	11.9
Granite	58	15
Limestone	27	7
Dolomite	78	20
Sandstone	39	10
Shale	31	8
Quartzite	174	45

Sources: Johnston (1981); Williams and Smith (1989).

If there is a change in the climatic conditions at the ground surface, the thickness of the permafrost will change accordingly. For example, an increase in mean surface temperature will result in a decrease in permafrost thickness, while a decrease in surface temperature will give the reverse. For example, Equation (5.2) can be used to calculate

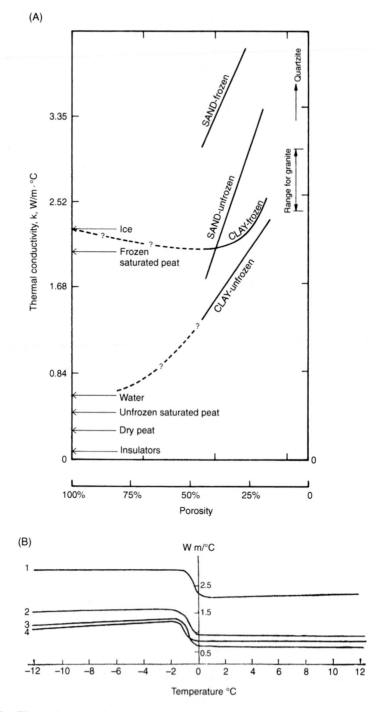

Figure 5.2. Thermal and physical properties of sand and clay when subject to freezing and thawing temperatures. (A) Relationship between thermal conductivity, porosity, and degree of saturation. From Johnston (1981). (B) Temperature dependence of thermal conductivity of materials of different grain size. Legend for B: 1, fine sand; 2, loess-like silty clay; 3, clay; 4, peat. From Yershov (1990). Reproduced by permission of Cambridge University Press.

the thickness of permafrost by letting $Tz = 0\,°C$. For the above case, the thickness of permafrost would be 200 m. If one assumes a constant geothermal gradient of 1 °C/50 m (Gold et al., 1972) and stable surface temperatures, it is possible to predict the approximate thickness of permafrost at any locality, given the mean annual temperature. This is done by multiplying the negative of the mean annual air temperature by the geothermal gradient. Thus, a locality with a mean annual air temperature of –10 °C might be expected to have a permafrost thickness of approximately 500 m.

When known data on permafrost thickness are tabulated against annual air temperatures (see Table 5.3, p. 97), it is clear that the rules-of-thumb outlined in the previous paragraph are not completely adequate. Several additional factors must be considered.

First, as Lachenbruch (1957) demonstrated, large bodies of water exert a distinct warming effect upon adjacent landmasses. Permafrost is generally absent from beneath the oceans and many of the larger water bodies. This explains why permafrost thickness values at many coastal locations are often less than predicted. Second, the effects of past climate change must be considered. For example, even where present-day mean annual surface temperatures exceed –15 °C, the thickness of permafrost may be so great (sometimes in excess of 800 m) that it must reflect colder climatic conditions during the Pleistocene. Third, the assumption of past and present climatic stability is questionable. For example, in parts of Siberia, permafrost extends to great depths (~1600 m) and varying geothermal gradients at varying depths indicate past climatic fluctuations. It seems reasonable to assume that much permafrost is relict and unrelated to present climatic conditions. A fourth factor, especially appropriate in the case of the Tibet Plateau, is that thermal discontinuities related to faults, sand dunes, lakes, and hot springs combine with normal altitudinal variations in the mean annual air temperature (MAAT) to produce a situation where it is difficult to generalize about the distribution of permafrost.

5.2.2. Physical Properties

For those interested in permafrost-related landforms, the most relevant physical properties of frozen ground are: (1) moisture content (i.e. total water content including both ice and unfrozen water), (2) density (bulk and dry), and (3) porosity. These are defined below.

The gravimetric water content (W) is the percentage of the weight of water (Ww) divided by the weight of dry soil (Ws). For frozen soils, $W = Wi + Wu$, where Wi is the ice content and Wu is the unfrozen water content:

$$W = Ww/Ws \tag{5.3}$$

Bulk density (ρ) is the total weight (Wt) divided by the total volume (V):

$$\rho = \text{weight/volume} = Wt/V \tag{5.4}$$

Dry density (ρd) is the dry weight divided by the total volume (V):

$$\rho d = \text{dry weight/volume} = Ws/V \tag{5.5}$$

Porosity (e) is the volume of pores (Vp) divided by the total volume (V):

$$e = \text{pore volume/total volume} = Vp/V \tag{5.6}$$

5.2.3. Thermal Properties

The thermal properties of frozen ground include thermal conductivity (K), mass and volumetric heat capacities (Cm and Cv), thermal diffusivity (α), and latent heat of fusion (Qi). Each of these is defined below.

Thermal conductivity (K) is a measure of the rate at which heat is driven through a unit area of a medium (soil or rock) per unit time, under a temperature gradient. It is usually measured in W m^1 K^1. The mass heat capacity (Cm), usually measured in J kg^{-1} K^{-1}, is a measure of the amount of heat that must accumulate in a unit weight (1 kg) to produce a change in temperature (1 K). The volumetric heat capacity is obtained by multiplying the mass heat capacity, Cm, by the bulk density, ρ, i.e.:

$$Cv = \rho \times Cm \tag{5.7}$$

The thermal diffusivity (α), measured in units of cm^{-2}s^{-1}, is the thermal conductivity divided by the volumetric heat capacity (Cv):

$$\alpha = K/Cv = K/Cm \times \rho \tag{5.8}$$

The latent heat of fusion (Qi), measured in J cm^{-3} or J g^{-1}, is the quantity of heat released per unit of volume or weight of a medium (soil or rock) when frozen:

$$Qi = L \times \rho d \times (W - Wu) \tag{5.9}$$

where ρd is dry density, L is latent heat of fusion of water (80 cal g^{-1} or 80 kcal kg^{-1}), W is total water content, and Wu is the unfrozen water content.

All these properties and relationships are important in understanding permafrost conditions and permafrost-related landforms.

5.3. HOW DOES PERMAFROST AGGRADE?

It is sometimes useful to think in terms of permafrost being either epigenetic or syngenetic in nature. Epigenetic permafrost occurs when soil or rock, already in existence, is subject to perennial freezing and permafrost forms by downward penetration from the ground surface. Syngenetic permafrost refers to permafrost that forms contemporaneous with the deposition of the sediment that is being frozen. For example, permafrost that is forming today in Paleozoic-age rock is obviously epigenetic in nature since it is younger than the host material. By contrast, permafrost that is forming today in deltaic or alluvial sediments that are currently being deposited must be regarded as largely syngenetic. This distinction is somewhat artificial since it can be argued that all permafrost is syngenetic in nature. However, the distinction is especially useful when dealing with relict permafrost (see pp. 103–4) and when undertaking Pleistocene periglacial investigations (Part III).

Permafrost-related landforms can be divided into two broad categories: those that form when permafrost aggrades and those that form as permafrost degrades. The latter are often associated with melt of ground ice and subsidence and erosion of thawed sediment. These phenomena are treated in Chapters 6, 7, and 8. Here, we examine the manner in which epigenetic permafrost forms.

5.3.1. General Principles

The growth of permafrost can be examined in terms of heat-conduction theory. Assuming an idealized homogeneous crust in thermal equilibrium, the distribution of ground temperature is a linear function of depth. Thus, the temperature regime in permafrost can be treated in a simple one-dimensional model:

$$Qg = K(dT/dz) \tag{5.10}$$

where Qg is heat conduction into the ground, K is thermal conductivity, and dT/dz is the thermal gradient. Obviously, different earth materials have different thermal conductivities (see Table 5.1).

Perhaps the simplest illustration of the application of heat-conduction theory is in the simulation of the geothermal disturbance that might result from the presence of a water body (Smith, 1977; Smith and Hwang, 1973). Several models are available. One which calculated the thermal contribution of a thaw-lake to the ground temperature regime in the Sachs River lowlands of southwest Banks Island, Western Canadian Arctic, is illustrated in Figure 5.3. An aerial photograph (see Figure 8.12) illustrates the nature of the terrain and the size and shape of the numerous lakes that are present. Given a mean annual air temperature of approximately −14 °C (Ikaahuk: Sachs Harbour; see Table 3.1), the model predicts maximum permafrost thickness to be 450–500 m beneath land and 50–60 m in the immediate offshore. However, the permafrost thickness on land varies greatly due to the influence of the numerous water bodies. For instance, beneath lakes approximately 50 m in radius (Lakes A and C in Figure 5.3), the thermal disturbance extends downwards for only 10–15 m. As a result, only shallow taliks develop. However, beneath Lake B, which has a radius of approximately 250 m, the upper permafrost boundary is depressed to form a 40 m deep talik. The thermal influence of this lake is also sufficient to raise the lower permafrost boundary. Finally, the large lakes greater than 1.0 km in radius (Lakes D and E in Figure 5.3) are predicted to form through-going taliks. As a

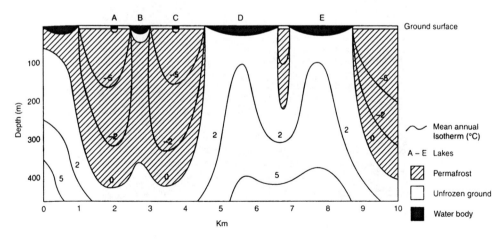

Figure 5.3. Simple two-dimensional numerical simulation of geothermal disturbances resulting from water bodies in the Sachs River lowlands, southwest Banks Island, Canada.

first approximation, lakes with minimum widths of twice the permafrost thickness will always be underlain by through-going taliks, and most lakes >2.0 m deep will possess some sort of talik.

Numerical models are limited by their inability to predict disequilibrium permafrost conditions. The downward penetration of a thermal wave is such that a lag may develop with respect to surface temperature. One estimate is that, with a mean surface temperature of $-10\,°C$, a thaw period of 10000–20000 years is required to form a through-going talik in a permafrost body 500 m thick (Mackay, 1979a, p. 30).

5.3.2. The Illisarvik Drained-Lake Experiment

The ideal way to investigate the growth of permafrost is via a three-dimensional conductive–convective heat transfer approach. However, this is not easy to model and field data are not available. This was why a novel and multidisciplinary field experiment on the growth of permafrost was initiated in August 1978 in the Western Canadian Arctic by J. R. Mackay (1981b, 1997). The site is termed Illisarvik (Inuit: "the place of learning"). Essentially, the experiment involved the artificial drainage of a shallow tundra lake (Figure 5.4). A bowl-shaped talik (unfrozen basin) with a maximum depth of 32 m underlay the lake bottom prior to drainage. Following drainage, the growth of permafrost has been

Figure 5.4. Oblique air photograph taken of drained Lake Illisarvik, August 1979, one year after lake drainage. Two shallow ponds remain, the larger of which (nearer the outlet) is underlain by a talik. The plunge pool, eroded to a depth of 3 m on August 13, 1978, is on the seaward side of the outlet. Several survey grid lines are visible on the drained lake bottom. The photo is supplied courtesy of Professor J. R. Mackay.

monitored (Mackay, 1997; Mackay and Burn, 2002). During the last 25 years, downward freezing has started on the exposed lake-bottom while upward freezing has developed from the permafrost beneath the talik. It is important to stress that the unfrozen sediments in the drained lake-basin were water-saturated sand and that the surrounding terrain was underlain by cold permafrost. Thus, freezing took place in a closed system. Figure 5.5 summarizes schematically what happens to a saturated unfrozen zone when sandwiched between downward- and upward-aggrading permafrost. As explained earlier in Chapter 4, the progressive freezing of a non-frost-susceptible, water-saturated soil can result in pore-water expulsion and the development of artesian pressures in the unfrozen zone. By contrast, in a frost-susceptible soil, the growth of ice lenses may result in water loss and consolidation of the unfrozen zone.

Measurements made in the years that followed drainage have included the detailed leveling of benchmarks inserted in the lake bottom, the recording of ground temperatures, pore-water salinities, and electrical freezing potentials, and analyses of surface water quality (Mackay, 1982, 1983a, 1984b, 1986d, 1997, 2000, pp. 62–63; Mackay and Burn, 2002; Parameswaran and Mackay, 1983). Collectively, they indicate uplift (heave) of the lake bottom, fluctuations in the amount of heave relative to a fixed datum (indicating artesian pressures), increasing values of pore-water salinity and freezing-point depression, a decrease in active-layer thickness in association with the establishment of a vegetation cover, and the growth of ice at the top of newly-formed permafrost. In April 1995, a small ice mound, approximately 1 m in height and 30 m in diameter, formed, indicating ground-water flow that previous winter. All these phenomena result primarily from permafrost

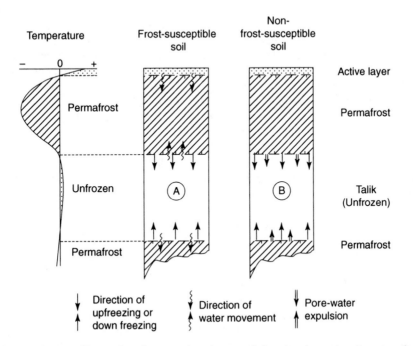

Figure 5.5. Diagram illustrating downward and upward freezing in a closed system in frost-susceptible and non-frost-susceptible soils. In frost-susceptible soil (A), the growth of ice lenses results in water loss and consolidation of the unfrozen zone. In non-frost-susceptible soil (B), pore-water expulsion results in water gain and artesian pressures in the unfrozen zone. Modified with permission from Mackay (1984a).

aggradation in saturated sand, pore-water expulsion in an open hydrological system, and intrapermafrost groundwater flow of progressively increasing salinity and decreasing temperature (Mackay, 1997, p. 31). By 1995, 17 years after drainage, freezing had penetrated more than 7 m downwards from the drained lake-bottom surface and, by 2005, the maximum thickness of the progressively shrinking talik was about 30 m.

The Illisarvik experiment is highly informative. For instance, in the first few years, the saturated lake-bottom sediments gradually hardened from water loss by drainage and evaporation in summer and repetitive freeze–thaw. However, after the first 5 years, the magnitude of frost heave and summer subsidence stabilized at about 3 cm/year (Mackay and Burn, 2002). A second observation is that, following upon the decrease in thickness of the active layer after the initial 5 years, ice began to accumulate in the upper 40 cm of the newly-formed permafrost. This ice is termed aggradational ice (see pp. 157–158). Other observations made at the Illisarvik drained-lake site have direct relevance to pingo growth, thermal-contraction cracking and ice-wedge growth, active-layer development, cryoturbation and patterned ground formation, and the growth and significance of tundra lakes (see below, and Chapter 4, 6, 7, and 8).

5.4. DISTRIBUTION OF PERMAFROST

Permafrost occurs in two contrasting but sometimes overlapping geographical regions, namely, high latitudes and high altitudes. Accordingly, permafrost can be classified into one of the following categories: (1) latitudinal, or polar, permafrost (i.e. permafrost in arctic regions), (2) alpine permafrost (i.e. permafrost in mountainous regions), and (3) plateau or montane permafrost, (i.e. areas of permafrost at high elevation, such as on the Qinghai-Xizang Plateau of China). In addition, sub-sea permafrost exists on the continental shelves of the Laptev, Siberian, and Beaufort Seas, and other permafrost bodies occur in terrestrial sub-arctic locations that bear no relationship to current climatic conditions. The latter constitute relict permafrost.

The importance of permafrost is best appreciated when it is realized that approximately 23–25% of the land surface area of the northern hemisphere is underlain by permafrost (Table 5.2A). This does not include the permafrost in the ice-free areas of Antarctica, the possible existence of permafrost beneath the Antarctic ice sheet, and the alpine permafrost that occurs in the Andes of South America. Excluding areas of frozen ground lying beneath glaciers and ice sheets, the former Soviet Union (primarily Russia) possesses the largest area of permafrost, followed by Canada, then China.

Permafrost varies in thickness from a few centimeters to several hundreds of meters. In parts of Siberia and interior Alaska, permafrost has existed for several hundred thousand years; in other areas, such as the modern Mackenzie Delta, permafrost is young and currently forming. According to J. Brown et al. (1997) about 72% of the northern hemisphere's permafrost occurs in mountains, uplands, and plateaus (Table 5.2B). Much of this is low in ice content. The remaining 28% occupies lowlands, highlands, and inter-montane depressions characterized by thick overburdens.

Permafrost is usually classified in its extent as being continuous (90–100%), discontinuous (50–90%), sporadic (10–50%), or isolated (0–10%) (Table 5.2B). In areas of continuous permafrost, frozen ground is present at all localities except for locally unfrozen zones, usually existing beneath lakes and river channels. In discontinuous permafrost terrain, bodies of frozen ground are separated by areas of unfrozen ground. Where permafrost is sporadic or isolated, it is usually restricted to isolated "islands," often occurring beneath peaty organic sediments.

Table 5.2.   Global distribution of permafrost. (A) According to Baranov (1959) and Shi (1988). (B) The extent of permafrost in the northern hemisphere based on continuity and ground ice content, as portrayed in the IPA Circum-Arctic map of permafrost and ground ice (Brown et al., 1997).

(A)

Northern Hemisphere (million km²)		Southern Hemisphere (million km²)	
Russia (Former Soviet Union)	11.0	Antarctica	13.5
Mongolian People's Republic	0.8		
China	2.1		
North American continent			
(a) Alaska	1.5		
(b) Canada	5.7		
Greenland	1.6		
Total	22.7		13.5

Total for both hemispheres	36.2 million km²
Total land area for both hemispheres	149.0 million km²
Area occupied by permafrost	24 per cent

(B)

	Area, 10⁶km² (%)					
Extent of permafrost	Lowlands/uplands with thick overburden			Mountains with thin overburden		
	High >20%	Med 10–20%	Low 0–10%	High >10%	Low 10%	Total
Continuous (90–100%)	1.49 (5.9)	1.31 (5.1)	0.38 (1.5)	2.14 (8.4)	5.66 (22.2)	10.98 (43.1)
Discontinuous (50–90%)	0.08 (0.3)	0.87 (3.4)	0.38 (1.5)	0.75 (2.9)	2.34 (9.2)	4.42 (17.3)
Sporadic (10–50%)	0.11 (0.4)	0.31 (1.2)	0.56 (2.2)	0.32 (1.3)	2.66 (10.4)	3.96 (15.5)
Isolated (0–10%)	0.34 (1.3)	0.07 (0.3)	0.60 (2.4)	0.03 (0.1)	2.85 (11.2)	3.89 (15.3)
Relict						0.12 (0.5)
Total area	2.02	2.56	1.92	3.24	13.51	23.37
Permafrost (%)	(8.6)	(11.0)	(8.3)	(14.1)	(58.2)	(100)

Source: *Frozen Ground*, 22, 1998 (back cover).

5.4.1. Latitudinal Permafrost

The broad distribution of latitudinal permafrost is relatively well known. In addition to the IPA Circum-arctic Map of Permafrost and Ground Ice (Brown et al., 1997), a number of more detailed permafrost maps are available for Alaska (Ferrians, 1965), northern Canada (Natural Resources Canada, 1995; Heginbottom and Radburn, 1993), and Russia

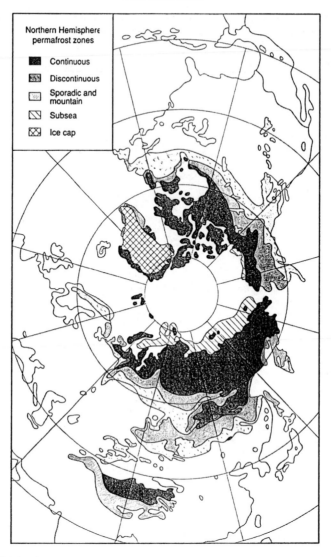

Figure 5.6. Distribution of permafrost in the northern hemisphere. From Brown et al. (1997), courtesy of the United States Geological Survey.

(Kudryavtsev et al., 1978). In the northern hemisphere, nearly one-half of Canada and 80% of Alaska are underlain by permafrost (Figure 5.6). In Russia, nearly 50% is underlain by permafrost, most occurring in the forest zone east of the Yenesei River. In China, permafrost occurs on the Tibet Plateau and in northeast China.

Typical permafrost thickness and mean annual air temperatures data are indicated in Table 5.3. In North America, the permafrost distribution is broadly governed by climate; the southern limit of continuous permafrost generally coincides with the position of the −6°C to −8°C mean annual air temperature isotherm (Brown, 1960, 1966, 1967a; Péwé, 1966b). The southern boundary of continuous permafrost extends east from the Seward

Table 5.3. Permafrost depths and mean annual air temperatures at selected locations in the northern hemisphere.

Locality	Latitude	Permafrost Zone	Mean Air Temperature (°C)	Permafrost Thickness (m)
Canada				
Resolute, NWT	74° N	Continuous	−12	390–400
Inuvik, NWT	69° N	Continuous	−9	100
Dawson City, YT	64° N	Discontinuous	−5	60
Yellowknife, NWT	62° N	Discontinuous	−6	60–100
Schefferville, PQ	54° N	Discontinuous	−4	80
Thompson, Man	55° N	Discontinuous	−4	15
Alaska				
Barrow	71° N	Continuous	−12	304–405
Umiat	69° N	Continuous	−10	322
Fairbanks	64° N	Discontinuous	−3	30–120
Bethel	60° N	Discontinuous	−1	13–184
Nome	64° N	Discontinuous	−4	37
Russia				
Nord'vik	72° N	Continuous	−12	610
Ust'Port	69° N	Continuous	−10	455
Yakutsk	62° N	Continuous	−10	195–250
Qinghai-Xizang (Tibet) Plateau				
Fenghuo Shan	34° N	Widespread	−6	110
Wudaoliang	35° N	Widespread	−5	40

Sources: Brown (1970), Ferrians (1965), Brown and Péwé (1973), Washburn (1979), Wang and French (1994).

Peninsula in Alaska and then south and east through Canada to the north of the Slave Lakes. Continuous permafrost reaches its most southerly extent at latitude 55°N, where it fringes the southern shore of Hudson Bay. East of Hudson Bay, continuous permafrost reappears in the Ungava Peninsula at latitude 60°N. The reason for this latitudinal jump is because the higher snow accumulations to the east of Hudson Bay keep ground temperatures relatively high. The southern limit of discontinuous permafrost roughly coincides with the −1°C mean annual air temperature isotherm. Southwards of this isotherm, permafrost becomes restricted to peatlands, north-facing slopes, and shady river banks.

The known thickness of permafrost in North America varies from 15–30m at the southern limit of the continuous zone to depths of over 500m in the Canadian Arctic Archipelago and northern Alaska.

As a generalization, permafrost in Russia is thicker than in North America. It increases in thickness northwards from 300m at the southern limit to over 600m along the Siberian coastal plain. A thickness in excess of 500m also occurs in central Yakutia, while in other areas, such as the Tamyr peninsula, permafrost reaches 400m in thickness (Baranov, 1959, pp. 15–19). The thickest known permafrost exists in northern Yakutia at a depth of ~1600 m, probably caused by super-cooled brine (Katasonov, in Brown, 1967b, p. 742).

Discontinuous permafrost is encountered in limited areas in northern Scandinavia, the Kola Peninsula, and the tundra and boreal forest areas between the White Sea and the Ural Mountains. East of the Urals, a broad zone of discontinuous permafrost exists across western and eastern Siberia. Here, the transition from discontinuous to continuous permafrost coincides approximately with the northern boundary of the boreal forest (taiga) and is accompanied by a sharp increase in thickness of permafrost. For example, within the boreal forest (taiga) discontinuous zone, an average thickness of 25–30 m is typical, as compared with 300 m in the forest-tundra and 400 m in the tundra.

Given similar mean annual temperatures at the depth of zero-annual amplitude, the Siberian permafrost is also colder than that in North American. This can be illustrated by the geothermal gradient. In Yakutia, values of the geothermal gradient range from 40 to 178 m/°C in permafrost and from 30 to 135 m/°C in non-frozen rock (Melnikov, in Brown, 1967b, p. 742). Observations in North America indicate values ranging from 20 m/°C at stations adjacent to large rivers or oceans to 55 m/°C at interior locations.

The greater extent, thickness, and coldness of permafrost in Siberia than in North America reflect differences in Quaternary glacial histories. During much of the Pleistocene, ice sheets covered the majority of Arctic North America but in Siberia ice sheets only formed in the principal mountain belts and uplands, leaving lowlands largely ice-free. An additional factor is that retreat of Late-Pleistocene ice sheets in North America was accompanied by development of extensive postglacial lakes and marine inundations. These limited the land areas exposed to low sub-aerial air temperature and partially explains the less-well developed Pleistocene periglacial zone in southern Canada and the northern United States (see Part III). By contrast, ice-free areas occurred widely in Siberia throughout the Pleistocene, and deep and continuous permafrost developed in response to the low air temperatures in the interior of the continent (Gerasimov and Markov, 1968). The lack of extensive glaciation in Siberia seems best explained by aridity because the mountain belts and plateaus of central Asia and Tibet, and the vastness of the landmass, effectively prevented the penetration of moisture-laden winds from either the Pacific or Atlantic Oceans.

The importance of glacial limits as regards permafrost distribution and thickness can be demonstrated with reference to known permafrost thicknesses in glaciated and unglaciated terrain that currently experience similar mean annual air temperatures. For example, 60 m of permafrost exist at Dawson City, in unglaciated Yukon Territory, Canada. The mean annual air temperature is −5 °C. A similar permafrost thickness occurs in southeast Siberia at Chita and Bomnak, which also experience mean annual air temperatures of −5 °C (Brown, 1967b, p. 746). However, at Fort Simpson, NWT, and at Thompson, Manitoba, both localities where air temperatures are similar to the other stations but which are located in recently-glaciated terrain, the permafrost is only about 15 m thick (Brown, 1970, p. 10).

5.4.2. Alpine (Mountain) Permafrost

Permafrost exists at high elevations in both mid- and low latitudes. However, since mountains also exist in polar regions and some mid-latitude mountain chains extend into polar regions, such as the Western Cordillera of North America and the Ural Mountains in Russia, it is sometimes difficult to distinguish between alpine and polar (high-latitude) permafrost. Although the term "mountain permafrost" is increasingly used to refer to both alpine permafrost and polar mountain permafrost (Cheng and Dramis, 1992), for simplicity, the term "alpine" permafrost is used here.

Alpine (mountain) permafrost is not to be confused with plateau (montane) permafrost, a term reserved exclusively for permafrost on the Tibet Plateau (see below). This is because many of the controls over the latter are quite different to those over alpine permafrost. Moreover, alpine permafrost implicitly involves consideration of steep slopes, solar radiation and aspect, and snow. Like latitudinal permafrost, the most important control over alpine permafrost is climate; this depends on the geographic location of the mountains in question (i.e. latitude, altitude, and distance from ocean).

Zonation of alpine permafrost is not unlike that of high-latitude (polar) permafrost; namely, a zone of continuous permafrost that occurs below the snow (firn) limit is replaced by a zone of discontinuous permafrost at lower elevations. This, in turn, is replaced by a zone of sporadic permafrost.

In the North American Cordillera, the distribution of permafrost varies both with altitude and latitude (Péwé, 1983a). Field observations suggest that the lower altitudinal limit of permafrost rises progressively from approximately 1000 m a.s.l. at 60°N to over 3000 m a.s.l. in Colorado and New Mexico at latitudes 39°N–33°N. In Central Mexico, isolated occurrences of frozen ground have been reported from the summits of volcanic peaks at elevations of 4500–4800 m a.s.l. (Figure 5.7A). In the European mountains, a number of recent mapping and monitoring studies have improved our understanding of alpine permafrost (Haeberli and Burn, 2002; Harris et al., 2001; Isaksen et al., 2001; King, 2000). The nature of permafrost in the Japanese Alps has also been subject to detailed investigation (Matsuoka and Ikeda, 1998).

In general, the relationship between the lower altitudinal limit of alpine permafrost and latitude can be expressed by a Gaussian curve (Figure 5.7B) (Cheng, 1983; Cheng and Dramis, 1992; Corte, 1988). This resembles, in broad outline, the snowline, the alpine cold-desert zone, and the timberline plotted against latitude. Regional controls over alpine permafrost are usually related to an increase in continental conditions away (inland) from nearby oceans. For example, in Scandinavia, as the snowline or glacial limit rises, the lower limit of permafrost decreases (King, 1986). Similar changes occur in the Cordillera of North America, with greater snowfall in the coastal ranges causing the lower limit of permafrost to decrease eastwards (Harris and Brown, 1978, 1982).

The extent and lower altitudinal limit of alpine permafrost is complicated by the relationships that exist between glaciers, timberline, and permafrost in the mid-latitude alpine environment (Figure 5.8). In humid-maritime environments, the lower parts of temperate glaciers often extend into forested valleys where summer warmth and winter snow accumulation prevent permafrost formation. As a result, no tundra or periglacial zone exists. By contrast, in dry-continental environments, as in many central Asian mountain chains, the glacier equilibrium line may be far above timberline. The occurrence of cold-air drainage, or temperature inversions and ice-caves (Harris, 1979, 1983), further complicates the situation. Surface features, such as rock glaciers, are sometimes used to infer the lower limit of alpine permafrost, but this can be controversial because some rock glaciers may be inactive while others may be glacial in origin (Barsch, 1988).

One method used to predict the occurrence of permafrost in the humid mountains of Europe is the bottom temperature of the winter snow cover. This is called the BTS method. W. Haeberli (1973, 1978) and others (Ikeda and Matsuoka, 1999; King, 1983) have demonstrated that permafrost is usually present in the Alps and in Scandinavia if the BTS beneath at least 80–100 cm snow cover during February and March is less than −2 °C. According to M. Hoelzle (1992), the following "rules of thumb" apply: values of

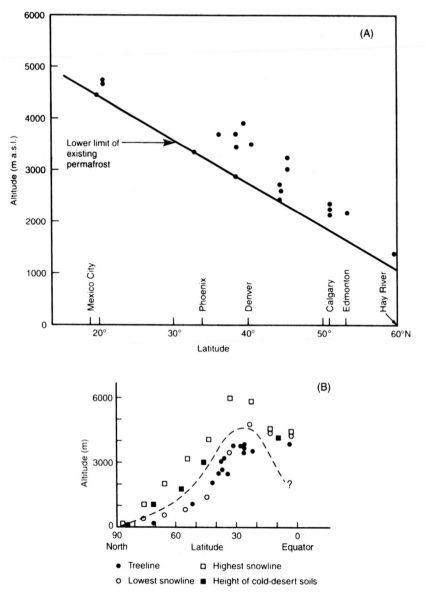

Figure 5.7. Alpine permafrost. (A) Relationship between latitude and altitude of alpine permafrost in the Cordillera of North America. Points indicate localities where permafrost has either been
measured by ground temperatures or inferred from either local micro-climatic data or from geomorphic phenomena (ice-cemented rock glaciers, blockfields). From Péwé (1983a). (B) Altitude of
selected geographical boundaries plotted against latitude in the northern hemisphere. Line is visual
best fit. From Cheng and Dramis (1992). Reproduced by permission of John Wiley & Sons Ltd.

$<-3\,°C$ indicate that permafrost is "probable," values of $-2\,°C$ to $-3\,°C$ that permafrost is
"possible," and values of $>-2\,°C$ that permafrost is "improbable." However, these categories have not been quantified. Thus, "probable" does not imply 100% likelihood of permafrost and "improbable" does not mean 0% likelihood.

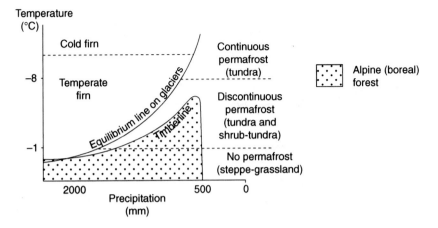

Figure 5.8. Schematic diagram illustrating the relationships between glaciers, permafrost, and timberline as a function of mean annual air temperature and average annual precipitation. From Haeberli and Burn (2002).

This empirical method has now been tested elsewhere. In the mountains of Hokkaido, Japan, the diagnostic BTS value of –2 °C for "permafrost occurrence" disagreed with the results from DC resistivity soundings (Ishikawa and Hirakawa, 2000). A more promising field test was undertaken in a mountainous basin in southern Yukon Territory, Canada (Lewkowicz and Ednie, 2004). Here, the distribution of frozen ground at 200 sites was compared with modeled BTS values using logistic regression. Analysis showed all categories of BTS-predicted permafrost were present at statistically significant levels of probability but that permafrost occurrence in the area varied markedly over short distances in response to snow-pack depth. Whether the BTS method is applicable to the arid permafrost of the mountain ranges of central Asia, such as the Tien Shan (Gorbunov, 1988a), is much more problematic.

5.4.3. Montane Permafrost of Central Asia and China

The so-called "alpine" permafrost of Central Asia (Qiu and Gorbunov, 1993) combines many of the characteristics of both alpine and latitudinal permafrost, as described above. For simplicity, it is referred to as "montane" permafrost. The main difference is that montane permafrost forms in cold, dry, continental climates. It is also spatially extensive. According to Qiu Guoqing (in Qiu and Gorbunov, 1993, 75), the total area of high-altitude permafrost in central Asia is 1,760,000 sq km, of which the vast majority is in the Qinghai-Xizang Plateau. The main difference from latitudinal permafrost is in terms of its high elevation and its low latitude.

The majority of montane permafrost occurs in China, with smaller amounts in Mongolia, Kazakhstan, and neighboring areas (Figure 5.9). Here, the controls over permafrost are quite different to those of latitudinal permafrost and, as such, demand further explanation. Central to any discussion is low latitude, high elevation, and relative aridity.

Chinese scientists map permafrost on the Tibet Plateau as being either "predominantly continuous" (70–80% of area underlain by permafrost) or "isolated" (less than 40–60% of area underlain by permafrost) (Shi, 1988). Equivalent terms in the North American literature are "widespread" and "sporadic" permafrost. What is important to stress is that

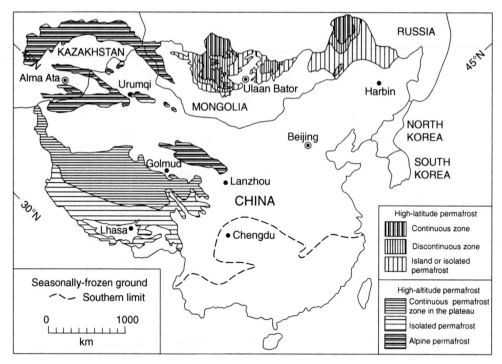

Figure 5.9. Permafrost distribution in China and adjacent areas of Mongolia and Kazakhstan. Compiled from Shi (1988), Brown et al. (1997), Brown (2001), and Aubekerov and Gorbunov (1999).

continuous permafrost, as defined in North America, does not exist on the Tibet Plateau. Equally, montane permafrost differs from alpine permafrost because alpine permafrost is characterized by steep slopes and bedrock terrain. By contrast, the Tibet Plateau has relatively low surface relief in between the various mountain ranges and can be underlain by as much as 1000m of unconsolidated sediments (Guo et al., 1982). Large mountain ranges separate many of the basins. Thus, the vastness and unusual nature of this terrain defies easy classification in terms of either alpine or latitudinal permafrost.

As a generalization, we can characterize montane permafrost as being a few meters to over 100m in thickness. The mean annual ground temperature (at the depth of zero-annual amplitude) varies between 0°C and –3.5°C. Therefore, it is similar in temperature to much alpine permafrost but considerably warmer than most latitudinal permafrost.

Some studies report that the thickness of Tibet Plateau permafrost increases at a rate of between 10m and 30m for each one degree increase in latitude (Tong and Li, 1983). This is not supported by ground temperature measurements, because air temperature varies irregularly with latitude and is dominated by altitudinal effects. Permafrost is known to thicken with increasing altitude at a rate of 15–20m for each 100m increase in elevation (Shang, 1982; Tong and Li, 1983; Zhou and Guo, 1982). A unique characteristic of Tibet Plateau permafrost is the increase in its lower altitudinal limit from north (at 4200m a.s.l.) to south (at 4800m a.s.l.), at a rate of about 110m per one degree of latitude. In general, the lower altitudinal limit of permafrost on the Plateau is usually 800–1100m lower than the snowline (Zhou and Guo, 1982). The mean annual air temperature at the lower limit is in the range of –2.5°C to –3.6°C (Tong and Li, 1983). This is considerably lower than that of alpine permafrost in the Rocky Mountains in North America (0 °C to −1 °C) (Péwé, 1983a) and the southern boundary of latitudinal permafrost in Canada (about −1.1 °C) (Brown and Péwé, 1973). The high average geothermal gradient on the Plateau (22 m/°C in permafrost; 16 m/°C below permafrost) is due, almost certainly, to the high geothermal heat flux associated with the recent tectonic uplift of the Tibet Plateau.

The regional distribution of permafrost in central Asia is complicated not only by the mountain ranges and adjacent inter-montane plateaus and basins but also by faulting, lakes and springs, and highly mobile sand dunes. The effect of snow cover upon the ground thermal regime is negligible because snow accumulation is often slight (<7 cm). Along the northern flank of the Tien Shan, in southern Kazakhstan, boreal forest occurs, but elsewhere, and at higher elevations in China, the vegetation is sparse.

5.5. RELICT PERMAFROST

Since permafrost is defined exclusively upon the basis of temperature, permafrost may occur on the sea floor wherever mean annual sea-bottom temperatures are below 0 °C. Moreover, because of the lower freezing point which exists under saline conditions, sediments are not necessarily frozen. The term "saline permafrost" is sometimes used. In addition to sub-sea permafrost, much terrestrial (land) permafrost is unrelated to the present climate (see earlier).

5.5.1. Sub-Sea Permafrost

Large areas of sub-sea permafrost occurs beneath the waters of the Laptev and East Siberian seas (Baranov, 1959; Danilov et al., 1998). Permafrost also occurs beneath the sea floor of the southern Beaufort Sea and the Western Canadian Arctic, and may extend as far as 100 km from shore (Mackay, 1972b; Dyke, 1991). Areas of sub-sea permafrost are usually identified either by either drilling or acoustic geophysical surveys. The latter approach only recognizes ice-bonded permafrost; that is, permafrost which is cemented together by ice crystals, and therefore underestimates the existence of unfrozen, saline permafrost bodies (Hunter et al., 1976).

Sub-sea permafrost can be in either thermal equilibrium or disequilibrium. In areas where sea-bottom temperature is negative, and has remained so over a considerable time period, shallow (<150 m thick) equilibrium permafrost may exist. Its thickness will depend upon such factors as mean sea-bottom temperature, soil properties, geothermal heat flux, and sedimentation rates. Most sub-sea permafrost probably developed during the colder periods of the Quaternary when eustatic sea level fluctuations exposed large areas of the continental shelves to cold, sub-aerial conditions (see Chapter 12). Today, this permafrost is relict and slowly degrading. Thawing of ice-bonded sediment below the sea floor in the southern Beaufort Sea is largely controlled by rapid shoreline retreat and the influx of warm spring and summer discharge of the Mackenzie River. This can affect sea-bottom temperatures to a depth of 10–20 m.

5.5.2. Relict (Terrestrial) Permafrost

Much relict terrestrial permafrost is Late-Pleistocene in age. In the western Arctic of Canada, glacially-deformed ground ice and icy sediments are thought to be at least 40000 years old (Mackay et al., 1972), and in central Alaska there is evidence that permafrost has been in existence for at least 125 000 years (Péwé et al., 1997). Evidence from Siberia also suggests that much permafrost originated during the Pleistocene (Gerasimov and Markov, 1968). Perhaps some of the most dramatic evidence is the presence of limbs of Pleistocene mammoths and other animals preserved in permafrost.

Pleistocene-age permafrost, and its implications for paleo-environmental reconstruction, is discussed in more detail in Part III.

5.6. PERMAFROST HYDROLOGY

It is generally assumed that groundwater movement in permafrost is restricted by perennially- or seasonally-frozen ground. However, this assumption is only partly true because it is not uncommon for unfrozen zones to exist in which normal groundwater movement may occur. Following early Russian literature associated with the development of gold mining in Siberia at the end of the nineteenth century, these unfrozen zones are now generally termed taliks (Muller, 1943; Nekrasov, 1963; Sumgin, 1927). The diagrammatic occurrence of groundwater in permafrost regions is illustrated in Figure 5.10.

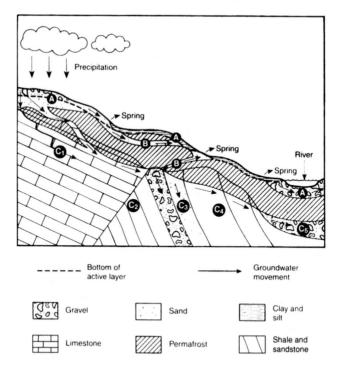

Figure 5.10. Occurrence of different types of groundwater in permafrost areas. Legend: A, suprapermafrost water; B, intra-permafrost water; C, sub-permafrost water; C1, karst water in solution channels; C2, fissure water in fault; C3, aquifer in porous rock; C4, fissure water in bedrock joint; C5, alluvial water in alluvial deposits. From Cederstrom et al. (1953), courtesy of the United States Geological Survey.

5.6.1. Aquifers

Because permafrost acts as an impermeable layer, groundwater movement is restricted to taliks (see Figure 5.10). These may be of three types. First, a supra-permafrost talik may exist immediately above the permafrost table but below the depth of seasonal frost. In continuous permafrost, supra-permafrost taliks are rare, but in discontinuous permafrost the depth of seasonal frost frequently fails to reach the top of permafrost since the latter is often relict. In these areas, the supra-permafrost talik is a residual thaw layer (see Figure 7.10) which may be several meters or more thick. Second, intra-permafrost taliks are unfrozen zones confined within permafrost. Third, sub-permafrost taliks refer to the unfrozen zones beneath permafrost.

Taliks may be further distinguished by the mechanism responsible for their unfrozen condition (Table 5.4). It is important to remember that taliks may be either cryotic (i.e. below 0 °C) or non-cryotic (i.e. above 0 °C), or either open or closed depending upon whether the talik reaches to the seasonally-thawed zone or not. Open taliks are common even within continuous permafrost, the majority resulting from local heat sources such as lakes, river channels, and other standing water bodies. Closed taliks usually result from a change in the thermal regime of permafrost, such as might occur following lake drainage and the downward aggradation of permafrost (as at Illisarvik; see pp. 92–94). Where

Table 5.4. The types of taliks that occur in permafrost regions.

Closed talik
a non-cryotic talik occupying a depression in the permafrost table below a lake or river (also called "lake talik" and "river talik"): its temperature remains above 0 °C because of the heat storage effect of the surface water.

Hydrochemical talik
a cryotic talik in which freezing is prevented by mineralized groundwater flowing through the talik.

Hydrothermal talik
a non-cryotic talik, the temperature of which is maintained above 0 °C by the heat supplied by groundwater flowing through the talik.

Isolated talik
a talik entirely surrounded by perennially-frozen ground; usually cryotic but may be non-cryotic (see transient talik).

Lateral talik
a talik overlain and underlain by perennially-frozen ground; can be non-cryotic or cryotic.

Open talik
a talik that penetrates the permafrost competely, connecting suprapermafrost and subpermafrost water (e.g., below large rivers and lakes). It may be non-cryotic (see hydrothermal talik) or cryotic (see hydrochemical talik).

Thermal talik
a non-cryotic talik, temperature of which is above 0 °C due to the general thermal regime. It includes the seasonally-thawed ground in the active layer.

Transient talik
a talik that is gradually being eliminated by freezing, e.g., the initially non-cryotic closed talik below a small lake which, upon draining of the lake, is turned into a transient isolated talik by permafrost aggradation

---

Source: Sloan and van Everdingen, 1988; van Everdingen (1990).

closed taliks occur at depth in thick permafrost bodies, they probably reflect long-past fluctuations in regional climate. In the discontinuous zone, where permafrost is often thin, the unfrozen zones that perforate permafrost form "through" taliks linking unfrozen ground above and below the permafrost layer.

5.6.2. Hydrochemistry

The low ground temperatures associated with permafrost regions means that reaction and dissolution rates are reduced. At the same time, because of the increased solubility of carbon dioxide at low temperature, the solubilities of calcites, dolomites, and gypsum are increased.

The chemical composition of intra- and sub-permafrost waters will depend largely upon the residence time in the subsurface, and the mineral composition of the aquifer. Low dissolved solids are usually associated with rapidly moving groundwater, as might be found in karst terrain or in fractured non-soluble rock. Sub-permafrost waters range from being freshwater of the $Ca(mg)$-HCO_3 type, to saline, to calcium/sodium brines. Needless to say, the latter may often be at temperatures below $0\,°C$, emerging from cryotic taliks. On the other hand, in some places, such as Engineers Creek on the Dempster Highway, northern Yukon, Canada, exceptionally iron-rich, sulfurous, and acidic springs occur, with pH values as low as 2.8, temperature of $+9\,°C$, and dissolved solids as high as $1080\,mg/1$ (Harris et al., 1983, pp. 75–79). In the same general vicinity, freshwater springs discharge at a rate of $1.4\,m^3/s$, with a temperature of $+4\,°C$ and a dissolved solids content of only $362\,mg/1$. Data such as these illustrate the importance of local geological conditions upon the hydrochemistry and water quality of permafrost waters.

Intra-permafrost groundwater may be highly mineralized. For example, in central Siberia, the mineralization of water in freezing taliks that still exist beneath previously drained lakes may be as high as $60\,g/1$ (Anisimova et al., 1973, p. 17). On the other hand, certain perennial springs originating from sub-permafrost waters are known for their excellent drinking quality. This is probably because of their low temperature and, in the absence of CO_2, the low solubility of carbonates in the host aquifer.

Perennial springs in areas of thick permafrost are relatively few but noteworthy. In central Yakutia, perennial springs are regarded with almost religious significance by local Yakut residents. There, some springs have discharges that exceed $70\,l/sec$ with dissolved solids concentrations as high as $0.34\,g/1$ (Anisimova et al., 1973). Those on Axel Heiberg Island, Canadian High Arctic, are associated with gypsum and, being highly mineralized (Na, Cl), give rise to a freezing-point depression of at least $7–10\,°C$. Spring discharge temperatures range from $>+5.0\,°C$ to $<-4.0\,°C$ (Beschel, 1963; Pollard, 2005; Pollard and McKay, 1997). Hot and/or highly mineralized springs are also known to occur in central Alaska (Grantz et al., 1962) and in the Mackenzie Mountains of Northern Canada (e.g. Takhini hot springs, southern Yukon Territory).

5.6.3. Groundwater Icings

Icings are sheet-like masses of ice which form at the surface in winter where water issues forth from the ground (Carey, 1970). They are also known as aufeis (Washburn, 1979, p. 44) or naledi (Brown, 1967b, pp. 74–79). Sometimes, the water source is of a sub- or intra-permafrost nature, in which case the spring is usually a perennial one.

Maximum groundwater icing accretions involve sub- or intra-permafrost waters. They usually occur in late winter when the ground is fully frozen and before the rise in air

Figure 5.11. The Babbage River icing, Barn Mountains, northern Yukon, Canada, is associated with perennial discharges through Triassic-age sandstone and limestone. The icing was ablating and about 3–4 m thick at the time the photo was taken in mid-July 1979.

temperature in the spring (van Everdingen, 1982, 1990). During the early summer months, icings usually melt completely. Sometimes, the thickness of the icing can be inferred from scars on the barks of trees adjacent to the icing location. Others, such as that which forms on the Babbage River, northern Yukon, Canada (Figure 5.11), are favorite gathering places for caribou to escape heat and mosquitoes in the summer.

Groundwater icings associated with perennial springs may assume considerable dimensions. In central Yakutia, four large icings associated with perennial springs are located in tributary valleys of the Lena River. In the Ulakhan-Taryn spring, the average maximum icing accretion volume exceeds 200 000 m^3 (Anisimova et al., 1973, p. 42). In the Momskaya Depression of northeast Yakutia, complex icings as large as 62 km^2 are known to occur (Nekrasov and Gordeyev, 1973, pp. 37–40). In Alaska, an icing over 10 m thick and more than 1 km wide and 2 km long formed in the St John River valley in 1969 (Ferrians et al., 1969, p. 346).

5.7. PERMAFROST AND TERRAIN CONDITIONS

Although the broad controls over permafrost distribution are climatic in nature, local variations in permafrost conditions are determined by a number of terrain and other factors (Kudryavtsev, 1965; Brown, 1973a, b). Of widespread importance are the effects of relief and aspect, and the physical properties of soil and rock. More complex are controls exerted by vegetation, snow cover, water bodies, drainage, and fire. Generally speaking, the most complex permafrost–terrain relationships occur in discontinuous permafrost.

In the following discussion emphasis will be placed upon the latitudinal permafrost of North America and Eurasia.

5.7.1. Relief and Aspect

Relief influences the amount of solar radiation received by the ground surface and the accumulation of snow. Slope orientation also influences the amount of solar radiation received. The effects of insolation are particularly clear in mountainous regions, such as northern British Columbia and Yukon Territory, where permafrost occurs on north-facing slopes and not upon adjacent south-facing slopes. Similarly, in continuous permafrost, the active layer is usually thinner on north-facing slopes. However, in certain instances, exposure to local weather conditions assumes greater importance. For example, on the Beaufort Plain of northwest Banks Island, Arctic Canada, the active layer is thinnest on southwest-facing slopes (French, 1970). This is attributed to the influence of the dominant southwest winds in this part of the Arctic, which promote evaporation and latent heat loss from exposed slopes during the summer months.

5.7.2. Rock Type

Variations in the nature of rock and soil express themselves in differing albedo and thermal conductivity values. These controls attain their greatest significance in the continuous permafrost zone, where climate is sufficiently cool to produce permafrost regardless of the type of terrain. Average albedo values for bare rock and soil can also vary between 10% and 40%. Thus, significant variations in active-layer thickness and the permafrost thermal regime can be expected in different rock and soil types. Table 5.1 provides typical thermal conductivity values for various materials. Because the thermal conductivity of ice is much higher than for water, frozen icy soils have higher thermal conductivities then unfrozen soils. Another important observation is that loose fresh snow and dry, organic (peaty) material have low thermal conductivity values, and hence are good insulators of the ground.

5.7.3. Vegetation

Probably the most complex terrain factor is vegetation. It affects permafrost in a variety of ways and is significant in all areas of discontinuous and continuous permafrost with the exception of the vegetation-free polar deserts. Its most fundamental influence is to shield the underlying permafrost from solar heat. The insulating property of vegetation and organic (peaty) material (see Table 5.1) is probably the single most important factor in determining the magnitude of the thermal offset (see Chapter 3) and the thickness of the active layer. Numerous observations from a wide variety of permafrost environments indicate that the active layer is thinnest beneath poorly-drained and well-vegetated areas, and thickest beneath well-drained bare soil or rock.

Much permafrost lies within the boreal forest and taiga regions of North America and Eurasia. In these regions, trees are important controls over local permafrost conditions because they shade the ground from direct solar radiation and intercept snowfall in winter. Thus, the winter cold often penetrates more deeply beneath trees than beneath areas of thick snow cover, and the amount of summer solar radiation received at the surface is reduced. Different tree species further complicate the situation. For example, the Siberian taiga is composed predominantly of pine (*Pinus silvestris*) and tamarack (*Larix dahurica*) whereas the spruce (*Picea glauca* and *Picea mariana*) is more common in North America. Since the spruce forest provides a denser canopy and more shade, a surface moss cover is widely developed; other things being equal, this promotes a thinner active layer in the North American boreal forest than in the Siberian taiga. In northern Finland, where birch (*Betula*) dominates the northern forest, the episodic occurrence of insect infestations, such as *Oporinia autumnata*, can reduce the leaf canopy, thereby increasing summer solar radiation and causing change in the near-surface thermal regime. In North America, the actions of rodents, such as the beaver (*Castor canadensis*) in building dams and raising water levels, may also cause trees to die and permafrost conditions to change (Lewkowicz and Coultish, 2004).

In those areas where the treeline and the southern boundary of continuous permafrost are in close proximity, as is the case for much of North America, the presence or absence of trees may become critical as regards permafrost occurrence. In Alaska, for example, even isolated white spruce may influence the energy exchange at the ground surface sufficiently for a small permafrost body to exist (Viereck, 1965).

5.7.4. Snow Cover

Snow cover, like vegetation and organic material, influences local permafrost conditions because it is also a good insulator (Table 5.1). The snowfall regime, type of snow, and length of time snow lies on the ground are critical factors. In general terms, a heavy snowfall in autumn or early winter inhibits frost penetration, while a winter of low snowfall does the reverse. Also, if snow persists late into spring, ground thawing will be delayed. In detail, variations in snow cover are controlled by site characteristics such as microrelief, vegetation, and direction of the dominant snow-bearing winds.

In areas north of treeline, snowfall amounts are more limited and the effects of snow cover are generally regarded as less important. However, significant differences in active-layer conditions still occur. For instance, it is not uncommon for upland surfaces and interfluves to be blown clear of snow for much of the winter while large snow banks accumulate in gullies and on lee slopes. Deeper frost penetration may occur on the uplands and interfluves, therefore, than in depressions and on lee slopes. On the other hand, ground thawing at snow-bank localities may be delayed until late summer when the snow bank finally disappears. As a consequence, the active layer is often thinner at snow-bank localities than on uplands. The role of snow banks as regards slope hydrology, mass wasting, and other processes in areas north of treeline is discussed more fully in Chapter 9.

South of treeline, especially in the discontinuous permafrost zones, snow cover frequently assumes great importance in terms of permafrost distribution. For example, a study at Schefferville, northern Québec, indicated that the pattern of accumulation of the seasonal snow cover is the controlling factor in the distribution of permafrost in that area (Granberg, 1973; Nicholson and Thom, 1973). Permafrost occurs only in uplands where an absence of trees prevents snow from accumulating. By contrast, in adjacent lowlands, where a thick snow cover accumulates each winter, permafrost is generally absent. It appears that a winter snow depth of 65–70 cm is sufficient to prevent the development of permafrost in the Schefferville region (Nicholson and Granberg, 1973).

To illustrate the complexity of the climate–vegetation–snow interaction, Table 5.5 summarizes ground temperature data from near Churchill, Manitoba, where permafrost occurs at three adjacent sites but not at a fourth. All are located within a 2 km^2 area. Churchill is at the border between continuous and discontinuous permafrost and is a few km north of treeline. The mean annual air temperature is −7.3 °C and the average permafrost thickness, where present, is between 40 and 60 m. Table 5.5 convincingly demonstrates that the progressive increase in average ground temperature at sites 2, 3, and 4 correlates with an increase in peat thickness and snow cover, and a decrease in snow density. The absence of permafrost at site 4, a depression adjacent to site 3, is related to waterlogging, enhanced by the accumulation of snow in the depression.

Similar subtle differences in terrain conditions occur widely throughout the discontinuous permafrost zones in both North America and Eurasia. As a consequence, it

Table 5.5. Ground temperatures and environmental conditions at four adjacent sites near Churchill, Manitoba (58°45′N), 1974–1976.

Site:	Average hole temperature (°C)	Active-layer thickness (cm)	Organic layer (cm)	Snow depth (1974/75/76) (cm)	Average snow densities (1974/75/76)	Permafrost Yes/No
Quartzite bedrock	−2.9	−750		61/51/20	0.23/0.32/0.26	Yes
Marine deposits overlying till	2.6	75–90	23	46/36/33	0.27/0.31/0.34	Yes
Palsa	−0.9	50	40	56/53/38	0.22/0.33/0.22	Yes
Depression	+0.4	75 (seasonal frost)	150	69/102/58	0.23/0.24/0.19	No

Source: Brown (1973a, 1978).

is often difficult to predict permafrost conditions in these areas without detailed site investigations.

5.7.5. Fire

Wildfires in the taiga and boreal forest are surprisingly common. Many start by lightning and much, if not all, of these regions have been burned over at least once. In Alaska, over 400 000 hectares burned between 1940 and 1969 (Viereck, 1973a, b) and in northwestern Arctic Canada, approximately one million hectares burned between 1962 and 1971 (Brown and Grave, 1979, p. 10). In many ways, the boreal forest can be regarded as a fire climax (Payette et al., 1989).

Tundra fires are less frequent on account of the relative absence of woody materials and lower summer temperatures. However, they are not unknown (Shilts, 1975; Wein, 1976; Wein and Shilts, 1976). In particular, the tussock-forming cotton grass, *Eriophorum vaginatum*, is especially prone to fire (Wein and Bliss, 1973). When the dry top of the tussock is burnt, the darkened surface promotes increased thaw. Where fire occurs in areas at the northern boundary of the boreal forest, and trees are killed, open tundra may develop. In Siberia, these areas are sometimes termed pyrogenic tundra (Kriuchkov, 1968).

The effect of fire upon permafrost depends upon the nature and dampness of the vegetation, and the speed at which the fire passes through the area. If the fire passes rapidly, and if the surface cover is peat, moss, or lichens, only the trees may burn and the ground beneath 2–3 cm may remain untouched. In this case, little change will occur to the permafrost. However, if the surface vegetation is exceptionally dry, and if the fire moves slowly, considerable changes in permafrost may result. At Inuvik, NWT, Canada, for example, the effects of a 1968 forest fire led to destruction of much of the vegetation, thaw of ice-rich sediments, rapid gullying and thermal erosion, and numerous earth flows (Heginbottom, 1973). The more long-term effect has been an increase in active-layer thickness in the burned-over area (Mackay, 1977b, 1995a) (see Table 7.1). Elsewhere in the Mackenzie Valley, forest fires have triggered permafrost degradation and slope instability (Harry and McInnes, 1988). More recently, the long-term degradation of permafrost following a fire in 1958 in a spruce forest in the southern Yukon Territory has been documented by C. R. Burn (1998a). An area which escaped burn in 1958 indicates that, under prevailing climatic conditions, permafrost has a thickness of about 17 m and an active layer of about 1.5 m. Measurements, made over a 6-year period (1991–1996) in the burn areas, indicate that the upper 2.4 m of permafrost had degraded by 1997 and ground temperature throughout the permafrost had warmed by approximately −0.2 °C. A simplified numerical model suggests that approximately 1000 years would be required for complete degradation of the permafrost.

5.7.6. Lakes and Surface Water Bodies

Water has a high volumetric heat capacity (see Table 5.1). It is not surprising, therefore, that lakes create the greatest natural thermal departure at the ground surface from conditions determined by climate (Brewer, 1958; Lachenbruch, 1957; Mackay, 1962, pp. 33–48). Numerous field observations, together with theoretical considerations (Brown et al., 1964; Johnston and Brown, 1964; Kudryavtsev, 1965, pp. 25–27) indicate that an unfrozen layer, or talik, exists beneath water bodies that do not freeze to their bottoms in winter. The size of talik varies with area and depth of the water body, water temperature, thickness of winter ice and snow cover, and nature and compaction of bottom sediments. Lakes are widespread in many permafrost regions and especially common in deltaic, coastal, and lowland tundra environments, where they may occupy as much as 15–20% of the total land area.

Along the southern fringes of the discontinuous permafrost zone, especially in wetlands, drainage conditions assume great importance in determining the presence or absence of permafrost (Brown, 1973b; Zoltai, 1973). This is discussed more fully in Chapter 6 in the context of palsas and organic terrain.

5.8. THE ACTIVE LAYER

The active layer refers to the layer of ground in areas underlain by permafrost which thaws during summer. As a rule, it is thinnest in polar regions (as little as 15 cm) and becomes thicker in sub-arctic regions (as much as 1.0 m or more). In the continuous permafrost zone, it generally reaches the permafrost table. In the discontinuous permafrost zone, it may be separated from underlying permafrost by a talik, or residual thaw layer. The thickness of the active layer varies from year to year depending on controls such as ambient air temperature, slope orientation and angle, vegetation, drainage, snow cover, soil and/or rock type, and water content.

The active layer includes the uppermost part of the permafrost wherever either the salinity or clay content of the permafrost allows it to thaw and refreeze annually, even though the material remains cryotic (i.e. below 0 °C). The term "depth to permafrost" as a synonym for thickness of the active layer is misleading, especially in areas where the active layer is separated from the permafrost by a thawed, or non-cryotic (i.e. above 0 °C), layer of ground.

The introduction of geocryological, or so-called "cryotic," terminology has confused the definition (see above) of the active layer. As shown in Figure 5.1A, the active layer, defined traditionally, includes a near-surface layer that is seasonally cryotic and a zone immediately beneath which is subject to thaw in spite of the fact that the temperature is below 0 °C (i.e. cryotic). This zone is termed "seasonally-active" permafrost. Because the active layer, so defined, varies consistently both in time and space, it is difficult to apply to operational procedures in the field (Burn, 1998b). The CALM monitoring program uses the traditional definition of the active layer.

5.8.1. The Transient Layer

The transient layer is the near-surface ice-rich layer marking the long-term position of the contact between the active layer, as traditionally defined, and the upper part of permafrost (Shur, 1988a, b; Shur et al., 2005). Figure 5.12 shows a three-layer system of the active layer that incorporates the concept of a transient layer.

The utility of the transient-layer concept follows upon an appreciation of the importance of the ice-rich layer that typically characterizes the base of the active layer and the upper part of permafrost. In years of exceptional summer warmth, the active layer deepens beyond its normal (average) thickness and part of this ice-rich zone thaws. This promotes slope instability, rapid mass movements (active-layer-detachment failures), and enhanced rates of solifluction activity. All these processes are described in Chapter 9. Continued global climate warming will also mean that this ice-rich layer will continue to thaw as the active layer progressively thickens in years to come (see Chapter 15).

The ice-rich layer exists because of temperature gradients that are set up within the active layer and the upper part of permafrost that allow moisture to migrate upwards during winter and downwards during summer. As explained earlier in Chapter 4, it is known that, in certain frost-susceptible soils, as much as 40% of the water content may

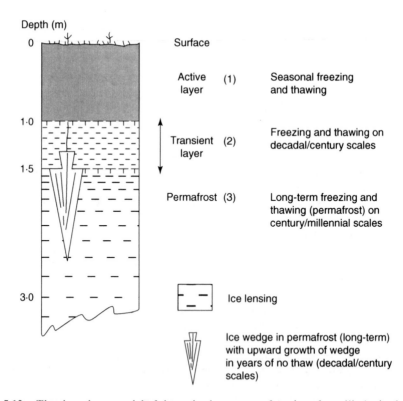

Figure 5.12.   The three-layer model of the active layer-permafrost interface. (1)-Active layer (seasonal freezing and thawing); (2)-transient layer (high ice content and freezing/thawing at decadal to century scales; (3)-long-term permafrost (freezing and thawing at century to millennial scales). Modified from Nelson (2004) and Shur et al. (2005).

remain unfrozen at −1.0 °C (Williams, 1976, 1977; see Figure 4.2) and that unfrozen water moves in the direction of colder temperature (the concept of capillarity). Therefore, the ice (water) content in the upper part of the active layer increases when a positive ground temperature gradient occurs in winter (i.e. the ground is warmer than the air) (Figure 5.13A). Under negative ground temperature gradients in summer, unfrozen water migrates downwards and the ice content in the upper part of permafrost increases. Late-summer heaving of the active layer is the result (Figure 5.13B).

In summary, the concept of the transient layer recognizes the different periodicities at which the near-surface of permafrost terrain cycles through 0°C (Nelson, 2004, p. 258), and the relative abundance, morphology, and distribution of ice contained within the active layer and the upper part of permafrost.

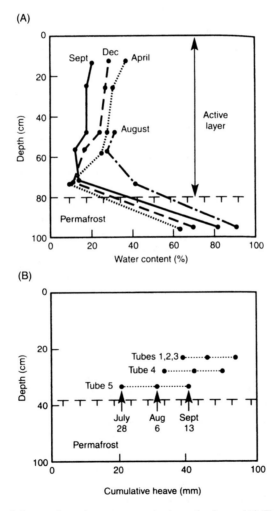

Figure 5.13. Seasonal changes in moisture content in the active layer. (A) Changes in water content for the active layer and top of permafrost in Siberia. (B) Summer (1978) movement (heave) of active layer at Garry Island, Mackenzie Delta region, Canada. From Mackay (1983a).

5.8.2. The Stefan Equation

There are a number of equations that can be used to calculate the thickness of the active layer. The most common is the Stefan equation:

$$Z = \sqrt{2TK/Qi} \tag{5.11}$$

where Z is the thickness of the active layer (m or cm), T is the ground surface temperature during the thaw season (°C), K is thermal conductivity of unfrozen soil (W/m K or kcal/m °C h), t is the duration of the thawing season (day, hour, second), and Qi is volumetric latent heat of fusion (kJ/m³). Qi is expressed as:

$$Qi = Lpd(W - Wu) \tag{5.12}$$

where $L = 80$ cal/g or 8 kcal/kg (latent heat of ice), pd is the dry density of soil (kg/m³), W is total moisture content (%), and Wu is unfrozen water content (%). Gold and Lachenbruch (1973) provide a further approximation of active-layer thickness, x (cm), with the following:

$$x = \sqrt{\alpha} \, P/\pi \log e |A_0/T_0| \tag{5.13}$$

where α is soil thermal diffusivity, P is the period of the temperature cycle, A_0 is surface temperature amplitude, and T_0 is mean annual surface temperature.

Typical values of thermal conductivity and thermal diffusivity are given in Table 5.1. A simple example illustrates use of the Stefan equation. If mean surface temperature during the thawing season is 6.5°C and thaw duration is approximately 150 days (~3620 hours), and if one assumes the thermal conductivity of unfrozen soil (K) is 1.25 kcal m h, dry density (pd) is 1334 kgm³, water content (W) is 18%, and unfrozen water content (Wu) is 3%, the thickness of the active layer will approximate 1.92 m. It is important to note that Equation (5.11) can also be used to calculate the depth of frost penetration. In this case, time t is the duration of the freezing season ($T < 0$°C) and K represents the thermal conductivity of frozen soil. The Stefan equation can also be used to calculate the previous surface temperatures that would have been responsible for the formation, at depth, of a thaw unconformity (see Chapters 7 and 11).

5.8.3. Active-Layer Thermal Regime

Ground climates have been discussed in general in Chapter 3 within the context of n factors and the thermal offset. Here, two further aspects of the active-layer ground-thermal regime are briefly mentioned.

First, freezing and thawing of the active layer occurs either on a diurnal basis, as in many temperate and subtropical regions, or on a seasonal basis, as in high latitudes. It is important to note that thawing is one-sided, from the surface downwards. Autumn freeze-back is a more complex process because in regions underlain by permafrost freezing is two-sided, occurring both downwards from the surface and upwards from the perennially-frozen ground beneath. Second, active-layer thermal regimes are closely related to snow thickness and density. This has been illustrated earlier with data from Churchill, Manitoba (see Table 5.5).

Other aspects of the ground-thermal regime are discussed elsewhere in this chapter and in Chapters 4, 9, 10, and 15.

ADVANCED READING

Ershov, E. D. (1990). *General Geocryology* (English translation, P. J. Williams, ed., 1998, Cambridge University Press, Cambridge, pp. 39–156, 276–303, 346–372).

Johnston, G. H., ed. (1981). *Permafrost: Engineering Design and Construction.* John Wiley & Sons, New York, 340 pp.

Muller, S. W., (1945). Permafrost or permanently frozen ground and related engineering problems. United States Engineers Office, Strategic Engineering Study, special report no. 62, 136 pp. (Reprinted in 1947, J. W. Edwards, Ann Arbor, Michigan, 231 pp.)

DISCUSSION TOPICS

1. What are the controls over the distribution of permafrost?

2. How does permafrost aggrade?

3. What are the thermal and mechanical properties of frozen ground?

6 Surface Features of Permafrost

Permafrost gives rise to a number of unique landforms. Some result from the growth of discrete ice bodies within permafrost, some reflect the unusual groundwater hydrology that characterizes permafrost terrain, some reflect the thermal properties of earth material when subject to freezing, and some reflect the ability of warm, ice-rich permafrost to creep and deform under its own weight. The most widespread permafrost-related landforms are the polygons associated with thermal-contraction cracking of the ground. A number of small-scale features are characteristic of the active layer; some are diagnostic of permafrost terrain while others occur in environments of both seasonal and perennial frost. Landforms associated with the thaw of ice-rich permafrost are discussed in Chapter 8.

6.1. INTRODUCTION

In broad terms, permafrost-related landforms can be divided into those associated with either the growth (aggradation) or thaw (degradation) of permafrost. Those associated with the thaw, subsidence, and erosion of ice-rich permafrost are examined in Chapter 8, after discussion of ground ice in Chapter 7. In this chapter, we consider landforms associated with permafrost aggradation.

6.2. THERMAL-CONTRACTION-CRACK POLYGONS

Thermal-contraction-crack polygons are the most widespread, most visible, and most characteristic feature of permafrost terrain (Figure 6.1). They are variously referred to as tundra polygons, frost-fissure polygons, ice-wedge polygons, sand-wedge polygons, or "Taimyr" polygons.

6.2.1. Coefficients of Thermal Expansion and Contraction

It is well known that different rocks, being composed of different minerals, possess different coefficients of expansion and contraction, and that these are temperature dependent. This has been discussed earlier in the context of frost shattering of rock (Chapter 4). It is also well known that the lowering of temperature of ice-rich frozen soil can lead to thermal contraction of the ground and the formation of fissures. These develop because pure ice has a coefficient of linear expansion of 52.7×10^1 at $0\,°C$ and only 50.5×10^1 at $-30\,°C$. It is generally assumed that the rates of expansion and contraction of ice-rich unconsolidated sediments are probably little different to those of pure ice. Thermal-contraction cracking also occurs in bedrock. Here, the coefficients of linear and

Figure 6.1. Oblique view of Sachs River Lowlands, southern Banks Island, Canada, showing high- and low-centered ice-wedge polygons and themokarst lakes.

volumetric expansion reflect the varying mineral composition of rocks. In consolidated sedimentary rocks, expanded bedrock joints are often the result (see Figure 7.6D).

6.2.2. Ice, Sand, and Soil Wedges

Thermal-contraction cracks can be filled with either ice, mineral soil, or a combination of both.

Ice wedges are wedge-shaped bodies of ice, composed of foliated or vertically-banded ice. They form when hoar frost develops in the open crack in winter and when water from melting snow penetrates the crack in the early spring. Ice wedges are best observed in unconsolidated sediments but they may also occur in bedrock and on sloping terrain. Favored environments for their formation are poorly-drained tundra lowlands underlain by continuous permafrost. In more arid polar deserts, such as the High Arctic islands, Antarctica, and the Qinghai-Xizang (Tibet) Plateau, ice wedges are not so well developed, undoubtedly reflecting lack of moisture. Many large ice wedges known to exist in central Alaska, northern Canada, and Siberia are probably Pleistocene in age and inactive (Péwé, 1966b). Ice wedges are discussed further in Chapter 7 in the context of ground ice.

The term "sand wedge" is somewhat of a misnomer, especially in the Pleistocene context, because the typical mineral infill of these wedges ranges from loess through to medium and coarse sand, and even to more locally-derived sandy infill. Usually, stratification is minimal although there may be a tendency for particle size to decrease in size with increasing depth. Most actively-forming sand wedges and sand veins that have been described are from the extremely cold, arid, and ice-free areas of Antarctica (Black, 1973; Péwé, 1959; Sletten et al., 2003) and Greenland (Djikmans, 1989; Péwé, 1974). Almost certainly, sand wedges also form in the polar deserts of the Canadian High Arctic (Hodgson, 1982; Pissart, 1968). In all the regions mentioned above, strong winds and a relative absence of moisture allow fractures to be filled with wind-blown sediment and other material (Figure 6.2A, B).

A literature review (Murton et al., 2000) suggests that many sand wedges are either relict (inactive) features or Pleistocene-age pseudomorphs (casts). Both are examined in more detail in Part III. Here, only actively-forming sand wedges are discussed.

(A)

(B)

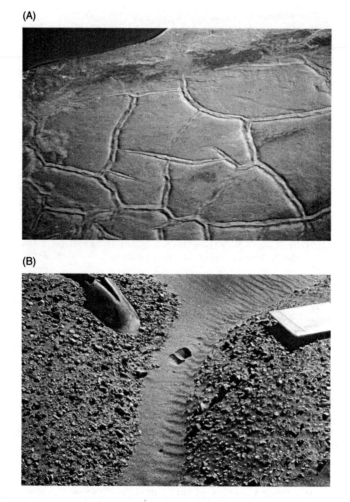

Figure 6.2. Sand wedges. (A) Oblique aerial view from 200 m of double-rimmed sand-wedge polygons formed on glacial outwash sediments, Peary Land, North Greenland. (B) Wind-blown sand infilling a thermal-contraction crack, Taylor Dry Valley, Southern Victoria Land, Antarctica. Both photos were supplied courtesy of the late Professor T. L. Péwé.

Fractures that develop primarily in the seasonally-frozen layer have been described from Siberia (Danilova, 1956; Dylik, 1966; Katasonov, 1973; Katsonov and Ivanov, 1973). They appear to be typical of cold continental climates and probably reflect the thickness of the zone of seasonal freezing and thawing. Apparently similar structures are described from Iceland (Friedman et al., 1971) and Svalbard (Jahn, 1975, pp. 74–78, 1983). Some transliterated Russian texts refer to these structures as "initially-ground wedges" (Romanovskii, 1974, pp. 71–74, 1977b, 1978; Melnikov and Spesitvsev, 2000, Figure 4-25), implying that they eventually evolve into other types of frost fissures. However, because their origin is unclear, the more descriptive term "soil wedge" is preferred here. In all cases, these cracks rarely exceed 1.5 m in depth, are filled with mineral soil, and usually occur in sandy loam and gravel. Sometimes, the enclosing sediments show deformation, often in the form of a "turning down" of the sediment layers. Relict soil wedges are discussed further in Chapter 13.

6.2.3. Development of the Polygon Net

Thermal-contraction cracks form polygonal, chiefly tetragonal, nets that cover extensive areas of the Arctic and sub-arctic. Typically, the average dimensions of the polygons that develop in unconsolidated sediments range from 15 m to 40 m. In bedrock, the polygons or nets are less well developed but, where they do occur, they are smaller, usually 5–15 m in diameter.

The theoretical development of the polygonal pattern was examined first by A. Lachenbruch (1962, 1966) and then by S. E. Grechishchev (1970). It was concluded that the angular intersection of the polygonal net should exhibit a preferred tendency towards an orthogonal (i.e. right angle) pattern (Figure 6.3). However, this contrasts with the many field descriptions of polygonal ground in which hexagonal or angular junctions dominate (Black, 1952; Leffingwell, 1919). One implication of the hexagonal net, and of

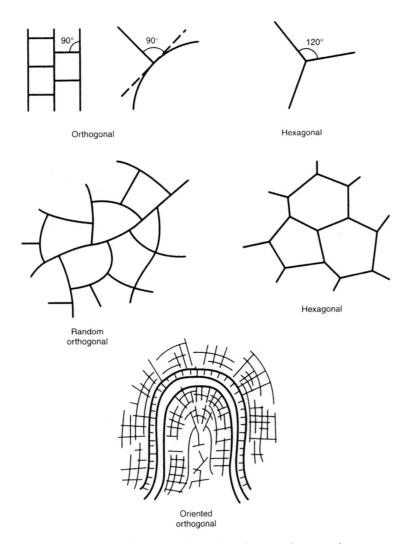

Figure 6.3.   Types of polygonal nets in permafrost terrain.

angular intersections of 120°, is that frost cracks develop at a series of points and that each crack develops more or less simultaneously. It is also suggested that the orthogonal pattern infers an evolutionary sequence in which primary cracks are followed by secondary cracks which progressively divide up an area. In this case, there is a tendency for an orthogonal intersection pattern. Thus, Lachenbruch (1966) classified the resulting polygonal network as a "random-orthogonal system" (Figure 6.4A). This contrasts with an "oriented-orthogonal system" (Figure 6.4B). The latter is commonly observed in the vicinity of large water bodies. In the case of "oriented-orthogonal" systems, one set of cracks is thought to develop normal to the water body and the other at right angles.

Numerical modeling techniques have also been applied to the initiation and development of thermal-contraction-crack networks. Whereas the earlier analyses by both Lachenbruch and Grechishchev examine a single idealized fracture and the dynamic pattern that is inferred from the stresses, L. J. Plug and B. Werner (2001, 2002) adopt a "complex-systems approach." They suggest that changes in the horizontal position of

(A)

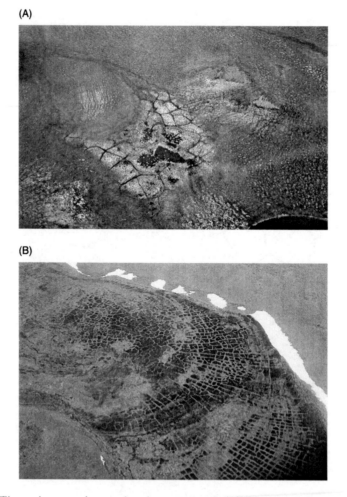

(B)

Figure 6.4. Thermal-contraction-crack polygon patterns. (A) Random-orthogonal pattern, Peel River floodplain, Central Yukon Territory, Canada. (B) Oriented-orthogonal pattern in an abandoned river channel, Mackenzie Delta region, Canada. In (B) the thermal-contraction cracks developed normal to the retreating water line, to be followed by cracks parallel to the water line.

fractures and the migration of fracture interactions cause orthogonal intersections to evolve towards interangular interactions. Analogy is made with basalt rock that forms hexagonal columns upon cooling. The complex-systems approach recognizes that spatial variability in snow cover and vegetation also influences fracture dynamics; in particular, the snow–stress feedback is complicated, site-specific, and not universal.

In the ice-free areas of Southern Victoria Land, Antarctica, evolution of the polygonal net has been examined on land surfaces that range in age from 10^{-3} to 10^{-6} years by R. L. Sletten, B. Hallet and R. C. Fletcher (2003). Several stages of polygon evolution are identified. In the mature or end phase, as typified by Beacon Valley (age 10^{-6} years), sand-wedge polygons are 10–20 m in dimensions. If an average growth rate of 0.6 mm (see below) is then applied, the ratio of polygon size to wedge-growth rate suggests that the entire land surface has been reworked or recycled by sand wedges on time scales of 10^{-4} to 10^{-6} years. Although such a dramatic interpretation of landscape appears unrealistic, it must be remembered that the ice-free terrain of Beacon Valley is probably the oldest land surface on Earth that has experienced continuous cold-climate conditions for several millions of years. It appears that, over time, there is a progressive change from curvilinear and quasi-orthogonal intersections towards 120° intersection angles and, eventually, to regular five- and six-sided polygons. Some support is given, therefore, to the numerical modeling approach described earlier. It may be that hexagonal patterns develop best in homogeneous material subject to long periods of uninterrupted and uniform cold-climate conditions, as in the Antarctic Dry Valleys, while orthogonal patterns are immature and develop in heterogeneous materials that experience changing environmental conditions.

6.2.4. Polygon Morphology

Typical of many polygonal systems is a raised rim on either side of the fissure. This is true for both ice- and sand-wedge polygons. Sometimes the rim may be as much as 0.5 to 1.0 m high. Even in newly-forming permafrost, as on the recently-drained lake bottom at Illisarvik, a shallow ridge a few centimeters high formed adjacent to the initial frost cracks during the first winter (Mackay, 1980a).

A commonly held view is that the double-raised rims are caused by the accumulation of either ice or mineral soil within the crack, thereby forcing adjacent frozen material upwards. This must be questioned because it is now thought that lateral thermally-induced movement of active-layer material occurs from the polygon center to the periphery. This is inferred from measurement of the distance and tilt of steel rods inserted into permafrost on either side of the fissure (Figure 6.5A) (Mackay, 1980a, 2000). The movements reflect summer warming and expansion of the active layer outwards from the middle of the polygon. In the Mackenzie Delta region of Canada, a movement rate of 0.25 cm/yr was estimated for one polygon, implying a coefficient of thermal expansion of about 1.7×10^{5} °C. The tilt of spruce trees adjacent to polygon rims (Kokelj and Burn, 2004) also suggests this sort of movement. The implication is that shearing occurs at the active layer-permafrost interface. At the same time, the progressive increase in width of the wedge itself must be accompanied by some deformation not only of the wedge but also of the enclosing ground.

This must also be the case for the shallow ridges that typically border many of the sand-wedge polygons in Antarctica (Sletten et al., 2003). Here, the development of the polygon net involves long-term convection-like cycling of material through the polygon (Figure 6.5B). Measurement of the spacing between steel rods hammered into the permafrost on either side of contraction cracks indicates average rates of surface widening of wedges of between 0.1 and 2 mm/year. A systematic tilt of the rods is interpreted to reflect

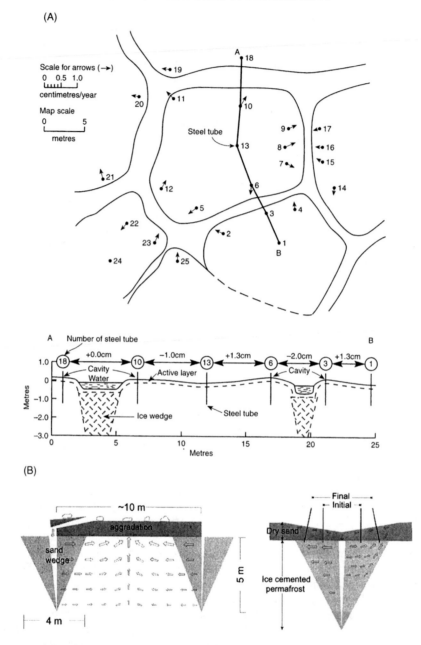

Figure 6.5. Polygon deformation. (A) Garry Island, Mackenzie Delta region, Canada. The plan sketch shows the location and direction of movement of steel rods between 1966 and 1978. The cross-section between rods 1 and 18 indicates changes in the distance between rods, 1966–1978. From: Mackay (1980a). (B) Hypothetical long-term soil motion in a sand-wedge polygon in Antarctica. The left-hand diagram shows the assumed displacement field and aggradation of polygon surface due to wedge growth. There is equality between the volume of the inflated surface and that of the contributing halves of the wedge. Wind erosion forms a gravel lag at the surface; some material is moved into the wedge. The right-hand diagram shows the typical displacement of steel rods on either side of a sand wedge; the inner pair shows the initial position in 1962–1963 and the outer pair shows the current (2002) position and inward tilt. The displacement fields are indicated by arrows. From: Sletten et al. (2003).

motion of loose soil towards the crack relative to ice-cemented soil at depth. It is also clear that recurrent cracking and continued addition of wind-transported sand into the crack causes long-term deformation of the surrounding permafrost. This must take the form of a net aggradation, or "inflation," of the ground surface within the polygon. The inferred displacement fields are indicated by arrows in Figure 6.5. According to Sletten et al. (2003, p. 15-4), net ground surface aggradation could average 0.05 to 0.1 mm/year.

Additional studies are needed in order to substantiate this inferred model of Antarctic landscape recycling. This may be difficult because few areas have experienced the same long period of uninterrupted cold-climate conditions as the Dry Valleys.

6.2.5. Controls over Cracking

Undoubtedly the most comprehensive investigations into the nature of thermal-contraction cracking have been undertaken at a number of sites in the Western Canadian Arctic by J. R. Mackay (1974a, 1975a, 1978a, 1984c, 1986d, 1992a, 1993a, 1993b, 2000; Mackay and Burn, 2002). Breaking cables and electronic crack detectors have established the timing, frequency, and direction of cracking. In addition, the speed and sound of cracking has been investigated, as have the relationships between cracking, snow cover, air and ground temperatures, and creep of frozen ground.

It appears that cracking is not related simply to a rapid drop in air temperatures in early winter, as first suggested by Lachenbruch (1962) and others. The best correlation between air temperature and cracking occurs in localities of thin snow cover. The favored duration and rate of temperature drop that results in cracking is about 4 days, at a rate of about 1.8 °C/day (Mackay, 1993b). It follows that tensions which cause cracking originate not at depth but either at the top of permafrost or in the frozen active layer. Proof of this can be seen by the repeated cracking that can be observed beneath any shallow tundra pond (Figure 6.6). Because water does not possess a "memory" which would permit cracking to occur in exactly the same location the following year, cracking must commence in frozen ground and then propagate both upwards and downwards.

Observations also indicate that less than half the fissures in any given area crack annually (Mackay, 1975a, 1989b; Harry et al., 1985). The frequency of cracking is also site-specific; for example, observations indicate that while a fissure may crack nearly every year at one locality, it may crack only once in every ten years or so at a site just several meters away. Moreover, years of exceptionally heavy snowfall inhibit cracking; along the western Arctic coast, an average snow depth of 60+cm is thought to be sufficient.

In reflecting upon his data, Mackay (1992a, p. 244) points out that theoretical analyses of thermal-contraction cracking (e.g. Grechishchev, 1970; Lachenbruch, 1962) assume uniform conditions. However, the reality is that crack formation is accompanied by changes in micro-relief associated with polygon development, and by associated vegetation and snow-cover changes. Cracking may be regarded as a random process, and Mackay (1992a) refers to "chaos" theory, and the complexity which follows growth, to explain the discrepancies between theoretical considerations and the field situation.

Few comparable data sets exist. Most reflect either one or two years of semi-quantitative observations inferences. For example, crack frequencies at Barrow, Alaska, ranged from 37% to 64% (Black, 1974) and only 38% of ice wedges examined in coastal exposures along the northern Yukon coast showed signs of cracking during the previous winter (Harry et al., 1985).

Several recent studies (Allard and Kasper, 1999; Christiansen, 2005; Fortier and Allard, 2005) permit a slightly more precise determination of the temperature requirements for

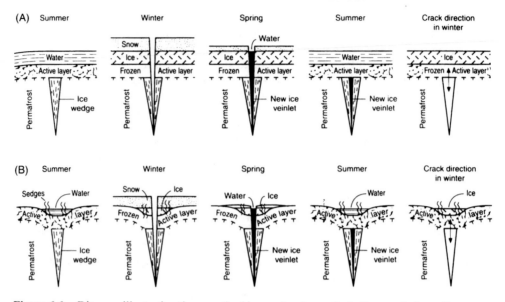

Figure 6.6. Diagram illustrating the growth of ice wedges beneath shallow pool of standing water (A) and beneath a water-filled ice-wedge trough (B). Crack initiation must commence at the top of permafrost and propagate upwards and downwards. Neither the pool of water above the ice wedge (case A) nor the semi-liquid active layer (case B) can memorize the location of the previous year's crack. From Mackay (1989b). Reproduced by permission of the Canadian Association of Geographers.

cracking (Table 6.1). Cold air temperatures are the primary requirement; these are usually between −25 °C and −40 °C with ground surface temperatures ranging from −15 °C to −25 °C. Given that cracking originates from the top of permafrost, attention must also focus upon cooling rates at both ground surface and top of permafrost. In all three field studies, the TTOP temperature at the time of cracking was between −10 °C and −20 °C and the ground cooling rate ranged between +0.1 °C and −0.6 °C/day.

Much still remains unclear about the mechanism of thermal-contraction cracking. For example, the early stages of permafrost aggradation at Illisarvik were characterized by unusually large (wide) thermal-contraction cracks (Mackay, 1986d), the cause of which is unclear. Second, the development of the polygonal net is not fully understood; lateral propagation is often short, unconnected, and frequently offset, and few cracks extend horizontally for more then 5 m. Finally, the influence which polygon topography exerts upon snow distribution, ground temperatures, and the cracking process needs further clarification.

6.2.6. Climatic Significance

Thermal-contraction-crack polygons are landforms of climatic and paleoclimatic significance. This is because (a) cracking requires permafrost and (b) cracking occurs when air temperatures drop well below 0 °C. Based upon the distribution of active cracking in Alaska and elsewhere, T. L. Péwé (1966b) concluded that thermal-contraction cracking occurs where the mean annual air temperature (MAAT) is −6 °C or colder. This simple threshold value must be treated with extreme caution because it is now clear that cracking is largely controlled by ground temperature (TTOP), the ground-thermal gradient, and snow cover.

Table 6.1. Temperature conditions for frost cracking.

Source	MAAT (°C)	Tair (°C)	ACR (°C/hr)	Tsurf (°C)	GCRsurf (°C/day)	Ttop (°C)	GCRtop (°C/day)	Gradient (°C/m)
Mackay, various papers, 1973–2000	−7 to −9	−29 to −34	−0.1	<−20	−0.5 to −0.9	−18 to −20	−0.1 to 0.4	−10 to −15
Allard and Kasper (1998)	−8	−25 to −43	−0.2 to −0.6	−21	−0.9 to −1.7	−15 to −20	−0.6	≥−10
Fortier and Allard (2005)	−15	−25 to −40	−0.2 to −0.9	−15 to −19	+0.1 to −1.1	−13 to −24	+01. to −0.3	−10.9
Christiansen (2005)	−6	>−25	n/a	−15 to −20	n/a	−15	−0.3 to −0.4	n/a

MAAT – Mear Annual Air Temperature
Tair – Air temperature at time of frost cracking
ACR – Atmospheric cooling rate of cold spell
Tsurf – Temperature at or near the surface
GCRsurf – Ground cooling rate at surface prior to frost-cracking
Ttop – Ground surface temperature at time of frost cracking
GCRtop – Ground cooling rate at top of permafrost prior to frost cracking
Sources: Fortier and Allard (2005), Christiansen (2005). Reproduced by permission of John Wiley & Sons Ltd.

An earlier suggestion that the size of the polygonal net reflects the severity of the climate (Dostovalov and Popov, 1966) appears, at first sight, to be logical, with smaller polygons reflecting increasingly severe winters (Figure 6.7A). However, observations made upon the lateral propagation of cracks that developed on the newly-exposed lake floor at Illisarvik do not support a fissure hierarchy (Figure 6.7B). As Mackay (1986d, p. 1784; Mackay and Burn, 2002) points out, the continued development of the crack is limited by the influx of vegetation to the site and the trapping of snow such that, with 60 cm snow cover, the main and lateral crack systems became inactive after the fourth winter.

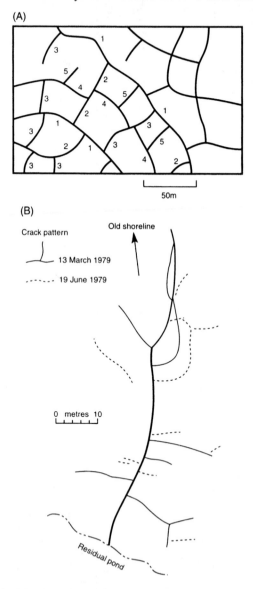

Figure 6.7. The growth of the polygonal net. (A) According to Dostovalov and Popov (1966), there is a successive growth of fissures of higher orders. (B) Plan of the new thermal-contraction-crack system at Illisarvik, one year after lake drainage, does not support the simple hierarchical model. From Mackay (1986d).

The climatic significance of cracking is further complicated by the nature of the substrate in which cracking occurs. Ground temperatures are generally lower in clay than in sand and gravel. As a result, N. N. Romanovskii (1985) suggests that cracking in clay substrate may occur with mean annual air temperatures as high as $-2\,°C$, but in gravel, the minimum temperature required may be $-6\,°C$. Further caution is provided by C. R. Burn (1990a), who reported upon elevated tritium concentrations in ice within frost cracks near Mayo, Yukon Territory. These values mean that, despite a mean annual air temperature of $-4\,°C$, thermal-contraction cracking must have been active at some time during the previous 30 years. One possibility is that contraction cracking at Mayo is related to below-normal winter temperatures, perhaps associated with cold-air drainage.

Because cracking is sensitive to a number of climate-related factors, it follows that a change in crack frequency may reflect a change in climate. In the western North American Arctic there is a body of anecdotal evidence that supports this. For example, the sound of cracking was reported in early literature (e.g. Leffingwell, 1915, pp. 638–639) from along the northern Alaska coast. However, on Garry Island in the outermost Mackenzie Delta, where cracking is known to be occurring today (Mackay, 1992a), no audible cracking or ground tremors resulting from cracking have been recorded during more than 30 years of monitoring. Today, audible cracking appears only to be reported from the High Arctic islands (Mackay, 1993a). Likewise, at Inuvik, 150 km south of the Arctic coast, winter temperatures are similar to those on Garry Island yet no cracking has been observed during the 1980–1988 period. The ice wedges are inactive. The critical factor that allows cracking to occur on Garry Island and not at Inuvik is that the former has less than half the winter snowfall of the latter. Therefore, the presence of inactive ice wedges at Inuvik and the absence of audible cracking today along the mainland Arctic coast suggest an increase in winter snowfall and higher winter temperatures during the last hundred years.

In summary, any climatic interpretation of thermal-contraction cracking must focus upon winter air temperatures, local site conditions, and winter snowfall amounts. In the Pleistocene context, the use of ice-wedge casts to reconstruct paleo-temperatures, as often attempted in the European periglacial literature, is fraught with uncertainty. A recent questioning of this approach (Murton and Kolstrup, 2003) is highly appropriate. The paleo-environmental significance of thermal-contraction cracking is discussed more fully in Part III.

Soil wedges pose a number of specific problems. First, it is not known what type of frost-action environment is necessary for them to form and whether perennially-frozen ground is necessary. Instances of seasonal-frost cracking have been reported from middle latitudes (Svensson, 1977; Washburn et al., 1963). Second, as stressed by J. Dylik (1966, p. 260), the reason for downturning of sediments adjacent to soil wedges is not well understood, and certain Siberian structures developed in silty alluvial terrace sequences may be subaqueous water-escape structures. Elsewhere, apparent frost cracks may be seismic-related (Leshikov, 1999). Third, the close spacing of many soil wedges is difficult to explain. Clearly, our understanding of soil wedges is incomplete and their use as indicators of perennially-frozen ground not justified.

6.3. ORGANIC TERRAIN

Organic material is particularly effective in protecting the ground beneath from atmospheric heat. This is most clearly demonstrated towards the southern limit of the discontinuous permafrost zone where permafrost is commonly restricted to peaty organic material. This is due to the thermal properties of peat (see Table 5.1). Dry sphagnum has thermal conductivity values approximately one order of magnitude lower than the lowest value for mineral soil. During summer, therefore, when surface peaty layers are dry, warming of underlying soil is limited. As a result, the depth of seasonal thaw in peat is considerably less than in other earth materials. During autumn and early winter, however, peat becomes saturated with moisture as rain and melting snow percolate downwards. When peat freezes, therefore, interstices become filled with ice and thermal conductivity increases dramatically. This leads to lower mean annual ground temperatures under peat than in adjacent areas without peat.

Permafrost-related landforms that relate to these unusual thermal properties are palsas and peat plateaus. Lithalsas are also discussed here because, although not associated with peaty materials, the principles involved are the same.

Palsas, lithalsas, and peat plateaus occur frequently in the sub-arctic regions of northern Canada, Scandinavia, and Russia. At the extreme southern limits of permafrost, they often form "islands" of permafrost (Zoltai, 1971) and many show signs of either past or present collapse (Delisle et al., 2003; Thie, 1974; Westin and Zuidhoff, 2001; Zuidhoff and Kolstrup, 2000). This reflects either natural causes, such as vegetation succession or forest fires, or may be an early sign of global climatic warming.

6.3.1. Palsas

A palsa is a peaty permafrost mound containing a core of alternating layers of segregated ice and peat or mineral soil material (ACGR, 1988). Typically, a palsa may vary between 1.0 m and 7.0 m in height and is less than 100 m in diameter (Figure 6.8). Usually, palsas occur in bogs and wetlands, where they protrude as low hills or knolls. The term is of Fennoscandian origin (Auer, 1927, pp. 33–41), originally meaning a "hummock rising out of a bog with a core of ice" (Seppälä, 1972b). Implicit in this definition are (i) their constructional nature, (ii) their origin in wetlands (fens or peat bogs), and (iii) that ice segregation in mineral soil beneath peat is the primary process responsible for growth. Once

Figure 6.8.   Abraded mature palsas on Luovdijeaggi palsa mire, western Utsjoki, Finland. In the foreground are several small vigetation-covered mounds (pounus) and a thermokarst pond. The photo is supplied courtesy of Professor M. Seppälä.

initiated, an absence of snow from the surface of the palsa allows deeper frost penetration and further growth. Implicit in the definition is the fact that permafrost extends beneath the organic surface layer into underlying fine-grained mineral soil.

The initial growth (heave) of a palsa or peat plateau is highly problematic. Random conditions, such as the growth of a spruce tree on a sedge tussock in a bog, may be sufficient to reduce the annual snow cover at a locality, thereby initiating deeper frost penetration. Another hypothesis is that palsa formation is triggered when wind turbulence causes a thinning of the snow cover on certain parts of a bog, also enabling frost to penetrate more deeply. This hypothesis was tested over a 3-year period in Finnish Lapland (Seppälä, 1982, 1995a) when snow was systematically removed from an experimental plot each winter. The result was the formation of permafrost and the growth of a small artificial palsa some 30 cm high.

The somewhat contradictory term "mineral palsa" (Pissart and Gangloff, 1984) has now been replaced by the less ambiguous term "lithalsa" (Pissart, 2000, 2002) to describe a shallow ice-segregation mound not covered with peat. The process of lithalsa growth is similar to that of a palsa but ice segregation is favored not by organic material but by the relatively high thermal conductivity of bare mineral soil, typically silty in nature.

Palsas are sometimes confused with seasonal-frost mounds. They differ in two important ways: (1) growth is primarily by ice segregation, with water moving towards the palsa from the surrounding wetland by cryosuction (see Chapter 4), and (2) palsas grow gradually over time, and may then persist for many decades. Also, although palsas are a type of frost mound, they are quite distinct from pingos. Likewise, various earth- or vegetation-covered mounds, locally referred to as "thufurs" or "pounus," and other small permafrost-related mounds, exist in permafrost terrain. All these forms are described later in this chapter. A. L. Washburn (1983a, b) attempted to bring order to the literature on palsas but it is now clear that a number of different mechanisms result in similar mound-like forms of varying sizes.

6.3.2. Peat Plateaus

Peat plateaus are flat-topped expanses of frozen peat, elevated above the general surface of a wetland. They are the simple result of freezing of peat with formation of segregated ice lenses and consequent uplift of the surface (Zoltai, 1972; Zoltai and Tarnocai, 1975). Segregated ice lenses may, or may not, extend downwards into underlying mineral soil. This is probably the main difference, genetically, between palsas and peat plateaus. Buoyancy effects may also initiate "floating" palsa-like elevations within water-saturated bogs (Outcalt et al., 1986).

Peat plateaus may be several square kilometers in extent. Towards the northern limit of the discontinuous permafrost zone, peat plateaus become increasingly more common as surrounding unfrozen zones become increasingly rare.

In summary, the recognition of palsas, lithalsas, and peat plateaus rests not only upon the nature of the surface material, which promotes deeper frost penetration than elsewhere, but also upon the presence of segregated ice, which causes the uplift.

6.4. ROCK GLACIERS

A rock glacier is a lobate or tongue-shaped body of frozen debris, with interstitial ice and ice lenses, which moves down-slope or down-valley by deformation of the ice contained within it (Figure 6.9A). Although some investigators regard rock glaciers as simply

(A)

(B)

Figure 6.9. Rock glaciers. (A) Muragl rock glacier, Switzerland. This well-investigated rock glacier is deforming at surface velocities of up to 50 cm per year (see Kaab and Kneisel, 2006). (B) Clast-supported angular rock debris in an ice matrix in the basal part of the Larsbreen rock glacier in Longyeardalen, Svalbard. This photo is supplied courtesy of Professor O. Humlum.

debris-covered glaciers (Johnson, 1974; Potter, 1972), the majority of rock glaciers discussed in the literature are essentially periglacial, or permafrost-related, features (Barsch, 1978, 1988; Haeberli, 1985).

Rock glaciers represent an extreme example of permafrost creep that occurs when relatively warm, ice-rich debris (Figure 6.9B) deforms on relatively steep mountain slopes. Rock glaciers are also thought significant because they are often used to infer the lower altitudinal limit of alpine permafrost.

6.4.1. Creeping Permafrost

It is well known that frozen ground can deform under the influence of gravity (Johnston, 1981, pp. 81–99; Yershov, 1990, pp. 145–156). Deformation is due mainly to the creep of pore ice and the migration of unfrozen pore water. Thus, permafrost creep will be most effective in ice-rich soils on steep slopes in areas of warm permafrost. This is because of the large amount of unfrozen water existing at these temperatures and the increasingly plastic nature of the ice. It follows that, ground temperature and ground-ice amount are the two obvious controlling variables; the warmer the permafrost and the greater the amount of ground ice, the greater will be the deformation.

Several field studies have documented the nature and magnitude of permafrost creep on gentle slopes (Table 6.2). In relatively warm permafrost, such as in the Mackenzie Valley and in Tibet, annual rates of deformation of 0.1–0.4 cm/yr appear typical, while in colder permafrost of the High Arctic, movement is one order of magnitude less, in the vicinity of 0.03–0.05 cm/year. Permafrost creep is discussed further in Chapter 9.

6.4.2. Types and Distribution

There are two types of periglacial rock glaciers: (a) "talus-rock glaciers" occur below talus slopes, and (b) "debris-rock glaciers" occur below glaciers (Figure 6.10). Active rock glaciers are probably best developed in continental and semi-arid climates since, under these conditions, the extent of the mountain periglacial zone is greatest (i.e. the snowline is highest).

Rock glaciers form where permafrost is present and where there is an adequate supply of debris. Two supply mechanisms are common: first, from talus slopes and their associated mechanical weathering, and second, from adjacent moraines, where the moraine forms the debris supply. Given an adequate accumulation of coarse clastic debris, percolating snowmelt infiltrates and freezes to form an ice–debris matrix which then deforms. Typical movement rates vary from several centimeters to several meters per year, and a typical surface relief consists of arcuate ridges and furrows aligned, in general, perpendicular to the flow direction.

Rock glaciers are reported from most of the major mountain regions of the world, including the European Alps (Baroni et al., 2004; Guglielmin and Smiraglio, 1997; Haeberli, 1985), the central Asian mountains (Cui Zhijui, 1983; Gorbunov, 1988b; Gorbunov and Tytkov, 1989), and the Cordillera of both North and South America (Corte, 1978; Brenning 2005; White, 1971). In high latitudes, talus rock glaciers have been described from Svalbard (André, 1994; Berthling, et al., 1998, 2000; Sollid and Sörbel, 1992), Greenland (Humlum, 1996, 1997, 1998a, 2000), the Faeroe Islands (Humlum, 1998b), and the Canadian Arctic (Evans, 1993). Relict rock glaciers have been described from numerous locations, including the northern Yukon (Harris et al. 1983, p. 69; Vernon and Hughes 1966), Gaspésie, Québec (Hétu et al., 2003), the Italian Alps and Apennines (Carton et al., 1988; Dramis and Kotarba, 1992), the Pyrenees (Chueca, 1992), and the southern Carpathian Mountains (Urdea, 1992).

6.4.3. Origin

The key to their origin lies in their internal structure. This is often difficult to ascertain without costly and sometimes impossible drilling. As a result, geophysical (seismic) methods are frequently used. Referring specifically to the Galena Creek rock glacier (Potter, 1972), D. Barsch (1988) comments that this rock glacier is probably composed of a mixture of debris and ice since P-wave velocities of 2400–4000 m/sec are typical, whereas

Table 6.2. Permafrost creep data from a number of localities in permafrost of varying temperatures.

Locality	Mean Annual Ground Temperature (°C)	Slope Angle	Movement Rate (cm/year)	Source
Cold permafrost:				
Rea Point, Eastern Melville Island, NWT	−16.0	5°	0.03–0.05	Bennett and French (1988, 1990)
Tuktoyaktuk, Mackenzie Delta	−6.0	Variable	0.03–0.04	Dallimore et al. (1996)
Warm permafrost:				
Great Bear River, Mackenzie Valley, NWT	−2.0	15–31°	0.0–0.3	Savigny and Morgenstern (1986a, 1986b, 1986c)
Fenghuo Shan, Qinghai-Xizang (Tibet) Plateau	−3.0	9–25°	0.44 (1.6 m depth) 0.16 (2.8 m depth)	Wang and French (1995b)
Alpine permafrost:				
Switzerland (Alps)	−1.0 to −3.0	(Rock glaciers)	5.0–6.0	Wagner (1992); Haeberli (1985)

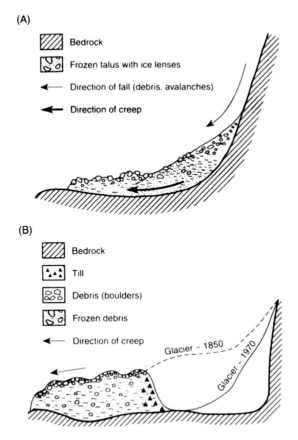

Figure 6.10. Types of rock glaciers. (A) Talus-derived rock glaciers are formed beneath scree slopes and are composed of frozen ice-cemented debris. They move by creep and deformation of the ice-rich rock debris. (B) Debris-derived rock glaciers form below glaciers and consist of a core of glacier ice overlain and subsequently buried beneath ablation till. They are regarded as simple debris-covered extensions to a glacier. Confusion arises when a buried body of glacier ice creeps and deforms, producing a feature similar to a permafrost (i.e. periglacial) rock glacier. From Barsch (1988). Reproduced by permission of John Wiley & Sons Ltd.

a core of pure glacier ice should have P-wave velocities of about 3600 m/sec. Observations such as these, and those made by co-workers elsewhere (e.g. King, 1986), led Barsch (1988, p. 79) to conclude that "the evidence to date fails to support the views that rock glaciers are debris-covered glaciers."

Rock glaciers continue to attract detailed study. No attempt is made to summarize these investigations except to say that they fall into two broad categories. First, rock glaciers are being used as a possible means of inferring the lower climatic and altitudinal limits of either present or past permafrost (Hauck et al., 2004; Hauck and Vonder Muhll, 2003; Humlum, 1998a, 1999; Whalley and Martin, 1994). Second, rock glaciers are being studied to better understand the creep and deformation of ice-rich permafrost bodies (Kaab and Reichmuth, 2005; Lambiel and Delaloye, 2004; Odegard et al., 2003). Methods include detailed field monitoring and geodetic surveys, geoelectrical measurements using DC resistivity and ground-penetrating radar, and the application of high-resolution digital-elevation models using advanced photogrammetric techniques (Hanson and Hoelzle,

2004; Kaab et al., 1997; Kneisel, 2004; Lambiel and Delaloye, 2004). Rheological models are also being examined (Whalley and Azizi, 1995). Currently, the study of rock glaciers constitutes a major component of alpine (mountain) permafrost study.

6.5. FROST MOUNDS

Various types of frost mounds occur in permafrost regions. They can be distinguished on the basis of their structure and duration, and by the character of the ice contained in them. Figure 6.11 summarizes the different types in terms of their uplift source and ice type. Certain types of frost mounds are of special interest because they are uniquely permafrost features. They are also of Pleistocene significance because collapse structures, interpreted as frost-mound remnants, have been described from many non-permafrost areas of the world. The latter are discussed more fully in Chapter 12.

6.5.1. Perennial-Frost Mounds

Pingos are perennial, intrapermafrost, ice-cored hills, typically conical in shape, that can grow and persist only in a permafrost environment (Mackay, 1998). The word "pingo" is of Inuit origin, used to describe an ice-cored conical hill in the Mackenzie Delta, Canada. The Russian equivalent is bulganniakh.

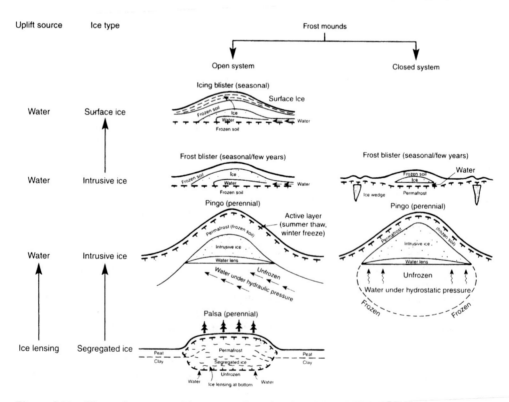

Figure 6.11. The various types of frost mounds and their origin. Modified from Mackay (1986c). Reproduced by permission of the Canadian Association of Geographers.

A well-developed pingo can be a striking geomorphic form (Figure 6.12). However, pingos are not common features and their existence is usually the result of a number of distinct and limiting geomorphic and hydrologic conditions. Pingos vary from a few meters to over 60 m in height and up to 300 m in diameter. They range in form from symmetric, to asymmetric, to elongate. Not all pingos have a typical conical form. Their one common characteristic, however, usually concealed by 1.0–10.0 m of overburden, is a core of massive ice or icy sediments. The ice may be remarkably pure, sometimes with seasonal bubble-rich and bubble-poor banding, or it may consist of layers of icy sediments. Fractures and faults are sometimes seen within the pingo ice core. Frequently, pingos portray dilation cracks and are ruptured near the summit. This is the first stage in decay, melt of its ice core, collapse of the mound, and ultimate formation of a shallow-rimmed depression (Figure 6.13).

Pingos were first described from northern Canada by Dr John Richardson (1851). They were later described in more detail from northern Alaska (Leffingwell, 1919) and from Siberia (Tsytovich and Sumgin, 1937). A. Leffingwell (1919) was the first to suggest hydraulic pressure as the cause while the botanist, A. E. Porsild (1938) was the first to suggest freezing in a closed system. Although the terms "open-system" and "closed-system" explain pingo growth, the words "hydraulic" and "hydrostatic" better identify the source of the water pressure that initiates and sustains pingo growth (Mackay, 1979a). These terms are used here. Put simply, hydraulic-system pingos derive their water pressure from a topographic gradient and hydrostatic-system pingos derive their water pressure from pore-water expulsion beneath aggrading permafrost in saturated sand.

Figure 6.12. Large hydraulic (open) system pingo ("Inner Pingo") in Adventdalen, Svalbard. Water issuing from the summit crater has formed an icing which extends down the flank of the hill. The photo, taken April 2006, is supplied courtesy of Professor O. Humlum.

Figure 6.13.   Collapsed hydrostatic (closed) system pingo, Sachs River lowlands, Banks Island, Canada.

6.5.2. Hydraulic (Open) System Pingos

Hydraulic (open) system pingos are hydrological phenomena that develop at sites where intrapermafrost or subpermafrost groundwater, under artesian pressure, reaches the surface (Müller, 1959). In general, they occur as isolated features, or as small groups within the same locality. A few attain considerable dimensions (~20–50 m in height) (see Figure 6.12). In nearly all instances, hydraulic (open) system pingos form in areas with topographic relief, such as lower hillslopes, alluvial fans, or valley bottoms. The majority is clearly associated with groundwater seepage. Typically, springs and icings form on their flanks in winter (see Figure 6.12). In East Greenland, groundwater flow and discharge characteristics suggest that structural weaknesses control the taliks that are associated with the intra-permafrost hydrology (Worsley and Gurney, 1996). Likewise, on Svalbard, many of the larger pingos are related to geological faults or are fed by high artesian pressures that are constantly being recharged by the melt of ice from the temperate basal zones of glaciers (Liestol, 1976; Yoshikawa, 1993; Yoshikawa and Harada, 1995). In Tibet, the large pingo at the head of the Kunlun Shan Pass is thought to have formed by groundwater rising to the surface along a fault (Wang and French, 1995c).

Despite numerous descriptive studies, there are no known surveyed growth data for hydraulic (open) system pingos. It is sometimes assumed they grow solely by ice injection. However, the role of artesian pressure is not to force overlying sediment upwards but rather to ensure a steady and slow movement of groundwater towards the surface. Pingo growth solely from injection represents an unstable condition that requires an unlikely long-term balance between three independent variables: (i) water pressure, which is determined by conditions external to the pingo, (ii) overburden strength, which varies with time of year, and (iii) the rate of freezing, which depends upon temperature (Mackay, 1973a, p. 1000). Since all three may change independent of the others, this balance will rarely be maintained for the total growth period of a pingo. The implication is that hydraulic (open) system pingos probably require a certain amount of ice segregation in addition to ice formed by intrusion of free water. In Yakutia, for example, where both "flat" bulganniakhs (dome-like elevations 2–5 m high) and upstanding large bulganniakhs (hills 10–50 m high) exist in close juxtaposition, it is thought that the ice core is of several origins, produced both by injection from groundwater under pressure from below and by segregation (Soloviev, 1973b, p. 148–151). The flat bulganniakhs are primarily of a segregated nature, composed of icy sediments, while the larger forms possess massive ice cores 5–10 m thick, formed through the repeated injection of water.

Approximately 500 hydraulic (open) system pingos occur in central Alaska and interior Yukon Territory (Holmes et al., 1968; Hughes, 1969). They are preferentially located on lower south- and southeast-facing valley-side slopes. Theoretical calculations by Holmes et al. (1968) suggest that the average pressure required to overcome the tensile strength of frozen ground and to subsequently maintain a 30 m high pingo is considerably higher than most artesian pressure measured in central Alaska. This probably explains why the majority of open-system pingos in Alaska and Yukon Territory never attain a fully-domed state but persist largely as doughnut-shaped, semi-circular, or circular ramparts. It may also explain why many of the hydraulic (open) system pingos on Svalbard and East Greenland are much larger because relative relief is greater and the hydraulic head, provided by adjacent sub-glacier melt water, is higher.

6.5.3. Hydrostatic (Closed) System Pingos

Hydrostatic (closed) system pingos result from pore-water expulsion caused by permafrost aggradation beneath the bottoms of drained lakes that are underlain by saturated sand (Mackay, 1962, 1985b, 1998). The highest concentration of this type of pingo occurs in the Tuktoyaktuk Peninsula area of the Pleistocene Mackenzie Delta region of Canada, but others occur elsewhere in northern Canada (Craig, 1959; French, 1975b; Pissart and French, 1976; St-Onge and Pissart, 1990; Tarnocai and Netterville, 1976; Zoltai, 1983), northern Alaska (Leffingwell, 1919; Walker et al., 1985), and central Siberia (Soloviev, 1973a).

The high concentration of hydrostatic-system pingos in the Tuktoyaktuk Peninsula area is the result of several favorable physical conditions. These include: (i) the occurrence of thick permafrost, (ii) large areas underlain by coarse-grained sediment, and (iii) numerous thermokarst lakes that drain frequently and easily, either by coastal erosion or by fluvio-thermal erosion along ice-wedge polygons (Mackay, 1998, p. 275). Typically, hydrostatic-system pingos occur within shallow lakes or former lake beds where both upward and downward permafrost growth occurs in the previously-unfrozen saturated sediment which comprises the sub-lake talik (Figure 6.14). They usually occur singly and not in groups, although at least one drained-lake basin is known to contain at least three actively-growing pingos (Mackay, 1973a, 1979a). On Banks Island, certain hydrostatic system pingos appear to have formed following the freezing of localized taliks that must have formed beneath the deeper sections of now-abandoned river channels (Pissart and French, 1976).

Detailed long-term field studies upon the growth of hydrostatic (closed) system pingos in the Tuktoyaktuk area have been undertaken by J. R. Mackay (1973a, 1979a, 1981c, 1986b, 1988b, 1990b, 1998).

The birth of a small pingo called Porsild Pingo (Mackay, 1988a) is typical of the early growth cycle. This pingo has grown in a lake which drained catastrophically about 1900. Birth probably took place between 1920 and 1930 when newly-aggrading permafrost ruptured and water was intruded into the unfrozen part of the active layer. A small mound, approximately 3.7 m high, was photographed by A. E. Porsild in May 1935 and subsequently described as part of a paper on "earth mounds" (Porsild, 1938, p. 53). Since then, Porsild Pingo grew steadily until 1976, at a growth rate approximately linear with height. After 1976, the growth rate has fallen. A similar pattern of rapid early growth (~1.5 m/yr) was monitored in a former lake bed that was drained by coastal erosion sometime between 1935 and 1950 (Mackay, 1973a, 1979a, pp. 14–18). Although quantitative observations on

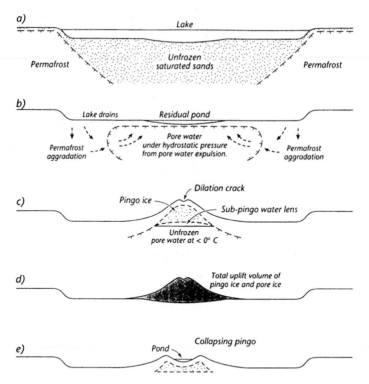

Figure 6.14. Genesis and growth of hydrostatic (closed) system pingos. (A) A large lake is underlain by unfrozen saturated sand. (B) Rapid lake drainage causes permafrost aggradation, pore-water expulsion, and development of hydrostatic pressure beneath a residual pond where permafrost is thin. (C) A growing pingo is underlain by a sub-pingo water lens whose downward freezing results in intrusive ice. (D) The total volume of ice required to grow a pingo is equal to the pingo volume above that of the bottom of the residual pond in which growth commenced. (E) Pingo collapse from partial thaw of pingo ice beneath the central pond that is surrounded by a pingo rampart. From Mackay (1998). Reproduced by permission of Les Presses de L'Universite de Montreal.

bulganniakh growth in Siberia are lacking, similar growth patterns are observed. For example, some hydraulic (open) system bulganniakhs have developed in recently-drained thermokarst (alas) depressions within the last 50 years. Eye-witness accounts indicate early growth rates of approximately 0.5–2.0 m/year (Soloviev, 1973a).

As pingos become older, their growth rates decrease. Some of the largest pingos, such as Ibyuk Pingo, are over 1000 years old and growing at a rate of only 2–3 cm/year (Mackay, 1986b). As a rough estimate, Mackay suggests that, for the Mackenzie Delta region, possibly 15 pingos may commence growth in a century. Probably, only approximately 50 are actively growing today.

Hydrostatic (closed) system pingos often exhibit pulsating patterns of heave or growth (Mackay, 1977a). This is caused by the build-up of water lenses that develop under pressure beneath the growing pingo (Mackay, 1978b). These pressures are released when water escapes to the surface, usually towards the periphery of the pingo where overburden strength (i.e. permafrost thickness) is least. This may result in seasonal frost mounds forming on the flanks of growing pingos (Mackay, 1979a, pp. 18–24). If penetrated by drilling, these water lenses can cause temporary artesian flow and geysers several meters high.

All pingos contain pore ice and varying proportions of intrusive and segregated ice. According to Mackay (1985a), the majority of growth is by ice segregation. In the early stage of pingo growth, as the recently-drained lake bottom begins to freeze, pore ice forms and the entire lake bottom heaves upwards. This is what was observed at the Illisarvik drained-lake site during the first 6 years of observation (see Chapter 5). Then, as permafrost continues to aggrade in the coarse-grained water-saturated sediment that underlies the basin, pore-water expulsion occurs (see Chapter 4). Segregated ice forms when hydrostatic pressure equals or exceeds overburden pressure, and uplift of the pingo, senso stricto, commences. Intrusive ice is associated with the freezing of the sub-pingo water lens. At that stage, pore water in the unfrozen and unbonded sand beneath the water lens remains at or below 0 °C because of a freezing-point depression. Using the opportunity of a rare pingo-ice exposure, Mackay (1990b) has described how seasonal-growth bands, consisting of alternating clear and bubble-rich bands, form in the intrusive ice (Figure 6.15). These

Figure 6.15. Seasonal-growth bands in pingo ice. (A) Oblique air view of pingo 20, Tuktoyaktuk Peninsula, Canada (see Mackay, 1990b, 1998), showing exposure of the ice core. (B) Seasonal banding in the ice core. Note the person for scale. Both photos were taken in August 1988.

result from the downward propagation of the cold and warm seasonal temperature waves. The gradual transition from clear to bubble-rich ice reflects the change from winter to summer, and the abrupt transition from bubble-rich to clear ice reflects the change from summer to winter.

In summary, hydrostatic (closed) system pingo growth is an example, albeit dramatic, of lake-bottom heave and aggrading permafrost.

6.5.4. Other Perennial-Frost Mounds

Pingo-like mounds occur in other geomorphic settings. For example, groups of pingo-like mounds have been described from the Canadian Arctic islands (Balkwill et al., 1974; Pissart, 1967a). On Prince Patrick Island, shallow mounds, between 1 m and 13 m in height and with average dimensions of about 60 m, are formed within thick sand formations of Tertiary age that rest discordantly upon impervious Paleozoic-age rocks. None appears to be growing today. Their location appears to coincide with deep-seated geological discontinuities that suggest the mounds are related to ancient groundwater movement along faults in underlying bedrock. The pingos probably developed when permafrost first began to form sometime during the early or middle Pleistocene. Those on Amund Ringnes Island are equally problematic.

A second geomorphic setting is provided by elongate and partially-collapsed mounds of varying sizes and shapes that occur in river valleys and on low fluvial terraces of Banks Island (French, 1975b, 1976c; French and Dutkiewicz, 1976). Most are less than 3 m in height and all appear relict. Especially puzzling are remnants of small, mutually-interfering mounds that occur on the broad fluvial surfaces of Central Banks Island. It is hypothesized that these may be seasonal-frost-mound remnants of Late-Pleistocene age that formed in shallow sections of braided channel systems.

6.5.5. Seasonal-Frost Mounds

Where freezing of the active layer restricts perennial discharge from intrapermafrost or subpermafrost aquifers, a variety of seasonal-frost mounds (frost blisters, icing blisters) may develop at the site of groundwater discharge. These have been described from northern Canada, Alaska, northern Scandinavia, Tibet, and Siberia. Typically, seasonal-frost mounds range between 1.0 m and 4.0 m in height. They form by the upheaval of seasonally-frozen ground brought about by the subsurface accumulation of water under high hydraulic potential. This occurs during progressive freezing of the active layer. Figure 6.16 illustrates the formation of frost blisters, one of the most common types of seasonal-frost mounds.

Seasonal-frost mounds are sometimes confused with palsas. The basic difference is that the former result from ice injection while the latter result from ice segregation. Thus, the interior of a frost blister, for example, is usually characterized by a core of pure ice with ice crystals aligned in a vertical columnar fashion that reflects the freezing of free water (Pollard and French, 1983, 1984, 1985).

Most seasonal-frost mounds are destroyed completely by thawing and collapse during the first summer after their formation. However, others may be preserved through one or more summers, depending on the insulating quality of their soil cover. As a result, they may assume the morphology and time duration of a palsa. It is also possible that some mounds combine both palsa and seasonal-frost-mound growth mechanisms. For example, palsa-like mounds have been described from a number of damp valley-bottom

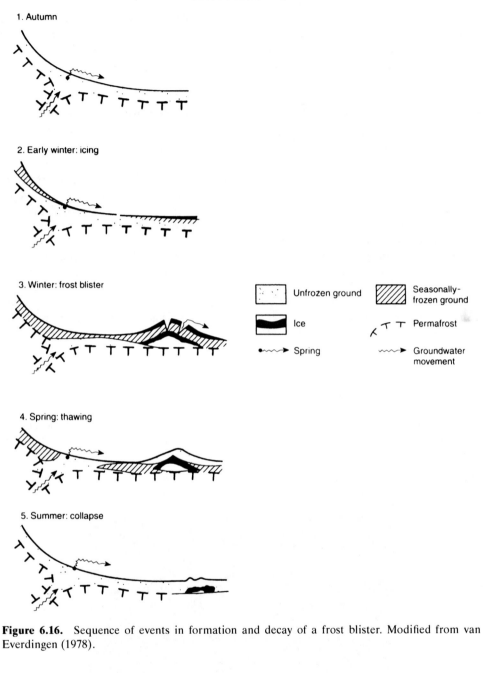

Figure 6.16. Sequence of events in formation and decay of a frost blister. Modified from van Everdingen (1978).

locations in the Mackenzie Mountains and southern Yukon Territory, Canada (Coultish and Lewkowicz, 2003; Kershaw and Gill, 1979; Lewkowicz and Coultish, 2004). Because high hydraulic potentials must exist in such hilly topography, the possibility exists that ice injection, rather than ice segregation, is the primary cause of uplift.

6.5.6. Hydrolaccoliths and Other Frost-Induced Mounds

A feature of many tundra landscapes is the presence of small mounds, many with either ice cores or ice lenses within them (Figure 6.17). These mounds are not regarded as pingos, mainly on account of their size and their location primarily within the active layer. They are also not usually regarded as a form of patterned ground since they are larger than the earth hummocks or non-sorted circles normally associated with patterned ground. Many function as owl perches and stand out as relatively dry sites. Their peaty soil frequently promotes the development of ice lenses immediately beneath the vegetation mat.

The variety of features suggests that they are not all of the same origin. In the Russian literature, the term "bugor" has been used to describe these small, gently-rising, and oval-shaped mounds (Dostovalov and Kudryavtsev, 1967). The North American literature (Sigafoos, 1951; Sharp, 1942a; Bird, 1967, p. 203; Porsild, 1955; French, 1971a; Washburn, 1983b) describes low circular mounds, rarely exceeding 2m in height, and usually between 15m and 50m in diameter. The origin of these features is not clear. Some are probably the result of localized ice segregation that has occurred in response to subtle thermal differences in soil and vegetation cover.

6.6. ACTIVE-LAYER PHENOMENA

The active layer (see Chapter 5) gives rise to a number of small-scale features of the permafrost landscape. These include disrupted bedrock and soil, and various patterned-ground phenomena. These are described below.

Figure 6.17. Small hydrolaccolith associated with poorly-drained low-centered ice-wedge polygons, Masik Valley, southern Banks Island, Canada. The mound consists of a body of pure ice immediately beneath the organic mat. The raised rims of the polygons create a closed system during annual freezing.

6.6.1. Bedrock Heave

The interaction between bedrock and groundwater controls the nature of bedrock heave (Figure 6.18). Usually, upward displacement is the result of excess water pressures created in the zone between the permafrost table and the downward-advancing freezing front. Where the saturated zone in the active layer becomes confined, the attempted expulsion of water supplies the heaving force. This is usually relieved along joints and bedding planes. Bedrock heave is particularly favored where the water table lies close to the surface in jointed granite, gneiss, quartzite, and other hard bedrock (Dredge, 1992; A. S. Dyke, 1978; L. S. Dyke, 1984). In such terrain, it is not uncommon for the active layer to be several meters in thickness.

Depending upon pre-existing fracture characteristics, bedrock heave varies from single ejected blocks to dome-shaped accumulations up to several meters in diameter. According to L. S. Dyke (1984) yearly movements may be as much as 5 cm, horizontally and vertically. As such, even compact sedimentary strata or igneous bedrock may experience deformation sufficient to damage man-made structures preferentially located upon these (assumed) firm foundations.

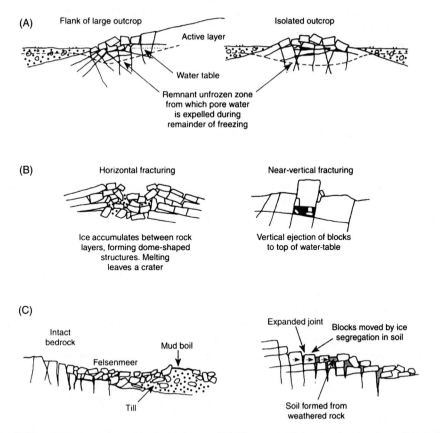

Figure 6.18. Different types of bedrock heave. (A) Water-expulsion mechanism in well-indurated rock with relatively-thick active layer (mainland, Canadian Shield). (B) Influence of fracture fabric on heaving style. (C) Ice-segregation mechanism displaces rock either by freeze–thaw creep of rock transported to a soil surface or by direct application of pressure applied by ice segregation in weathering products. Modified from Dyke (1984).

The main resistance to heave is offered by the weight and shear resistance of any over-lying ice-bonded rock mass. The deformation behavior is regarded as plastic in nature. Laboratory experiments (Michaud and Dyke, 1990) suggest that the vertical displacement of bedrock blocks is characterized by progressive and slow movement on a monthly time scale.

6.6.2. Needle Ice

An interesting, but small-scale, heave phenomenon produced by diurnal one-sided freez-ing at or just beneath the ground surface is needle ice. Delicate vertical ice crystals grow upwards in the direction of heat loss and range in length from a few millimeters to several centimeters. Occasionally, they may lift small pebbles or, more commonly, soil particles. Needle-ice formation is particularly common in wet, silty, and frost-susceptible soil. The importance of needle ice as a soil-disruptive agent has probably been underestimated, especially if it exposes soil to wind action, deflation, and cryoturbation activity. It may also damage plants by causing mechanical stresses within the root zone (Brink et al., 1967). In the coast range of British Columbia, needle ice occurs in oriented stripes (Mackay and Mathews, 1974), and both wind direction and solar radiation have been sug-gested as explanations. However, it is not clear whether oriented needle ice is primarily a shadow effect, developed by thawing, or a freezing effect. In the French Alps, certain micro-forms of patterned ground are thought to result from processes associated with the thawing and collapse of needle ice (Pissart, 1977). These include frost sorting, frost creep, and the differential down-slope movement of fine and coarse material.

6.6.3. Cryoturbation and Frost Heave

Cryoturbation is a collective term used to describe all soil movements due to frost action. The term is also used in the plural sense to refer to irregular structures formed in soils by deep frost penetration and frost-action processes. It is the first (singular) meaning that is used here. Thus, cryoturbation includes frost heave, thaw settlement, and all differential movements that include contraction and expansion due to temperature changes and the growth and disappearance of (segregated) ice bodies. According to ACGR (1988), the water–ice phase change is necessary for cryoturbation, and this distinguishes cryoturba-tion from other soil-movement processes. Therefore, frost sorting is part of this complex process by which migrating particles are sorted into uniform particle sizes. The process is still not fully understood.

Frost heaving on an annual basis is intimately associated with the freezing of moisture in the active layer. Annual ground displacements of several centimeters (Table 6.3) with cyclic differential ground pressures of many kilopascals per square centimeter are common. Two types of heave are recognized: primary (i.e. capillary) and secondary heave. These are discussed in Chapter 4. The engineering hazards associated with these displacements and pressures, together with the adverse effects of segregated ice formation, are frequently encountered in the construction and maintenance of roads, buildings, and pipelines in cold environments. These aspects are considered in Chapter 14.

The progressive upward movement of stone and objects is the direct result of frost heaving in the active layer. This movement, especially common in heterogeneous uncon-solidated sediments, is called upfreezing. The mechanics are not fully understood. At least two different hypotheses are suggested. The "frost-pull" mechanism assumes that the top of a pebble or coarser particle is gripped by the advancing freezing plane and raised in

Table 6.3. Some frost heave values recorded in unconsolidated sediments and soils in various periglacial environments.

Climatic Type	Location	Year of record	Site Characteristics		Total Heave, per year (cm)		Source
High Arctic:	Mesters Vig, Greenland 72° N	1958–64	Slopes	(a) "Wet" (b) "Dry"	Depth 10 cm 20 cm	0.0–1.0 1.5–5.8	Washburn (1969)
	Cape Thompson, Alaska 70° N	N/a	Frost boil; highly frost-susceptible. Surface			32.5	Everett (1965)
	Mackenzie Delta, Canada, 69° N	1976–78	Mud hummocks	(a) Undisturbed (b) Disturbed		10.3 14.0	Mackay et al. (1979) Smith (1985b) Burn (1989)
Alpine:	Colorado Front Range, USA, 39° N	1969–70	Frost boil; highly frost-susceptible. Surface			25.0–29.5	Fahey (1974)
	Alberta Rockies, Canada, 50° N	1980–82	Sloping terrain; non-sorted circle			2.0–4.5	Smith (1987)
Low Temperature Range:	Signy Island, South Orkney, 61° S	1964	Sorted circle, highly frost susceptible (a) Surface (b) Buried (c) On stones (d) At edge			4.0 0.4 3.6 2.0	Chambers (1967a)
Qinghai-Xizang (Tibet) Plateau:	Fenghuo Shan, 34° N	1991–92	Sloping terrain (4 sites)			4.3–7.7	Wang and French (1995a)
Antarctica:	Sor-Rondane Mountains, 72° S	1987	Ridge site		Diurnal heave >0.2 mm < 1.8 mm occurred 12 times		Matsuoka et al. (1988)

conjunction with the overall heave that is associated with freezing of the enclosing sediments. Upon thawing, the pebble is unable to return to its initial position because the space originally occupied by the pebble has been compressed by lateral frost heaving (frost thrusting) during the freezing process and, during thaw, material has slumped into the hollow.

A second hypothesis, termed "frost-push," relies upon the greater thermal conductivity of the stone resulting in the formation of ice preferentially around and beneath it, thereby forcing the pebble upwards. Upon thawing, as in the frost-pull mechanism, the infill of fines beneath the pebble prevents its return to its original position. Both hypotheses assume that ice lenses grow on the cold (freezing) side of the object with water migration from the warm (unfrozen) side. However, two-sided freezing of the active layer means this mechanism cannot be directly applied to upward freezing. Also, the possibility of downward water movement in summer from the thawed active layer into the lower portion of the frozen active layer, where refreezing and associated ice lensing and heaving may occur, complicates the process. J. R. Mackay (1984a) concludes, somewhat unsatisfactorily, that the unfreezing of any object must be considered within the context of several variables. These include (a) the direction of freezing, either upwards or downwards, (b) the degree of frost-susceptibility of the enclosing soil, and (c) the degree of frost-susceptibility of the object concerned.

A related characteristic is that frost action commonly leads to the tilting of stones, the more angular of which may become aligned on-edge (Figure 6.19A). This phenomenon is thought to result from differential frost heave at the top and bottom of the stone that results in the rotation and tilting of its axis. The probable mechanism, illustrated in Figure 6.19B, assumes that "frost-pull," by the downward-advancing freezing plane, grips the top of the stone while the lower part remains within the unfrozen zone. Depending upon the strength parameters of the unfrozen material, the axis of the stone undergoes rotation. When all of the stone is totally within the frozen layer, rotation and tilting ceases. Thus, with repeated annual cycles of freezing and thawing, there is a progressive increase in angle of the axis of the stone towards the vertical, as well as a progressive upward movement of the stone in general.

6.6.4. Frost Sorting

Under laboratory conditions, at least three types of sorting mechanisms have been simulated: (a) sorting by uplift (i.e. frost heave), when freezing and thawing occur from the top; (b) sorting by preferential migration of finer particles ahead of a moving freezing plane, when freezing and thawing occur from either the top or the sides; and (c) mechanical sorting, when larger particles migrate under gravity when mounds and frost-heaved structures are produced. Experiments have shown that fine particles migrate under a wider range of freezing rates than coarser particles (Corte, 1966, 1971). This means that a heterogeneous material inevitably becomes sorted by freezing.

The effects of frost sorting can be demonstrated both in the field and in the laboratory. For example, on Banks Island, numerous complex deformations can be observed in the near-surface sediments (French, 1986, p. 172; Pissart, 1975). Similar phenomena can be produced by subjecting shallow trays filled with sediment layers of different particle size and frost susceptibility to repeated freeze–thaw cycles (Corte, 1971). For example, in one experiment when a relatively non-frost-susceptible material was the middle layer, this layer became convoluted. In a second experiment, when frost-susceptible material was

(A)

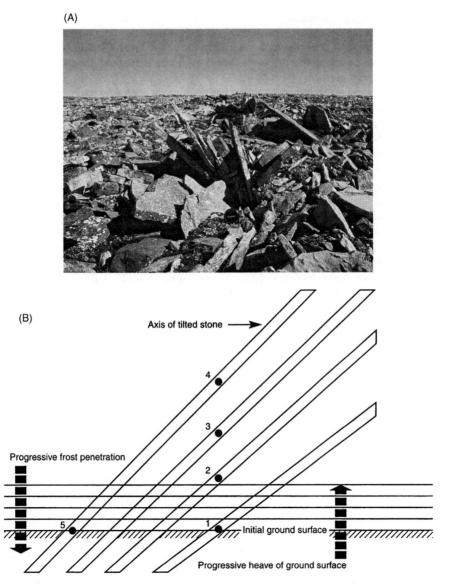

(B)

Axis of tilted stone →

Progressive frost penetration

Initial ground surface

Progressive heave of ground surface

Figure 6.19. Stone tilting. (A) Uptilted siltstone and sandstone slabs of Weatherall Formation (late Devonian age), near Rea Point, Eastern Melville Island, Arctic, Canada. (B) Illustration of the mechanism by which a stone becomes tilted through repeated freezing and frost heaving of the ground, according to Pissart (1970).

the middle layer, a single domed structure developed at the surface by extrusion of finer particles from below, to form a simple circle. A number of other experiments emphasize the importance of negative frost-susceptibility gradients in producing the upward injection of finer materials to the surface to form mudboils or coalescing sorted circles (van Vliet-Lanoë, 1988, 1991).

6.6.5. Patterned Ground

Most patterned-ground phenomena form within the active layer. Circles, earth hummocks, and mudboils are undoubtedly the most common. Some patterned ground may also form in seasonally-frozen ground.

Patterned ground attracted much attention from early explorers and scientists working in the tundra and Arctic regions, who often traveled on foot or by dog team. The relative lack of vegetation makes the various patterns distinctive, and early botanical explorers quickly identified the intimate relationships that exist between patterned ground and vegetation. A voluminous literature describes patterned ground phenomena, and A. L. Washburn (1956) lists no fewer that 19 hypotheses for its origin. Today, more recent studies suggest more universal explanations (Gleason et al., 1986; Hallet, 1990; Hallet et al., 1988; Krantz et al., 1988; Mackay, 1980b; Werner and Hallet, 1993).

The main geometric forms that are recognized include circles, polygons, and stripes. Where circles interact with each other, the resulting pattern is sometimes called a net. Patterned ground features are also described in terms of whether they are sorted or non-sorted as regards the material in which they are formed. Circles, nets, and polygons usually occur on flat or nearly-flat surfaces. Then, as slope angle increases, these features become elongate and irregular and, depending upon local conditions, may change to stripes further down slope.

Undoubtedly, the most common form of patterned ground is the non-sorted circle (Figure 6.20A) sometimes termed a "plug circle" (Washburn, 1997). If the feature assumes a raised topography, it is termed a "hummock" (Mackay, 1979b, 1980b). Hummocks grade from those which are completely bare (mud hummocks) to those which are vegetated (earth hummocks) (Figure 6.20B). Non-sorted circles, so defined, occur singly or in groups, and commonly vary from 0.5 m to 3.0 m in diameter and up to 0.5 m in height. Beneath the circle or hummock, the late-summer frost table is gently bowl-shaped.

These circular forms of patterned ground are likely the result of cryoturbation (see above). Typically, they are composed of fine-grained, frost-susceptible, soil. The upward displacement of material is caused by the freeze and thaw of ice lenses at the top and bottom of the active layer, with a gravity-induced, cell-like movement (Mackay, 1979b, 1980b). Evidence of cell-like circulation is deduced from the grain-size distribution of the hummocky soil, radiocarbon dating of organic materials intruded into the hummock center from the sides, and from upturning tongues of saturated soil observable in late summer (Figure 6.20B) (Tarnocai et al., 1973; Zoltai and Tarnocai, 1974; Zoltai et al., 1978). An equilibrium model is inferred in which cell-like movement occurs because the top and bottom of the freeze–thaw zones have opposite curvatures (Figure 6.20C).

Support for some sort of circulatory movement within the active layer is provided by measurements of soil movement within sorted circles (Hallet and Prestrud, 1986; Hallet et al., 1988) (Figure 6.21). These indicate periods of activity in the summer separated by long quiescent periods. One possibility is that the density difference that exists between water at ~0 °C (at the base of the active layer) and ~3–4 °C (at the near surface) may be sufficient to induce free convection of water in the active layer (Gleason et al., 1986; Ray et al., 1983). This may lead to spatially non-uniform thawing at the top of frozen ground caused by convective heat transfer. Such a process would explain the concave upward curvature of the frost table beneath circles, a phenomenon not adequately explained in the equilibrium model (see Figure 6.20C).

Even though soil circulation can be demonstrated, it still does not completely explain the origin of all circular forms. For example, circulation may evolve subsequent to a protoform, such as a diapir (Washburn, 1989, pp. 953–954). It has also been suggested that

(A)

(B)

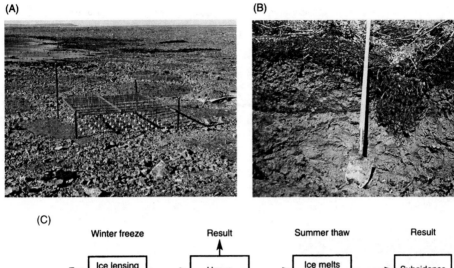

(C)

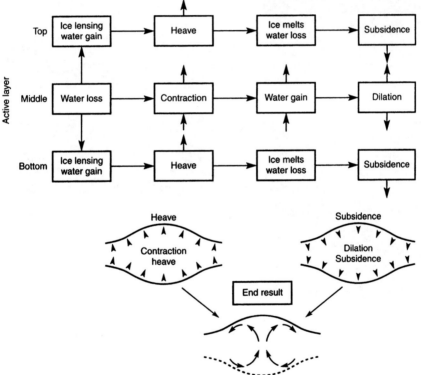

Figure 6.20. Patterned ground. (A) Plug circles, Cornwallis Island, Arctic Canada. The photo shows instrumentation used to measure vertical displacements at depth within a circle by free-moving rods. Photo taken: August, 1988 (see Washburn, 1989, 1997). (B) Section through an earth hummock showing cryoturbated organic material, Hume River region, near Fort Good Hope, NWT. Photo was supplied by the late S. C. Zoltai. (C) The equilibrium model of earth-hummock formation, inferred by J. R. Mackay (1980b) following field observations on Garry Island and at Inuvik, western Canadian Arctic.

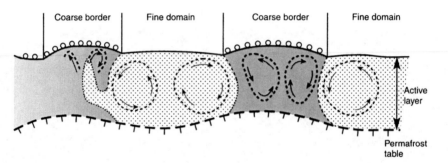

Figure 6.21. Illustration of the surface relief and inferred soil circulation patterns of non-sorted circles developed in raised beach gravels, western Spitsbergen. From Hallet et al. (1988).

Figure 6.22. Examples of mudboils and non-sorted circles (hummocks), Arctic Canada. (A) Mudboil in silty sand, Eastern Melville Island. (B) Slope hummocks at a snow-bank locality, Prince Patrick Island.

differential heave and ice segregation combine with density instability in thawed soils to explain cryoturbation structures and, by extension, hummocks (van Vliet-Lanoë, 1988, 1991). Finally, locally-high hydraulic potentials may develop in the active layer from water confinement between the underlying permafrost and an overlying semi-rigid carapace. The latter may be produced by desiccation and hardening of the surface in late summer; the former may result from hydrostatic- or artesian-induced pressures on slopes, often generated by summer rain. The result is the formation of mudboils (Figure 6.22A). These features are ubiquitous over extensive areas of Keewatin and northern Canada and are typical of poorly-sorted sediments (mud or diamict) with significant silt and/or clay content (Dyke and Zoltai, 1980; Egginton and Dyke, 1982; Egginton and Shilts, 1978; Shilts,

1978). The moisture contents of such sediments are close to their liquid limits so the muds liquefy and flow readily. Where internal stresses cannot be relieved, the mud bursts through the semi-rigid surface layer to create the mudboil. The process occurs primarily in late summer when the active-layer thickness is greatest, and is often associated with excess pore-water pressures created by summer rain and/or thawing of the ice-rich layer (the so-called "transient" layer; see Chapter 5) at the top of permafrost. Sometimes, mud-bursts produce small-scale mudflows which extend down slope for several meters.

Polygonal patterned ground can be either small (diameter >1.0 m) or large (diameter <1.0–2.0 m). The larger forms result from thermal-contraction cracking and have been discussed earlier. The smaller forms constitute nets that are delineated by furrows and narrow cracks to give a hummocky micro-relief of between 0.1 m and 0.3 m (Figure 6.22B). Many are as small as 5 cm but most are 20–50 cm in diameter. Terms such as "tundra

(A)

(B)

Figure 6.23. Examples of sorted and non-sorted stripes, Arctic Canada. (A) Oblique air view of non-sorted stripes, eastern Banks Island. (B) Sorted stripes formed in shale, Eastern Prince Patrick Island.

hummock," "earth hummock," "slope hummock," "thufur," and "pounus" have been used to describe these features (Beschel, 1966; Corte, 1971; Lewkowicz and Gudjonsson, 1992; Raup, 1966; Seppälä 1998; Schunke, 1977; Van Vliet-Lanoë and Seppälä, 2002).

A polygenetic origin is most likely. Certainly, not all are of the same origin. Desiccation or seasonal-frost cracking, or a combination of the two, is probably the initial cause. Desiccation can occur for a number of reasons: wind may promote evaporation from the ground, the ground may be subjected to a drainage change, and desiccation occurs during freezing and ice segregation. Some, such as thufur and pounus, occur in areas of seasonal rather than perennial frost (Luoto and Seppälä, 2002). Where small polygons form distinct nets or hummocks, it is likely that other processes, such as differential frost heave and snowmelt erosion, are also important (Lewkowicz and Gudjonsson, 1992; Seppälä, 2005).

Stripes occur on sloping terrain. Most non-sorted stripes consist of bare areas, commonly 0.3–1.0 m wide (Figure 6.23A), that are separated by parallel lines of vegetation-covered ground. In all probability, they reflect the same processes of cryoturbation that form circles but modified by the effects of slopewash and other mass-wasting processes (see Chapter 9). For example, the vegetation lines often merge, suggesting seepage and/or surface wash. Sorted stripes are less common and show a marked differentiation between lines of coarser and finer material oriented parallel to the gradient (Figure 6.23B). Usually, the coarser stripes are narrower than the fine stripes, and stones and boulders are commonly on edge with their long axes parallel to the line of movement. As with non-sorted stripes, it is possible that wash, by preferentially removing fines, may be a factor in their development.

ADVANCED READING

Haeberli, W. (1985). Creep of mountain permafrost: internal structure and flow of alpine rock glaciers. *Mittgeilugen der Versuchsanstalt fur Wassenbau Hydrologie und Glaziologie*, no. 77, 142 pp.

Mackay, J. R. (1998). Pingo growth and collapse, Tuktoyaktuk Peninsula area, Western Arctic Coast, Canada. *Géographie physique et Quaternaire*, **52**, 271–323.

Mackay, J. R. (2000). Thermally-induced movements in ice-wedge polygons, Western Arctic Coast: a long-term study. *Géographie physique et Quaternaire*, **54**, 41–68.

Washburn, A. L. (1997). Plugs and plug circles: a basic form of patterned ground, Cornwallis Island, Arctic Canada – origin and implications. *Geological Society of America*, memoir 190, 87 pp.

DISCUSSION TOPICS

1. What surface features are typical of permafrost terrain?

2. What are the unique hydrological characteristics of permafrost?

3. Differentiate between the various forms of frost mounds.

4. What are the controls over thermal-contraction cracking in ice-rich sediments?

7 Ground Ice

Many periglacial landforms result from either the formation or melt of ground ice. This chapter describes the nature, amount, and distribution of ground ice and the basic principles of cryostratigraphy and cryolithology. The landforms associated with the aggradation of permafrost and formation of ground ice bodies have been discussed in Chapter 6. The processes and landforms associated with the thaw (degradation) of permafrost (thermokarst) are dealt with in Chapter 8.

Cryostratigraphy is useful in reconstructing Quaternary environmental history. The recognition of ground-ice pseudomorphs in non-permafrost regions today provides evidence for the former existence of permafrost. Understanding ground ice is important because the thaw of ice-rich permafrost has applied and geotechnical significance.

7.1. INTRODUCTION

Ground ice is the general term used to refer to all types of ice formed in freezing and frozen ground. Ground ice occurs in the pores, cavities, voids, or other openings in soil or rock (ACGR, 1988, p. 46). Although traditional North American usage of the term excludes buried surface ice (Mackay, 1972a), a recent trend includes such ice (i.e. buried glacier, lake, river, and snow-bank ice). Ground ice may be present in both unconsolidated sediments and in bedrock. Relatively little is known about ground ice in bedrock.

Two quantitative parameters are used to describe ground ice conditions in unconsolidated materials. First, the "ice content" of a soil is defined as the weight of ice to dry soil, and is expressed as a percentage. For example, if a soil sample weighed 100 g when frozen and 40 g when oven dried, then the weight of ice (i.e. moisture) in that sample was 60 g. Therefore, the ice content is 150%. Low ice-content soils are generally regarded as those having ice contents less than 40–50%. Soils with high ice content are usually fine grained with values that commonly range between 50% and 150%.

It is important to determine the Atterberg Limits for the sediment content of permafrost. It is not uncommon for the natural water (ice) content of frozen soil to exceed its liquid limits. If this were the case, the ground will pass upon thawing from a solid state possessing considerable strength to a lower state strength. In this case, thawed sediments may become relatively mobile, capable of flowing on slight inclines.

The amount of "excess ice" is a second parameter commonly used in the description of ground ice. Excess ice refers to the volume of supernatant water present if a vertical column of frozen sediment were thawed. In this case, the sample is allowed to thaw and the relative volumes of supersaturated sediment and standing water (i.e. excess ice) are noted. The volume of supernatant water is then expressed as a percentage of the total volume of sediment and water. For example, if upon thawing the relative volumes of supersaturated sediment and supernatant water were 300 cm and 500 cm respectively, then the excess ice value would be 62.5%. The advantage of this index is that it provides some indication of the potential morphological change, or volumetric ground loss, consequent upon thaw.

Visual estimates of the amount of excess ice present in frozen soil are deceptive because sediment may not necessarily contain excess ice even though it may contain visible ice lenses. This is because the sediment between the lenses may not be saturated with ice and, upon thawing, all water is retained in the voids. Sediments that contain excess ice are referred to as "ice-rich" or "icy." In unconsolidated materials, excess ice values of between 15% and 50% are common although exceptionally icy sediments may have values as high as 70–80%. Frozen sediments containing excess ice are "thaw-sensitive" and may be contrasted with "thaw-stable" materials which contain no excess ice. The latter are not subject to thaw settlement and retain much of their mechanical strength when thawed.

The terms "massive ice" and "massive-icy bodies" are usually reserved for relatively pure ice bodies whose ice content averages at least 250% for a thickness of several meters.

7.2. CLASSIFICATION

Ground ice may be either epigenetic (i.e. develops after the enclosing sediment has been deposited) or syngenetic (i.e. forms at, or almost at, the same time as the enclosing sediments are deposited).

Many attempts have been made to classify ground ice. The earliest were developed in Russia (Shumskii, 1959; Shumskii and Vtyurin, 1966; Solomatin, 1986; Vtyurin, 1975). Although these systems are not widely adopted in North America, several comments are appropriate. First, as many as 20 different ground-ice types are recognized by Russian scientists. P. A. Shumskii (1959) lists segregated ice, injection ice, vein ice, recurrent-vein ice, cave ice, thermokarst-cave ice, karst cave ice, and buried ice as the major types, each with several subtypes. Second, Russian classifications often include the burial of surface ice as a ground-ice category. Surface ice includes the following: (i) river, lake, and sea ice, (ii) bottom (anchor) ice, and (iii) surface icings. It also includes snow banks and glacier ice, both of which experience recrystallization and other changes following burial. Third, many Russian scientists consider the composition, genesis, and initial moisture content on freezing (syngenetic or epigenetic) (Kudryavtsev, 1978; Melnikov and Spesivtsev, 2000). For example, A. I. Popov (1973) identified ground ice as existing within three "permafrost zones": for simplicity, these are described as follows: (i) the "horizon of discontinuous (seasonal) cryohypergenesis" (i.e. the active layer), (ii) the "horizon of active cryodiagenesis" (i.e. the permafrost that exists above the depth of zero-annual amplitude), and (iii) the "horizon of passive cryodiagenesis" (i.e. the permafrost below the depth of zero-annual amplitude). This three-fold division is useful in that it focuses attention upon the movement of unfrozen moisture within frozen ground in response to the temperature gradient (see Chapter 4).

In North America, the approach is simpler and more general. For example, an early non-genetic classification was developed by field engineers in Canada in order to describe ground ice conditions (Pihlainen and Johnston, 1963). Ice was classified as being in one of three categories: either "not visible," "visible – less than 1" thick," or "visible – greater than 1" thick." Each category was further subdivided and given a symbol designation for easy use. The scheme is presented in Table 7.1.

Table 7.1. Summary of a ground-ice classification system according to Pihlainen and Johnston (1963).

A. ICE NOT VISIBLE

Group Symbol	Subgroup	
	Description	Symbol
N	Poorly bonded or friable	Nf
	No excess ice	Nbn
	Well-bonded	Nb
	Excess ice	Nbe

B. VISIBLE ICE – LESS THAN 1 INCH THICK

Group Symbol V	Subgroup	
	Description	Symbol
	Individual ice crystals or inclusions	Vx
	Ice coatings on particles	Vc
	Random or irregularly-oriented ice formations	Vr
	Stratified or distinctly-oriented ice formations	Vs

C. VISIBLE ICE – GREATER THAN 1 INCH THICK

Group Symbol	Subgroup	
	Description	Symbol
ICE	Ice with soil inclusions	ICE + soil type
	Ice without soil inclusions	ICE

Note: The impression received by the unaided eye is that none of the frozen water soils in the N group occupies space in excess of the original voids in the soil. The opposite is true of frozen soils in the V group.

While easy to use, this descriptive system provides no indication of genesis; predictions as to the extent and possible occurrence of ground ice are difficult. Today, a more commonly used classification is based upon (a) the source of water immediately prior to freezing, and (b) the principal transfer process which moves water to the freezing plane (Johnston, 1981; Mackay, 1972a). The result is seven mutually-exclusive ground-ice types (Figure 7.1). It must be noted that the various types of buried ice are excluded. The advantage of this classification, in addition to its clarity, is that it emphasizes the variety and complexity of the transfer process, recognizing sublimation, gravity transfer, and soil-water expulsion, together with the effects of soil-water pressure and soil-water gradient as mechanisms. Although the principles behind these transfer processes are beyond the scope of this book, the classification has the added advantage that it focuses attention upon four broad types of ground ice: (1) vein (wedge) ice, (2) segregated ice, (3) intrusive ice, and (4) pore ice. The first three types are important from a geomorphological

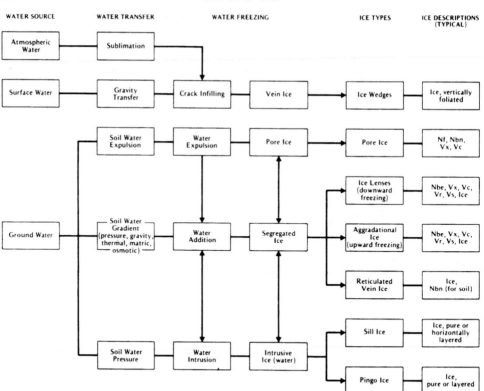

Figure 7.1. Classification of ground ice based upon the water source and the principal transfer processes at the time of freezing (modified from Mackay, 1972b), together with ice descriptions according to Pihlainen and Johnston (1963).

viewpoint because their occurrence gives rise to landforms such as ice-wedge polygons, frost mounds, and ice-cored topography. In addition, pore ice and segregated ice not only determine the amount of frost heave (see Chapter 4) but are also major determinants of cryotextures and cryostructures (see below) and are essential elements in the thermokarst process (see Chapter 8).

These four major types of ground ice and their characteristics require brief description.

7.2.1. Pore Ice

Pore ice, sometimes termed interstitial or "cement" ice, is the bonding cement that holds soil grains together. It is ubiquitous wherever moisture is present within permafrost and occupies, in varying amounts, voids in the material concerned. It is widespread in near-surface permafrost and in the active layer. The distinction between pore and segregated ice relates to the water content of the soil. This is best determined by thawing the soil and noting the presence or absence of excess ice or supernatant water. If supernatant water is present, this indicates the frozen soil was supersaturated and segregated ice was present. Pore ice has been discussed earlier in Chapter 4. In the Russian literature, pore ice give rise to what are termed "massive" cryotextures (see below).

7.2.2. Segregated Ice

Segregated ice is a broad term for soil with high ice content. Segregated ice forms in a variety of materials but water-saturated fine-grained sediments are especially suitable. Usually, ice lenses are visible to the naked eye; they vary in thickness from layers a few millimeters thick to massive ice bodies, sometimes tens of meters thick. In the case of closed-system freezing, as in the freezing of a sub-lake talik, lenses may be inclined to reflect the position of the freezing plane (Figure 7.2A). In theory, segregated ice can be distinguished from intrusive ice (see below) on account of the relative purity of the latter, and the stratification and presence of soil particles and air bubbles oriented normal to the freezing plane in the former. However, when dealing with massive icy bodies, this distinction is often difficult to make.

(A)

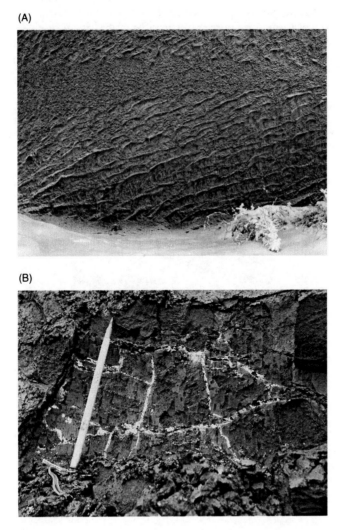

(B)

Figure 7.2. Examples of cryostructures in frozen ground. (A) Inclined segregated ice lenses, 30–80 cm long and 5–10 cm thick, formed by subaqueous syngenetic freezing of glaciolacustrine silty clay near Mayo, Yukon Territory, Canada. (B) Network of reticulate vein ice formed in silty clay diamicton, Pelly Island, Pleistocene Mackenzie Delta, NWT, Canada.

Segregated ice has been discussed earlier in the context of downward freezing (see Chapter 4). However, upward freezing may also occur when, for example, the upper surface of permafrost rises in response to sediment accretion, the accumulation of organic material, or climate change. Under these conditions so-called "aggradational" ice is formed (Mackay, 1972a, p. 10). A second sub-type of segregated ice is reticulated-vein ice (Figure 7.2B). This develops in fine-grained sediment in response to desiccation and shrinkage upon freezing with water being derived from sediment in a semi-closed freezing system rather than from an upward migration of water in an open system (Mackay, 1974b).

7.2.3. Intrusive Ice

Intrusive ice is formed by water intrusion, usually under pressure, into the seasonally- or perennially-frozen zone. Sill and pingo ice are two types of intrusive ice which are usually identified. Pingo ice, formed by pore-water expulsion, has been discussed in Chapter 6. Sill ice forms when water is intruded under pressure into a confining material and then freezes in a tabular mass. Sill ice may be stratigraphically controlled, as in sandstone and coal bedrock on the Fosheim Peninsula of Ellesmere Island (Pollard, 2000b; Robinson and Pollard, 1998), or it may cross-cut sediment boundaries to form ice "dikes" in softer sediments (Figure 7.3) (Mackay, 1975b; Mackay and Dallimore, 1992a, p. 1237). Typically, sill ice forms a sharp contact with enclosing sediment, is relatively clear, and fractures conchoidally, indicative of its formation under pressure. Although groundwater is usually the source of sill ice, it may also form adjacent to river banks by flood water being injected laterally along the base of the active layer and parallel to the permafrost table (Lewis, 1962; Mackay, 1972a).

Figure 7.3. Injection (i.e. intrusive) ice, Pullen Island, Pleistocene Mackenzie Delta, Canada. The ice bed continues for more than 100 m to the left of the photo (see Mackay, 1975c). The hypsithermal thaw unconformity is parallel to the top of the snowbank. Photo is supplied courtesy of Professor J. R. Mackay.

7.2.4. Vein Ice

Vein ice is formed by the penetration of water into open fissures developed at the ground surface. The origin of the water prior to freezing is atmospheric, usually meltwater from snow and summer rain. Vein ice can be distinguished from segregation ice on account of its vertical foliation and structures.

There are two types of vein ice. First, single-vein ice develops in small cracks, usually less than 0.2 cm thick and less than 70 cm deep, that form by thermal-contraction cracking at the ground surface. Second, repeated-vein ice forms in a thermal-contraction crack that opens in approximately the same place for a number of years. At depth the result is a complex of ice veins that mirror the different cracks; nearer the surface, the result is a vertical, or near-vertical, sheet of foliated ice that is termed an ice wedge. The latter are discussed later in this chapter (Section 7.5, pp. 176–181).

7.2.5. Other Types of Ice

Other types of ground ice occur in response to specific situations. These often reflect the complexities of thermokarst processes (see Chapter 8). For example, the term "thermokarst-cave ice" was initially used by Russian scientists to describe bodies of near-pure ice that lack foliation, may contain organic fragments dispersed throughout the ice, and that posses ice fabrics and bubble patterns that indicate inward freezing (Gasanov, 1969; Shumskii, 1959; see the English translation by Leuchtenberg, 1964, pp. 45–46). In North America, the more colloquial term "pool ice" is used to describe this type of ice (Mackay, 1988b, p. 87, 1997). Typically, pool ice forms in gullies or tunnels that are preferentially eroded along ice wedges. In ice-rich unconsolidated sediments, the rapidity of gully erosion along tundra polygons is well known, and can lead to lake drainage (see Chapter 8). As thaw progresses during the summer, fallen masses of soil and slumped material plug the tunnel or allow standing water to accumulate in the channel floor. These water bodies subsequently freeze during the following winter. On poorly-drained tundra, pool ice may be more common than generally thought (Shur et al., 2004). For example, pool ice was present in the near-surface of at least three of the seven ice wedges that were exposed in the gully that was eroded when Lake Illisarvik was artificially drained (Mackay, 1997, p. 22). It is exposed also in the walls of the CRREL permafrost tunnel in Alaska (Shur et al., 2004).

Ice within caves and tunnels in bedrock must also be considered a type of ground ice. Karst geomorphologists report ice forming when either groundwater seeps into a cave system and freezes (Marshall and Brown, 1974) or when warm and humid air circulates within a cave system (Lauriol et al., 1988). On Svalbard, the ice present within expanded bedrock joints beneath warm-based glaciers (see Figure 7.6C) reflects the refreezing of downward percolating glacial meltwater (Christiansen et al., 2005).

7.3. ICE DISTRIBUTION

7.3.1. Amounts

The total volume of ground ice present within permafrost varies from negligible, as in certain igneous and metamorphic rocks, to considerable, as in unconsolidated, fine-grained Quaternary-age rocks and unconsolidated sediments. For example, Table 7.2 summarizes the typically high ground-ice volumes which exist in the upper 5.0 m of permafrost at three

Table 7.2. Typical ground-ice volumes, Arctic Canada. (A) In the upper 5 m of permafrost in tundra lowlands of the western Arctic. (B) In the upper 7 m of permafrost on the Fosheim Peninsula, Ellesmere Island.

A. Upper 5.0 m of permafrost, unconsolidated sediments, Western Arctic

	Pore/Segregated ice (%)	Wedge Ice (%)	Total ice (%)
King Point, Yukon	43 (79)	11 (21)	44 (100)
Richards Island, NWT	28 (79)	7 (21)	36 (100)
Southwest Banks Island, NWT	44 (77)	21 (37)	56 (100)

B. Upper 7.0 m of permafrost, bedrock, Fosheim Peninsula, Ellemere Island, Nunavut

	Vol of ice (km³)	Ice Content (%)	Total ice (%)
Massive ice	0.179	69.0	6.8
Pore/segregated ice	1.970	48.6	74.5
Bedrock	0.407	9.7	15.4
Wedge ice			
High density	0.028	3.5	1.0
Low density	0.060	1.8	2.3

Note: Values in parentheses in (A) indicate percentage contribution of ground ice to total ice volume. In (B), ice content (%) is that of ice volume (km³) and total ice (%) is the percentage of total ice volume by ice type.
Sources: Harry et al. (1985), Pollard and French (1980), French and Harry (1983), Couture and Pollard, (1998).

lowland localities in the Western Canadian Arctic. The total volumetric ice content varies between 35% and 60%, of which the majority (66–80%) is either segregated or pore ice. Similar conclusions were reached by J. Brown (1967) for the northern Alaskan Coastal Plain. Clearly, in regions such as these, ground ice is an important component of permafrost.

Even where extensive natural exposures of permafrost are present, or where detailed borehole data are available, most estimates of ground ice volume in the upper 1–10 m of permafrost are gross approximations. The problem is one of generalizing from a relatively small number of observation points, often with an irregular spatial distribution, to the landscape at large. Bearing this in mind, several attempts have been made to characterize ground-ice volumes in near-surface permafrost. For example, using information from man-made trenches and large excavations (sumps) associated with the drilling of explora- tory boreholes on Melville Island, H. M. French et al. (1986) concluded that mean values of ice volume ranged between 30% and 70% in the upper 7.0 m. Similarly, on the un- dulating terrain of the Fosheim Peninsula of Ellesmere Island, the mean ice content of 154 cores drilled to 1.5 m depth was 53% (Hodgson and Nixon, 1998), but with additional observations upon massive ground ice exposures and soil moisture measurements at 53 sites, a slightly lower value of 48% for the upper 7.0 m of permafrost was subsequently obtained (Couture and Pollard, 1998; see Table 7.2B).

Little information is available about total ground ice amounts in igneous and meta- sedimentary rocks. However, it seems reasonable to assume that ice volumes are low and vary from 1–2% to 10–15%.

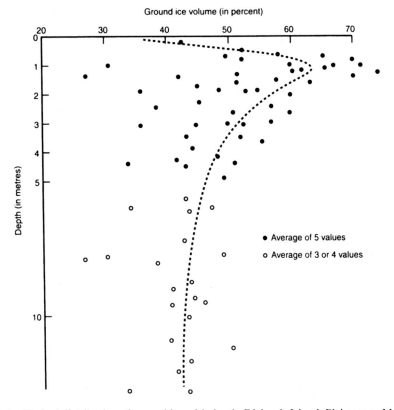

Figure 7.4. Typical distribution of ground ice with depth, Richards Island, Pleistocene Mackenzie Delta, Canada. Note: the curve represents a visual best fit line. From Pollard and French (1980).

7.3.2. Distribution with Depth

The majority of ground ice is concentrated in the upper 1–3 m of permafrost, in the zone immediately below the permafrost table (Figure 7.4). J. Büdel (1977; see the English translation by Fischer and Busche, 1982, pp. 103–105) was one of the first to highlight this, following observation of shattered ice-rich bedrock beneath stream channels on Svalbard. As a consequence, he formulated the "ice-rind" concept in which this near-surface ice-rich layer favored rock shattering, and thus stream incision.

The "ice-rind" concept is not particularly satisfactory in explaining what Büdel termed "excessive valley deepening" in permafrost. The deep valley incision on Svalbard is best explained in terms of deglaciation and recent isostatic uplift. The area is in paraglacial transition (see Chapter 2). On the other hand, Büdel's observation concerning the ice-rich nature of near-surface permafrost on Svalbard is a valid generalization. In northern Canada, for example, ice-rich near-surface permafrost has been documented in a number of environments (Mackay, 1981a; Pollard, 2000a; Stangl, et al., 1982; Williams 1968). There is also a relationship between near-surface ice content and vegetation. For example, Table 7.3 gives mean ice content in the top 50 cm of permafrost for six major ecological terrain types in the Mackenzie Delta. The volume of ice in the top 50 cm of permafrost ranges from ~15% at a spruce/feathermoss forest site to almost 50% at a spruce/crowberry-lichen site (Kokelj and Burn, 2005).

Table 7.3. Excess ice content (%) of the upper 50 cm of permafrost beneath point-bar willow and alder communities, spruce forests, and lake-side alder and sedge communities, southern, central, and northern Mackenzie Delta, Canada. From Kokelj and Burn (2005, table 1). Reproduced by permission of John Wiley & Sons Ltd.

	Willow	Alder	Spruce/alder-bearberry	Spruce/feathermoss	Spruce/crowberry-lichen	Lakeside alder/sedge
Cores (N)	14	10	10	18	10	11
Mean	0.2	6.6	6.0	27.3	41.1	33.6
Median	0.0	7.3	5.6	26.8	40.9	32.8
Minimum	0.0	0.0	0.0	14.8	35.0	19.6
Maximum	3.2	11.8	12.5	43.2	48.6	46.9

Our understanding of moisture migration within freezing and frozen soils now permits explanation of this ice-rich zone in the upper 0.5–2.0 m of the permafrost profile. In Chapter 4, it was explained how unfrozen water moves in the direction along which the ground temperature decreases in response to an imposed thermal gradient. In summer, moisture migrates downwards in the near-surface permafrost. Later, in Chapter 5, the concept of the transient layer, an ice-rich zone at the base of the active layer and at the top of permafrost, was explained. In many ways, the transient-layer concept is the modern equivalent of the ice-rind concept of Büdel, but based upon a better understanding of ground-ice dynamics.

7.3.3. Ice in Bedrock

Relatively little information is available concerning ice in bedrock. Theoretically, ice amounts should vary in accordance with rock porosity and permeability, and the presence or absence of discontinuities such as faults, joints, and bedding planes. The Russian literature recognizes a number of distinct ice-distribution patterns ("cryogenic textures"; see below) that characterize solid or semi-solid rock (Figure 7.5).

As a generalization, fine-grained rocks that are both permeable and porous contain higher ground ice amounts than coarse-grained rocks. For example, on Melville Island, fine-grained shale of Mesozoic age is more ice-rich that older and more consolidated sandstone and siltstone of Paleozoic age (French et al., 1986) (Figure 7.6A, B). Likewise, on the Fosheim Peninsula of Ellesmere Island, areas underlain by well-lithified sandstone and siltstone have a mean ice content of 30%, while fine-grained (disaggregated) shale and siltstone have a mean ice content of 44% (Hodgson and Nixon, 1998). The presence of joints, bedding planes, faults, and other structural discontinuities in bedrock complicates any assessment of ground ice. For example, bodies of intrusive ice and ice-expanded joints are common in sedimentary bedrock sequences (Figure 7.6C, D) (French, 1981, p. 12; Christiansen et al., 2005; Robinson and Pollard, 1998; Wang, 1990; Wernecke, 1932).

7.3.4. Ice in Unconsolidated Sediments

The occurrence of ground ice in unconsolidated sediments is much better understood than in bedrock. First, many regional surficial geology mapping programs in northern Canada demonstrate that the majority of heterogeneous, fine-grained surficial materials, be they of alluvial, fluvial, glacial, lacustrine, or marine origin, typically possess relatively high

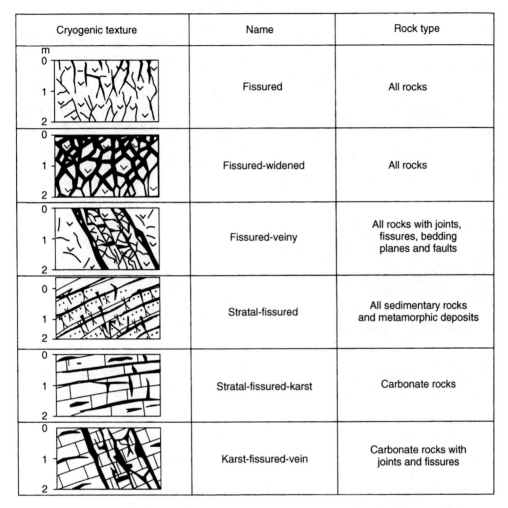

Cryogenic texture	Name	Rock type
	Fissured	All rocks
	Fissured-widened	All rocks
	Fissured-veiny	All rocks with joints, fissures, bedding planes and faults
	Stratal-fissured	All sedimentary rocks and metamorphic deposits
	Stratal-fissured-karst	Carbonate rocks
	Karst-fissured-vein	Carbonate rocks with joints and fissures

Figure 7.5. Cryogenic textures associated with solid and semi-solid rocks, according to Kudryavtsev (1979). From Melnikov and Spesivtsev (2000). Reproduced by permission of SB RAS Publishing House.

ground-ice amounts (Dredge et al., 1999; Hodgson and Nixon, 1998; Rampton, 1982, 1988; St-Onge and McMartin, 1995). Second, if a weathered mantle overlies bedrock the weathered mantle commonly possesses higher ice content than the unweathered bedrock from which it is derived (French et al., 1986; Stangl et al., 1982). Third, marine sediments that lie below the maximum level of marine submergence are usually more fine-grained than sediments that lie above the marine limit (Pollard, 2000b) and, as a result, commonly possess higher ice contents. These generalizations can be applied to most other permafrost areas, including Alaska, Scandinavia, and northern Russia.

In spite of the previous comments, it should not be assumed that coarse-grained unconsolidated sediments are always ice-poor. An increasing body of evidence suggests that high ice contents and bodies of ice may occur in a variety of granular materials (Dallimore and Wolfe, 1988; Gowan and Dallimore, 1990; Hyatt, 1998). Pore-water expulsion during freezing in a closed system, the burial of glacier ice, and the injection of groundwater

Figure 7.6. Ground ice in bedrock. (A, B): Ground ice in shale bedrock, Melville Island, exposed in section excavated during the winter of 1983–1984. The ice crystal structure has been destroyed by blasting. (A) cryoturbated ice-rich shale in upper 0.6–1.25 m; (B) brecciated ice-rich shale at 1.7–3.0 m depth. A thaw unconformity separates these two cryostratigraphic units. See Figure 7.8A. From French et al. (1986). (C) Ice fills a 50 cm wide expanded joint in the main tunnel of the Gruve-7 coal mine, Adventdalen, Svalbard . The ice is clear and contains mineral inclusions. Oxygen-isotope values suggest the ice is basal meltwater/subglacial regelation ice. See Table 7.6. From Christiansen et al. (2005). (D) An expanded joint, previously filled with ice, occurs in shale bedrock and is exposed here in a borrow pit, kilometer 366, Dempster Highway, Yukon Territory, Canada.

under pressure are all possible reasons why apparently non-frost-susceptible materials may, in fact, possess high ice content. This is also true for deltaic sediments, where high ice-content permafrost may be associated with a range of grain sizes. For example, in the Mackenzie Delta, the sand and clay content of ice-rich sediments (>70% moisture content) was between 5% and 45%, and 7% and 24%, respectively (Kokelj and Burn, 2005). It is unwise to assume that all coarse-grained sediments are ice-poor.

7.4. CRYOSTRATIGRAPHY AND CRYOLITHOLOGY

Cryostratigraphy refers to study of frozen layers in the Earth's crust. Central to cryos-tratigraphy is the fact that ice within perennially-frozen sediment imparts structures distinct from those found in other sedimentary environments. These structures are termed "cryostructures." Intimately related to cryostratigraphy is cryolithology, a sub-branch of geocryology (see Chapter 1). Cryolithology refers to the relationship between the lithological characteristics of rocks and their ground ice amounts and distribution. The standard texts are those by Popov et al. (1985) and Melnikov and Spesivtsev (2000).

7.4.1. Cryostructures, Cryotextures, and Cryofacies

Cryostructures are determined by the amount and distribution of ice within pores (i.e. pore or cement ice), and by lenses of segregated ice. The type and arrangement of ice within frozen material will depend upon the initial water content and the nature of water migration during freezing. Figure 7.2 illustrates two common cryostructures observed in northern Canada: (a) a lenticular cryostructure formed in frozen lacustrine clay near Mayo, Yukon Territory, and (b) a reticulate cryostructure formed in clay diamicton in the Mackenzie Delta region. Cryotextures must be distinguished from cryostructures. The former refers to the grain and/or ice crystal size and shape, and the nature of the contacts between grains and ice crystals in frozen earth materials. Both cryostructures and cryo-textures are useful in determining the nature of the freezing process and the conditions under which frozen sediment accumulates.

The first to systematically identify cryotextures and cryostructures were Russian permafrost scientists (Katasonov, 1969, 1975; Kudryavtsev, 1978). Unfortunately, Russian classifications tend to blur the distinction, made above, between texture (i.e. grain size) and structure (i.e. aggregate shape). Traditionally, Russian scientists used the term texture to describe what North Americans denote as structure (see comment, ACGR, 1988, p. 24). This confusion was perpetuated in one of the few early English-language transla-tions on this subject (Demek, 1978, pp. 139–153). Russian geocryologists recognize at least ten "cryogenic textures" (i.e. cryostructures); these are termed massive, massive-porous, basal, basal-layered, crust-like, porphyry-like, massive agglomerate-lens type, lattice type, layered, and lattice-block type (Kudryavtsev, 1978, pp. 301–304) This classification of "cryotextures" is primarily one of cryostructures (Murton and French, 1994).

Problems with Russian cryostructural classifications arise primarily from their complex and unwieldy nature. For example, E. M. Katasonov's (1969) classification involves 18 different cryostructures and A. I. Popov et al.'s (1985) classification has 14, excluding those that are composite. A second limitation is that these classifications apply primarily to permafrost containing little excess ice, detailing the distribution of ice within sediment and neglecting that of sediment within ice. Thus, V. A. Kudryavtsev's (1978) classification contains seven cryostructural terms that describe frozen ground whose ice content is $\leq 50\%$ by volume and only one term (basal layered: ataxitic or breccia-like; see Cheng, 1983; Shur, 1988a) that describes permafrost that is very ice-rich.

A simplified North American cryostructural classification encompasses the range of ice contents found within permafrost (Figure 7.7) (Murton and French, 1994). Several Russian terms are transliterated but the cryostructures proposed can all be recognized by the naked ice. A structureless cryostructure refers to frozen sediment in which ice is not visible and consequently lacks a cryostructure. Lenticular cryostructures, by compar-ison, can be of several types; they are described by inclination, thickness, length, shape, and relationship to each other. That shown in Figure 7.2A is best described as "lenticular, parallel, curved" according to this classification. In the case of lenticular cryostructures, the orientation of ice lenses reflects the orientation of freezing fronts and/or the struc-tural properties of the sediment (e.g. bedding; Smith and Williams, 1990). Layered

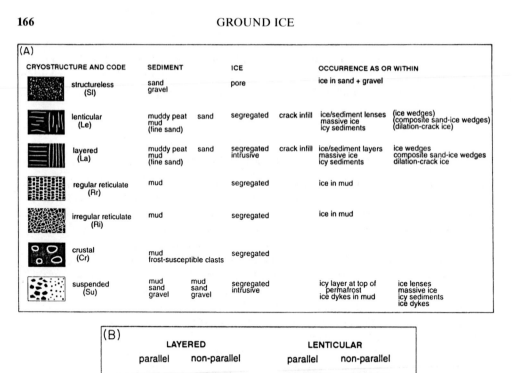

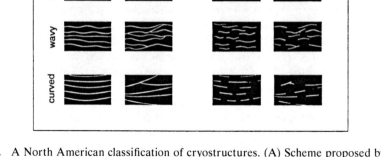

Figure 7.7. A North American classification of cryostructures. (A) Scheme proposed by Murton and French (1994). Ice is shown in white and sediment in black. In lenticular and layered cryostructures, lenses and layers may comprise either ice or sediment. (B) Terms and illustrations used to describe layered and lenticular cryostructures. Structureless (SI) and reticulate (Rr, Ri) cryostructures result from epigenetic freezing. Layered (La) and lenticular (Le) cryostructures result from syngenetic freezing.

cryostructures are continuous bands of ice, sediment, or a combination of both. The reticulate category, described earlier (see Figure 7.2B), is a three-dimensional net-like structure of ice veins surrounding mud-rich blocks. The best known, but probably least common, cryostructure is that of an ice crust or rim around a rock clast. This is termed crustal; it occurs commonly just beneath the permafrost table where ice crusts up to a few centimeters thick envelop pebbles and wood fragments, typically within silt-rich facies. Crustal cryostructures most likely form by localized ice segregation around frost-susceptible clasts. The suspended category refers to grains, aggregates, and rock clasts suspended in ice. By definition, it refers to icy sediment and massive icy bodies. The aggregates are typically mud (silt), and range in diameter or length from ≤1 mm to several

Table 7.4. Cryofacies types applicable to ice-rich sediments in the Pleistocene Mackenzie Delta, Canada.

Cryofacies type	Volumetric Ice Content (%)	Cryofacies	Code	Cryostructures
Pure ice	100	Pure ice	I	Le, Le
Sediment-poor ice	>75	Sand-poor ice	SPI	Le, La, Su
		Aggregate-poor ice	API	
Sediment-rich ice	>50 to ≤75	Sand-rich ice	SRI	
		Aggregate-rich ice	ARI	Le, La, Su
Ice-rich sediment	25 to ≤50	Ice-rich sand	IRS	SI, Le, La
		Ice-rich mud	IRM	Le, La, Rr, Ri, Cr
		Ice-rich diamicton	IRD	
Ice-poor sediment	≤25	Ice-poor mud	IPM	SI; various non-ice
		Ice-poor sand	IPS	sedimentary structures
		Ice-poor gravel	IPG	
		Ice-poor diamicton	IPD	
		Ice-poor peat	IPP	

Source: Murton and French (1994).

centimeters or more. In segregated ice, angular mud aggregates are abundant in the ice-rich layer commonly observed at the top of permafrost (e.g. Cheng, 1983; Burn, 1988; Mackay, 1972a; Shur, 1988b). Rounded mud aggregates have been reported from debris-rich glacier ice (Boulton, 1970; French and Harry, 1988).

Many cryostructures are composite or transitional, either merging into adjacent categories or mixing two categories. They may also be hierarchical, although such hierarchies do not necessarily imply genesis.

Cryostratigraphy adopts many of the principles of modern sedimentology. For example, "cryofacies" are defined according to volumetric ice content and ice-crystal size, and then subdivided according to cryostructure. One such system is illustrated in Table 7.4, where five cryofacies are distinguished according to arbitrarily-defined volumetric-ice-content values. Although this particular facies classification was developed specifically for ice-rich sediments, facies can be modified to meet other conditions.

Cryofacies can also be described according to cryostructures. Where a number of cryofacies form a distinctive cryostratigraphic unit, these are termed a cryofacies assemblage.

Two examples illustrate the application of cryostructural analysis. First, Figure 7.8A summarizes the ground ice conditions observed in shale bedrock on Eastern Melville Island. The cryostratigraphy is indicated together with ice types according to the Pihlainen and Johnston (1963) classification (see earlier) and the Murton and French (1994) cryostructural classification. A more sophisticated example (Figure 7.8B) describes an 8 m high exposure of massive ice overlain by sandy diamicton. Two cryofacies assemblages are recognized together with a secondary thaw contact (see below).

7.4.2. Epigenetic and Syngenetic Cryostructures

Certain cryostructures are diagnostic of certain freezing conditions. Here, we ignore the details of the various Russian classifications (Melnikov and Spesitvsev, 2000) and attempt to generalize in terms of the cryostructures listed in Figure 7.7A.

(1)

(2)

Unit	Depth (m)	Description	Ice-type	Cryostructures
1	0–0.35	Active layer: desiccated; blocky	–	Not described
2	0.35–0.60	Ice-rich, silty colluvium: blocky; 12% excess ice	Vr, Vs	Lenticular Irregular reticulate (incipient)
3	0.60–1.25	Shale-rich (-poor) ice: cryoturbated, angular shale fragments in ice matrix; 72% excess ice	Ice	Suspended (Irregular reticulate)
4	1.25–1.35	Ice-rich shale	Vs, Vr	Lenticular (Irregular reticulate, incipient)
5	1.35–1.70	Shale-rich ice: angular, displaced and reoriented shale fragments in ice matrix; 30% excess ice	Ice	Irregular reticulate Suspended
6	1.7 1.7–3.0	Thaw unconformity Ice-rich shale: brecciated, *in situ*, with ice lenses	Vs	Regular to irregular reticulate Lenticular Suspended
7	3.0–5.0	Ice-rich to ice-poor shale: lithified shale containing pore ice and occasional ice lenses	Vr, Vs	Lenticular (Irregular reticulate, incipient)

Figure 7.8. Examples of cryostratigraphy. (A) Observed in Mesozoic-age shale bedrock (Christopher Formation), Sherrard Bay F-34 wellsite, eastern Melville Island, Canada. (1) Stratigraphy; Figures 7.6A and 7.6B illustrate units 3 and 6, respectively. (2) Description using Pihlainen and Johnston (1963) ice Types and Murton and French (1994) cryostructures. Modified from French et al. (1986).

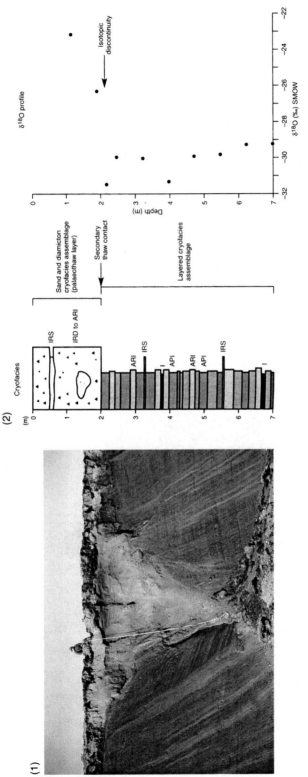

Figure 7.8. (*Continued*)   (B) Observed at Crumbling Point, Pleistocene Mackenzie Delta, Canada. (1) Large sand wedge penetrates a sand and diamicton cryofacies assemblage that overlies a layered cryofacies assemblage consisting of massive icy beds. The person in the photo is ~1.8 m. (2) The generalized cryofacies log and δO^{18} profile obtained from both cryofacies assemblages. At the contact of the two cryofacies assemblages is an isotopic (thaw?) discontinuity. SMOW is standard mean ocean water. See Table 7.4 for explanation of cryofacies types indicated. From Murton and French (1994).

Two common examples of epigenetic freezing are the structureless (Sl) and reticulate (Rr, Ri) cryostructures. The structureless cryostructure reflects in-situ freezing of pore water. It is termed "massive" in the Russian transliterated literature. The reticulate (regular, irregular, or "chaotic') cryostructure, (see Figure 7.2B) reflects either the migration of water into shrinkage cracks in a semi-closed freezing system (Mackay, 1974b) or hydraulic fracturing in unfrozen soil ahead of the advancing freezing front (McRoberts and Nixon, 1975).

Layered, lenticular-layered and micro-lenticular cryostructures indicate syngenetic freezing. The various sub-types recognized in the Russian literature are shown in Figure 7.9. It also shows how the inclined cryostructures shown in Figure 7.2A would have formed during the freezing of lake-bottom sediments. Based upon observations in the CRREL permafrost tunnel, Alaska, layered and lenticular cryostructures are regarded as diagnostic indicators of syngenetic permafrost (Shur et al., 2004; Bray et al., 2006).

(A)

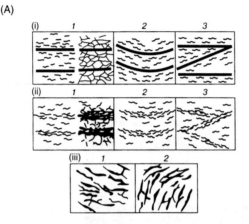

(B)

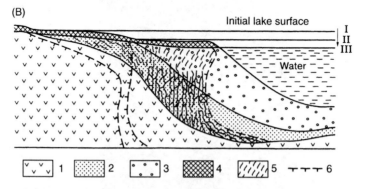

Figure 7.9. The different cryostructures that form during syngenetic freezing of subaqueous lake-bottom sediments. (A) Those associated with freezing of the seasonally-thawed layer from below (i and ii), and in subaqueous situation (iii). (B) Schematic illustration of cryostructures that form during freezing of lake sediments. Legend: 1, thawed material beneath original lake; 2/3, lake-bottom sediments; 4, seasonally-thawed layer; 5, reticulate ice cryostructure; 6, permafrost boundary at different stages of lake infilling and water-level lowering. From Kudryavtsev (1978) and Melnikov and Spesivtsev (2000).

7.4.3. Thaw Unconformities

Discontinuities in the nature and distribution of ground ice bodies are the result of either thawing of frozen material or subsequent refreezing of previously-thawed material. The significance of these discontinuities, termed "thaw unconformities," is that they allow inferences to be made as to past permafrost conditions. This is explained below.

Figure 7.10 shows two typical permafrost conditions that might exist in continuous permafrost of high latitudes and in the zone of deep seasonal frost overlying relict permafrost in parts of the sub-arctic. In Figure 7.10A the active layer is shown as the near-surface layer that thaws during the summer. A "relict active layer," shown as ground immediately below the modern active layer, was once part of the active layer but is now perennially frozen. A "paleo-active layer" is also indicated as the horizon between the ground surface and the base of the relict active layer. In Figure 7.10B, a "residual thaw layer" is shown as referring to an unfrozen layer, formerly permafrost, lying between the modern depth of seasonal frost penetration and an underlying (relict) permafrost body. The base of the active layer, shown in Figure 7.10A, is the simplest and most obvious example of a thaw unconformity.

It is possible to distinguish between primary (i.e. present-day) and secondary (i.e. paleo-) thaw unconformities. Both are shown in Figure 7.10A but in Figure 7.10B the paleo-thaw unconformity overlies relict permafrost that is unrelated to present surface conditions.

The manner in which permafrost degrades and subsequently forms again, and the cryostratigraphic evidence which it leaves, is illustrated in Figure 7.11. The diagrams depict an initial permafrost profile that is subject to thaw-degradation from above, possibly caused by regional climate warming. As thaw proceeds, a primary thaw unconformity

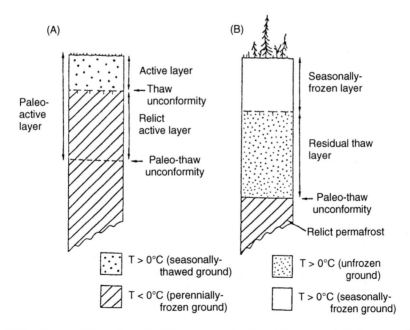

Figure 7.10. Types of ice discontinuities commonly found in perennially-frozen sediments. (A) Arctic regions of continuous permafrost. (B) Sub-arctic regions of discontinuous permafrost or deep seasonal frost.

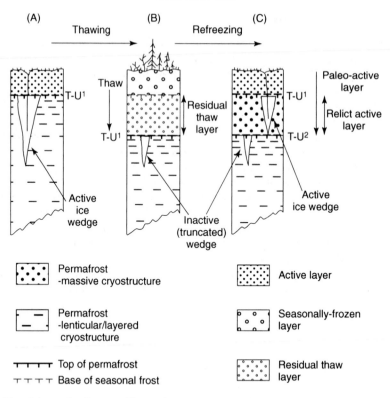

Figure 7.11. Schematic diagram illustrating the cryostratigraphic evidence associated with the degradation and subsequent aggradation of permafrost. In (A) and (B) an initial permafrost sequence is subject to downwards thawing, possibly as the result of climate warming. The cryostructure of the permafrost is indicated as being lenticular-layered. As thaw proceeds, a primary thaw unconformity (T-U^1) forms at depth below a residual thaw layer. In (C), the climate subsequently deteriorates, permafrost aggrades, and the base of the active layer again becomes the primary thaw unconformity. In (C) a massive cryostructure is illustrated.

(T-U^1) forms at depth below a residual-thaw layer. At this time, the ground surface experiences only seasonal freezing and thawing. In the process, an ice wedge is truncated and is no longer active. When climate subsequently deteriorates, permafrost aggrades and the base of the active layer again becomes the primary thaw unconformity. Renewed thermal-contraction cracking at the ground surface permits a new ice wedge to form. During this process, the original thaw unconformity at depth becomes a secondary (i.e. paleo-) thaw unconformity (T-U^2). The latter can be recognized by both the truncated ice wedge and by different ice structures (cryostructures) in sediment above and below.

Besides the truncation of ice bodies and differences in ice contents, a thaw unconformity might be recognized by differences in stable isotope values (see Figure 7.8B), heavy mineral and pollen assemblages above and below the unconformity (Burn et al., 1986), and horizons of enhanced micro-organisms (Gilichinsky and Wagener, 1995). The recognition of thaw unconformities is a valuable line of evidence in reconstructing past permafrost histories. In currently-unfrozen Pleistocene sediments of middle-latitudes, thaw unconformities are sometimes inferred from the study of soil microfabrics and structures (Derbyshire et al., 1985; van Vliet-Lanoë, 1988).

7.4.4. Ice Crystallography

The petrofabric analysis of ground ice is not only useful for descriptive purposes but, like the study of cryostructures, helps to infer growth processes and conditions. This is because the crystal size, shape, boundary characteristics, and c-axis orientations are directly related to the direction and speed of the freezing process. Ice crystals normally grow at right angle to the direction of freezing, and crystal size varies inversely with the rate of freezing. In recent years, petrofabric analyses of ground ice bodies have become standard procedures (French and Pollard, 1986; Pollard, 1990; Solomatin, 1986). The techniques and procedures follow closely those developed for glacier ice (Ostrem, 1963).

Although different ice types may display a characteristic range of fabric and texture patterns, the reality is that a wide range of fabrics exist. Without good cryostratigraphic control, ice fabrics do not permit unambiguous identification of ice types. However, segregated ice tends to be composed of large equigranular anhydral crystals whose c-axes form a loose girdle oriented normal to the plane of the ice layer. By contrast, buried snow-bank ice is composed of small enhedral equigranular crystals with a high concentration of vertically-oriented inter-crystalline bubble trains and tubular bubbles. The petrography of intrusive ice reflects the groundwater transfer mechanism and freezing conditions. For example, in seasonal-frost mounds, the ice mass is composed of large tabular crystals oriented normal to the freezing direction with c-axes forming a horizontal girdle normal to the long axes of the crystals (Pollard and French, 1985).

7.4.5. Ice Geochemistry

Standard chemical analyses, including conductivity and cations (Ca, Na, Mg, and K), can characterize ground ice (water). Although such determinations usually reflect local geologic and/or hydrologic conditions, they are useful for comparison purposes and for differentiating between ice bodies.

More inferential from the viewpoint of cryostratigraphy is the use of isotopic data (e.g. δO^{18}, deuterium, and tritium). For example, when water freezes, δO^{18} is preferentially incorporated into the ice, which becomes isotopically heavier. Usually isotopic values are compared to standard mean ocean water (SMOW) values and expressed in ‰. There is also a positive linear relationship between temperature and δO^{18}, as demonstrated from ice cores from Greenland. Finally, there is also a relationship between δO^{18} and δO^{16}: the colder the climate, the lower the $\delta O^{18}:\delta O^{16}$ ratio becomes. It follows that the isotopic analysis of ground ice is not only a useful descriptive tool but also allows inferences to be made concerning the approximate temperature of the water prior to freezing and, by comparing the isotopic signatures with those from adjacent groundwater, the water source.

A simple example of the effects of freezing upon the oxygen-isotope composition of groundwater is provided by data in Table 7.5. Two sites are compared; one a large drained lake where aggrading permafrost is 20–35 m in thickness, the other a more-recently drained lake where permafrost is currently only 15–20 m thick. It is assumed that initial groundwater conditions were similar. The ice (water) samples from within newly-formed permafrost is −16‰ while the subpermafrost waters range from −27‰ to −29‰, indicating

Table 7.5. Isotopic and geochemical composition of water (ice) in mg/L at two drained lake sites in the Mackenzie Delta region, Canada. Site 1: lake drained about 150 years ago, permafrost is 20–35 m thick. Site 2: lake drained between 1935 and 1950, permafrost is 15–20 m thick.

	Site 1			Site 2	
	Surface water (residual lake)	Permafrost (at 20.5 m)	Sub-permafrost water (at 21.5 m)	Surface water	Sub-permafrost water (at 17 m)
$\delta H_2{}^{18}O$ (in ‰)	−21.6	−16.2 (at 14.5 m)	−28.8	−19.7	−27.7
Specific conductance (μmho\cm)	188	190	1224	134	1714
Chloride	16	–	106	18	141
Magnesium	7	2.3	49	4	148
Potassium	2.4	2	5.6	1.5	7.3
Calcium	14	17	93	10	54
Sodium	2	4	83	9	92

Source: Mackay and Lavkulich (1974).

Table 7.6. Typical δO^{18} values reported from different ground ice types in western Arctic Canada and Siberia, together with typical values of basal meltwater/regelation ice from Svalbard and the Swiss Alps.

	δO^{18}	Reference
Basal meltwater/regelation ice:		
Tranfeuron Glacier, Swiss Alps	−10 to −15‰	Lemmens et al. (1982)
Gruve 7, Adventdalen, Svalbard	−15‰	Christiansen et al. (2005)
Longyearbreen, Svalbard	−12‰	O. Humlum (pers. communication)
Ground ice:		
Massive ice	−28 to −35‰	Mackay (1983b)
Buried glacier ice	−26 to −35‰	Lorrain & Demeur (1985) French & Harry (1990)
Modern ice wedges:	−22 to −26‰	Mackay (1983b), Michel (1990), Lauriol et al. (1995)
Late-Pleistocene ice wedges:	−28 to −34‰	Mackay (1983b) Vasil'chuck and Vasil'chuck (1997)

that significant fractionation has occurred during freezing. The values for surface water compare favorably with annual SMOW values.

δO^{18} values are useful not only in inferring thaw unconformities but, because different ground-ice types have different isotopic signatures, can characterize ground-ice bodies. For example, Table 7.6 provides typical δO^{18} isotope values for massive ground ice of intra-sedimental origin, massive icy sediments of presumed buried glacier-ice origin, modern and Pleistocene-age ice wedges, and basal meltwater/regelation ice.

Tritium (H^3) is a radioactive isotope of modern age with a short half-life (12.43 years). Its detection in the upper layers of permafrost, in the active layer, and in seasonal ice bodies is useful in determining recent water migration into permafrost (Burn and Michel, 1988; Michel and Fritz, 1982), the study of seasonal-frost mounds (French and Guglielmin, 2000; Pollard and French, 1984), or recent ice-wedge growth (Lewkowicz, 1994).

It is clear, therefore, that isotopic compositional differences in groundwater and ground ice can assist in the interpretation of the history and stability of permafrost, such as whether permafrost grew in open or closed systems, in the recognition of thaw unconformities, in understanding certain geomorphic processes, and in the characterization of ground ice bodies.

7.4.6. Cryostratigraphy and Past Environments

Cryostratigraphic observations from the lowlands of the Western Canadian Arctic indicate that during the early Holocene the climate ameliorated, causing permafrost to partially thaw but to then subsequently refreeze towards the end of the Holocene. The evidence consists of a widespread paleo-thaw layer (Burn 1988, 1997; French et al., 1986; Harry et al., 1988; Murton and French 1994). It can be recognized by distinct cryostructural contrasts and, in places, by truncated ice bodies. A good example is from Pullen Island (see Figure 7.3), where the regional hypsithermal thaw unconformity (a "paleo-thaw layer") occurs at a depth of 125–150 cm. This is approximately 2.5 times thicker than the present active layer. Radiocarbon dating of organic material just above the unconformity suggests thaw was greatest about 8.0–9.0 ka.

This regional thaw unconformity in the Western Arctic can be used to infer the relationship between permafrost and past climate by application of the Stefan equation (see Chapter 5). If one assumes that an increase in the active layer is the result of summer thaw and that thaw is linked to thawing degree-days (TDD), the Stefan equation indicates that a doubling of active-layer thickness corresponds to a fourfold increase in thawing index and that an active layer 2.5 times as thick as today implies an increase in the thawing factor by 6.25.

This type of analysis can be applied to the climatic records from five settlements that represent a north–south transect, covering 16 degrees of latitude, across the Western Canadian Arctic (Table 7.7). From north to south, the five localities each typically record thawing indices of approximately 1900, 1200, 800, 400, and 300 degree-days per year, respectively. The thickness of the active layer at each locality is approximately

Table 7.7. Data showing average active-layer depths, the depth of the Early Holocene thaw unconformity (where recognized), typical annual thawing degree-days, and bio-climatic zonations for five localities in the Western and High Arctic of Canada.

Locality	Latitude	Thawing Degree-days (°C)	Active Layer (cm)	Early-Holocene thaw unconformity	Ecozone
Whitehorse	61°N	1900	125–150		Boreal forest
Inuvik	68°N	1200	100		Treeline
Tuktoyaktuk	69°N	800	50	125–150*	Tundra
Sachs Harbour	72°N	400	30–50		Tundra
Eastern Melville Island	77°N	300	25–50	113**	Polar semi-desert

*Thaw unconformity at 125–150 cm depth corresponds to ~1800 thawing degree-days.
**Thaw unconformity at 113 cm depth corresponds to ~600 thawing degree-days.
Sources: French et al. (1986), Burn (1998a).

Figure 7.12. Large epigenetic ice wedges, near King Point, Yukon Coast, Canada.

150 cm, 100 cm, 50 cm, 30–50 cm, and 25–50 cm, respectively. When the maximum depth of the thaw unconformity observed at Pullen Island is used in the Stefan equation, the resulting thawing index value (approximately 1800) is similar to that currently experienced by Whitehorse. Thus, during the early Holocene, the summer climate at the location of the present Arctic coast was probably typical of that which exists today near Whitehorse, in central Yukon Territory. Likewise, the thaw unconformity that occurs on Melville Island corresponds to a summer climate experienced by Tuktoyaktuk and Sachs Harbour (Ikaahuk).

7.5. ICE WEDGES

Their widespread occurrence and distinctive surface manifestation makes ice wedges some of the most characteristic features of the periglacial landscape (see Figure 6.1). After pore and segregated ice, ice-wedge ice constitutes the third most important ground-ice type in terms of volume (Figure 7.12). Some Russian authorities regard wedge ice as the dominant type of underground massive ice in central and northern Siberia (Dostovalov and Popov, 1966; Popov, 1962; Shumskii and Vtyurin, 1966). More typically, ice-wedge ice constitutes between 20% and 35% of the total ice volume in the upper 5–10 m of permafrost (see Table 7.2A).

Ice wedges are best developed in unconsolidated sediments but they also occur in bedrock and on sloping terrain. The most favorable environments for their formation are poorly-drained tundra lowlands underlain by continuous permafrost.

For several reasons, the growth of an ice wedge must involve deformation of both the ice and adjacent ground. First, the ice must deform because the length of the initial crack depth is always less than the length of the side of the wedge (Figure 7.13A). Second, because wedge ice is less dense than the surrounding frozen ground, there is a tendency for ice wedges to move upwards in diapiric fashion (Figure 7.13B). Third, the top of an actively-forming ice wedge may be affected by thaw as the wedge rises into the base of the seasonally-thawed zone (the active layer) (Figure 7.13C). In like fashion, because of the volume addition to the growing ice wedge, the enclosing sediments adjacent to the wedge typically exhibit upward bending, and there is a tendency for deformation to occur upwards and away from the wedge (Figure 7.13D). Finally, if ice wedges form on sloping

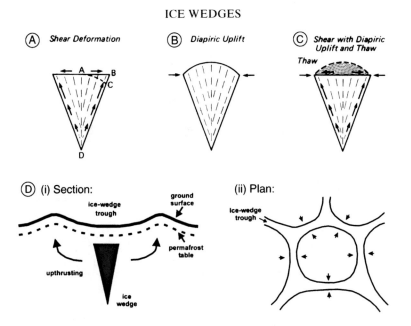

Figure 7.13. Ice-wedge deformation patterns (A–C) and mode of polygon deformation pattern (D) according to Mackay (1990a) and Kokelj and Burn (2004). Reproduced by permission of John Wiley & Sons Ltd.

terrain, the creep of near-surface permafrost means that wedges will deform in the down-slope direction.

Ice wedges can be classified in several ways. Based upon the frequency of occurrence of cracking, they can be active, inactive, or relict. They may also be classified in terms of their infill composition. This can range from pure ice to icy silt and sand. To complicate matters further, pseudomorphs and casts of Pleistocene-age wedges can be classified according to whether the wedge infill is either primary or secondary in nature, or if it is composite (see Chapter 12). When discussing actively-forming wedges in present-day permafrost environments, as in this chapter, a useful approach is to consider the direction of wedge growth relative to the land surface. Following J. R. Mackay (1990a, 2000), ice wedges can then be classified as being epigenetic, syngenetic, or anti-syngenetic (Figure 7.14).

7.5.1. Epigenetic Wedges

Epigenetic ice wedges grow in pre-existing permafrost and are younger than the host sediment. With repeated cracking and infilling, the ice wedge grows progressively wider but not deeper (Figure 7.14A). At the same time, there is a net outward movement of the active layer from the polygon center towards the bordering ice-wedge trough. The end result is the highly distinctive "double-ridge" ice-wedge polygon topography. The top of an actively-forming epigenetic wedge commonly shows a small ice nipple extending towards the base of the active layer. Where there has been recent formation of near-surface organic material, the active layer will thin and wedge "rejuvenation" will occur (Figure 7.15A). In a sense, this is the beginnings of syngenetic growth. Sometimes several episodes of rejuvenation can be related to vegetation-succession sequences and the associated thinning of the active layer (Figure 7.15B). Epigenetic ice wedges typically occur on flat surfaces that are experiencing neither erosion nor sediment accretion. In poorly-drained

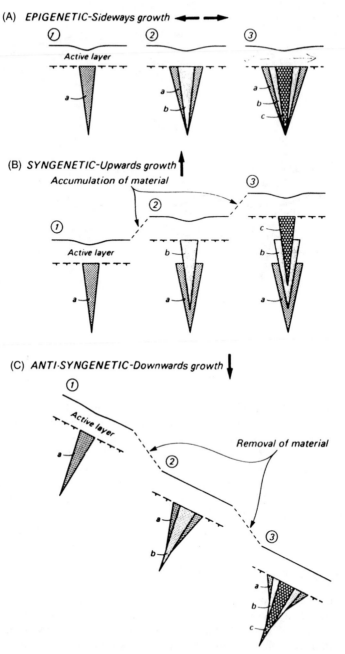

Figure 7.14. Schematic diagram showing the growth of epigenetic, syngenetic, and anti-syngenetic wedges. Three growth stages are indicated for each type; a–c indicate stages of ice-wedge growth. From Mackay (1990a). Reproduced by permission of John Wiley & Sons Ltd.

(A)

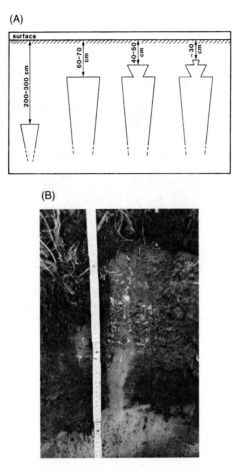

(B)

Figure 7.15. Ice wedge rejuvenation. (A) Schematic diagram showing a buried (i.e. inactive) wedge and modern wedges of single-, two-, and multi-stage nature. (B) Rejuvenated ice wedge, southern Banks Island, Canada.

areas, they may attain widths of 1.0–2.0 m near the surface but are usually much smaller. Epigenetic wedges rarely exceed 3.0–4.0 m in depth.

7.5.2. Syngenetic Wedges

Syngenetic ice wedges form at approximately the same time as the enclosing sediments are laid down. In other words, as the ground surface rises, so the wedge grows upwards in response to surface aggradation (Figure 7.14B). The typical locations for syngenetic wedges are on floodplains, beneath peat, as in tundra polygons, and beneath gelifluction deposits, as at the bottom of a slope. The size and shape of a syngenetic wedge is a function of both horizontal and vertical growth rates. If sedimentation is high and ice-vein accretion is low, then the wedge becomes thinner; if ice-accretion rate is high and sediment accumulation is low, then the wedge becomes wider.

Syngenetic ice wedges appear to reach their greatest thickness and vertical dimensions on ancient alluvial surfaces and river terraces in central Siberia (Dostovalov and Popov,

(A) (B)

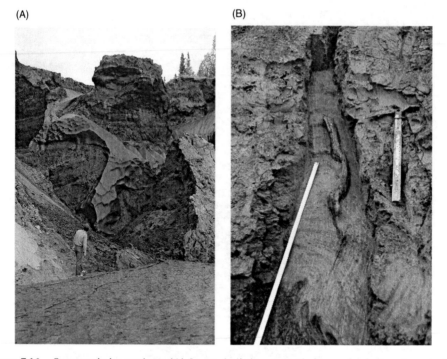

Figure 7.16. Syngenetic ice wedges: (A) Large (relict) syngenetic ice wedge in loessic materials, Fairbanks, Alaska. (B) Small (relict) syngenetic ice wedge (vein) in outwash sands, Sachs River lowlands, Banks Island, Canada (see French et al., 1982).

1966; Gasanov, 1978; Vasil'chuk and Vasil'chuk, 1997). These high-ice-content sediments are described as "ice-complex" or "Yedoma" in the Russian literature (Romanovskii et al., 2000). Large inactive (Pleistocene) syngenetic ice wedges also occur in the muck deposits of central Alaska (Figure 7.16A; also see Chapter 11). A characteristic of these wedges, apart from their size, is that their internal near-vertical foliations are truncated along the sides of the wedges.

In most other permafrost areas, smaller syngenetic wedges and veins are more common (Figure 7.16B). Certainly, most Pleistocene pseudomorphs of syngenetic wedges that can be observed today in mid-latitude are also small (French and Gozdzik, 1988; see also Chapter 12). It appears that the growth of large syngenetic wedges requires exceptionally long periods of uninterrupted surface-sediment accretion under cold-climate conditions. Only depositional surfaces of either alluvial or wind origin, located in areas that have experienced cold non-glacial climatic conditions throughout much of the Pleistocene, are capable of fulfilling these requirements. Vasil'chuk and Vasil'chuk (1997) use radiocarbon dating and oxygen-isotope determinations to show that many of the large syngenetic ice wedges in Northern Siberia have grown vertically at rates varying between 1.0 m and 2.7 m per 1000 years throughout the Late-Pleistocene.

7.5.3. Anti-Syngenetic Wedges

Anti-syngenetic wedges form on receding surfaces that are experiencing net surface removal (i.e. erosion) (Figure 7.14C). If thermal-contraction cracking and ice-vein forma-

Figure 7.17. Photo showing the surface expression of an anti-syngenetic ice wedge on sloping terrain near Illisarvik, Pleistocene Mackenzie Delta, Canada. Excavation indicated the wedge width to exceed 8.5 m and an adjacent wedge to exceed 6.5 m. There was no proof of winter cracking. The individuals are standing on either side of the wedge. Photo and caption information supplied courtesy of Professor J. R. Mackay.

tion keep pace with removal of surface material, the ice veins penetrate progressively to greater depths. Therefore, anti-syngenetic wedges grow downwards normal to the ground surface and not upwards. The top of an anti-syngenetic wedge continues to be truncated by thaw as the active layer descends in response to surface lowering. These wedges were first described from the Mackenzie Delta region, Canada (Mackay, 1990a, 1995b).

Favorable places for the growth of anti-syngenetic wedges include convex sites near hilltops where the rate of downslope sediment movement is high, on actively receding or fluvially-undercut slopes, and in badland thermokarst terrain. However, their surface expression is masked by gravity-controlled mass movements in the active layer, such as gelifluction. As such, they are rarely recognizable in the landscape (Figure 7.17). Their identification is further complicated by the lack of natural exposures on slopes far removed from actively eroding stream banks or coastal bluffs. It can be theorized that the shape of anti-syngenetic wedges will depend largely upon the rate of surface lowering and the time over which the wedge has formed. One must suspect that anti-syngenetic wedges are a most common type of ice wedge.

7.6. MASSIVE ICE AND MASSIVE-ICY BODIES

Thick, often bedded, and sometimes deformed layers of massive ground ice and icy sediments are the most spectacular of ground-ice forms (Figure 7.18). These icy bodies are important not only because of their origin, and the light this may throw upon permafrost histories, but also because of the thaw-settlement properties of terrain underlain by such material.

7.6.1. Nature and Extent

Bodies of massive ground ice and icy sediments exist in parts of Western Siberia (Astakhov, 1992; Astakhov and Isayeva, 1988; Astakhov et al., 1996; Dubikov, 1982, 2002;

Figure 7.18. Photo showing a deformed massive ground ice body exposed near Nicholson Point, Pleistocene Mackenzie Delta, Canada. Note also the large epigenetic ice wedge in the silty diamicton above the massive ice.

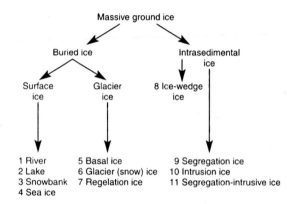

Figure 7.19. A classification of massive ground ice proposed by J. R. Mackay at a Geological Survey of Canada seminar in Ottawa in 1989.

Kaplanskaya and Tarnogradsky, 1986) and the Western Canadian Arctic (e.g. French and Harry, 1988, 1990; Mackay, 1971, 1973b; Mackay and Dallimore, 1992; Murton et al., 2005). Many show glaciotectonic structures or deformations (see Figure 7.18; Lokrantz et al., 2003; Murton et al., 2004). Others exhibit anticlinal structures apparently related to diapiric uplift and pressure release following removal of overlying material (Mackay and Dallimore, 1992).

The two main explanations advanced for the origin of these icy bodies are: (1) segregated ice which, with increasing importance of water-injection processes, grades into intrusive ice, and (2) buried glacier ice, without a clear distinction being made between glacier ice derived from snow and sub-glacier regelation ice. A more comprehensive classification of massive ground ice bodies, based on these two different origins, is given in Figure 7.19. Massive segregated ice bodies are regarded as "intra-sedimental."

Massive icy bodies of both segregated and buried glacier-ice origin may appear similar in the field. Both may contain considerable quantities of sediment, often stratified (Klassen and Shilts, 1987, pp. 15–22). Moreover, basal glacier ice that has experienced regelation is essentially segregation ice. A further complication arises because, in both cases, the ice may have existed beneath an overburden for several thousands of years and may have experienced major post-formational or post-burial alteration (moisture migration, thermal and geochemical changes, and loading/unloading).

Several observations may assist in differentiating between intra-sedimental and buried basal glacier ice. First, gradational contacts should rarely exist with buried ice, and, in the case of intrusive ice, suspended fragments might occur just below the contact. Second, if the contact between overlying material and underlying ice is of a thaw or erosional nature, this indicates the ice predates burial and must be buried. Third, features suggestive of ice origin may be found within the ice itself. For example, matched soil fragments indicate segregated and/or intrusive ice, ice coatings beneath clasts within the ice indicate segregation, striated clasts and pods of diamicton within the ice suggest glacier ice, and lumps of clear ice frozen (i.e. suspended) within silty bubbly ice suggest regelation ice. Finally, ice dykes formed from water intruded upwards under pressure into permafrost along fissures are associated with segregated and/or segregated-intrusive ice (Mackay, 1989a; Mackay and Dallimore, 1992).

7.6.2. Intra-Sedimental Ice

Undoubtedly, the strongest evidence in favor of an intra-sedimental (ice segregation) origin for massive icy bodies is provided by over 600 drill hole records from both the western Canadian Arctic (Mackay, 1973b) and northern Siberia (Dubikov, 1982, 2002). These are summarized in Mackay and Dallimore (1992, figure 3). They indicate two salient facts about the occurrence of massive ice bodies: (1) in the majority of instances where massive ice is encountered, it is overlain by clay-grade sediment (diamicton) and underlain by sand-grade sediment; and (2) a significant number of drill holes encounter massive ice at depths in excess of 30 m, sometimes at depths of 100–200 m; in some holes two or more layers of massive ice, usually separated by sand, are penetrated by the same drill hole.

These observations support a segregation or segregation-injection origin and fit a model of ice growth with fine-grained sediment underlain by coarse-grained sediment. Supporting evidence includes ice dykes which penetrate the overlying diamicton and possess isotopic signatures similar to the massive ice. These structures are proof of the high water pressures that must have been involved. Moreover, the contact between the massive ice and the overlying diamicton is always conformable and the continuity of δO^{18} and δD profiles from the top of the massive ice downwards into underlying sand indicate a common water source for the massive ice and the interstitial ice in the underlying sand. Finally, a network of reticulate ice veins in the overlying diamicton and bubble trains which originate at the upper ice–sediment contact are evidence of downward freezing. More than 40 years ago, J. R. Mackay concluded "The ice-sheets are believed to have formed during progressive downward aggradation of permafrost in fine-grained soils in an open system, where fresh water moved to the bottom of the freezing plane, as new ice formed." He also added "some ice sheets may be sill-like injection features, formed similar to pingo ice" (Mackay 1963, p. 63). These statements must remain the standard against which other interpretations are judged.

7.6.3. Buried Glacier Ice

Notwithstanding the previous conclusion, it is probable that certain bodies of massive ice and icy sediments are of buried glacier-ice origin. This hypothesis was initially suggested by Russian Quaternary geologists (Astakhov and Isayeva, 1988; Kaplanskaya and Tarnogradskiy, 1986). Terms such as "ice/sediment complex," "sheet-ice complex," and "deposit-forming ice" are used to describe these icy sediments.

Several studies in the western Canadian arctic suggest that at least some massive icy bodies may also be buried basal glacier ice (Dallimore and Wolfe, 1988; French and Harry, 1988, 1990; Lorrain and Demeur, 1985). Several lines of evidence support the buried-glacier ice mechanism: (1) deformation structures and folds of varying dimensions and intensities are typical of glacier ice; (2) sediment-rich layers often resemble shear planes (Rampton and Walcott, 1974); (3) there is a range of grain and clast sizes contained within the massive ice and icy bodies, from clay to pebbles to striated boulders; and (4) the overlying soil–ice contact is often unconformable. It should also be pointed out that many massive icy bodies in both Siberia and northern Canada that have been ascribed a glacial origin occur in regions that were glaciated during the Pleistocene.

Recent stratigraphic investigations provide further support for a buried glacier ice hypothesis. For example, at two locations in the Tuktoyaktuk Coastlands, features characteristic of both basal glacier ice and intra-sedimental ice occur (Murton et al., 2004, 2005). The basal-ice features include (a) ice facies and ice-crystal fabrics similar to basal ice in contemporary glaciers and ice-cored moraines, (b) an erosional unconformity that truncates the top of the massive icy bodies, and (c) a covering of glacigenic or aeolian sediment. A similar interpretation has been proposed for massive icy sediments in the Kara Sea region of northern Russia. There, it is hypothesized that relict glacier ice, buried in sediment and incorporated within permafrost, may survive for many tens of thousands of years and even several interglacial and stadial events (Henricksen et al., 2003; Lokrantz et al., 2003).

7.6.4. Other Mechanisms

In addition to the two main hypotheses discussed above, several others have also been proposed. For example, K. Fujino et al. (1983, 1988) concluded that the massive ground ice at Peninsula Point near Tuktoyaktuk, Canada, was regelation ice formed from the episodic inundation of a perennial snow bank. However, none of the ice at that location originated as firn, and chemical and isotopic data indicate it grew in-situ with ground water as the source (Moorman et al., 1996, p. 265; 1998, p. 761). Massive tabular ice bodies are also known to occur on Ellesmere Island, where the ice is interbedded with fine-grained marine sediments lying below the maximum limit of marine submergence (Pollard, 2000b). The ice is interpreted as intra-sedimental in origin, the result of permafrost aggradation into recently-emerged marine sediments.

ADVANCED READING

Astakhov, V. I., Kaplyanskaya, F. A., Tarnogradsky, V. D. (1996). Pleistocene permafrost of West Siberia as a deformable glacier bed. *Permafrost and Periglacial Processes*, **7**, 165–191.

Mackay, J. R., Dallimore, S. R. (1992). Massive ice of the Tuktoyaktuk area, Western Arctic Coast, Canada. *Canadian Journal of Earth Sciences*, **29**, 1235–1249.

Melnikov, V. P., Spesivtsev, V. I. (2000). Cryogenic formations in the Earth's lithosphere. Novosibirsk Scientific Publishing Center UIGGM, SB RAS Publishing House, 343 pp. (in Russian and English).

Murton, J. B., French, H. M. (1994). Cryostructures in permafrost, Tuktoyaktuk coastlands, Western Arctic, Canada. *Canadian Journal of Earth Sciences*, **31**, 737–747.

Murton, J. B., Waller, R. I., Hart, J. K., Whiteman, C. A., Pollard, W. H., Clark, I. D. (2004). Stratigraphy and glaciotectonic structures of permafrost deformed beneath the northwest margin of the Laurentide ice sheet, Tutktoyaktuk Coastlands, Canada. *Journal of Glaciology*, **50**, 399–412.

DISCUSSION TOPICS

1. How can we describe ice within permafrost?

2. What is the origin of massive ground ice?

3. Can ice wedges be used to make paleo-climatic and paleo-environmental inferences?

8 Thermokarst

Thermokarst processes achieve their greatest importance in terrain underlain by ice-rich and unconsolidated sediments. In alpine regions, in areas underlain by consolidated and resistant bedrock, and in the extremely arid polar deserts, thermokarst modification is less apparent. Thermokarst-affected landscapes occur on a regional level in central and northern Siberia and the western North American Arctic.

The thaw of permafrost, consequent upon climate warming, poses special problems at the southern fringes of discontinuous permafrost. Human-induced thermokarst continues to present problems as regards settlement and resource development in permafrost regions.

An understanding of thermokarst is essential to the correct interpretation of structures and sediments that resulted from the degradation of permafrost in mid-latitudes during the Pleistocene.

8.1. INTRODUCTION

The term "thermokarst" was first used by M. M. Ermolaev (1932a, b) to describe irregular, hummocky terrain due to the melt and thermal abrasion of "ice-complex" sediments exposed along the coastal lowlands of the Laptev Sea, northern Siberia. Subsequently, the term has been applied to the processes associated with the thaw of permafrost that lead to local or widespread collapse, subsidence, erosion, and instability of the ground surface. Although J. Dylik (1968) argued that the term should be reserved for the melt of underground ice as opposed to buried glacier or surface ice, it is now accepted that the term applies generally to the thaw of icy permafrost, irrespective of the origin of the ice contained within it. In fact, the term "thermokarst" now encompasses the whole range of geomorphic effects resulting from subsurface water on landforms in permafrost regions.

Thermokarst is not a variety of karst. The latter is a term which is applicable to limestone areas where the dominant process, solution, is a chemical one. Underlying the development of thermokarst is a physical, i.e. thermal, process, namely, that of ground ice melting. Thermokarst is peculiar to regions underlain by permafrost.

There is a large Russian literature on thermokarst processes and phenomena (Grave and Nekrasov, 1961; Kachurin, 1938, 1955, 1962; Popov, 1956; Popov et al., 1966 Romanovskii, 1977a; Romanovskii et al., 2000; Shamanova, 1971; Shur, 1977, 1988a, b; Tolstov, 1961). In parts of central and eastern Siberia, it is thought that over 40% of the land surface has been affected, at some time or another, by thermokarst processes (Are, 1973; Czudek and Demek, 1970; Soloviev, 1962, 1973b). In the permafrost regions of North America, thermokarst phenomena were first described from central Alaska (Hopkins, 1949; Péwé, 1954) and then from the lowlands of northwestern Arctic Canada

The Periglacial Environment, Third Edition   Hugh M French
Copyright © 2007 John Wiley & Sons Ltd

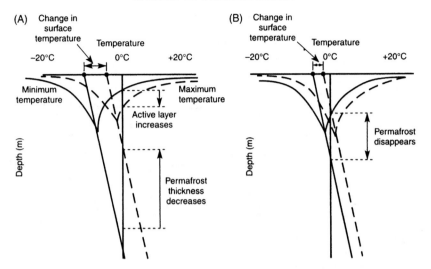

Figure 8.1. Graph showing the long-term effect of a 4 °C warming upon the ground temperature profile. (A) Continuous permafrost zone; (B) discontinuous permafrost zone. In both cases, the active layer thickens and near-surface permafrost thaws. In discontinuous permafrost, where ground temperatures are close to 0 °C, thaw degradation results in disappearance of permafrost. In situation (A), permafrost degrades in the near-surface and decreases in overall thickness.

(Brown, 1974; French, 1974b, 1975a; French and Egginton, 1973; Mackay, 1970; Rampton, 1974).

The importance of thermokarst is commonly underestimated. Furthermore, as will be explained in Part III, thermokarst has been neglected in the study of Pleistocene cold-climate phenomena (Dylik, 1964b). A full discussion of Pleistocene thermokarst is given in Part III.

8.2. CAUSES OF THERMOKARST

The development of thermokarst is due primarily to the disruption of the thermal equilibrium of the permafrost consequent upon an increase in surface temperature and a corresponding increase in the depth of the active layer. The geothermal effect is illustrated in Figure 8.1.

8.2.1. General

There are many reasons why thermal disequilibrium and permafrost degradation may occur. These are summarized in Figure 8.2. They are classified as being geomorphic, vegetational, or climatic, and either natural or human-induced in origin. Typical trigger mechanisms are also indicated.

With respect to regional climate change, thermokarst occurs when either the mean annual temperature of the soil rises or when there is an increase in the amplitude of ground temperature. The latter implies increased continentality of climate while the former implies amelioration (overall warming) of climate. We must also differentiate

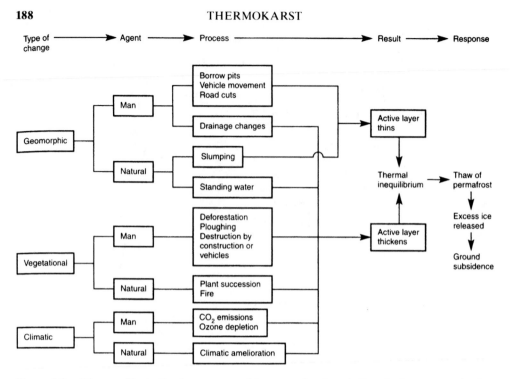

Figure 8.2. Diagram illustrating how geomorphic, vegetational, and climatic changes may lead to permafrost degradation.

between regional climate warming that occurs on a time scale of 10^{-4} to 10^{-6} years from that which occurs on a time scale of 10^{-1} years. For example, many periglacial environments are paraglacial in nature and, as such, have experienced an overall progressive amelioration of climate during the past 10000–15000 years following retreat of the Late-Pleistocene ice sheets. Equally, ongoing global warming trends of the last 100 years, irrespective of whether they are natural or anthropogenic, are capable of increasing thermokarst activity. Finally, economic activities, such as construction, land clearance, or terrain disturbance, may initiate thermokarst activity on a time scale of less than a decade.

In reality, any increase in mean annual temperature probably leads to a diminution of heat exchange in the soil and a lower summer soil temperature. This is because it would most likely be associated with an increase in precipitation and cloud cover. As a result, there would be a decrease in thaw depth. This is the exact opposite to what is required for thermokarst development. Therefore, the simplest condition for the onset of thermokarst is a progressive increase in continentality of climate. This results in a greater range of soil temperatures and summer thaw depths. However, recent global warming scenarios suggest that increases in winter temperatures and snowfall amounts will be greatest in the high latitudes. As a result, any predictions concerning an increase in thermokarst activity in response to global climate change are problematic.

In central and northern Siberia thermokarst has occurred on a regional scale (Czudek and Demek, 1970; Romanovskii et al., 2000) and over extended periods of fluctuating cold-climate conditions during the Quaternary and throughout the Holocene. This reflects the never-glaciated history of much of the area. The distinctive thermokarst terrain of central Yakutia is discussed later in this chapter while that of coastal lowlands is discussed in Chapter 11. By contrast, much current thermokarst activity in Arctic North America is largely the result of local, non-climatic factors. However, a widespread regional thaw unconformity in the western Canadian Arctic does suggest a period of regional thermokarst activity between approximately 9000 and 4500 years ago (Burn, 1997; Burn et al., 1986; French, 1999; Mackay, 1975b). Since then, climate has cooled and regional thermokarst has been limited. For example, on Banks Island, the freezing of taliks and the growth of pingos, between 4000 and 2500 years ago (Pissart and French, 1976), indicates that climate deteriorated at that time. The impact of the recent climate warming of the last 30 years in the northern polar region has yet to be fully appreciated or understood.

It is difficult to generalize about the global distribution of thermokarst phenomena since the climatic, permafrost, and ground ice controls vary from area to area. However, several general observations are appropriate. First, thermokarst develops best in unconsolidated ice-rich sediments rather than in bedrock. Obviously, this reflects the structural coherence of bedrock and the fact that fine-grained sediments promote ice segregation. Thus, in certain areas, thermokarst processes may assume dramatic regional importance. Second, thermokarst is rarely reported from alpine and mountainous regions. Clearly, this observation is related to the first. Third, thermokarst is largely absent from many of the ice-free areas of the extreme polar latitudes. This is largely because the aridity of these regions leads to low ground-ice amounts in the near-surface sediments. An exception to this particular generalization, the fine-grained sediments of the Fosheim Peninsula of northern Ellesmere Island, has been discussed in Chapter 7. In the case of "ice-free" Dry Valleys of Southern Victoria Land, Antarctica, buried glacier ice is widespread, and sublimation, rather than melt, of buried ice occurs. This must be regarded as a special and unusual form of thermokarst activity. Fourth, thermokarst is obviously favored in areas of warm permafrost, where near-surface ground temperatures are close to 0°C. For example, in discontinuous and sporadic permafrost, the vegetation cover and organic mat are crucial to permafrost preservation and, if disturbed for any reason, thermokarst may be initiated. By contrast, in terrain underlain by cold permafrost, thermokarst may be less important because the active layer is shallow, the period of summer thaw is short, and the thermal change required to initiate thermokarst is large.

8.2.2. Specific

It must be stressed that thermokarst can develop in a stable (i.e. unchanging) cold climate in response to a variety of situations that may be either natural or human-induced. This was first emphasized by S. P. Kachurin (1955). The process is sometimes described as "self-developing thermokarst" (Aleshinskaya et al., 1972).

The widespread presence of ice-wedge polygons in many periglacial landscapes (see Figure 6.1) means that, in summer, water accumulates not only in the trough above the ice wedge, but also in the depression formed at the junction of ice wedges and within the depressed centers of low-centered polygons. These shallow bodies of standing water invariably favor preferential thaw during summer and impede freeze-up in the autumn. For example, Table 8.1A provides data that demonstrate the thickness of the active layer in typical lowland terrain of central Siberia beneath 20–25 cm of standing water is twice as great as beneath adjacent tundra. Thus, localized thermokarst is a natural and ubiquitous process in many poorly-drained tundra lowlands. Once polygons have developed their low-center characteristic, further thaw is promoted without any supplementary agent (see Figure 8.8B).

Table 8.1. Active-layer thickness under different terrain conditions. (A) Depth of the active layer in ice-wedge polygon terrain in Yana-Indigirka lowland of northern Siberia. (B) The increase in active-layer depth, thickness of permafrost thawed, and amount of ground subsidence at the site of the 1968 forest fire near Inuvik, NWT, Canada.

(A) Moisture conditions in central part of polygon

	Depth of active layer (cm)	
	Latitude 70°N	Latitude 71°N
Relatively dry	30–34	23–25
Moist	38–41	31–33
Water layer, 7–10 cm deep in polygon centre	52–56	39–42
Water layer, 20–25 cm deep in polygon centre	62–69	51–53
Lake, 30 × 40 m with depth 25 cm or more	75+	

(B) Site	Mean increase since 1960 in active-layer depth (approx)	Mean thickness of permafrost thawed (cm) (approx.)	Mean ground subsidence (cm) (approx.)
Unburned	From 42 cm to 52 cm in 1979	15	5
Burned	From 35 cm to 72 cm 1973	55	18
Burned:			
Ridges	From 60 cm to 137 cm in 1988	116	39
Depressions		117	39
Unburned:			
Hummocks	From 115 cm to 132 cm in 1981	26	9
Depressions	From 76 cm to 108 cm in 1981	48	16
Burned:			
Hummocks	From 116 cm to 136 cm in 1979	30	10
Depressions	From 68 cm to 102 cm in 1979	51	17

Source: (A) Czudek and Demek (1970). (B) Mackay (1995a).

Another cause of thermokarst is associated with stream incision acting preferentially along ice wedges causing the undermining and slumping of overlying material, tunneling, and piping. Sometimes, lateral stream erosion may initiate thermokarst activity. For example, retrogressive thaw slumps occur mainly upon steeper west- and southwest-facing slopes of asymmetrical valleys on eastern Banks Island (French and Egginton, 1973). The thermokarst activity appears to be triggered by the lateral migration of the stream towards the base of the steeper slope.

A third cause of thermokarst is the destruction of surface vegetation and any organic cover by fire. For example, Table 8.1B demonstrates the increase in active-layer thickness and the amount of near-surface permafrost that thawed following the 1968 forest fire near Inuvik, Canada. At some ridge sites, the active-layer thickness doubled in 20 years. As explained in Chapter 4, lightning-induced forest fires are characteristic of the boreal forest and not uncommon on the tundra. Other local triggers for thermokarst include ice-push and scour along coasts, cyclical changes in vegetation, slope instability, and deforestation or disruption of the surface by human activity.

8.3. THAW-RELATED PROCESSES

Of the various processes associated with thermokarst, a basic distinction should be made between those associated with subsidence and those associated with erosion.

8.3.1. Thermokarst Subsidence

Thermokarst subsidence is associated with a loss of water (excess ice) upon thawing and its removal by either evaporation or drainage. Thermal melting depends upon heat conduction from, for example, a pool of water directly overlying icy soil, or through an intervening layer of unfrozen soil. Therefore, quite unlike thermal erosion (see below), flowing water is not required. It follows that thermokarst subsidence can operate just as efficiently upon flat and well-drained uplands as in poorly-drained valley bottoms. To illustrate, simple thermokarst subsidence over a 5-year period at an experimental plot near Mayo, Yukon Territory, amounted to 35 cm. The active layer increased from 33 cm to 90 cm (Figure 8.3) and total permafrost degradation was 92 cm, of which 35 cm resulted from melt of excess ice and 57 cm was due to active-layer development.

8.3.2. Thermal Erosion

Thermal erosion refers to the complex of erosional processes that are associated with running water acting upon ice-rich permafrost (Romanovskii, 1961; Shamanova, 1971). Typically, thermal erosion results when surface runoff, from snowmelt, summer precipitation, or thawing permafrost, becomes concentrated along ice wedges, causing preferential thaws. It is sometimes referred to as "fluvio-thermal" erosion. The gullies that result (Figure 8.4) are often characterized by an inverted "T" cross-profile because water first erodes vertically and then, as the bed becomes armored with transported sediment from up-gully, erodes laterally to leave organic-mat overhangs. Slumping, piping, and the creation of small tunnels above and adjacent to the partially-eroded ice wedge are all common (French, 1975b; Mackay, 1974c; Murton, 2001, pp. 185–186; Seppälä, 1997). Standing water bodies may accumulate in the channel floor behind slumped masses to form "pool"

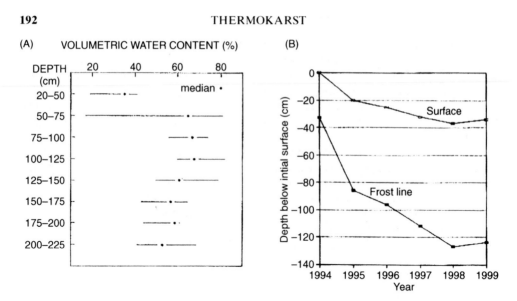

Figure 8.3. Thermokarst subsidence at Mayo, Yukon Territory, Canada. (A) Typical volumetric water content in the upper 2.25 m of glaciolacustrine sediment. (B) Ground subsidence (1994–1999) observed following surface disturbance. From Haeberli and Burn (2002). Reproduced by permission of CABI Publishing.

Figure 8.4. Photograph showing thermal erosion along an ice-wedge system near Sachs Harbour (Ikaahuk), Banks Island. Gullying developed followed vehicle disturbance and snowmobiles in late spring 1973. Photograph taken in July 1975.

ice the following winter. Many tundra lakes that drain rapidly are thought to have been effected by thermal erosion operating along ice wedges (Mackay, 1988b). Beaded drainage patterns may also result from channel enlargement located at the intersection of ice wedges.

8.3.3. Other Processes

Numerous other processes that operate in cold-climate environments are related to thermokarst activity. For example, thermal abrasion or thermo-erosional "wash" are considered by some to be important components of periglacial slopewash ("ruisellement") (Czudek and Demek, 1973; Dylik, 1971, 1972; Kachurin, 1962, p. 52). In fact, thermal erosion, in all its forms, is intimately linked to the whole question of slope and valley development in periglacial environments. This topic is discussed in Chapter 10. Likewise, thermal abrasion and fluvio-thermal erosion are important causes of rapid coastal erosion and river-bank retreat wherever ice-rich unconsolidated sediments are exposed (Are, 1988; Romanovskii et al. 2000; Walker and Arnborg, 1966). These aspects are discussed more fully in the context of cold-climate fluvial and coastal processes (Chapter 11).

8.4. THERMOKARST SEDIMENTS AND STRUCTURES

Thermokarst sediments and structures are little studied, yet they are widespread in many areas of permafrost terrain and, almost by definition, are ubiquitous in areas in which Pleistocene-age permafrost no longer exists.

Thermokarst sediments form when ice-rich permafrost degrades. By definition, excess ice is involved. Typically, thermokarst sediments mantle slopes and infill valley bottoms and depressions. They are colluvial in nature and consist of a range of locally redeposited and heterogeneous materials, or diamictons, which often incorporate clumps of organic materials. When refrozen, thermokarst sediments typically contain cryostructures indicative of epigenetic freezing (see Chapter 7). Where thaw lakes and basins form, re-sedimentation may occur and where the original permafrost was exceptionally icy, debris-flow deposits associated with retrogressive-thaw-slumps may result.

Frequently, thermokarst sediments are stratigraphically confused and often indicate differential loading and density readjustment in water-saturated sediment (Murton and French, 1993c). Ice-wedge and composite-wedge casts must also be regarded as thermokarst structures. Likewise, thermokarst-cave ice ("pool" ice), and other ice, silt, and gravel pseudomorphs, all associated with erosion, re-sedimentation and refreezing of material within permafrost, are especially complex thermokarst structures (Bray et al., 2006; Shur et al., 2004).

8.4.1. Involuted Sediments

Thermokarst involutions take the form of load casts, pseudo-nodules, ball-and-pillow structures, and diapirs. The usual mechanism is loading, buoyancy, and water-escape although fluidization may also be involved. In dimensions, thermokarst involutions vary from a few to 90 cm in width and from a few centimeters to several meters in depth. There is nothing sedimentologically unusual about these involutions, similar structures having been observed in other depositional environments throughout the geologic record (Murton and French, 1993c).

To form thermokarst involutions, ice-rich permafrost must thaw, drainage conditions must be poor, and sediments must vary in texture or composition. In addition, the sediments should be susceptible to fluidization, liquefaction, or hydroplastic deformation. Those formed by loading and buoyancy require a reversed density gradient; those formed by fluidization require either open-system groundwater conditions or associated water-saturated sediments susceptible to liquefaction.

Relatively few studies describe these structures from present permafrost environments. On Banks Island, the author (French, 1986) refers to "periglacial involutions," formed by loading and liquefaction in the active layer in late summer with the underlying permafrost acting as a confining layer. These are better described in the context of cryoturbation and other frost-induced movements that occur within the active layer. These are discussed in Chapter 6. By contrast, "thermokarst involutions" are structures that occur in a paleo-thaw layer. These have been described from the Tuktoyaktuk coastlands, where a paleo-active layer, partially refrozen in permafrost beneath the modern active layer, occurs widely (Burn, 1997; Mackay, 1975b; Murton and French, 1993c). The most common type of involution is a sand ball-and-pillow structure that varies in size upwards to over 1.0 m. The base is approximately 60–70 cm above the thaw unconformity that delimits the base of the paleo-active layer. Therefore, the stratigraphic location of these structures indicates that they are unrelated to the modern active layer because that varies between only 0.25 m and 0.9 m in thickness. According to J. B. Murton (2001), the soft-sediment deformations occurred during active-layer deepening with the sand sinking into a saturated diamicton of lower density, while the diamicton ascended into the overlying sand. Deformation probably occurred as pore-water pressures increased when water from the melt of excess ice was generated faster than it could drain.

Undoubtedly, the most detailed descriptions of thermokarst involutions are to be found in the Pleistocene periglacial literature, especially from northern Germany (Eissmann, 1978, 1994). There, the occurrence of thick lignite beds meant that large ice lenses would have formed in poorly-drained lowland terrain during permafrost conditions. Pleistocene thermokarst structures are discussed in Part III.

8.4.2. Retrogressive-Thaw-Slumps and Debris-Flow Deposits

Retrogressive-thaw-slumps (Figure 8.5) are short-lived but rapidly developing features, the majority of which become stabilized within 30–50 summers after their initiation (French and Egginton, 1973). They are important in that they represent one of the most rapid erosive processes currently operating in present-day periglacial environments. Maximum rates of headwall retreat of between 6.0 m and 8.0 m may occur (French, 1974b; Lewkowicz, 1987; Mackay, 1966). They occur widely and wherever ice-rich permafrost is exposed to thaw. Typically, the result is a semi-circular steep headwall, 1–2 m in height, beneath which there is a gently sloping floor. As the headwall retreats by thaw, sediments are redeposited by falls, slides, debris flows, meltwater, and sub-slump melt-outs. The result is a matrix-supported diamicton, a sediment type that must be regarded as typical of thermokarst activity. At the same time, sheetfloods may exit from the slump floor during the summer to form alluvial fans, braided channels, and braid-plains composed of well-stratified mud aggregates, pebbles and/or sand. Sometimes, debris-flow deposits are laid down over these alluvial sediments to form a complex stratigraphic unit. Many slumps are polycyclic. Accordingly, thermokarst sediments are complex, showing stratification and heterogeneity in close association.

There are several trigger mechanisms responsible for retrogressive-thaw-slumps. For example, on Banks Island, some slumps occur on the steeper slopes of asymmetrical

Figure 8.5. A retrogressive-thaw-slump, eastern Banks Island, Canada. Rates of headwall retreat averaged 3.0 m per year between 1972 and 1974.

valleys and are triggered by lateral stream erosion undercutting the steeper bank. Others are found adjacent to lakes or the coast, and probably owe their initiation to wave action or ice push. Finally, some appear unrelated to any obvious trigger mechanism and appear to occur randomly. These may have developed initially from simple active-layer slope failures. Such failures are not uncommon in permafrost regions (Bird, 1967, pp. 217–220; Harris and Lewkowicz, 1993a, 2000; Lewkowicz, 1992b) and are associated with the development of high pore-water pressures after either prolonged summer rain or during years of exceptionally high winter snowfall and/or rapid and late thawing. The nature of thaw consolidation, the stability of thawing slopes, and active-layer detachments are discussed in more detail in Chapter 9.

8.4.3. Ice-Wedge Pseudomorphs and Composite-Wedge Casts

Casts, or pseudomorphs, represent the previous shape of a structure and can be formed in material other than that which formed the original structure. Ice-wedge pseudomorphs are wedges of secondary mineral infilling and are thermokarst structures because they result from the thaw of excess ice. Composite wedges are slightly different because they comprise variable proportions and distributions of both ice and clastic sediment (Gozdzik, 1973; Romanovskii, 1973). Thus, they may or may not contain excess ice, and only certain forms can also be regarded as thermokarst structures.

Ice-wedge pseudomorphs form when the ice in the wedge slowly melts, usually as permafrost degrades. As this happens, there is a general collapse of sediment into the void that is created. The majority of pseudomorphs described in the literature are from mid-latitude regions where permafrost no longer exists. These are discussed in detail in Chapter 12. However, ice-wedge pseudomorphs also exist in present-day permafrost environments (Murton and French, 1993b). The latter provide the best indication of the range of processes that are involved in their thaw-transformation. These include thermal erosion, collapse, subsidence, refreezing, loading, buoyancy, spreading, folding, and shearing. Typical pseudomorphs above partially-thawed ice wedges in northern Canada are illustrated in Figure 8.6. Noteworthy is the bent, U-shaped involution, 60–70 cm deep and ~20 cm wide,

(A)

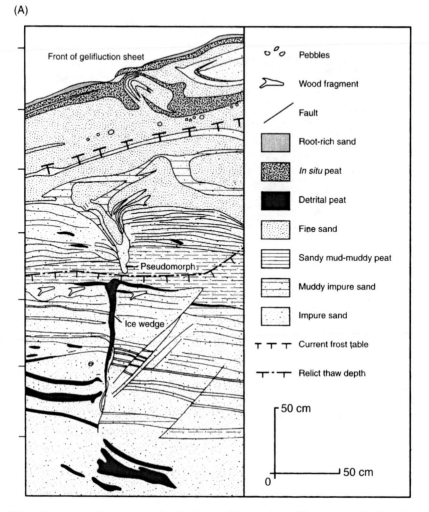

Figure 8.6. Pseudomorphs associated with thaw of ice wedges, Western Arctic Canada. (A) A pseudomorph above partially-thawed ice wedge. (B) A partially-thawed composite (sand) wedge comprises icy sand and sandy ice that is truncated along a planar, horizontal thaw contact. (C) An ice-wedge pseudomorph with a tunnel comprising the middle section. From Murton and French (1993b). Reproduced by permission of John Wiley & Sons Ltd.

that has formed above one wedge. This is a typical feature of the thaw-transformation from an ice wedge to an ice-wedge pseudomorph (Harry and Gozdzik, 1988).

A fully developed ice-wedge pseudomorph is characterized by the penetration of material into the fissure from above and from the sides, the downward inflection of layers if the enclosing sediments are stratified, and various systems of miniature faults and displacements. However, the majority of ice-wedge pseudomorphs are much more complex than this. It must be stressed that the melt of the wedge ice and the release of any excess ice (water) from the enclosing sediment may result in considerable deformation of the original structure. Thus, a shape far removed from the simple wedge-like form can result. A fuller discussion of ice-wedge pseudomorphs is given in Part III.

(B)

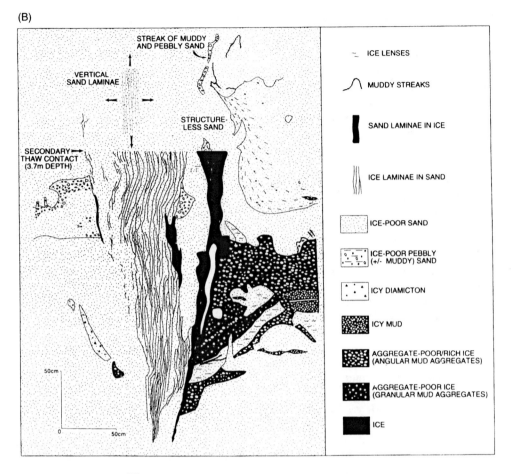

(C)

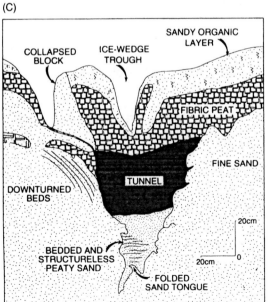

Figure 8.6. *Continued*

8.4.4. Ice, Silt, Sand, and Gravel Pseudomorphs

Ice wedges and composite wedges are often partially thawed because thermal erosion operates preferentially along ice-wedge troughs. As a result, piping and tunnels are common (see Figure 8.6C). Cast structures of partially-thawed wedges may reflect tunnel shape and the material that has either fallen from the roof or sides of the tunnel or been transported along the tunnel. If the initial wedge were of a composite nature, the infill may be a highly variable and chaotic mix of ice and mineral material.

Sometimes, pool ice ("thermokarst-cave ice") replaces wedge ice that has been eroded during a thermokarst, or underground erosion, episode. The result is the formation of an "ice cast," that is, an "ice pseudomorph" (Rozenbaum et al., 1978; Solomatin, 1986). Where sediment fills the void, formed either in ice or adjacent sediment, the result is a "soil pseudomorph" (Murton and French, 1993b, pp. 186–188). These are especially difficult to recognize because of the similarity of the fill material with enclosing material. Ice pseudomorphs are also difficult to recognize because they are often intersected by ice veins related to the continued growth of the wedge that they replace. Examples of various sand pseudomorphs that form above partially-thawed ice wedges are illustrated in Figure 8.6B and C. Ice pseudomorphs, usually composed of "pool" ice (see Chapter 7), are significant because the ice may extend laterally within an erosional niche beyond the extent of the original wedge, thereby producing an ice body that bears little or no relationship to the original wedge shape.

The manner by which ice wedges can be thaw-modified is illustrated schematically in Figure 8.7. The schematic diagram initially considers undisturbed permafrost that is penetrated by large ice wedges. The permafrost contains micro-lenticular cryostructures typical of syngenetic permafrost growth. Thaw-modification of the ice wedges could result in an ice-wedge pseudomorph (left), a partially-thawed (truncated) ice wedge (center), or an ice pseudomorph (thermokarst-cave ice) filling a tunnel void in the wedge (right). The extent of secondary (thaw-modified) deposits is indicated schematically. Expanded images are intended to represent micro-lenticular cryostructure in the permafrost, indicative of syngenetic growth and a reticulate-chaotic cryostructure adjacent to the ice pseudomorph indicative of epigenetic growth.

A fuller discussion of syngenetic and epigenetic permafrost and their associated cryostructures was given earlier in Chapter 7.

8.5. ICE-WEDGE THERMOKARST RELIEF

The surface morphology created by ice-wedge polygons is a distinctive characteristic of much lowland tundra terrain (Carson and Hussey, 1962; Hussey and Michelson, 1966; Mackay, 1963). However, the genetic relationship between this topography and thermokarst processes needs to be emphasized. Two types of polygon relief can be recognized. First, high-centered polygons develop by preferential fluvio-thermal erosion along ice wedges. Second, low-centered polygons result from the growth of ramparts adjacent to the thermal-contraction crack.

8.5.1. Low-Centered Polygons

Low-centered polygons are characteristic of poorly-drained tundra. They commonly possess a double raised rim, or rampart, often in excess of 50 cm in height, on either side of the ice-wedge trough. The depressed wet center contains sedges and grass. The raised

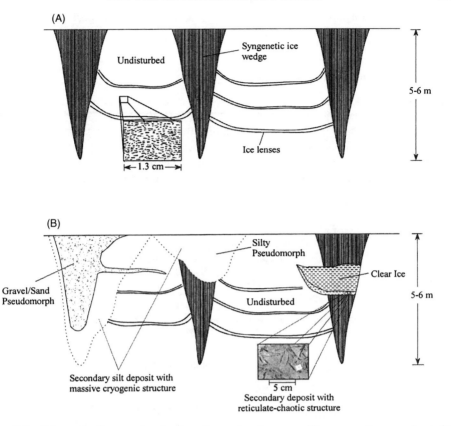

(A)

Undisturbed

Syngenetic ice wedge

Ice lenses

5-6 m

|←— 1.3 cm —→|

(B)

Gravel/Sand Pseudomorph

Silty Pseudomorph

Clear Ice

Undisturbed

5-6 m

Secondary silt deposit with massive cryogenic structure

5 cm

Secondary deposit with reticulate-chaotic structure

Figure 8.7. Schematic diagram showing how thermokarst may modify syngenetic permafrost. (A) Undisturbed syngenetic permafrost with large ice wedges. The permafrost is shown to contain micro-lenticular cryostructures. (B) Thermokarst modification. The three ice wedges have been thaw-modified to (i) an ice-wedge pseudomorph (left), (ii) a partially-thawed (truncated) ice wedge (centre), and (iii) an ice pseudomorph (thermokarst-cave ice) that fills a tunnel in the wedge (right). The extent of secondary (thaw-modified) deposits is indicated schematically. The expanded image shows the reticulate-chaotic cryostructure that would form adjacent to the ice pseudomorph. From Bray et al. (2006). Reproduced by permission of John Wiley & Sons Ltd.

rims are the result of lateral thermal expansion within the active layer moving material from the polygon center to the periphery, as explained earlier (see Chapter 6). Low-centered polygons are well suited, therefore, for the initiation of self-developing thermokarst. Ground subsidence in the polygon center is associated with standing water and the melt of pore and segregated ice. At the same time, standing water in the ice-wedge trough accelerates thawing along the line of the wedge. The most favored location for thaw is at the junction of two or more wedges, and small, deep pools of standing water may persist in such localities throughout the summer months. If integrated drainage occurs in such terrain, it can assume a "beaded" pattern.

An evolutionary sequence of low-centered polygon morphologies can be recognized, with the end phase being characterized by the formation of so-called "fortress," or thermokarst, polygons (Root, 1975) (Figure 8.8). The near-vertical walls of such extreme low-centered polygons may be as much as 1.0–1.5 m high. They have been little studied but they seem to form in poorly-drained areas that have recently experienced local lowering of the drainage base level.

8.5.2. High-Centered Polygons

The change from low-centered to high-centered polygons is usually brought about by an improvement in drainage, often associated with stream incision along the lines of the bordering ice wedges. On well-drained uplands, the preferential melt of ice wedges often leads to the formation of high-centered polygon terrain consisting of gentle convexities or mounds. In lowland areas, where low-centered polygons are more common, the raised rims progressively disappear as the trough broadens and the central depressed areas of

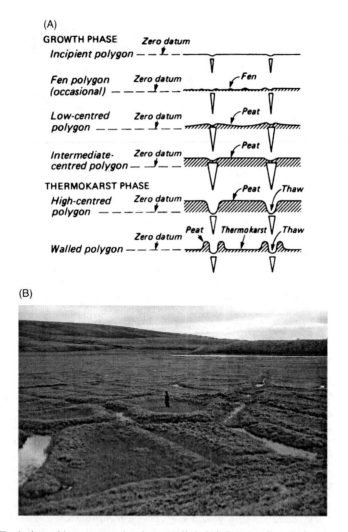

Figure 8.8. Evolution of low-centered polygon relief. (A) Stages of evolution towards formation of thermokarst ("fortress") polygons. (B) Tundra terrain showing "fortress" polygons, Sachs River valley, southern Banks Island, Canada.

sedge and tussock grass gradually disappear. An elevated peaty polygon finally emerges surrounded by troughs along the underlying ice wedges.

8.5.3. Badland Thermokarst Relief

Dramatic badland-type relief can occasionally develop as the result of thermal melting and erosion operating preferentially along large ice-wedge systems in unconsolidated sediments. For example, in an area drained by two streams on Eastern Banks Island a large amphitheatre-like hollow, over 0.5 km in diameter and several hectares in area (Figure 8.9), is being eroded within a thick sequence of glacigenic (lacustrine) silt that possesses 20–30% excess ice on average and contains ice wedges exceeding 6 m in depth (French, 1974b, pp. 791–793). Fluvial erosion, currently operating preferentially along the ice wedges, has led to the isolation of sharp conical mounds within the depression and a serrated edge to the amphitheatre. Mudflows and re-sedimentation go on within the depression and the mounds are degrading due to their exposure to thaw from all sides. This terrain appears similar to some of the thawing permafrost features first described by Ermolaev (1932b, pp. 7–10) from Lyakhovski Island in the Laptev Sea of northern Siberia.

Figure 8.9.   Oblique air view of badland thermokarst formed in ice-rich silty sand, eastern Banks Island. Thermal erosion is preferentially concentrated along ice wedges producing residual mounds over 8 m high. Remnants of the original polygon surface can be seen in the foreground and mid-distance.

8.6. THAW LAKES AND DEPRESSIONS

Thaw lakes are exceedingly common in arctic tundra lowlands. For example, in the Tuktoyaktuk Coastlands of the Western Canadian Arctic, over 10 000 lakes exist (Mackay, 1963), and many thousands more occur on Banks, Victoria, and the other Arctic islands. Likewise, thaw lakes are numerous on the Alaskan coastal plain (Hinkel et al., 2005; Sellman, et al., 1975), interior Yukon (Burn and Smith, 1988b), and on the alluvial lowlands of northern Eurasia and Siberia (Dostovalov and Kudryavtsev, 1967; Romanovskii et al., 2000; Tomirdiaro and Ryabchun, 1978). In fact, they are ubiquitous in flat-lying terrain wherever fine-grained ice-rich sediments are present (Figure 8.10A). Fluvial ter-races, outwash plains, coastal areas, and drained-lake basins are favored localities. Typi-cally, the outline shape of thaw lakes is smooth. Thermokarst ponds may also form in bedrock, provided it is poorly consolidated, fine-grained, and ice-rich (Figure 8.10B).

Intimately associated with thaw lakes are numerous shallow depressions that mark the existence of previous thaw lakes. Many depressions appear to overlap with each other, thereby intuitively suggesting a relatively rapid and progressive evolution of thaw lakes through initiation, expansion, capture, and eventual drainage.

In the North American literature, thaw lakes first attracted attention in northern Alaska (Hopkins, 1949; Wallace, 1948), where they are variously called tundra lakes, thaw lakes, thaw depressions, thermokarst lakes, and tundra ponds (Black, 1969).

Like many landforms, the initial cause of a thermokarst depression is problematic. It may be the quite random melting of ground ice, the subsidence of the ground, and the

(A)

(B)

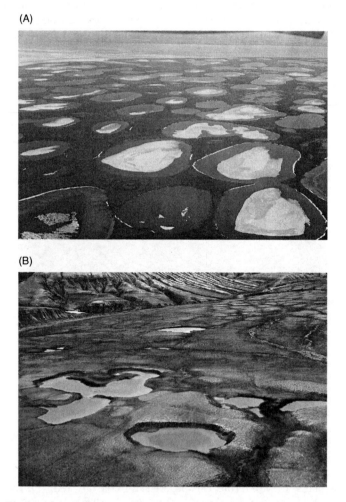

Figure 8.10. Thermokarst lakes in northern Canada. (A) Thaw lakes, Baillie Island, Bathurst Peninsula, western Arctic coastal plain. (B) Thermokarst ponds developed on ice-rich shale of Cretaceous age, Sabine Peninsula, Eastern Melville Island, High Arctic.

accumulation of water in the depression. For example, in the Seward Peninsula of Alaska, where permafrost is thin and discontinuous in nature, thaw depressions are thought to be the result of disruption of vegetation by frost heave, accelerated thaw beneath standing water pools occupying ice-wedge intersections, or accelerated thaw beneath pools in small streams. In some instances, where permafrost was particularly thin, the water "pierces," or thaws, the permafrost, allowing subterranean drainage to develop, thereby creating so-called "cave-in" lakes and thaw "sinks" (Wallace, 1948).

In areas of more continuous permafrost, many thermokarst lakes probably originate in poorly-drained low-centered ice-wedge polygons, as explained earlier. In such areas, the depressed polygon centers promote standing water bodies, beaded drainage develops at ice-wedge intersections, and surface runoff is concentrated along ice wedges. With time, the combination of fluvio-thermal erosion and thermokarst subsidence causes the coalescence of water bodies. Under this interpretation, thaw lakes represent a classic example of self-developing thermokarst.

8.6.1. Morphology

Thaw lakes vary considerably in size and shape. Some attain diameters of 1–2 km but the majority are much smaller, usually not exceeding 300 m in diameter. As a generalization, most thaw lakes are shallow and the majority possess smooth curving margins. Most of the larger lakes have a central pool, typically 2–4 m deep, surrounded by a littoral terrace, usually 0.3–0.5 m deep. Because the central pool is deeper than the thickness of winter ice cover, a talik usually underlies the larger lakes while permafrost is present beneath the littoral terraces. For example, on northern Richards Island, Western Arctic Canada, thaw lakes of average size 33 ha., occupy approximately 25% of the area. Annual mean lake-bottom temperatures in the central pool may exceed 3–4 °C (Burn, 2002, 2005), and geothermal modeling indicates that many sub-lake taliks penetrate the regional permafrost. By contrast, the annual mean lake-bottom temperature on the littoral terraces is approximately −2 °C to −3 °C.

Thermokarst basins constitute important depositional environments that allow inferences to be made as regards thermokarst processes and landscape modification. Based upon sections across thaw-lake basins that were truncated by coastal erosion, J. B. Murton (2001) recognizes several depositional stages in thermokarst lake-basin development. These are broadly similar to the sequence of thermokarst lake, alas, and talik formation in the Laptev sea area (Romanovskii et al, 2000, pp. 146–149). The principal thermokarst-lake-bottom sediment appears to comprise a diamicton that is the result of retrogressive-thaw-slumping around the lake margins and which transports upland sediment into the lake. This unit progrades across the basin floor through subaqueous debris flows while, at the same time, the lake edge recedes through retrogressive-thaw-slumping and sub-aerial debris flows. Eventually, as thaw-slumping stabilizes, sediment reduction into the basin allows reworking around the lake margins and suspension settling in the deeper parts.

8.6.2. Growth and Drainage

The drainage of lakes and the formation of thaw depressions are ongoing processes. For example, J. R. Mackay (1988b) estimates that about 65 lakes drained in the 1950–1986 period in the Tuktoyaktuk Peninsula. There, and elsewhere, many thaw lakes drain by the expansion of adjacent basins and fluvial "tapping" (i.e. erosion) along ice-wedge systems (Mackay, 1963, 1988b, 1992b; Walker, 1978). This can result in rapid, sometimes catastrophic, lake drainage and was the method chosen by Mackay to artificially induce the draining of Lake Illisarvik (see Chapter 6). In coastal areas of the western Arctic and northern Siberia, thaw lakes may drain when truncated by rapid coastal retreat associated with thermal abrasion and Holocene wave action (Mackay, 1986a; Romanovskii et al., 2000).

Their growth, expansion, and eventual drainage are mechanisms that are not fully understood. What is clear, however, is that thaw lakes evolve relatively rapidly through lateral bank erosion that may average as much as 15–25 cm per year. This makes thaw lakes an especially dynamic feature of the tundra landscape.

D. M. Hopkins (1949) was one of the first to outline a cyclic growth model that envisaged thaw lakes growing in size by coalescing with adjacent lakes and migrating across the tundra surface. With time, it was hypothesized that vegetation would grow upon the newly-exposed lake floor while the migrating lake would slowly infill with silt and organic matter. At the same time, permafrost aggradation and lake-bottom heave on the floor of the depression and mass-wasting upon the banks would lead to the ultimate obliteration of both lake and depression from the landscape. In subsequent years, the concept of a thaw-lake cycle, consisting of sequential stages of initiation, expansion, and drainage, became well established in the Alaskan literature (Billings and Peterson, 1980; Black, 1969; Britton, 1967; Tedrow, 1969). In support of this model, radiocarbon dates obtained from organic material within drained thaw-lake basins are used to argue that the growth, drainage, and rebirth of small thaw lakes and depressions is accomplished within a relatively short time span, approximately 2000–5000 years (Black, 1969; Tedrow, 1969).

While this model may be satisfactory for northern Alaska, where topography, climate, and geology are reasonably uniform, it is by no means universally applicable. It lacks a rigorous understanding of thermokarst processes and, specifically, of thermokarst lake-basin sedimentation. Furthermore, in spite of radiocarbon dating of organic material from within drained thaw-lake depressions, these dates merely indicate a minimal age for lake drainage and there is no evidence to indicate that the later stages of the cycle actually exist. It has also been suggested that the processes responsible for the hypothesized thaw-lake migration across the tundra are also the same as those responsible for thaw-lake orientation (Tedrow, 1969). However, no geomorphic or stratigraphic evidence for such migration can be found in many areas of oriented thaw lakes. It seems best to conclude, following J. R. Mackay (1963), that thaw lakes are quasi-equilibrium landscape elements.

The question of thaw-lake drainage is another interesting aspect of thaw-lakes. Many are either partially or completely drained. Two contrasting modes of lake drainage have been suggested: (1) gradual infilling and sedimentation, as described in the cyclic model above, and (2) catastrophic outflow following lake tapping or truncation by coastal retreat (Mackay, 1979a, p. 31; Walker, 1978; Weller and Derksen, 1979). It can be argued that the latter, a rapid permafrost-related process, is more appropriate to explain the transition from thaw lakes to drained thaw-lake depressions than gradual sedimentation, lake-bottom heave, and infill as suggested in the Alaskan thaw-lake cycle. In fact, tapping must be regarded as a ubiquitous process that occurs in poorly-drained tundra terrain wherever polygonal ice-wedge systems are well developed. Several examples of lake tapping can be seen in Figure 8.11 (see below).

8.6.3. Oriented Thaw Lakes

One of the more perplexing aspects of thaw lakes is that they are often elongate in shape, with a common and systematic orientation of their long axes (Figure 8.11). The thaw lakes near Point Barrow, northern Alaska, are exceptionally good examples of this phenomenon (Black and Barksdale, 1949; Carson and Hussey, 1962, 1963; Hinkel et al., 2005). Other oriented lakes are described from the Beaufort Sea coastal lowlands (Côté and Burn, 2002; Harry and French, 1983; Mackay, 1963), interior Yukon Territory (Price, 1968), other parts of Arctic Canada (Bird, 1967, pp. 212–216; Dunbar and Greenaway, 1956, pp. 132–134), and the alluvial coastal lowlands of northern Siberia (Tomirdiaro and Ryabchun, 1978).

A variety of oriented forms can be recognized. For example, in northern Alaska, the lakes are commonly elliptical or rectangular in shape, ranging in size from small ponds to large lakes 15 km long and 6 km wide (Black and Barksdale, 1949; Hinkel et al., 2005). In Alaska, the ratio of length to breadth varies from 1:1 to 1:5. Lake depths suggests two varieties: (a) with a shallow shelf surrounding a deeper central part, which may be 6–10 m deep, and (b) with a uniform saucer-shaped cross profile with depths of less than 2 m. In the Tuktoyaktuk Peninsula of the Mackenzie Delta, numerous oriented lakes are lemniscate, oval, triangular, or elliptical in shape (Mackay, 1963). On southwest Banks Island, oriented lakes assume a D-shape (Harry and French, 1983) while in western Baffin Island, on the Great Plain of Koukdjuak, many thaw lakes are clam-shaped with one straight edge and with length–breadth ratios varying between 1.5 and 2.5 to 1 (Bird, 1967, p. 215).

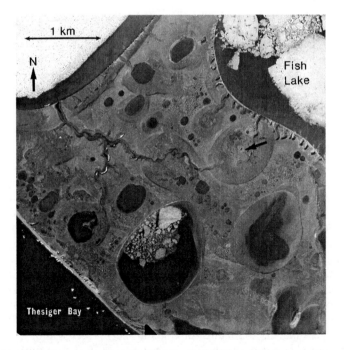

Figure 8.11. Vertical air photograph of part of the Sachs River lowlands, southern Banks Island, Canada, showing elliptical D-shape and preferred orientation of thaw lakes. There are also numerous drained lake basins, one of which contains a small pingo (arrow). Air Photograph A 15980-25, National Air Photo Library © 1996. Produced under licence from Her Majesty the Queen in Right of Canada, with permisssion from Natural Resources Canada.

The cause of these systematic orientations has attracted considerable controversy. In most cases, the long axes of the lakes are at right angles to prevailing wind. Therefore, some relationship between the two is clearly apparent but the exact mechanism remains obscure. One suggestion is that wind-induced littoral drift reaches a maximum at the corner of lakes and that eroded material is then distributed uniformly along the long-axis shorelines (Carson and Hussey, 1962). However, this explanation does not account for either the orientation of very small lakes where such circulation systems are not well developed or the lack of erosive currents at the ends until after the basin is elongate. In a recent attempt to explain the oriented thaw lakes of northern Alaska, J. D. Pelletier (2005) suggests that lake orientation is controlled by regional topographic aspect rather than predominant wind direction, and develops a model based upon thaw slumping of the downslope (lower) end of the lake basin. However, this hypothesis ignores the reality of the field situation and neglects the unique geomorphic history of the Barrow area, notably the Pleistocene-age sand sea, that formed linear dunes oriented parallel (E22N) to the paleowind (Hinkel, 2005). Much earlier, and in a more general fashion, J. R. Mackay (1963) concluded that oriented lakes represent an equilibrium condition in response to currently-operating processes. Other explanations are equally limiting. For example, H. M. French and D. G. Harry (1983) conclude simply that the strongest relationship exists between lake morphology and the storm wind regime during the summer period of open water conditions, and go no further in explanation, thus implicitly supporting earlier ideas of C. E. Carson and K. M. Hussey (1962). Clearly, it is difficult to distinguish between cause and effect in the circulation pattern of oriented thaw lakes. The problem is further complicated by the fact that oriented lakes are not solely a feature of permafrost regions. Many are found in non-permafrost regions (Price, 1968). At the same time, it is clear that oriented lakes are forming today because they are sometimes found upon plains and terraces which have only recently emerged from beneath the sea.

The oriented nature of thaw lakes remains a perplexing problem in periglacial geomorphology. More detailed studies are required concerning wave and current effects upon the thaw and transportation of sediment within the lake.

8.7. THERMOKARST-AFFECTED TERRAIN

Although thermokarst activity is especially favored in areas underlain by ice-rich sediments or bedrock, it is not a major characteristic of all periglacial environments. In fact, extensive areas of northern Canada, northern Scandinavia, and central and southern Siberia are underlain by highly resistant and ancient rocks associated with major tectonic plates. In other areas, such as the montane uplands and plateaus of central Asia, and at high elevations in the major mountain ranges, ground ice is limited in occurrence. In all these areas, thermokarst activity is conspicuous by its relative absence, being restricted to glacigenic materials, to localized ice-rich organic, lacustrine, or marine deposits, and to the numerous poorly-drained depressions within the largely bedrock-controlled topography. For example, in northern Québec, thermokarst activity is limited to the decay of ice wedges in emergent marine sediments, piping and thermal erosion in localized areas of low-centered ice-wedge polygons, and the degradation of peaty permafrost plateaus and mounds (Allard et al., 1996; Kasper and Allard, 2001; Seppälä, 1997; Seppälä et al., 1988; Séguin and Allard, 1984). Likewise, in northern Manitoba and northern Saskatchewan, thermokarst activity is only associated with site-specific features such as palsas and peat plateaus (Thie, 1974; Zoltai, 1971).

Regional-scale thermokarst activity is best developed in lowland environments underlain by ice-rich unconsolidated sediments or soft bedrock, and which often have extended histories of cold, non-glacial, conditions. Two such environments are described below.

8.7.1. The Lowlands of Central and Northern Siberia

Highly distinctive thermokarst terrain occurs in Siberia, where thermokarst phenomena are thought to develop in a predictable and sequential fashion. The type area is the terraces of the Lena and Aldan rivers of Central Yakutia (Soloviev, 1973a, b), but similar terrain occurs in much of the coastal lowlands of the Laptev and Kara Seas (Romanovskii et al., 2000).

As with almost any type area, a number of regional site-specific factors favor the distinctive nature of the thermokarst topography in Central Yakutia. First, the terraces are constructed from thick sequences of fine-grained alluvial sediments in which segregated ice constitutes up to 50–80% by volume. In addition, large syngenetic ice wedges, reported to exceed 50–60 m in vertical extent in some areas, underlie between 30% and 60% of the terrace surfaces. In terms of ground ice, therefore, the area is exceptionally well-suited for thermokarst development. Second, the geomorphic history which enabled such conditions to develop is uncommon. The central Yakutian lowland remained unglaciated for much, if not all, of the Quaternary and acted as a stable aggradational region in which alluvial sediments were deposited under cold-climate conditions over a long period of time. Third, the present climate is continental in nature; for example, the annual air temperature range at Yakutsk is 62 °C and summer temperatures commonly exceed 30 °C (see Chapter 3). As a result, the typical active-layer thickness may exceed 1.5–2.0 m. In general terms, therefore, one must regard the thermokarst terrain of central Yakutia as unusual, and it is unlikely that exactly similar conditions will be found elsewhere.

The range of thermokarst relief features in central Yakutia, initially described by P. A. Soloviev (1962), was subsequently expanded into a progressive (cyclic) sequence of landscape evolution (Czudek and Demek, 1970; Katasonov and Ivanov, 1973; Popov et al., 1966; Shumskii and Vtyurin, 1966; Soloviev, 1973a, b). It is summarized in Figure 8.12.

The first stage follows upon an increase in annual thaw depth that leads to preferential thaw-subsidence along large ice-wedge polygon systems. Continued ground subsidence leads to slumping and disruption of the vegetation cover and to the isolation of distinct conical mounds. These are termed "baydjarakhii," a local Yakut term used to describe silty or peaty mounds, and are commonly between 3 m and 4 m high and between 3 m and 15 m wide. From the air, baydjarakhii resemble a cobblestone surface, distributed in a checkerboard or rectilinear pattern. Similar features occur in the Tamyr Peninsula, near Dickson, and in many other parts of Northern Siberia. In the discontinuous permafrost zone of southern Siberia, where the active layer is thicker, and in areas where ice wedges have been consumed (thawed) by thermokarst-induced subsidence, baydjarakhii are more subdued in form and locally referred to as "graveyard mounds."

The second stage is characterized by the progressive collapse and decay of baydjarakhii as depressions develop in the center of the thermokarst terrain, often with a central hollow or sinkhole. With the linkage of these sinkholes, continuous depressions form with steep slopes and uneven bottoms. This hummocky depressed topography is locally called a "dujoda."

By stage three, a distinct depression with steep sides and a flat bottom has developed through the continued collapse of baydjarakhii on the side of the dujoda. This depression is locally called an "alas," a Yakut term that denotes a circular or oval depression with

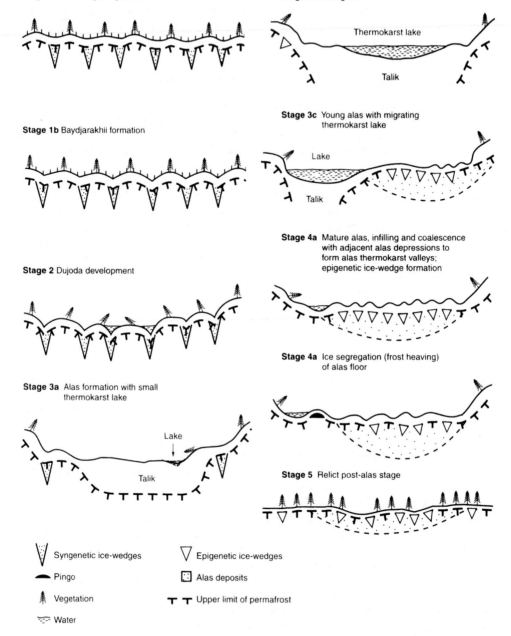

Stage 1a Ice-wedge degradation

Stage 1b Baydjarakhii formation

Stage 2 Dujoda development

Stage 3a Alas formation with small thermokarst lake

Stage 3b Young alas

Stage 3c Young alas with migrating thermokarst lake

Stage 4a Mature alas, infilling and coalescence with adjacent alas depressions to form alas thermokarst valleys; epigenetic ice-wedge formation

Stage 4a Ice segregation (frost heaving) of alas floor

Stage 5 Relict post-alas stage

Syngenetic ice-wedges

Pingo

Vegetation

Water

Epigenetic ice-wedges

Alas deposits

Upper limit of permafrost

Figure 8.12. Sequence of development of alas thermokarst relief in central Yakutia, according to Soloviev (1973b).

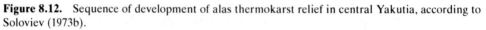

steep sides and a flat floor. In contrast to the surrounding boreal forest or taiga, an alas has no trees and usually contains a small thermokarst (thaw) lake. The latter, upon obtaining the minimum depth at which the bottom does not freeze in winter, promotes the formation of an unfrozen zone or talik. Thus, as thaw-consolidation continues, the lake progressively deepens and an open talik eventually develops.

In the fourth stage, alas lakes disappear, either by infilling with thermokarst (mass-wasted) sediments derived from the sides of the expanding alas or by drainage tapping towards an adjacent alas floor or stream channel. From this stage onwards, permafrost aggradation and lake-bottom heave occur and young epigenetic ice-wedge systems begin to form. Sometimes, perennial-frost mounds or pingos form. These are locally termed "bulgannyakhs." The larger of these are usually of the open (hydraulic) type since the unfrozen zone beneath the alas floor will often penetrate the permafrost. However, smaller bulgannyakhs may be hydrostatic (closed) in nature. The end result of this sequence of alas relief is the development of a depression with gentle slopes and an undulating floor.

In central Siberia, alas formation has greatly modified the lowland areas. In one region near to Yakutsk nearly 40% of the initial land surface has been destroyed by alas formation. Their coalescence may lead to the formation of depressions often in excess of $25\,km^2$. Complex thermokarst valleys also develop. These are irregular in plan and consist of wide sections (alas depressions) separated by narrow sections that cut through the intervening watersheds. Other characteristics of these valleys are right-angle turns, blind spurs, and a general misfit relationship with overall topography and drainage.

The rate of alas formation varies considerably since there are reports that some have developed within historic time, while others are obviously very old features. S. P. Kachurin (1962) regards the beginning of thermokarst development in Siberia as coinciding with the early Holocene warm period. However, many of the thermokarst features in the vicinity of Yakutsk are not active today and probably only 10% of the terrain is currently undergoing thermokarst modification.

Similar thermokarst terrain occurs on a regional scale in the "ice-complex" sediments that formed during the Late-Pleistocene regression (marine-isotope stages 3–5) on the drained Laptev Sea shelf and coastal lowlands of northern Siberia (Romanovskii et al., 2000). More than 50% of the area is, or has been, influenced by thermally-induced subsidence (Grosse et al., 2005). During the Late Pleistocene and Early Holocene, thermokarst processes began to destroy the ice-complex. Then, in the Middle and Late Holocene, and continuing today, thermokarst lakes and alas basins have been truncated by coastal erosion. Rising sea level has transformed the coastline into one of shallow headlands, bays, and lagoons. A fuller discussion of the role of thermokarst in the evolution of coastal topography, and the roles of thermal abrasion and coastal processes, is given in Chapter 11.

8.7.2. The Western North American Arctic

Most areas of arctic North America did not experience the same protracted periods of non-glacial cold-climate conditions that characterized central Siberia during the Quaternary. As a consequence, it is only in parts of the western North American arctic that one finds similar regional thermokarst terrain to that previously described from Yakutia. Here, certain aspects of Siberian alas thermokarst relief find North American analogues.

In central Alaska, W. A. Rockie (1942) first drew attention to the settling of ground caused by the thawing of ice wedges in recently cleared field systems near Fairbanks. T. L. Péwé (1954) and others subsequently referred to these hummocks as "thermokarst mounds." These are clearly analogous to the baydjarakhii or "graveyard mounds" of Yakutia. The thaw-sinks, funnels, and "cave-in" lakes reported from the Seward Peninsula and other areas (Hopkins, 1949; Péwé, 1948; Wallace, 1948) also find a clear counterpart in the "dujoda" stage. However, the alas depression, with bulgannyakh or pingo growth within it, does not occur. The closest comparable features are the shallow thermokarst depressions and lakes of the Mackenzie Delta region, in which closed (hydrostatic) system pingos form. However, few of these depressions have the same striking relief as the Yakutian depressions. Moreover, they are primarily the result of lateral growth rather than subsidence. Probably the most analogous terrain to the alas thermokarst relief is the ice-cored topography described from parts of the Canadian arctic coastal plain (Mackay, 1963; Rampton, 1974). For example, in the Tuktoyaktuk Peninsula, thermokarst development along small creeks has produced "macro-beaded" drainage systems where a series of depressions, partly lake-filled, are connected to each other by narrower sections or shallow interfluves.

Much of the lowland terrain of the Yukon Coastal Plain and Pleistocene Mackenzie Delta regions are essentially thermokarstic in nature and similar to that of the Laptev Sea coastal lowlands previously described. For example, extensive areas of ice-cored terrain near Sabine Point, Northern Yukon, are regarded as being remnants of an initial upland surface, greatly modified by thermokarst processes (Harry et al., 1988). The bodies of massive ice and icy sediments formed as permafrost aggraded following ice retreat from the maximum Late-Wisconsinan (Buckland) glacial limit. Then, multiple episodes of regional, climatically-induced thermokarst occurred between 14 000 and 8000 years ago that resulted in thaw settlement and surface subsidence in a series of coalescent and polycyclic thaw-lake basins. Thaw mobilization of sediment resulted in the transport of material from upland areas into the basins and to the progressive isolation of rolling ice-cored uplands areas draped with a variable thickness of surficial materials of thermokarst origin.

8.8. HUMAN-INDUCED THERMOKARST

In spite of our understanding of thermokarst processes in permafrost environments, numerous examples of human-induced thermokarst continue to occur today.

Russia, by virtue of its long history of northern exploration and settlement, has by far the greatest experience in this respect. As early as 1925, experiments were being undertaken to determine the effects of vegetation changes on the underlying permafrost, brought about either by deforestation or plowing (Tikhomirov, 1959; Tyrtikov, 1964) (Table 8.2). In Alaska, similar experimental studies were subsequently undertaken following the colonization and settlement of the 1920–1940 period. For example, one study in the Fairbanks region involved the cutting and/or stripping of surface vegetation by the US Army Corps of Engineers in 1946. In the stripped area, the active-layer thickness increased from 1.0 m to more than 3.0 m over a ten-year period. Subsequent studies emphasized the thermal role played by the surface organic layer and/or forest cover (e.g. Babb and Bliss, 1974; Brown et al., 1969; Haugen and Brown, 1970; Kallio and Reiger, 1969).

8.8.1. Causes

Without doubt, the most common cause of human-induced thermokarst is the clearance of vegetation for agricultural or construction purposes. However, it must be emphasized that even small disturbances to the ground surface may be sufficient to induce thermokarst activity. J. R. Mackay (1970), for example, describes how an Eskimo dog in the Mackenzie Delta was tied to a stake with a 1.5 m long chain. In the ten days of tether, the animal

Table 8.2.   Some examples of the effects of human-induced vegetational changes upon permafrost conditions in central Siberia, according to P. I. Koloskov (1925).

(A) Increase in average July soil temperature at 40 cm depth after ploughing, Yenesie region

Soil	Previous Vegetation Cover	Temperature Increase (°C)
Semi-bog	Forest	+14
Peat-bog	Grass	+12
Semi-bog	Grass	+9
Light sod, gravelly	Grass	+3

(B) Soil temperature changes as a result of deforestation, Amur Province

Depth of Measurement (m)	Thickness of Snow (cm)	Soil Temperature Changes within one Year (°C)
0.2	20	+0.5
	10	−1.6
0.3	20	+0.7
	10	−1.2
0.4	20	+0.6
	10	−0.6

Source: quoted in Tyrtikov (1964).

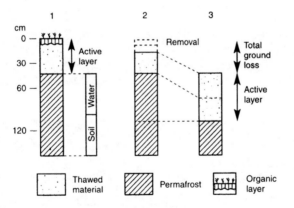

Figure 8.13.   Diagram illustrating how terrain disturbance of ice-rich soil can lead to thermokarst subsidence. From Mackay (1970).

trampled and destroyed the tundra vegetation of that area. Within two years, the site had subsided like a pie dish by a depth of 18–23 cm and the active layer thickness had increased by more than 10 cm within the depression. Figure 8.13 illustrates how the removal of the surface organic layer from an area underlain by ice-rich permafrost (composed of 50% excess ice (water) and 50% saturated soil if it were to thaw) will lead to both ground subsidence and thickening of the active layer.

The ice-rich zone in the transient layer at the top of permafrost (see Chapter 5) is the first to be affected. Typically, an irregular hummocky relief, similar to baydjarakhii or graveyard mounds, develops (Figure 8.14A, B). More widespread clearance, usually for agricultural purposes, leads to the formation of large thermokarst (alas) depressions. For example, one depression at Maya village, central Yakutia, is approximately 5–8 m deep and between 200 m and 300 m in diameter. It formed in historic times (the last 200 years) following deforestation associated with establishment of the agricultural settlement (Soloviev, 1973a). Similar human-induced thermokarst terrain has been described from the Fairbanks region of central Alaska, where extensive areas were cleared for agricultural purposes in the 1930s (Rockie, 1942). The following years saw the formation of mounds and depressions in the fields, the mounds varying from 3 m to 15 m in diameter and up to 2–4 m in height. Eventually many fields had to be abandoned because it was impossible to use mechanized equipment (Péwé, 1954).

(A)

(B)

Figure 8.14. Human-induced thermokarst. (A) In borrow pits adjacent to the Sachs Harbour airstrip, southern Banks Island, Canada, as seen in 1973, approximately 15 years after initial disturbance. (B) Baydjarakhii adjacent to the Maya-Abalakh road, Central Yakutia, Siberia, as seen in July 1973. The topography developed following disturbance in 1966.

A specific cause of thermokarst activity relates to the movement of vehicles over permafrost terrain. If this occurs in summer, when the thawed surface (active layer) is soft and wet, vegetation can be destroyed and trenching and rutting can result. Some of the worst examples of this sort of activity occurred on the Alaskan North Slope in the old US Naval Petroleum Reserve No. 4 during the late 1940s and early 1950s, when uncontrolled movement of tracked vehicles in summer led to widespread disruption of the surface organic layer. Subsequent thaw subsidence and thermal erosion along vehicle tracks formed trenches as much as 1 m deep and between 3 m and 5 m wide (Figure 8.15A). In Canada, a similar error was made when a summer seismic exploration program was authorized in the Mackenzie Delta in the late 1960s. Approximately 300 km of seismic line were bulldozed and long strips of vegetation and soil, approximately 4.2 m wide and 0.25 m thick, were removed. These lines were subsequently transformed into prominent trenches and canals over much of their length (Kerfoot, 1974). Another example of vehicle track disturbance occurred in 1970 on the Sabine Peninsula of Eastern Melville Island (Figure 8.15B). At that time a blow-out occurred at a wildcat well and vehicles were moved, of necessity, across tundra in summer. Sensitive lowlands underlain by soft ice-rich shale were crossed, and substantial and dramatic trenching occurred.

8.8.2. Case Studies

Two case histories from Banks Island, Arctic Canada, illustrate the nature and progression of human-induced thermokarst and provide insight into the time constraints involved. They suggest that terrain stabilization only begins 10–15 years after initial disturbance and probably is not complete until 30–50 years have passed. In both cases, disturbance was associated with the removal of surface material for construction purposes.

In the first example, disturbed terrain was monitored over a four-year period following initiation disturbance in August 1973 at an exploratory well site on Banks Island (French, 1978). Disturbance was caused during site restoration when surface material was scraped from an adjacent gravel ridge and used to infill a waste-disposal pit. A total of approximately 3000 m^2 was disturbed. Two years later, a crude polygonal system of gullies had developed at the site, reflecting erosion along underlying ice wedges. After two further years, the area consisted of unstable and irregular hummocks interspersed with standing water bodies.

The second example is the human-induced thermokarst terrain which formed adjacent to the airstrip, constructed during the summers of 1959–1962, at Sachs Harbour (French, 1975a). In order to grade the proposed strip, thawed material was removed each summer from adjacent terrain and transported, via access ramps, to the site. In all, a total of 50000 m^2 was disturbed and as much as 2.0 m of material removed in places. When first examined in the summer of 1972, the borrow pits portrayed actively subsiding thermokarst mound topography (see Figure 8.14A). By 1998, the terrain showed signs of stabilization, with vegetation beginning to re-establish.

Efforts to reduce terrain damage due to human-induced thermokarst activity focus upon regulatory procedures and geotechnical methods aimed at minimizing surface disturbance and preserving the thermal equilibrium of the permafrost. Canada was one of the first countries to implement such procedures through the Territorial Arctic Land Use Act and Regulations of the Federal Government in the early 1970s. Procedures include the introduction of vehicles with low-pressure tires, the restriction of heavy equipment movement to winter months, the implementation of terrain and biophysical mapping

(A)

(B)

Figure 8.15. Terrain disturbance associated with vehicle movement on tundra. (A) Old vehicle track, probably made in the early 1950s in the United States Navy Petroleum Reserve No. 4 (now NPR-4), northern Alaska, initiated thermokarst subsidence and became a canal, August 1977. (B) Gully erosion occurred along a vehicle track made in the summer of 1970 near the site of the Drake Point blow-out, Sabine Peninsula, Melville Island, NWT, Canada. The terrain is underlain by ice-rich shale of the Christopher Formation (Cretaceous). The photograph was taken in August 1976.

programs and environmental impact assessments, the timing of winter-road use, exploratory drilling and seismic work, and the mining of aggregate sources. In Canada, the right-of-way associated with the Norman Wells pipeline is monitored regularly to ensure that terrain stability is maintained (Burgess and Harry, 1990). Similar environmental procedures are adopted in Alaska and Svalbard, but their late introduction in Russia meant that significant terrain damage occurred in the Yamal oil and gas fields throughout the 1980s and early 1990s.

Today, all aspects of resource exploration and development in permafrost regions are closely regulated in order, among other things, to minimize thermokarst and environmental damage. Further discussion of the geotechnical aspects of human settlement and economic activity in periglacial environments is given in Chapter 14.

ADVANCED READING

Carson, C. E., Hussey, K. M. (1962). The oriented lakes of Arctic Alaska. *Journal of Geology*, **70**, 417–439.

Czudek, T., Demek, J. (1970). Thermokarst in Siberia and its influence on the development of lowland relief. *Quaternary Research*, **1**, 103–120.

Murton J. B. (1996). Thermokarst-lake-basin sediments, Tuktoyaktuk Coastlands, western arctic Canada. *Sedimentology*, **43**, 737–760.

Murton, J. B. (2001). Thermokarst sediments and structures, Tuktoyaktuk Coastlands, western Arctic Canada. *Global and Planetary Change*, **28**, 175–192.

Romanovskii, N. N., Hubberten, H.-W., Gavrilov, A. V., Tumskoy, V. E., Tipenko, G. S., Grigoriev, M. N. (2000). Thermokarst and land–ocean interactions, Laptev Sea region, Russia. *Permafrost and Periglacial Processes*, **11**, 137–152.

DISCUSSION TOPICS

1. What are the most common causes of thermokarst?

2. What happens when ice-rich permafrost degrades?

3. Discuss the origin of oriented thaw lakes.

4. Explain why thermokarst processes are some of the most important processes fashioning certain periglacial landscapes.

9 Hillslope Processes and Slope Evolution

A variety of slope forms exist in periglacial environments and none is exclusively "periglacial" in nature. Their morphologies often resemble those found in temperate, warm, and hot, semi-arid regions. Slopes evolve under cold-climate conditions primarily by mass-wasting processes. Flow includes permafrost creep, soil (frost) creep, gelifluction, slopewash, debris-flow, and slush-flow. Slip involves active-layer failures and ground-ice slumping. Fall involves avalanche and rockfall. Superimposed upon these processes is control exerted by the presence of frozen or thawing substrate.

Slope evolution in periglacial environments probably involves a progressive and sequential reduction of relief. Limited evidence suggests this is accomplished by replacement from below with Richter denudation slopes being ultimately replaced by low-angle (cryo) pediments.

9.1. INTRODUCTION

Our understanding of slopes in cold-climate environments is limited. Although significant advances have been made in recent years, the link between slope form and process is still unclear. Moreover, as argued in Chapter 2, many periglacial landscapes are in disequilibrium with current cold-climate conditions. There is the inevitable problem, therefore, of distinguishing between the effects of past and present processes. There is no slope form, or slope assemblage, that may be regarded as uniquely "periglacial" in nature. A general statement, made more than 40 years ago, remains valid today, namely: *"the similarities of form in diverse climatic regions and the differences of form in similar climatic environments emphasise the need not of classification but of understanding the interrelation of climate, lithology, and process"* (Leopold et al., 1964, p. 383).

In this chapter, the various slope assemblages that are commonly encountered in periglacial environments are described. The influence of lithology is explicitly considered. The various mass-wasting processes that operate on cold-climate slopes are described. The peculiarities of thawing and frozen slopes are outlined. A final section contains speculation concerning the nature of periglacial slope evolution.

9.2. SLOPE MORPHOLOGY

A number of cold-climate slope-form assemblages, identified largely on the basis of the author's personal experience, occur in periglacial environments (Figure 9.1). These are described below.

The Periglacial Environment, Third Edition   Hugh M French
Copyright © 2007 John Wiley & Sons Ltd

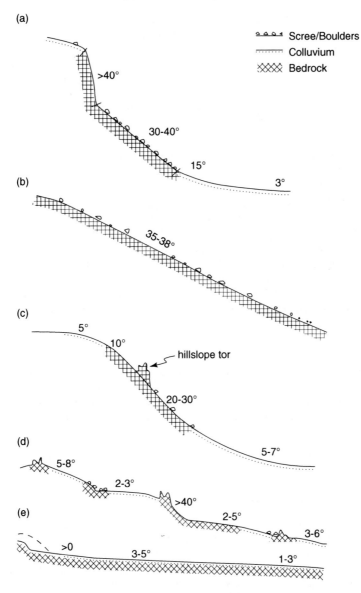

Figure 9.1. Typical slope forms found in present-day periglacial environments. (a) free-face (King model) slope, (b) rectilinear debris-mantled (Richter) slope, (c) convexo-concavo debris-mantled slope, (d) stepped (terraced) profile, and (e) pediment-like form.

9.2.1. The Free-Face Model

Probably the best-known slope form is one composed of a vertical or near-vertical rockwall below which is, first, a talus, or scree slope and, second, a footslope, or basal complex. This morphology is similar to the "standard" hillslope described by L. C. King (1953), in which there is a crest (waxing slope), a scarp (free-face), a debris slope (constant slope), and a pediment (waning slope).

In periglacial environments, the free face usually stands at angles in excess of 40°. The rock is subject to disintegration and there is cliff recession through rockfall. The talus slope beneath is at the repose angle of the coarser blocks, usually between 30° and 40°. The junction between the talus slope and the footslope (basal) complex may be abrupt or it may be a smooth concave profile. The upper part of the footslope varies in angle between ~25° and 5–10°; the lower end has inclinations of between 2° and 5°. Typically, the footslope is characterized by a micro-relief of lobes and terraces.

This slope assemblage is well developed in regions that have experienced recent deglaciation, such as Svalbard and northern Scandinavia (André, 1993; Jahn, 1960, 1976; Rapp, 1960a, b). This slope assemblage can be regarded as an inherited, glacial form that is being progressively modified.

The cold-climate processes operating on this type of slope assemblage fall into three groups: (1) rockfall and debris-cone (apron) accumulations, the result of mechanical weathering of the free face, (2) debris flows, the result of melting snow and summer rain on the talus slope, and (3) slow mass wasting (solifluction and slopewash) on the lower slope, the result of near-surface soil saturation. In terms of rapidity of profile change, the most important process appears to be either debris-flow activity associated with unusually heavy summer rain (Jahn, 1976) or snow avalanches (André, 1993; Rapp, 1960b). Examples of debris-flow activity on talus slopes within this slope-form assemblage are illustrated in Figure 9.2.

The visual impression of great thickness of talus is misleading because, except for cones or fans, the boulder cover mantles a bedrock surface. Moreover, talus profiles are usually concave, and not rectilinear as often thought, with higher angles occurring towards the top of the slope. Talus is also coarsely stratified, the result of either frost-coated clasts sliding over each other (Hétu, 1995), or from relatively dry debris flows, (van Steijn et al., 1995). Talus movement itself may be locally initiated by the impact of falling rock, or when individual rock particles expand upon heating during periods of strong solar insolation.

9.2.2. Rectilinear Debris-Mantled Slopes

These slopes develop upon consolidated bedrock and occur in regions characterized by extreme cold combined with extreme aridity. Typically, they are at the angle of repose, varying from 25° to 35–38°, and are covered with a thin veneer of loose material (Figure 9.3A). They are a kind of Richter denudation slope (Young, 1972, p. 107) in which debris supply and debris removal are in some form of equilibrium.

This slope form has been described in the periglacial context almost exclusively from the ice-free areas of Antarctica (Augustinus and Selby, 1990; French and Guglielmin, 1999; Iwata, 1987; Selby, 1971b, 1974; Souchez, 1966, 1967a). M. J. Selby (1971b, 1974) concluded that slope debris is produced by retreat of exposed bedrock subject to salt weathering enhanced by frost action. The debris is then removed by wind action. The balance between weathering supply and removal results in rectilinear slopes at repose angles. An essentially similar explanation is proposed by Iwata (1987), who observed that these slopes occur most frequently on north-facing sides of bedrock ridges that are subject to more frequent freeze–thaw action. In addition, they are best developed in gneissic rocks possessing intensive joint systems but less well developed in adjacent granitic rocks where joint density is lower (see Figure 9.3B).

(A)

(B)

Figure 9.2. Debris-flow activity on talus slopes. (A) At Longyearbyen, Svalbard, slopes are formed in near-horizontal beds of Tertiary-age sandstone, siltstone, and shale. The free face (upper part of the slope) is developed within massive sandstone while the talus slope (lower part) is within shale and siltstone (Dallmann et al., 2001). (B) At Sleepy Mountain, northern Yukon Territory, Canada, the free face is developed in Cretaceous-age sandstone and quartzite that overlies softer shale and siltstone, covered by talus, which forms the lower concavity (See also Figure 2.5).

9.2.3. Convexo-Concavo Debris-Mantled Slopes

A third slope form is characterized by a relatively smooth profile with no abrupt breaks of slope. There is a continuous or near-continuous veneer of rock debris but no widely developed free face or bedrock outcrop. Maximum slope angles range from 10° to as high as 25–30°, depending upon lithology. In areas of variable rock resistance, bedrock outcrops may be present in the upper sections and indistinct mounds of coarser, less weathered debris may constitute the summits of rounded interfluves (Figure 9.4). In form, the slope profiles extend over the complete range of convexo-concavo forms, from dominantly convex to dominantly concave.

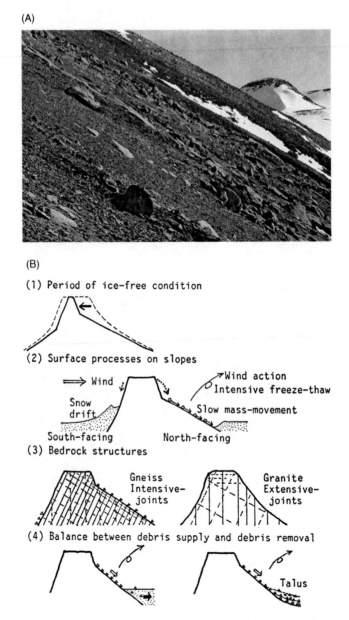

Figure 9.3. Rectilinear debris-mantled slopes, Antarctica. (A) Richter denudation slope formed in metagranites and high-grade metamorphic rocks, Terra Nova Bay, Northern Victoria Land. (B) Schematic illustration of debris-mantled rectilinear slopes in Sor Rondane Mountains, according to Iwata (1987) showing (1) time duration for slope formation, (2) surface processes acting on the slopes, (3) bedrock structures, and (4) assumed balance between debris supply and debris removal.

These slope forms have been reported from several localities (Büdel, 1960; Jahn, 1975, pp. 163–165; Pissart, 1966a; Rudberg, 1963). While the lack of detailed observations makes their interpretation difficult, several generalizations are possible.

First, convex and concave slopes usually form upon predominantly soft, unconsolidated bedrocks. For the most part, these form lowland, as opposed to upland, terrain. According

Figure 9.4. Smooth convexo-concavo debris mantled slopes with bedrock outcrop, in Cretaceous-age shale, Prince Patrick Island, NWT, Canada.

to J. Büdel (1960), the convexities and concavities reflect solifluction and slopewash processes while the steeper middle section is a backwearing debris slope subject to weathering and gravitational processes. The maximum slope angle, which may range from 10° to 25–30°, is essentially one of repose and reflects lithological factors. Thus, it can be regarded as a modified Richter slope. Second, many convexo-concavo slopes have an abrupt break of slope at the junction between the main part of the slope and the lower concavity. This phenomenon is similar to the "pediment junction" of hot arid regions. One explanation is that snow banks remain longest in depressions and at the foot of lee slopes (e.g. see Figure 2.4). Thus, in relation to the total slope profile, there is an increase in slopewash intensity at this location, which, in turn, leads to the progressive removal of fine material. This idea, sometimes termed "nivation," is discussed more fully later in this chapter. Third, it is not uncommon for more resistant bedrock to outcrop and disrupt the smooth convexo-concavo profile. These bedrock outcrops, surrounded by frost-derived debris, are commonly referred to as "tors." They occur upon a wide range of lithologies, including sandstone and dolerite (Derbyshire, 1972), gneiss (Dyke, 1976), shale (Figure 9.4), quartzite, and dolomite (see Figure 9.5A).

In view of their supposed cyclic (see below) and climatic significance (see Part III), it is useful to distinguish between tors that occur on slopes and those that form summit elevations. Usually, slope tors are surrounded by coarse debris and are structurally controlled. They form through enhanced disintegration of adjacent rock and corresponding slope retreat. However, other factors besides lithological variability may also be important. For example, in Southern Victoria Land, Antarctica, tors are developed predominantly upon west- and northwest-facing slopes (Derbyshire, 1972). This suggests they may also reflect relatively high solar radiation received by these slopes. However, differential mechanical weathering may not be the only process responsible. On Ellef Ringnes Island, in the Canadian High Arctic, tor-like features ("relief ruinifore") have developed in sandstone as the result of wind erosion (St-Onge, 1965).

9.2.4. Pediment-Like Slopes

Extensive low-angled surfaces are reported from many of the more continental and arid periglacial environments. These include the unglaciated regions of central Siberia, interior

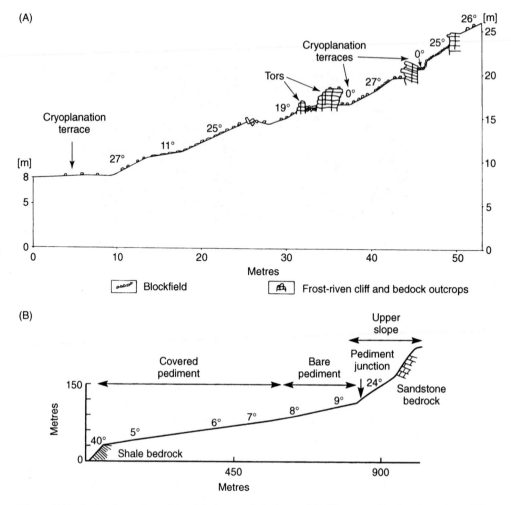

Figure 9.5. Irregular and straight debris-mantled slopes. (A) Slope profile showing "frostcliffs" (tors) and cryoplanation terraces developed upon dolomite of Cambrian age, Aldan River, Central Yakutia, Russia. From Czudek (1990). (B) Topographic profile of a pediment in the Barn Mountains, unglaciated Northern Yukon Territory, Canada. Reproduced with permission from French and Harry, 1992, Blackwell Publishing.

northern Yukon Territory, and central Alaska (Czudek and Demek, 1973; Demek, 1972a; French and Harry, 1992; Hughes, 1972b; Nelson, 1989; Priesnitz, 1981, 1988; Priesnitz and Schunke, 1983). In the periglacial literature, the surfaces are sometimes termed cryopediments, "gently-inclined erosion surfaces developed at the foot of valley sides or marginal slopes of geomorphological units developed by cryogenic processes in periglacial conditions" (Czudek and Demek, 1970, p. 101).

Cryopediments vary in angle from as much as 8–10° in their upper sections to as low as 1° in their lower sections (Figure 9.5B). Usually, only a thin veneer of surficial material mantles the slope (see Figure 2.6B). J. Dylik (1957) was one of the first to conclude that these forms are analogous to the pediments of tropical and sub-tropical regions.

Pediments in hot arid regions have been the subject of considerable controversy in the geologic and geomorphic literature; they are generally believed to be rock-cut transportation surfaces which truncate geologic boundaries. Some workers regard pediments as part of the semi-arid landscape evolution model proposed by King (1953) and earlier authors (e.g. Johnson, 1932). Others regard such surfaces as being essentially basal weathering surfaces, or "etchplains" (Büdel, 1970). The cause of planation is not easy to understand but most workers in hot semi-arid regions generally agree that pediments develop through the action of rills, gullying, rainwash, and sheetflooding coupled with backwearing of steep upper slopes (Leopold et al., 1964, pp. 496–498).

The traditional interpretation of pediments in cold-climates relies upon the assumed activity of cryogenic processes (Czudek and Demek, 1970, 1973; Priesnitz, 1981) whereby frost action extends the pediment upslope by a process of scarp retreat. However, in arid regions, such as central Siberia, interior Yukon Territory, and central Alaska, it seems prudent to downplay the role attributed to frost. A more reasonable assumption is that cold-climate processes act on these slopes within the constraints imposed by a pre-existing pediment morphology. One is tempted to believe that pediments are not typical cold-climate slope forms but are inherited elements in the landscape.

9.2.5. Stepped Profiles

Closely related to pediment-like surfaces are large terraces or steps carved in bedrock. They usually occupy higher elevations in the landscape (Figure 9.6; also see Figure 9.5A). Terms used to describe these features include "goletz" terraces, altiplanation terraces, nivation terraces, and equiplanation terraces. All are now generally referred to as cryoplanation terraces. Like cryopediments, these slope profiles occur mainly in the unglaciated regions of northern North America and northern Russia (Boch and Krasnov, 1943; Czudek, 1990; Demek, 1969a, b; Lamirande et al., 1999; Lauriol and Godbout, 1988; Lauriol et al., 1997; Reger and Péwé, 1976). Cryoplanation terraces have also been described from South America (Grosso and Corte, 1991). In Antarctica, they have not been explicitly recognized although the literature variously refers to "block terraces," "flat erosion surfaces," "pavements," "step-like topography," and "stepped slopes" (Hall, 1997b, p. 182; Matsuoka, 1995).

Figure 9.6. Bedrock (cryoplanation) terrace developed upon argillite and chert of Pre-Cambrian age, Buckland Hills, Barn Mountains, Northern Yukon Territory, Canada. The helicopter provides scale.

Cryoplanation terraces vary in form; they may be sickle-like, or elongate and relatively narrow in shape. Their dimensions also vary; the smallest may be less than 50 m in maximum dimensions while others exceed 400–600 m in length and 150–200 m in width. In central Siberia, some terraces are over 1 km in width and several kilometers in length, and some bevel upland summits (Demek, 1969a, p. 42). The height of the scarp that limits the upslope end of the terrace also varies. On gentle slopes the scarp height may be 1–2 m but, in areas with greater overall terrace dimensions, the scarp height may exceed 10–20 m. The angle of the terrace tread varies between 1° and 12°; usually the larger the tread and the lower the inclination of the original slope, the smaller the gradient.

Lithology must play an important role in the development of these stepped profiles. For example, on Ellef Ringnes Island, Canadian Arctic, D. A. St-Onge (1969) describes the effects of snow and the different "nivation benches" that develop in gabbro, sandstone, and shale bedrock. In areas of gabbro, the terraces are between 10 m and 15 m in width and the risers, with slopes of between 25° and 35°, are between 2 m and 5 m in height. The terraces form giant, near-horizontal steps and reflect the predominantly coarse boulders and relatively few fines which result from the disintegration of gabbro. Since the finer particles are quickly removed by snowmelt percolating through the boulders, the terrace is bounded by an apron of coarse angular debris. On adjacent sandstone the terraces are more subdued in form because sandstone weathers to silt, sand, and sandstone aggregates, all of which are relatively easily moved by wash. Thus, the terrace becomes an inclined surface of 6–8°. Finally, in shale and siltstone, a variety of features develop, ranging from large amphitheatre-like semi-circles to smaller hollows and ledges. These features reflect (i) the ease by which the soft shale is reduced to fine sand and silt, and (ii) the effectiveness of wash in removing such material. A second example is provided by K. Hall (1997b), who describes numerous small benches that occur on Alexander Island, in the humid Antarctic Peninsula. They are interpreted in the context of near-horizontally-bedded and extensively-jointed sedimentary rock that is being subject to dilation and thermal stress.

Cryoplanation is discussed further in the Chapter (pp. 244–246) and in Chapter 13 (pp. 332, 341).

9.3. MASS WASTING

Mass wasting is the term applied to the downslope movement of debris under the influence of gravity. For convenience, mass-wasting processes in cold environments can be divided into those that are slow and those that are fast. Slow mass-wasting processes are discussed under the general heading of solifluction, and then subdivided into frost creep and gelifluction. Rapid mass-wasting processes are discussed under the headings of active-layer failures, debris flows and avalanches, and rockfalls.

While mass wasting is not unique to cold climates, mass-wasting processes are especially effective under periglacial conditions. There are several reasons. First, frost action promotes rock disintegration and the resulting loose material is readily available for transport by mass-wasting processes. Second, diurnal and short-term freezing of the ground surface accelerates near-surface sediment movement. Third, the typically high moisture content of the thawed active layer favors gravity-induced downslope movement. Fourth, permafrost directly aids mass wasting by limiting the downward infiltration of water into the ground, thereby inducing high pore-water pressures in the near-surface. Fifth, the permafrost table acts as a water-lubricated slip plane for movement of thawed material. Finally, in many areas, the glacial legacy, consisting of over-steepened slopes and an abundance of loose glacial debris, is conducive to the active development of slopes.

9.4. SLOW MASS-WASTING PROCESSES

9.4.1. Solifluction

Solifluction is regarded as one of the most widespread processes of soil movement in periglacial environments. The term was first used by J. G. Andersson (1906, p. 95) to describe the "slow flowing from higher to lower ground of masses of waste saturated with water" that he observed in the cool, damp environment of the Falkland Islands. Since solifluction, so defined, is not necessarily confined to cold climates, the term "gelifluction" has been proposed to describe solifluction associated with frozen ground (Washburn, 1979).

Intimately associated with solifluction is frost creep. This is the ratchet-like downslope movement of particles as the result of the frost heaving of the ground and the subsequent settlement upon thawing, the heaving being predominantly normal to the slope and the settling more nearly vertical (Jahn, 1975; Washburn, 1979). In the Russian literature, the transliterated term "cryogenic deserption" is sometimes used for frost creep. When operating together, the two processes of gelifluction and frost creep constitute the movement that is generally termed solifluction in the modern sense.

There are three components to solifluction (Figure 9.7A): (i) potential frost creep (PFC) (the horizontal component of creep following heave), (ii) the horizontal component

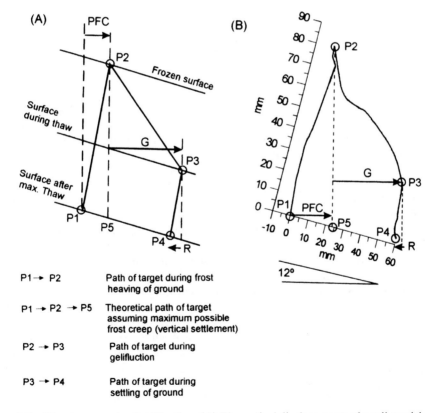

P1 → P2	Path of target during frost heaving of ground
P1 → P2 → P5	Theoretical path of target assuming maximum possible frost creep (vertical settlement)
P2 → P3	Path of target during gelifluction
P3 → P4	Path of target during settling of ground

Figure 9.7. The components of solifluction. (A) Theoretical displacements of a soil particle due to potential frost creep (PFC), gelifluction (G), and retrograde movement (R). (B) Interpretation of observed surface movement vector in a controlled large-scale laboratory simulation of gelifluction movement at CNRS, Caen, France. Data from Harris and Davies (2000).

Table 9.1. Some recorded rates of solifluction movement.

Locality	Reference	Gradient (degrees)	Rate (cm/year)
(A) Arctic:			
Spitsbergen	Jahn (1960)	3–4	1.0–3.0
Spitsbergen	Jahn (1961)	7–15	5.0–12.0
Svalbard	Akerman (1993)	2–25	
East Greenland	Washburn (1967)		0.9–3.7
Banks Island, NWT, Canada	French (1974a)	3	1.5–2.0
	Egginton and French (1985)	<10	0.6
(B) Sub-arctic:			
Kärkevagge, Sweden	Rapp (1960a)	15	4.0
Tarna area, Sweden	Rudberg (1962)	5	0.9–1.8
Norra Storfjell, Sweden	Rudberg (1964)	5	0.9–3.8
Okstindan, Norway	Harris (1972)	5–17	1.0–6.0
Garry Island, NWT, Canada	Mackay (1981a)	1–7	0.4–1.0
Ruby Range, YT, Canada	Price (1973)	14–18	0.6–3.5
(C) Alpine:			
French Alps	Pissart (1964)		1.0
Colorado Rockies	Benedict (1970)		0.4–4.3
Swiss Alps	Gamper (1983)		0.02–0.1

of gelifluction movement (G) (the downslope displacement in excess of maximum potential creep), and (iii) retrograde movement (R) (the horizontal movement due to resettlement perpendicular to the soil surface). Recent laboratory simulations add significantly to our understanding of the mechanisms involved in solifluction. For example, simulations conducted using frost-susceptible soils subject to one- and two-sided freezing over a number of freeze–thaw cycles, several lasting for as much as 30–60 days, closely mirror the theoretical components of solifluction (Figure 9.7B) (Harris and Davies, 2000; Harris et al., 1993, 1995, 1997, 2000).

Some of the most detailed field studies upon solifluction have been undertaken by N. Matsuoka (Matsuoka, 2001c; Matsuoka and Hirakawa, 2000; Matsuoka et al., 1997, 2003, 2005). Today, a large data base is available; some rates of solifluction movement published in the literature are summarized in Table 9.1.

The various types of frost heave and solifluction movement, and the typical velocity profile with depth that is associated with each, are summarized in Figures 9.8 and 9.9.

Under one-sided freezing, solifluction consists of only two components: (i) frost creep, usually associated with repeated diurnal frost throughout the year and/or needle ice, and (ii) gelifluction, occurring in the spring as the seasonally-frozen ground thaws from the surface downwards. Under two-sided freezing, solifluction comprises not only frost creep and gelifluction but also a movement which occurs in late summer when the thawed active layer is capable of sliding, en masse, across the lubricated slip plane provided by the ice-rich zone at the top of permafrost. The latter is termed "plug-like" flow (Mackay (1981a).

The relative importance of each component of movement varies depending upon site factors such as moisture availability or the ground temperature regime. For example, in

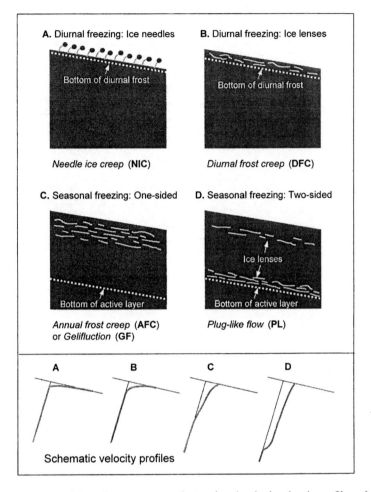

Figure 9.8. Summary of frost heave types and associated velocity-depth profiles of solifluction movement. From Matsuoka (2001c). Reprinted from Earth Science Reviews, with permission from Elsevier.

alpine regions with strong diurnal and seasonal temperature rhythms, frost creep and needle ice may be the dominant mass-wasting process for much of the year. On the other hand, observations from high-latitude environments indicate that either frost creep or gelifluction may dominate, not only in any given year but also in different sectors of the same slope depending upon moisture conditions. In areas where the base of the active layer contains icy lenses, an end-of-summer plug-flow movement may dominate.

9.4.2. Frost Creep

Frost creep is associated with either one-sided or two-sided freezing (see Figures 9.9 and 9.10). One-sided frost creep involves movement decreasing from the surface downwards and depends upon the frequency of freeze–thaw cycles, angle of slope, moisture available for heave, and frost susceptibility of the soil. Thus, if the slope angle is σ, the heave normal

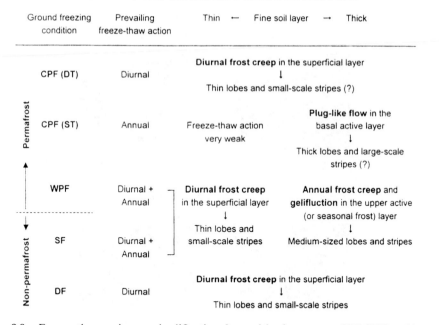

Figure 9.9. Freeze–thaw regimes and solifluction. Legend for frost types: CPF (DT), cold permafrost with diurnal thaw; CPF (ST), cold permafrost with seasonal thaw; WPF, warm permafrost; SF, seasonal frost; DF, diurnal frost. From Matsuoka (2001c). Reprinted from Earth Science Reviews, with permission from Elsevier.

to the ground surface is h, and soil settles vertically upon thaw, the potential frost creep (Δl) parallel to the surface is:

$$\Delta l = h \tan\sigma \qquad (9.1)$$

In the case of two-sided freezing, which is the situation when dealing with permafrost environments, frost creep may exceed the amount calculated from Equation (9.1). The growth of ice lenses at the top of permafrost during freeze-back means that the potential frost creep (Δl) varies from a maximum in the thickest part of the ice lens to zero at the periphery. Furthermore, towards the end of the thaw period, a residual ice lens in a soupy matrix may make a larger contribution to total movement than that of Δl due to vertical settlement. This causes plug-like flow, where hummocks slide over water-saturated sediment lying immediately above the permafrost table.

9.4.3. Gelifluction

Conditions suitable for gelifluction occur in areas where downward percolation of water is limited by frozen ground and where melt of segregated ice lenses provides excess water which reduces internal friction and cohesion in the soil. Locations of abundant moisture, such as below late-lying or perennial snow banks, are especially favored.

Gelifluction is essentially a process operating mainly during the thaw period. It is laminar in nature; movement decreases with depth in a linear fashion. Usually, movement is restricted to the uppermost 50 cm of the active layer.

If one assumes typical average surface movement rate of between 0.5 and 5.0 cm/year, and that movement in the upper 50 cm is, on average, one-quarter of this amount, it follows that gelifluction can transport between 6.0 and 60.0 cm^3 cm^{-1} year^{-1} of material (French, 1974a, 1976b). In plug-like flow the vertical velocity profile is convex downwards. Also, movement usually occurs in late summer when the active layer is at its thickest. Rates of plug flow appear similar to those of gelifluction. For example, J. R. Mackay (1981a) measured rates of movement that ranged between 0.2 and 1.0 cm/year, which indicate volumetric transport of between 7 and 52 cm^3 cm^{-1} year^{-1}. Small mudbursts and mudflows, sometimes observed in mid- and late summer on lower slopes (Egginton and French, 1985), are related to plug-like flow. They probably result from liquefaction associated with positive differential pore-water pressures that are released along shearing zones.

9.4.4. Solifluction Deposits and Phenomena

Solifluction deposits are heterogeneous, largely unstratified, sediments that have undergone transport and subsequent deposition by frost creep and gelifluction processes. They are commonly matrix-supported, the matrix consisting of fine-grained sediment that, typically, has low liquid limits and plasticity indices (Figure 9.10). A characteristic of solifluction deposits is that the larger clasts tend to become oriented with the longer axis parallel to the direction of movement (Harris, 1981, 1987). Referring specifically to solifluction deposits on Baffin Island, NWT, A. Cailleux and G. Taylor (1954) suggest that at least two-thirds of all elongate clasts in solifluction deposits lie parallel to the slope direction. Platy microfabrics are typical of solifluction deposits and reflect the previous existence of segregated ice lenses that promote localized slip during thaw (Harris, 1985; van Vliet-Lanoë et al., 1984). Where plug-like flow has occurred, soil pods may possess slicken-sided surfaces due to small-scale shears (Reanier and Ugolini, 1983).

Geomorphic features produced by solifluction include uniform sheets of locally-derived surficial materials and tongue-shaped lobes (Figure 9.11). Solifluction sheets are probably best developed where the absence of vegetation enables solifluction to operate uniformly. Tundra and forest-tundra favor more localized lobate movement. The correlation between

Figure 9.11.   A turf-banked solifluction lobe, Holman, western Victoria Island, NWT, Canada.

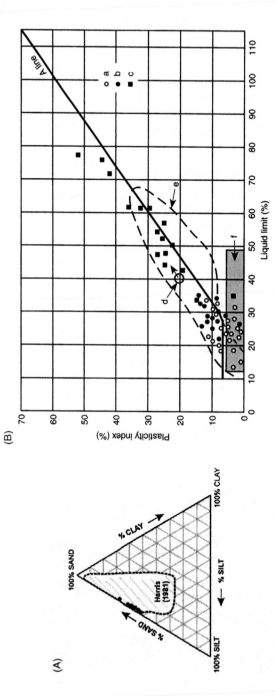

Figure 9.10. Sediment characteristics associated with solifluction activity. (A) Ternary plot of sediment matrix (<2 mm) showing the textural envelope for active solifluction identified by Harris (1981). From Kinnard and Lewkowicz (2005). (B) Typical plasticity values for solifluction sediments: (a) active solifluction sediments, Banks Island, Canada (French, 1974a), Greenland (Everett, 1967; Washburn, 1967), and Norway (Harris, 1977); (b) relict solifluction sediments, England (Harris and Wright, 1980; Mottershead, 1971); (c) relict planar slides in clay, England (Chandler, 1970a, b, Chandler et al., 1976); (d) minimum plasticity and liquid limit values, Garry Island, NWT, Canada (Mackay, 1981b); (e) range of values reported for near-surface sediments, District of Keewatin, NWT., Canada (Shilts, 1974); (f) range of values reported for solifluction lobes in the Swiss Alps (Furrer et al., 1971). From Harris (1987).

process and landform suggests that riser height of solifluction lobes is indicative of maximum depth of movement and prevailing freeze–thaw type (see Figure 9.9). For example, in tropical high mountains, where diurnal frost is limited to shallow but relatively rapid movements of the superficial 5–10 cm, thin stone-banked lobes are typical. In mid-latitudes (i.e. areas of annual frost), where either deep seasonal frost or warm permafrost occurs, solifluction affects a thicker soil layer and produces medium-sized lobes, often 0.3–0.5 m high at the riser. In colder regions of either high-latitude or higher elevation, the presence of permafrost results in two-sided freezing and plug-like flow that can affect a soil mass 60 cm or more in thickness. This can also produce well-pronounced lobate forms. Finally, in semi-arid polar deserts, thaw depths are shallow and movement is restricted to snow banks and other sites of localized soil humidity. As a result, riser height is often small, and solifluction sheets, rather than lobes, are better developed.

The site-specific controls over solifluction activity can be illustrated with two studies from the upland terrain of the southern Yukon Territory, Canada. In the Ruby Range, movement rates vary according to vegetation cover, aspect, and micro-climate (Price, 1973). Greatest rates (~2.4–2.7 cm year^{-1}) occurs on north- and east-facing slopes, where vegetation is poorly developed, while lower rates (~1.6 cm/year) occur on south- and east-facing slopes, where vegetation is well developed. Movement is least (~0.7 cm/year) on south- and west-facing slopes which, being exposed to prevailing winds, are drier than those of other orientations. A second study examined lobe movement in the Kluane Range (Kinnard and Lewkowicz, 2006). They observed that vegetated lobes progressively increase in size downslope. This happens either by the merging of adjacent lobes to form sheets or by an increase in riser height (Figure 9.12). It was inferred that lobe advance leads to a progressive steepening of the riser and its eventual rupture and collapse. Rebuilding of the lobe and the full cycle of development is estimated to take several hundred years.

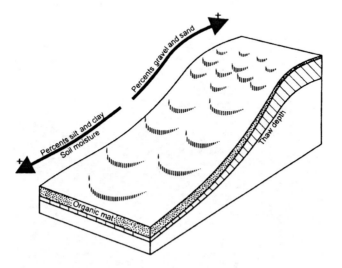

Figure 9.12.   Diagram illustrating typical downslope trends in soil moisture, sediment-size composition, and morphology of solifluction lobes, Kluane Range, Yukon Territory, Canada. The thickness of the organic mat and thaw depths are not drawn to scale. From Hugenholtz and Lewkowicz (2002) and Kinnard and Lewkowicz (2006). Reproduced by permission of John Wiley & Sons Ltd.

9.5. RAPID MASS WASTING

A number of high-magnitude, low-frequency, mass-wasting processes also fashion slopes in periglacial environments. Their geomorphic significance is best described in the context of their magnitude and frequency.

9.5.1. Active-Layer-Detachment Slides

Localized and small-scale slope failures that are confined to the active layer are common in regions underlain by perennially-frozen and ice-rich unconsolidated sediments. They are usually restricted to middle or upper slopes, with the entrance and exit of the failure plane well within the slope (Figure 9.13). These rapid mass movements are termed active-layer-detachment slides (Harris and Lewkowicz, 1993a, b, 2000; Lewkowicz, 1992). Failure involves the thawed or thawing active layer, together with any overlying vegetation mat, becoming detached from underlying frozen material.

The cause of failure is associated with a reduction in shear strength through time. This results from the progressive weathering of soil and a gradual increase in basal active-layer ice content (see Chapters 5 and 7). Because many active-layer-detachment failures occur in mid- and late summer, soil shearing during annual gelifluction movement is often the cause (Harris and Lewkowicz, 2000). The permafrost table acts as a lubricated slip plane and controls the depth of the failure plane. Thus, active-layer detachments are usually shallow, the slump scar or hollow often being no more than 1 m deep. In particularly fine-grained sediments possessing high liquid limits, failure may take the form of mudflows. Where icy sediments are exposed in the permafrost, regressive- thaw-slumping may follow (see Chapter 8).

In terrain underlain by ice-rich and poorly-lithified or unconsolidated sediments, active-layer detachments are especially frequent (Chatwin and Rutter, 1978; French, 1988, pp. 167–168; Harris and Lewkowicz, 1993a, b, 2000; Hodgson, 1982; Hughes et al., 1973;

Figure 9.13. Active-layer detachment, Ellesmere Island, Arctic Canada. This slide occurred approximately 24 hours prior to the photograph as a result of rapid thaw at the base of the active layer caused by warm temperatures and nearly continuous sunshine. Photo taken August 2005. Photograph and caption information are courtesy of Professor A. G. Lewkowicz.

Table 9.2. Detachment failure volumes, components of mass transfer, and long-term rates of mass transfer, Fosheim Peninsula, Ellesmere Island, Nunavut, Canada, 1994–1995.

	Year of study	Black Top Creek (3.6 km²)	Hot Weather Creek (0.82 km²)	Big Slide Creek (7.6 km²)
Detachment failure	1988	24–40	6.2–13.0	0.2–0.4
volume (10^3 m³ km⁻²)	1998	0.5–0.8	0.2–0.5	1.4–3.2
Downslope mass transfer	1988	3600–3200	240–280	8.2–8.9
(10^3 Mg m km⁻²)	1998	8.5–6.9	3.6–3.8	190–240
Long-term rates of downslope mass transfer (10^3 Mg m km⁻²) a⁻¹		40–84	5.9–12.0	17–38

Source: Lewkowicz and Harris (2005). Reproduced by permission of John Wiley & Sons Ltd.

Lewkowicz, 1990, 1992; Stangl et al., 1982). Often, failure is initiated during periods of rapid spring thaw and/or following periods of summer precipitation. Failures may also occur following the destruction of surface vegetation by forest fire (Zoltai and Pettapiece, 1973) or following human-induced terrain disturbance (Heginbottom, 1973).

The magnitude and frequency of active-layer-detachment failures occurring in relatively warm, discontinuous permafrost, and those occurring over cold, continuous permafrost terrain, has been analyzed by A. G. Lewkowicz and C. Harris (2005). They conclude that rates of geomorphic work, when considered over a time period of 100–200 years, are of the same order of magnitude. However, pre-conditioning of the active layer appears important because it was found that rapid thaw did not necessarily initiate activity. Most failures involved elements of both compression and translation of the soil mass. Table 9.2 summarizes some of the volumes and components of mass transfer that were calculated and from which long-term rates of unit mass transfer were approximated. These data illustrate not only the considerable amounts of material that can be moved by this process but also its extreme variability.

9.5.2. Debris Flows, Slushflows, and Avalanches

In environments characterized by abundant snowfall, rapid mass movement may occur through snow and debris avalanches. Such activity is especially favored in areas of glacially-oversteepened slopes. Most avalanches start as snow avalanches which then pick up varying amounts of rock debris en route, ultimately becoming debris avalanches or slides. These are sometimes termed "dirty avalanches" (Rapp, 1960a, p. 127) or "mixed avalanches" (Washburn, 1979, p. 193). More liquid forms such as slush avalanches and mudflows occur where excessively wet (ripe) snow is subject to rapid thaw.

Debris flows (see Figure 9.2A, B) are rapid movements of masses of rock and/or debris, gliding on slide planes causing considerable friction erosion (Rapp, 1985). They are characterized by a distinct slide scar, and an eroded slide track terminating in a slide tongue or lobe. Often, debris slides are heavily saturated with water, usually from melting snow, and quickly become viscous debris flows creating lateral debris-flow levees that terminate in debris fans.

The importance of debris flows and avalanches depends upon such factors as relief, climate, and lithology. One of the earliest studies to distinguish between the processes was undertaken in Kärkevagge, a deeply-eroded glacial valley in northern Lapland (Rapp, 1960a) (see Chapter 4, pp. 68–69). A specific conclusion was that, next to solution, rapid mass movements (debris flows and slides, slush avalanches, rockfalls) were the most important in terms of denudation (Table 9.3). However, as mentioned in Chapter 4, it is unwise to assume that the activity measured in the Kärkevagge is typical of all mountainous, periglacial environments. Much of the movement reported in 1960 resulted from a series of extreme events that occurred the previous year and which were regarded as "a centennial or probably even millennium maximum" (Rapp, 1960a, p. 185). It is important, therefore, to consider the frequency and magnitude of occurrence of episodic events when assessing their overall importance. For example, assuming a recurrence interval of 200 years for such major events, the rates of rock denudation from several documented debris slides and flows which occurred in Scandinavia and Spitsbergen over the previous three decades can be extrapolated (Table 9.4). This analysis demonstrates that the activity recorded at Kärkevagge is at the low end of the high magnitude scale, and that denudation accomplished by such activity is highly variable and largely dependent upon lithology and antecedent conditions.

In some instances, avalanche activity may be more important than debris-flow activity. For example, long-term studies by in the Canadian Rockies (Luckman, 1977, 1978) indicate mean avalanche debris-accumulation rates of 5 mm/year. In a second study, in northwest Spitsbergen (André, 1993, 1995b), the amount of debris deposited at the foot of avalanche paths was correlated with rockwall surfaces. Average denudation rates were obtained of 0.007 mm/year in massive gneiss and 0.08 mm/year in fractured mica-schist (Table 9.5). These values are considerably higher than comparable values from Kärkevagge.

Slushflows are a particular type of snow avalanche in which water-saturated snow masses are mobilized when rainfall, snowmelt, or a combination of the two increases the water content of the snow. When instability is reached, the pack is mobilized. They are widespread in arctic and sub-arctic environments (André, 1995b; Elder and Kattelman, 1993; Laroque et al., 2001; Nyberg, 1985; Onesti, 1985; Onesti and Hestness, 1989) and often occur on relatively gentle slopes. Slush avalanches probably have recurrence

Table 9.3. Ranked list of major slope process at Kärkevagge, Swedish Lapland, 1952–1960.

Process	Tonnes/ km/year	Denudation (mm/year)	Mass transfers (tonne-metres)	Remarks
1. Transport of solutes by running water	26	0.010	136 500	–
2. Debris slides and flows	49.4	0.019	96 300	Extreme event, October 1959
3. Slush avalanches, rock debris transport	14	0.005	20 000	Extreme events, 1956 and 1958
4. Rockfalls	8.7	0.003	19 600	Seasonal events of high frequency
5. Solifluction	5.4	0.002	5300–19 800	Based on 9 km of soliflucted slope length; material density of 1.8
6. Talus creep	1.5	0.001	2700–4700	Based on 6 km of talus slope length; material density of 1.8

Source: Rapp (1960a).

Table 9.4. Instances of alpine debris slides and flows in Scandinavia and Svalbard, and estimated indices of their denudation impact. From Rapp (1985).

Locality	Date	Catchment Area (km^2)	Rock Type	Volume of debris (m^3)	Denudation		
					Debris (mm)	Rock (mm)	Extrapolated rock[1] (mm)
Nissunvagge, N. Sweden	23.6.79	5.5	Amphibolite	85 000	15.5	10.8	54
Tarfala, N. Sweden	6.7.72	11	Amphibolite	55 000	5	3.5	18
Kärkevagge, N. Sweden	6.10.59	15	Mica-schist	4 600	0.3	0.2	1
Longyearbyen, Svalbard	11.7.72	4.5	Schist, sandstone	5 000	1	0.7	3.5
Ulvadal, W. Norway	26.6.60	7	Granite	3 000 000	43	30.1	150

[1] Extrapolated rates of rock denudation are given in mm per 1000 years assuming that the recurrence interval is 200 years.

intervals varying between 2 and 10 years, and, depending upon magnitude, can transport anywhere between 50 and 200 m^2 of rock debris. A recent case study from Mont Albert in the Gaspésie Mountains of eastern Québec, Canada, involved the use of dendrochronology and air photographs to date slushflow events (Laroque et al., 2001). It was concluded that slushflows occur at regular 20–30 year intervals and have contributed to the formation of a tongue-shaped accumulation of 17 900 m^2 at the base of one small first-order stream.

It is clear that slushflows and snow avalanches are geomorphic events that may undertake considerable geomorphic activity. However, it is difficult to assess their importance relative to debris flows. A general perspective is that debris-flow activity ranges from events of major geomorphic importance, with recurrence intervals of several hundreds of years, to smaller "spot" events occurring every 2–3 years within limited areas. As regards avalanches, the clear relationship between winter snowfall and spring temperatures is critical, and recent investigations suggest that avalanches are, in themselves, an important denudation process in snowy mountainous regions. Slush avalanches appear confined to smaller valleys where they occur in years of relatively rapid spring thaw. With all three processes, frequency and magnitude concepts are fundamental to any interpretation of their geomorphic significance.

9.5.3. Rockfall

Where bedrock outcrops in the form of vertical or near-vertical free-faces, rockfalls assume local importance and talus may accumulate below the free-face. Theoretically, if the scree is not renewed, the free-face will be progressively eliminated by the accumulation and upward growth of the talus. In many areas of Svalbard, northern Scandinavia, and the eastern Canadian Arctic, extensive and imposing rock faces have been inherited from previous glacial periods; in other areas, these slope forms occur in association with specific geologic structures (intrusive sills, dykes), with sea cliffs, or with deeply-incised stream valleys. The weathering and recession of these free-faces occurs primarily through the melting of interstitial ice and the loosening of rock particles by frost wedging in the

Table 9.5. Tentative rates of denudation due to avalanche activity in glacial cirques on Svalbard. From André (1993).

Avalanche Activity	Annual mean 1983–85			
	Number (%)	Surface (m^2)	Weight (kg)	Volume (m^3)
(a) Steep cirque wall in massive gneisses. Area: 145000 m^2				
Debris-free avalanches	21 (65%)	5800	0	0
Slightly dirty avalanches	9 (26%)	1300	390	0.12
Moderately dirty avalanches	2 (6%)	310	1530	0.5
Highly dirty avalanches	1 (3%)	50	1030	0.4
Total	33 (100%)	7460	2950	1.0
(b) Cirque wall in densely fractured mica-schist. Area: 135000 m^2				
Debris-free avalanches	8 (14%)	1400	0	0
Slightly dirty avalanches	26 (48%)	4200	1250	0.4
Moderately dirty avalanches	16 (29%)	2600	13000	4.8
Highly dirty avalanches	5 (9%)	760	15200	5.6
Total	55 (100%)	8960	29450	10.8

Rate of denudation in massive gneiss: 0.007 mm/year
Rate of denudation in fractured mica-schist: 0.08 mm/year

small joints and fractures that are inevitably present. Eventually, the shear strength of the material is reduced below the level at which it is capable of countering the stresses imposed by gravity.

Like the other rapid mass movement processes described earlier, rockfall is episodic and varies in magnitude from year to year, as well as spatially across the free-face. It is difficult to establish rates of debris production and rockwall retreat without extensive and long-term observations. Unfortunately, our understanding of rock weathering under cold-climate conditions is still incomplete (see Chapter 4). For example, a recent study attempted to measure weathering and rockfall occurrence on a cliff consisting of sandstone and shale near Longyearbyen, Svalbard (Prick, 2003). During 8 months of continuous monitoring, traps were placed beneath the free-face to collect rockfall debris. At the same time, sensors, inserted into bedrock to depths of 40 cm, measured rock temperatures and, at an adjacent locality, tablets of sandstone and limestone, exposed to the atmosphere, were weighed at regular intervals and non-destructive determinations of the Young's modulus of elasticity were carried out (Prick, 1997). It was found that the appropriate combinations of cold temperature and high rock moisture thought conducive to frost disintegration were largely lacking and there was little evidence for thermal shock. Moreover, the elasticity measurements suggested that the limited frost shattering that was observed was probably the result of simple wedging along joints in the sandstone because no changes in elasticity were observed in the sandstone control tablets. One is tempted to conclude that rockfall activity is largely controlled by the inherited mechanical properties of the rock in question.

A number of studies have attempted to measure the rate of rockwall retreat under cold-climate conditions (Table 9.6). Rates, which range from 0.003 to 2.50 mm/year, are all gross approximations and should be treated with caution. For example, R. Souchez (1967b) computed the volume of material lying downslope of a raised beach. This involved assumptions concerning the shape of the talus, its thickness, and the particle size and porosity of the debris. Thus, his volumes may be too large and retreat rates too high. On the other hand, A. Rapp (1960a) estimated the volume of material in fresh rockfalls and then averaged that amount over the whole area of the rockwall in question. This leads to an underestimate of retreat since rockfalls do not necessarily occur uniformly over the whole of the free-face. Finally, the methods that have been used to arrive at mean recession rates (mm/year) vary in nature, reliability, and precision. For example, M.-F. André (1993) derived rates at Kongsfjord from debris arriving at the foot of snow avalanches, while those from Wijdefjord and Ossian Sarsfjellet were computed using a standard lichen growth curve.

Bearing these considerations in mind, the available evidence suggests that rates of rockwall retreat of between 0.3 and 0.6 mm/year are probably typical for most lithologies. If correct, there is little support for the conventional view that weathering and slope retreat is appreciably faster in periglacial environments than in other environments. When recession rates are compared to those for humid temperate and subtropical semi-arid environments (see Table 9.6), it is clear that retreat rates in periglacial environments are, in general, at least one order of magnitude lower.

9.6. SLOPEWASH

Slopewash refers to a group of processes that include both surface wash (the downslope transport of weathered material over the ground surface by running water) and subsurface wash (the set of processes associated with water movement and sediment transport within the regolith) (Lewkowicz, 1988b, p. 354). In non-permafrost regions, the primary instigator of slopewash activity is snowmelt, rather than rainfall. In permafrost regions, the

Table 9.6. Summary data on rates of cliff recession under periglacial and non-periglacial conditions.

Location	Lithology	Recession (mm/year)	Source
(i) Periglacial – high latitude:			
Mt Templet, Spitsbergen	Limestone and sandstone	0.34–0.50	Rapp (1960b)
Mt Langtunafjell, Spitsbergen	Limestone and sandstone	0.05–0.50	Rapp (1960b)
Longyeardalen, Spitsbergen		0.3	Jahn (1976)
Spitsbergen, Kongsfjord	Massive gneiss	0.007	André (1993)
	Fractured mica-schist	0.08	
Spitsbergen, Wijdefjord	Amphibolite	0.03–0.11	André (1993)
Spitsbergen, Ossian Sarsfjellet	Quartzite	0.1–1.5	André (1993)
Northern Lapland, Karkevagge	Schists	0.04–0.15	Rapp (1960a)
Ellesmere Island, NWT, Canada	Dolomitic limestone	0.30–1.30	P. Souchez; personal communication (1971)
Yukon, Canada	Syenite, diabase	0.003–0.019	Gray (1973)
(ii) Periglacial – alpine:			
Austrian Alps	Gneiss, schist	0.7–1.0	Poser, in Rapp (1960a)
Swiss Alps		2.5	Barsch (1977)
Canadian Rockies	Limestone	0.06–0.26	Luckman (1972)
(iii) Non-periglacial:			
Mt St Hiliaire, PQ, Canada	Gabbro, breccia	0.02–0.04	Pearce and Elson (1973)
Brazil	Granite	2.0	Quoted in Young (1972)
South Africa	Granite	1.5	Quoted in Young (1972)
Southwest USA	Shale	2–13	Quoted in Young (1972)

effects of snowmelt are complicated by the presence of impermeable frozen sediments at the base of the active layer.

Slopewash in periglacial environments has often been linked to the concept of nivation (Matthes, 1900). The latter is a vague term associated with the frost shattering, gelifluction, and slopewash processes thought related to late-lying seasonal snow. In its simplest form, nivation was seen as a suite of snow-bank-induced erosional processes that are thought to erode hollows, cirque-like basins, or benches (Cook and Raiche, 1962; Kariya, 2002; Nyberg, 1991; St-Onge, 1969). This causal relationship between snow banks and landforms does not always stand up to rigorous analysis because many field measurements now demonstrate the lack of effective freeze–thaw cycles beneath snow banks (Thorn, 1976, 1979a, b; Thorn and Hall, 1980). As a result, modern usage of the term is not generally recommended (Thorn, 1988; Thorn and Hall, 2002). Having stated this, it remains true that an understanding of snow-bank hydrology is an essential requirement if one wishes to assess the significance of slopewash processes.

9.6.1. Snow-Bank Hydrology

The distribution of snow in any one area is largely controlled by prevailing winds and topography. It is also a major control over the ground-thermal regime (see Chapter 5) and snow-cover depth is often influenced by the presence or absence of vegetation. North of treeline and above timberline snow is blown clear of exposed surfaces and accumulates in lee-slope positions, hollows, and other topographic irregularities. Snow distribution is further complicated because, in the northern boreal forest, it may also depend on arboreal species present. For example, the Siberian taiga is composed predominantly of pine (*Pinus silvestris*) and tamarack (*Larix dahurica*) whereas the spruce (*Picea glauca, Picea mariana*) is more common in the North American boreal forest. Because more snow remains on spruce foliage than on more delicate larch branches, the snow cover is typically thinner beneath a spruce canopy than a larch canopy.

The pattern of snow melt varies from year to year and from locality to locality, depending upon climate (i.e. solar radiation, aspect, wind direction, etc.) and other site-specific details. Nevertheless, a few generalizations about snow melt are possible. To illustrate, the pattern of snow melt over a three-week period in a small drainage basin near Resolute, in the Canadian Arctic (Figure 9.14), permits the following observations. First, the primary control over snowmelt is clearly solar radiation. This is especially the case in high latitudes because of the near-continuous daylight in the summer months. There are also marked diurnal variations in runoff. These are related to solar radiation inputs. Second, as the snow pack ablates, water percolates downwards through the pack and refreezes. Thus, the formation of basal ice layers complicates the snowmelt–runoff relationship and prolongs the snowmelt season. Third, as the snow bank continues to ablate and progressively shrink in size, exposed ground starts to thaw. In its final stage, the snow bank consists almost entirely of basal ice.

9.6.2. Surface and Subsurface Wash

Depending upon the volume of snow, the rapidity of thaw, and the nature of the substrate, runoff can be both overland (surface) and subsurface in nature. Both are concentrated at the downslope edge of an ablating snow bank.

One of the earliest quantitative measurements of slopewash is provided by A. Jahn (1961) from Spitsbergen. Using simple sediment trays, it was calculated that approximately $12-18 \, \text{g m}^{-2} \, \text{year}^{-1}$ of sediment was being washed downslope beneath large perennial snow banks. This corresponds to a surface denudation rate of approximately 7 m/1000 years, a value considerably lower than comparable rates from non-permafrost regions (Young, 1974).

Subsequent studies by A. G. Lewkowicz and colleagues (Lewkowicz, 1983; Lewkowicz and French, 1982a, b; Lewkowicz and Kokelj, 2002) indicate that surface denudation due to suspended sediment removal varies between ~0.4 and ~2.6 mm/1000 years. By contrast, solute removal is 8–30 times greater (between ~2.0 and ~74.0 mm/1000 years). It has also been found that most surface wash is derived primarily from snowmelt and that summer rain rarely produces overland flow. Finally, average values of suspended sediment that are removed by slopewash processes can be as high as $1200 \, \text{g m}^{-2} \, \text{year}^{-1}$ on disturbed slopes (experiencing active-layer detachments) but, on undisturbed slopes, erosion rates are low in comparison to global rates.

Relatively little is known about subsurface wash and the role of seepage in slope modification. Solute concentrations probably increase throughout the summer in response to a higher residence time and a progressive desaturation of the active layer. Denudation from subsurface wash probably approximates that from surface wash. The Russian literature suggests that "thermo-erosional wash" exists, in which small mineral particles are liberated by melt of frozen ground. This may contribute to what is termed "thermo-planation"

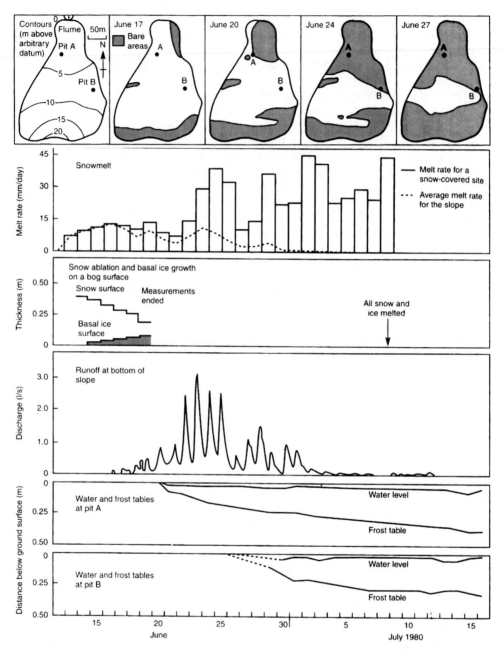

Figure 9.14. Typical pattern of snow ablation and associated runoff and active-layer conditions near Resolute Bay, Cornwallis Island, Nunavut, Canadian High Arctic. From Woo and Steer (1982, 1983).

(Dylik, 1972; Kachurin, 1962) but field studies have yet to substantiate this claim. If subsurface flow is concentrated at a locality, and if lithological conditions are favorable, collapse hollows or soil pipes may form by subsurface erosion and the transport of fines (Carey and Woo, 2000). This process also occurs in central Yakutia, where springs, emerging from the foot of terraces and fed by supra- and sub-permafrost waters, excavate large amounts of fine sand (Anisimova et al., 1973).

Few landforms can be attributed directly to slopewash activity but subtle features may be associated with seepage of supra-permafrost groundwater or subsurface flow. For example, on concave slopes underlain by coarse sediment possessing a high hydraulic conductivity and where the frost table is shallow, the topography may intersect the water table and seepage produces local saturated zones ("wet" spots) and seepage lines (Woo and Xia, 1995).

9.7. FROZEN AND THAWING SLOPES

Frozen terrain introduces at least two complications when considering slope evolution. These relate to the creep of frozen ground and thaw consolidation. The latter is often associated with thaw subsidence.

9.7.1. Permafrost Creep

Permafrost creep refers to the long-term deformation of frozen ground under the influence of gravity. Deformation is due mainly to the presence of pore ice, the migration of unfrozen pore water, and thaw consolidation (see below). Essentially, the warmer the permafrost and the greater the amount of ground ice, the greater will be the deformation. Fine-grained frozen sediments, such as silt and clay, which contains large unfrozen water content, are especially suited to frozen-creep deformation.

Early attempts to investigate frozen creep involved in situ experimental studies (Ladanyi and Johnston, 1973; Thompson and Sayles, 1972). Subsequently, a number of field studies have determined movement rates in several permafrost environments (see Table 6.2). These data illustrate that the magnitude reflects ice content, permafrost temperature, and slope angle. At one locality in the Mackenzie Delta, Canada (Dallimore et al., 1996b), a significant upslope movement was recorded during late winter, thought to be due to thermal contraction.

On slopes of moderate angle in discontinuous permafrost terrain, such as the Mackenzie Valley, Canada, and on the Tibet Plateau, China, rates of deformation of 0.1–0.4 cm/year appear typical. In colder permafrost terrain of the higher latitudes, movement rates are approximately one order of magnitude less, in the vicinity of 0.03–0.05 cm/year. Finally, in alpine environments where slope gradient is high, permafrost is marginal (warm), and ice content is high (as in rock glaciers), movement rates may be as great as 5.0–6.0 cm/year (Wagner, 1992).

Permafrost creep must be considered in any assessment of permafrost landscape evolution. Based on published data (Morgenstern, 1981), A. G. Lewkowicz (1988b) computes that downslope volumetric transport rates of 800–1000 cm^3 cm^{-1} year^{-1} are possible. More specific examples are that permafrost creep may result in the downslope curvature of ice wedges (Bozhinskiy and Konishchev, 1982) and it may complicate thermal-contracting cracking (Mackay, 1993b).

9.7.2. Thermokarst and Thaw Consolidation

An important cause of slope instability in permafrost environments is associated with the time-dependent compression of frozen ground that results from thaw and subsequent drainage of pore water. This process, which occurs annually in the active layer, is termed thaw consolidation (Morgenstern and Nixon, 1971). Many geotechnical engineers, working either in permafrost regions (McRoberts and Morgenstern, 1974) or upon slope stability problems of strata previously frozen (Hutchinson, 1974; Hutchinson and Gostelow, 1976), regard gelifluction as one form of thaw consolidation. In all probability, a process continuum exists between slow mass wasting and certain rapid mass movements with thaw consolidation being the central, common, mechanism.

The significance of thaw consolidation is only beginning to be appreciated by geomorphologists working in permafrost regions, and some additional words of explanation are appropriate.

During thaw, the flow of water from a thawing soil may be unimpeded. In this situation, the variation of settlement with time is controlled solely by the position of the thawing front. However, if flow is impeded, as is often the case in fine-grained sediments, the rate of settlement with time is controlled by the compressibility and permeability of the thawed ground. If the rate of thaw is sufficiently fast, water is released at a rate that exceeds that at which it flows from the soil. As a result, pore-water pressures increase in the thawed material and, if these exceed shear strength (i.e. the cohesion of the soil), instability will result.

The thaw-consolidation ratio, R, is commonly used to describe the relative rates of generation and expulsion of excess pore fluids during thaw (Morgenstern and Nixon, 1971; Nixon and McRoberts, 1973):

$$R = \alpha/2\sqrt{Cv}$$

and

$$\alpha = d/\sqrt{t}$$

where Cv is the coefficient of consolidation, and d is the depth of thaw in time t. Because the depth of slippage is limited by the frost table (i.e. top of permafrost), slope failures in permafrost terrain are generally planar in nature. The obvious illustration of slope failures in permafrost that can be explained in terms of thaw consolidation is the active-layer detachment described earlier in this chapter.

According to the thaw-consolidation ratio, a value greater than unity predicts the danger of sustained substantial pore pressure at the thawing front, and hence the possibility of instability due to reduction of shear strength at that plane. Usually, the permafrost table acts as a lubricated slip plane and controls the depth of the failure plane. Previously, such failures were colloquially termed "skin flows" (Capps, 1919). In all instances, the active layer and its vegetation mat detaches from the underlying permafrost surface (see Figure 9.13).

In areas of discontinuous permafrost, thawing of permafrost frequently involves adjacent non-permafrost sediments. The interactions are complex. For example, in the Mackenzie Valley, NWT, Canada, failure can occur through frozen soil with the base of the slide in unfrozen clay (McRoberts and Morgenstern, 1974). The permafrost temperatures were between $-2\,°C$ and $-4\,°C$ and high pore-water pressures controlled the available shear strength in the unfrozen clay. The long-term strength of the permafrost soil was governed by a frictional resistance factor. Since a considerable amount of unfrozen water was present in the permafrost, the frictional resistance of the frozen clay was significantly reduced and failure occurred.

9.7.3. Stability of Thawing Slopes

Permafrost slopes commonly experience failure at angles considerably lower than the equilibrium angle predicted by standard geotechnical analysis. Table 9.7 lists a number of studies where analysis indicates that the failures have occurred at angles below their predicted equilibrium angle. The table also includes data from several studies undertaken in non-permafrost environments where low-angled slopes have experienced unusual failure. These failures have been attributed to frozen ground conditions in the Pleistocene.

The geotechnical details of slope stability analysis are beyond the mandate of this text. However, so-called "back-analyses," involving assumptions as to the depth of thaw, and the bulk density and water content values that might have been involved in a particular slope failure, permit identification of the conditions necessary to initiate failure. For example, on Ellesmere Island, C. Harris and A. G. Lewkowicz (2000, pp. 457–460) concluded that, even taking residual shear strength of 25°, pore-water pressures associated with thaw and a progressive reduction in shear strength at the base of the active layer from gelifluction movement were sufficient to initiate failure on a 12° slope.

Similar slope failures have been described from the permafrost regions of western Siberia (Liebman, 1996; Liebman et al., 2003). In the boreal forest and taiga zones, it is not uncommon for failure to occur following the destruction of vegetation by fire (Zoltai and Pettapiece, 1973). Where substantial bodies of ground ice are present, an initial slope failure may be the trigger mechanism for subsequent retrogressive-thaw slumps (see Chapter 8).

With predicted global climate warming, it must be anticipated that permafrost slopes will experience substantial slope instability and thermokarst modification. This will be especially pronounced in areas of warm permafrost and at the southern (i.e. warm)

Table 9.7. Summary data on reported failures in thawing slopes.

Locality	Lithology	Soil residual strength ø (degrees)	Predicted angle (degrees)	Failed angle (degrees)	Source
Active slope failures in permafrost:					
Svalbard	Sandy clayey silt	36	20	6–12	Chandler (1972)
Mackenzie Valley, NWT	Clay	23	12.5	3–9	McRoberts and Mogenstern (1974)
Pleistocene slope failures:					
England	Soliflucted clay (Weald, Gault, and London clays)	12.4–15.5	6.8–8.1	3–7	Weeks (1969)
England	Sandy clay	23	12.0	6.8	Chandler (1970a)
	Sandy clayey silt	16	8.8	4.0	Chandler (1970b)

borders of the discontinuous and sporadic permafrost zones. In high mountains, the instability of rock slopes and the movement of ice-rich debris (rock glaciers) will undoubtedly increase (Haeberli and Burn, 2002). Slope failures in permafrost terrain are a geotechnical hazard for many of the construction activities described in Chapter 14.

9.8. COLD-CLIMATE SLOPE EVOLUTION

Periglacial slope evolution involves a progressive and sequential reduction of relief with the passage of time. Limited evidence suggests that this takes place by slope replacement from below, with the formation of Richter denudation slopes.

The Richter denudation slope is thought to represent a balance between debris supply and debris removal. Where the rate of weathering of a free-face and the debris at its base is less than or equal to the ability of transportational processes to remove weathered debris, a denudation slope forms below the free-face. Because weathering rates are assumed to be uniform over Richter slopes, they decline at a constant and inherited angle. Thus, over time, Richter slopes are replaced by (cryo) pediments as relative relief is progressively reduced.

The major complication to this simple model is that few regions exist where periglacial slope evolution has managed to run its full course. This has been discussed in Chapter 2. Nevertheless, two general and descriptive models are thought useful in our understanding of periglacial landscapes. These are outlined below.

9.8.1. Cryoplanation

The periglacial literature contains numerous references to cryoplanation as a process promoting low-angled slopes and level bedrock surfaces (Demek, 1978, pp. 148–149; Dylik, 1957; French, 1976a, pp. 155–165; Péwé, 1970; Washburn, 1979, pp. 237–243). According to Demek (1969a), cryoplanation terraces best develop in continental semi-arid periglacial environments. According to Reger and Péwé (1976), cryoplanation terraces require permafrost for their formation.

Following descriptions of bedrock ("goletz") terraces and apparent "mountain planation" in Siberia (Jorré, 1933), Russian geologists proposed a cyclic model for the formation of these flat bedrock surfaces (Boch and Krasnov, 1943) (Figure 9.15). This model was subsequently promoted in the European periglacial literature as "cryoplanation" (Demek, 1969a, b; Czudek, 1990; Czudek and Demek, 1973; Richter et al., 1963). The surfaces were thought to initiate below structural benches, or initial slope irregularities, all of which favor snow accumulation in lee positions. Then, frost action beneath the snow bank leads to steepening and retreat of the slope and the formation of a frost-riven "riser." Ultimately, this leads to the development of a summit flat as the riser is eventually consumed by an adjacent terrace. In the final stages, downwearing becomes the dominant mode of evolution.

Recent Russian texts rarely mention cryopediments or cryoplanation (for example, see Yershov, 1990; Kudryavtsev, 1978; Popov et al., 1985; Romanovskii, 1980). The reality is that features previously identified as cryoplanation terraces are probably controlled primarily by lithology. Although the significance of "cryoplanation" is debatable, the existence of these bedrock benches is not in question. In addition to Siberia, they also occur in Alaska and Central Yukon (Cairnes, 1912; Eakin, 1916). Most are certainly erosional

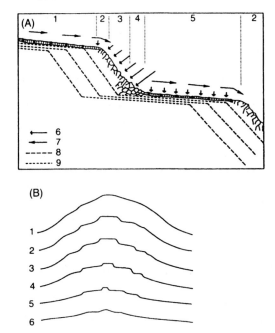

Figure 9.15. The altiplanation (cryoplanation) concept as proposed by S. G. Boch and I. I. Krasnov (1943) for the Ural Mountains, Russia. (A) Altiplanation terrace development. Legend: (1) flat terrace, or step (slope angle ~3–5°); (2) edge/border of terrace; (3) terrace riser with bedrock exposed; (4) frost-riven rock face; (5) lower level terrace; (6) triangular-headed arrows signify varying intensities of frost weathering; (7) pointed arrows indicate direction of material transport across the terrace; (8) successive stages of inferred retreat of the frost-riven rock face; (9) successive stages in downwearing of the terrace. (B) Proposed stages of landscape evolution through altiplanation: stage 1, initial topography prior to commencement of periglacial conditions; stages 2–5, development of flat summits and/or interfluves, and "stepped" lower profiles and/or tors; stage 6, ultimate cryoplanation of the landscape and formation of a "frost and solifluction peneplain." From Boch and Krasnov (1943); translated and reproduced in Evans (1994).

in origin. For example, on the Yukon–Alaska border, the benches truncate dip directions and attitudes in the underlying bedrock (French et al., 1983, pp. 61–63; Reger and Péwé, 1976). Moreover, most terraces that are described appear to be relict. Their age of formation is uncertain. In the Richardson Mountains of northern Yukon Territory, analysis of the silty veneer, and the weathering of quartzite bedrock on which bedrock terraces are formed, fail to demonstrate their active formation or the process(es) involved (Lamirande et al., 1999; Lauriol et al., 1997). On Svalbard, bedrock erosional surfaces initially attributed to "altiplanation" (Waters, 1962) were subsequently found to be raised beaches (Péwé et al., 1982).

Notwithstanding these problems, the cryoplanation concept is a useful working model for landscape evolution under cold, non-glacial conditions. It underlies the periglacial "cycle of erosion" first proposed by L. C. Peltier (1950). The initial assumption is that an existing non-periglacial landscape is subject to intense frost action and solifluction. An initial stage begins with downslope movement of soil and frost debris that exposes bedrock on upper slopes. The latter is then subject to further frost shattering, rockfall, and retreat.

On upland surfaces, frost action produces extensive areas of angular debris, forming blockfields. With time, continued retreat of the frost-riven "riser" leaves only residual bedrock outcrops at the summit. At this stage, so-called "maturity" is reached in which the pre-existing landscape has disappeared and the landscape is covered with a mantle of frost-derived material. In "old age," solifluction degrades and flattens the summits while adjacent valleys and lowlands are progressively "plugged" by the accumulation of soliflucted debris.

In terms of slope evolution, the Peltier periglacial cycle of erosion represents an intuitive synthesis since it emphasizes many of the typical slope forms described earlier in this chapter. It is too general, however, to provide anything but an overall framework within which to view slope evolution. For example, no quantitative parameters are given for the landscape changes which are thought to occur, there is no realistic discussion of the manner in which frost shattering and mass movement influence slope form, and there is a lack of attention paid to other processes, particularly running water.

9.8.2. Slope Replacement and Richter Denudation Slopes

In the extremely cold and arid regions of Antarctica, other landscape models have been proposed to explain the development of rock slopes. They envisage slope replacement and the formation of Richter denudation slopes (see Figure 9.3).

One of the first, developed by R. Souchez (1966) and based upon observations in the Sor Rondane Mountains of East Antarctica, emphasized the role of plastic deformation and shearing failure. Under conditions of plastic flow, regolith movement was assumed proportional to the angle of slope, and ground loss was proportional to convex curvature. The resulting model was one in which slopes decline over time but remain predominantly convex.

A second model, proposed by M. J. Selby (1974; Augustinus and Selby, 1990), is based upon observations in the Dry Valleys of Southern Victoria Land. On steep upper slopes, stress-release joints and slab failures are a consequence of gravitational loading. This leads to headwall retreat. On lower slopes, if the rate of weathering exceeds the rate of debris removal, talus cones or aprons accumulate. As the rock slope retreats, the rate of debris supply is progressively reduced since the size of the headwall progressively decreases. Ultimately, a Richter denudation slope extends upslope to eventually consume the upper slope.

9.8.3. Rapidity of Profile Change

It is sometimes assumed that slopes evolve more rapidly under periglacial conditions than under non-periglacial conditions (Tricart, 1963; see 1970, p. 112). This remains unproven, since the few data that are available suggest that weathering and slope processes do not operate at significantly faster rates in periglacial, as opposed to non-periglacial, environments. For example, the rate of rockwall retreat under periglacial conditions appears to be less than in rainforest, humid temperate, and tropical semi-arid conditions, and limestone solutional activity is no greater in arctic regions than in other arid or semi-arid areas. Indeed, although the impression gained from recently-glaciated regions is one of relatively dynamic slope evolution, if one considers the arid, cold, and never-glaciated regions of the world, a more subdued model of landscape evolution seems more appropriate.

9.8.4. Summary

A number of points summarize the preceding discussion.

First, slopes evolve in periglacial environments primarily through the combined action of mass movement and running water, and not through "unique" periglacial processes. Second, a variety of slope forms exist in periglacial environments and none is limited to periglacial environments. Third, slope form is primarily influenced by the lithological characteristics of the underlying rock. Fourth, the presence or absence of rock moisture, in determining the type and rate of weathering and weathering removal, is of prime importance. Fifth, periglacial slope forms exhibit many similarities with slope forms found in hot arid and semi-arid regions of the world. Sixth, many slopes are best interpreted as having been inherited from a previous glacial period, and are, therefore, in disequilibrium with the present cold-climate environment. Seventh, there is no evidence to assume that slopes evolve more rapidly under periglacial conditions than under non-periglacial conditions.

ADVANCED READING

Augustinus, P. C., Selby, M. J. (1990). Rock slope development in McMurdo Oasis, Antarctica, and implications of glacial history. *Geografiska Annaler*, **72A**, 55–62.

French, H. M., Harry, D. G. (1992). Pediments and cold-climate conditions, Barn Mountains, unglaciated Northern Yukon, Canada. *Geografiska Annaler*, **74A**, 145–157.

Harris, C., Lewkowicz, A. G. (2000). An analysis of the stability of thawing slopes, Ellesmere Island, Nunavut, Canada. *Canadian Geotechnical Journal*, **37**, 449–462.

Iwata, S. (1987). Debris-mantled rectilinear slopes in the western Sor Rondane Mountains, East Antarctica. Proceedings, NIPR, *Antarctic Geoscience*, **1**, 178–192.

Lewkowicz, A. G., Kokelj, S. V. (2002). Slope sediment yield in arid lowland continuous permafrost environments, Canadian Arctic Archipelago. *Catena*, **46**, 261–283.

McRoberts, E. C. (1978). Slope stability in cold regions. In: Andersland, O. B., Anderson, D. M., eds, *Geotechnical Engineering for Cold Regions*. McGraw-Hill, New York, pp. 363–404.

Rapp, A. (1960). Recent development of mountain slopes in Kärkevagge and surroundings, northern Scandinavia. *Geografiska Annaler*, **42**, 65–200.

Rapp, A. (1960). Talus slopes and mountain walls in Tempelfjorden, Spitsbergen. *Norsk Polarinstitutt Skrifter*, **119**, 96 pp.

DISCUSSION TOPICS

1. What are the mass-wasting processes which operate on slopes in periglacial environments?

2. How does the presence of frozen or thawing substrate influence slope processes?

3. How similar are the slope forms found in the hot and cold semi-arid regions of the world?

4. How do slopes evolve under cold non-glacial conditions?

5. Is there a periglacial cycle of landscape evolution?

10 Azonal Processes and Landforms

Fluvial processes assist in fashioning the periglacial landscape. Both large and small rivers exist, all dominated to varying extent by spring snowmelt-induced (nival) peak discharge. Freeze-up and break-up are distinct cold-climate characteristics, together with fluvio-thermal erosion. Although all types of stream channel morphology can be found in periglacial environments, braided channels are most common. The asymmetry of many small valleys reflects a complex interaction between slope and channel processes.

Wind action is enhanced by the lack of vegetation, and by the ability of wind not only to transport and redistribute snow but also to use snow as an abrading agent. Niveo-eolian deposits illustrate the intimate connection between wind and snow. Wind also influences, via the snow cover and direct evaporation, microclimates and ground-thermal regimes. Wind also influences ice movement and wave action in both marine and lake environments.

Beach processes are restricted by the presence of sea ice and beach-fast ice for parts of the year. Coastlines formed in ice-rich permafrost sediments are subject to thermo-abrasion and rapid modification.

10.1. INTRODUCTION

Many azonal processes are either enhanced or inhibited by cold-climate conditions while others assume special characteristics. This is the case for fluvial, eolian, and coastal processes and justifies their collective treatment in this chapter. The intent is not to present a systematic treatment of each since other geomorphology texts are readily available. Instead, the approach highlights those aspects that are peculiar to, and achieve their greatest importance in, periglacial environments.

10.2. FLUVIAL PROCESSES AND LANDFORMS

The large-scale organization of periglacial terrain is not unlike that of other environments. A well-developed drainage network exists, even in those areas that have recently emerged from beneath Quaternary ice sheets and glaciers. It is most striking in the extreme high latitudes, where, in spite of aridity, and in the absence of vegetation, the intricacies of the network are clearly visible (see Figure 2.4). In broad terms, the fluvial network is composed of two types of rivers. First, large rivers, such as the Mackenzie and Yukon rivers in North America, and the Ob, Yenisei, Kolyma, and Lena rivers in Siberia, originate in non-periglacial regions and gain additional discharge from springs in the discontinuous permafrost zone. These are some of the larger rivers of the world, flowing for several thousands of kilometers from their headwaters to the Arctic ocean. To some extent, their discharge and sediment characteristics are independent of the terrain and climate of the region through which they flow. Second, innumerable smaller rivers of varying sizes constitute the overwhelming majority of the fluvial network. Their drainage originates totally within the periglacial domain and their discharge and sediment characteristics more truly reflect the environment in which they flow.

Most high-latitude stream channels are occupied by running water for only a few summer months when air temperatures rise above freezing. For most of the year there is little surface or subsurface water movement. Yet running water is capable of significant denudation and transportation activity in cold environments, especially when compared with other geomorphic agents. The spring snowmelt is rapid, usually occurring over a period of 2–4 weeks, and gives rise to a prominent freshet, or flood. Thus, although precipitation may be slight (<200 mm per year), between 25% and 75% of total runoff can be concentrated in only a few days during the spring thaw. At that time, the active layer is shallow and permafrost inhibits downwards percolation.

10.2.1. Major Rivers

Before proceeding any further, it is appropriate to consider the very large northern rivers that cross the vast periglacial domains of Eurasia and North America. Their channel hydrology and fluvial runoff characteristics have a significant impact upon transportation, shipping, and human settlement.

The Mackenzie River of northern Canada is typical of these rivers (Figure 10.1A). The simplicity of the flow hydrograph (Figure 10.1B) masks the complexity of the watershed basin and the different effects of both mountain and northern hydrology. The Hay River (drainage area 47 900 km^2) drains the lowlands south of Great Slave Lake and has the normal attributes of winter low flow, a brief spring-snowmelt high flow, and a gradual summer recession interrupted by rainstorms. The Liard River (drainage area 33 000 km^2) drains the Cordillera and also has low winter flow. However, high runoff is sustained for a longer period by snowmelt from the range of elevations within the basin. Large runoff per unit area is another feature of this mountainous basin. By contrast, the Camsell River (drainage area 30 900 km^2) drains the undulating Canadian Shield, where runoff is impounded by the myriad of small lakes. As a consequence, lake storage evens out flow over the year such that snowmelt runoff is withheld and a moderate level of winter flow is maintained. Thus, these three sub-basins have distinct hydrological regimes of either nival, mountain, or Shield character. The annual discharge regime of the Mackenzie River is the sum total of these tributary contributions combined with regulation of flow provided by dammed and natural lakes.

The Mackenzie River is navigable throughout its length from Hay River on the south shore of Great Slave Lake to Tuktoyaktuk on the coast of the Beaufort Sea. The river plays a vital role in the annual re-supply of settlements in the western Canadian Arctic. During the navigation season (Table 10.1), barges frequent the waterway. However, the season is short: break-up begins in May just north of the Liard River junction, and progressively moves down-river, while freeze-up begins during early October in the Mackenzie Delta and advances up-river. The Yukon, the other large river in northern North America, has similar discharge characteristics, and plays a similar transportation role.

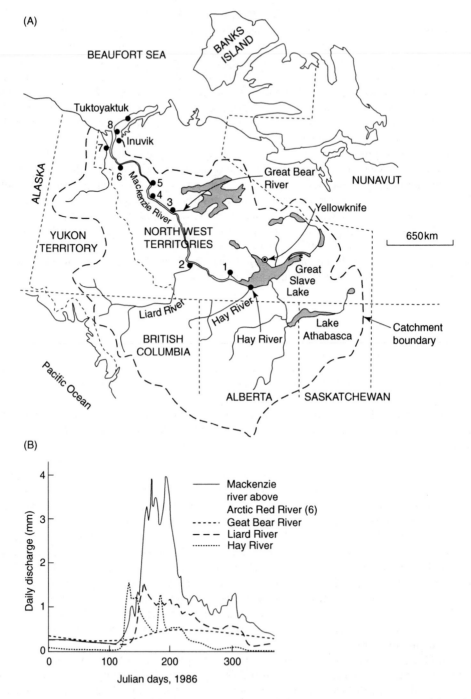

Figure 10.1. The Mackenzie River, northern Canada. (A) Extent of the catchment with major tributaries and settlements indicated. (B) Daily discharge at Tsiigehtchic (Arctic Red River), and of Great Bear, Liard and Hay River tributaries, in Julian days (1986). From French and Slaymaker (1993). 1-8 = localities listed in Table 10.1. Reproduced by permission of McGill-Queens Univeristy Press.

Table 10.1. Mean freeze-up and break-up dates on the Mackenzie River system, NWT, Canada, 1946–1955. *See Figure 10.1 for locations.

Location	Distance from Great Slave Lake (km)	Freeze-up	Break-up
Fort Providence*[1]	80	24 Nov.	18 May
Fort Simpson*[2]:			
(i) Mackenzie above Fort Simpson	335	27 Nov.	15 May
(ii) Mackenzie below Fort Simpson	351		11 May
Tulita (Fort Norman)*[3]	825	15 Nov.	14 May
Norman Wells*[4]	909	10 Nov.	15 May
Fort Good Hope*[5]:			
(i) Ramparts	1094	15 Nov.	22 May
(ii) Settlement	1101	12 Nov.	15 May
Tsiigehtchic (Arctic Red River) settlement*[6]:			
(i) Arctic Red River	1445	8 Oct.	25 May
(ii) Mackenzie River	1445	1 Nov.	24 May
Aklavik*[7]	1607	9 Oct.	28 May
Reindeer Station (north of Inuvik)*[8]	1615	18 Oct.	27 May

Source: French and Slaymaker (1993). Reproduced by permission of McGill-Queens Press.

In Siberia, a number of similar large rivers flow northwards (Ob, Yenisei, Lena, Indigirka, and Kolyma rivers), with the added advantage of a northern sea route along the Siberian coast from Kolymsk (Cherskiy) in the Kolyma Delta in the Far East to Archangel'sk on the White Sea in the west.

All these large rivers usually retain water in the main channel under the ice throughout the winter. The river ice can be as much as 2.0 m thick and the channels are surprisingly deep. Break-up usually begins with the flow of locally-derived meltwater over the ice surface. Then, the ice rises as flow increases in the channel beneath and as shore leads develop. Break-up is usually rapid, often accompanied by flooding. For example, in the Mackenzie River basin, a major reason for the creation of Inuvik as a regional administrative center in the mid-1950s was because Aklavik, the original center, was subject to periodic flooding. Ice-jams, caused by large pieces of river ice becoming concentrated at localities where the ice has yet to break-up, or where there are obstructions to flow, as at bridge crossings, can accentuate flooding, causing serious damage (Gerard, 1990). Table 10.2 lists the direct costs associated with ice jams on the Mackenzie and Yukon River systems between 1960 and 1990. In recent years, the continued relocation of settlements and structures away from localities prone to flooding, and improved geotechnical engineering has minimized this hazard (see Chapter 14). Comparable data for the large Siberian rivers are not easily available but it is reasonable to assume that similar hydrograph patterns and freeze-up/break-up conditions prevail. On the Lena and Kolyma Rivers, annual ice damage to wharf facilities and settlements is an ongoing hazard.

10.2.2. Freeze-Up and Break-Up

We can distinguish between stream channels which are fed by both surface runoff and groundwater and those which are fed almost entirely by surface runoff. The latter occur in areas of continuous permafrost while the former occur in sub-arctic locations where discontinuous permafrost permits springs to emerge from taliks and sub-permafrost aquifers. In both types of channel the winter cold and abundance of snow and/or ice in the

Table 10.2. Ice-jam damage on Yukon and Mackenzie Rivers, Canada, 1960–1985.

	Year	Cost of flooding (Can. $)
Yukon		
Yukon River, Dawson City	1979	2 910 136.00
Yukon River, Dawson City	1967	182 912.00
Yukon River, Dawson City	1965	158 573.00
Yukon River, Dawson City	1963	27 994.00
Stewart River, Mayo	1973	23 041.00
Stewart River, Mayo	1972	14 890.00
Stewart River, Mayo	1965	103 833.00
Ross River, Ross River	1973	45 824.00
Northwest Territories		
Mackenzie River System:		
Hay River, Hay River	1985	695 736.00
Hay River, Hay River	1963	3 953 374.00
Liard River, Fort Simpson	1963	3 294 479.00

Source: Van der Vinne et al. (1991).

channels intensifies freeze-up and break-up. The broader issues of groundwater hydrology in permafrost regions have been discussed in Chapter 6.

The process of river-ice formation is described in the northern hydrological and geotechnical literature (Gerard, 1990) and is illustrated in Figure 10.2A. In the initial stages in autumn, turbulence generated by stream flow is sufficient to produce supercooled water. Hence, ice forms first over quiet waters along the banks. This ice is termed "sheet" ice. Elsewhere, floating ice crystals ("frazil" ice) form in deeper water. Where this comes in contact with the bed, as in shallow reaches, it forms "anchor" ice. As freezing continues, the frazil particles form slush, which in turn agglomerates into frazil pans. At the same time, ice grows outwards from the river bank and eventually frazil pans lodge against the bank-fast ice. As more frazil pans arrive from upstream, the initial pack grows in the upstream direction. At this point, water level begins to rise in response to increasing resistance to flow beneath the ice and the necessity for the channel to carry the discharge with increased resistance. The depth increase is typically greater than 30% of the mean depth of the open-water situation. It is further enhanced by the fact that the ice cover, or pack, floats with more than 90% of its thickness submerged; this necessitates an additional rise in water level. River geometry and weather conditions then influence the final stages of freeze-up. If discharge is slow and air temperatures are low during pack formation, the pack will be thin and frazil production will be high. In this case, the pack forms as one pan and progression upstream is rapid. If discharge is high, and air temperatures are mild during pack formation, the pack will be thick and the rate of frazil-pan production will be slow. In this case, the increase in water level will be large but pack progression upstream will be slow.

River icings are further phenomena which may occur during freeze-up and in winter (Figure 10.2B). They should be distinguished from the groundwater icings described in Chapter 6. River icings are attributed either to a reduction in the cross-sectional area of an ice-covered channel as freezing advances or to an increase in snow load on an initial ice cover thus raising the hydrostatic head beneath the ice to an elevation higher than the ice surface (Wankiewicz, 1984). Because the water cannot escape from the banks due to freezing of the active layer, fractures in the ice allow water to escape over the ice cover, to subsequently freeze as an icing. In shallow braided streams, small icing mounds, 1–3 m high, may develop in response to localized restrictions of flow by ice freezing to the bed

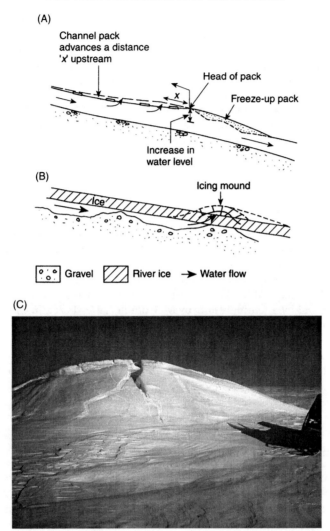

(A)

Channel pack
advances a distance
'x' upstream

Head of pack

Freeze-up pack

x

Increase in
water level

(B)

Icing mound

Ice

Gravel   River ice   → Water flow

(C)

Figure 10.2. River freeze-up processes. (A) Growth of ice cover at freeze-up and its associated water-level changes. (B) Formation of a river icing in a shallow section of a multiple-channel system. (C) Photograph showing river-icing mound formed in winter 1978–79 on the flood plain of Big River, Central Banks Island, Arctic Canada. The mound is 2.2 m high. The photograph, taken in April 1979, is courtesy of D. Nasagaloak (Sachs Harbour Hunters and Trappers Association). Diagrams A and B are modified from Gerard (1990).

of the braid bars. The icing mound shown in Figure 10.2C is one of several that occur annually in approximately the same location on the braided channel of the Big River in central Banks Island, Canada. These river-icing mounds are a type of seasonal-frost mound discussed earlier in Chapter 6 (see Figure 6.11).

Break-up usually occurs rapidly, often with considerable flooding. The problem is accentuated in rivers that have headwaters in the south where snow and ice melt begins earlier. Ice jams frequently occur, as described above for the Mackenzie River. On smaller rivers, where ice may have frozen to the bed, break-up sometimes involves avulsion, flow is diverted around ice obstructions, and local bank erosion may occur.

In extreme high latitudes, rivers are generally small, and river ice and river icings are rare. Instead, the stream valleys are usually infilled with snow that accumulated by wind-drifting during the winter months (Pissart, 1976b). The break-up sequence occurs when the snow reaches saturation point through the arrival of runoff from snowmelt on adjacent slopes. As ripening of the pack continues, slushflow may briefly be initiated, quickly followed by stream flow, which rapidly carves unstable channels or tunnels in the snow. In places, large snow drifts may dam sections of small valleys to impound water up valley, only for this to be released when the snow jam breaks (Woo and Sauriol, 1980). Thus, the drainage network often opens up in segments, and until all segments are linked, the basin is incapable of coherent transport of meltwater runoff (Woo, 1986).

10.2.3. Basin Hydrology

Because permafrost restricts downward percolation, runoff in basins underlain by permafrost responds quickly to snowmelt and rainfall events. Vegetation, if present, promotes a longer recession period, especially in wetlands.

One of the first basin hydrology studies in Arctic Canada was undertaken by F. A. Cook (1967), who measured the flow of a small stream on Cornwallis Island, NWT. This was shortly followed by a seminal study by M. Church (1972, 1974), in which the different hydrologic regimes of northern rivers were summarized (Figure 10.3).

To varying extents, all runoff regimes are dominated by rapid melt of snow and ice in the short winter–summer transition period. In the Canadian Arctic, for example, this occurs in late June or early July. Then, during the rest of the summer, the runoff steadily decreases as less and less snow remains to be melted. This progressive decrease in runoff is periodically interrupted by subsidiary runoff peaks related to summer storms and direct surface runoff. Such single-peak runoff regimes are termed nival. They may be either sub-arctic or Arctic in nature depending upon whether or not base flow is maintained throughout the winter. A proglacial regime occurs in watersheds where permanent snow or ice fields exist. Here, melt occurs throughout the summer whenever warm and/or overcast conditions develop. Peak runoff is often delayed until late July or early August, and the nival peak is not so important. A third regime is the wetland type. Because of the water-retaining capacity of organic terrain and tundra, and the relatively high resistance to runoff presented by such terrain, peak flows are attenuated in such basins. A final category is the spring-fed regime, often associated with basins underlain by carbonate rocks in the discontinuous permafrost zone. These streams have relatively stable discharges because the primary source of water is groundwater.

The runoff season in high latitudes can be divided into four seasons: (1) break-up, (2) the snowmelt period (the "nival flood"), (3) late summer, and (4) freeze-back. In detail, however, each season is determined by the pattern of local summer weather for each year. Usually, break-up begins in late spring. Although air temperatures are still below 0 °C at this time, local snowmelt is brought about by direct solar radiation. Meltwater percolates to the base of the snowpack and into snow-choked stream channels, where it refreezes. It is not until two or three weeks later that sufficient melt has occurred for flowage of saturated snow to occur. The river channels turn to slush and runoff begins over the snow and ice on the stream bed. Usually after two or three days of intense runoff, the winter ice on the stream bottom has been eroded and runoff continues on the stream bed proper. Sometimes, a trigger, commonly a period of warm weather or a heavy storm which flushes the snow out of the stream channel, initiates runoff.

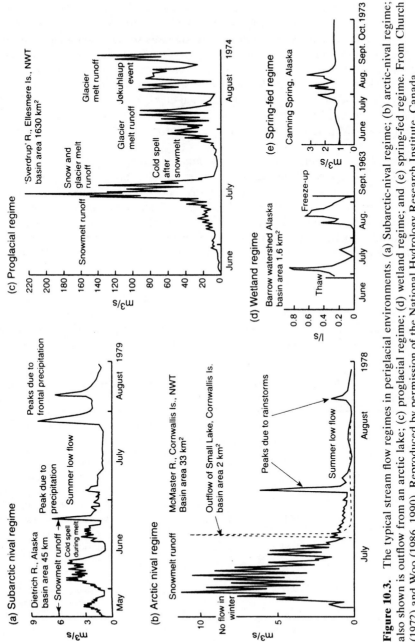

Figure 10.3. The typical stream flow regimes in periglacial environments. (a) Subarctic-nival regime; (b) arctic-nival regime; also shown is outflow from an arctic lake; (c) proglacial regime; (d) wetland regime; and (e) spring-fed regime. From Church (1972) and Woo (1986, 1990). Reproduced by permission of the National Hydrology Research Institute, Canada.

After the snowmelt period, runoff progressively decreases and discharge becomes dominated by storm runoff. The latter occurs after heavy cyclonic activity or following periods of prolonged overcast and drizzly conditions. The hydrologic response is rapid, reflecting the presence of permafrost and the absence of vegetation. Superimposed upon these short-term fluctuations are diurnal fluctuations; discharge usually increases in late afternoons and early evenings, or following periods of uninterrupted solar radiation.

During late summer, many small streams are virtually without flow and, by freeze-back time at the end of the summer, there is little or no runoff. Any stream flow that does exist is maintained almost entirely by recession flow from either groundwater discharge or from seepage through the active layer. Icing ablation augments stream flow, particularly after snowmelt (Clark and Lauriol, 1997; Kane and Slaughter, 1972; Reedyk et al., 1995; Sokolov, 1978; van Everdingen, 1987, 1990). In general, icing ablation in warmer regions generates high flows for a short period of time while icing ablation in colder regions sustains a longer period of flow.

It is difficult to generalize about the magnitude and frequency of flow events because there is a lack of long-term data. However, a number of catchment studies, undertaken in northern Canada by governmental and other regulatory bodies (Table 10.3), suggest that annual peak flows and flow variability are highest for mountainous catchments fed by glaciers and for catchments underlain by continuous permafrost. They are least for lowland catchments in discontinuous permafrost terrain that include numerous lakes and extensive wetlands. Almost certainly, similar general relationships exist for the northern rivers of Eurasia.

Long-term flow measurements are usually accompanied by observations upon precipitation, snow cover, and evaporation. Together, these data can be analyzed within a water-balance context in which the various water gains and losses within the basin are examined. Typically, snowfall constitutes the bulk of annual precipitation in northern catchments, most runoff results from snowmelt, and the importance of snow declines in areas where

Table 10.3. Mean annual discharge (Q) and standard deviation of flow (S) of selected rivers in northern Canada. Q and S are measured in m^3 s^{-1} km^{-2}

River	Drainage area (km^2)	Mean discharge (Q)	Standard deviation (S)	Years of Record	Environment
Kluane	4950	0.0148	0.0027	30	Mountainous, glacerized, discontinuous permafrost (below lake outlet)
Klondike	7800	0.0078	0.0016	18	Mountainous, discontinuous permafrost
Kakisa	14900	0.0026	0.0010	19	Interior plain, discontinuous permafrost (below lake outlet)
Thoa	9630	0.0037	0.0010	15	Canadian Shield, discontinuous permafrost
Dubwant	67600	0.0049	0.0008	15	Canadian Shield, continuous permafrost (below lake outlet)
Freshwater	1490	0.0050	0.0013	8	Arctic Islands, continuous permafrost

Source: Woo (1986).

rainfall contributes a larger proportion of total precipitation. In fact, runoff consumes the majority of total precipitation, with runoff ratios (R/P) as high as 0.7–0.8 for most basins excluding wetlands. A problem with these sorts of analyses is that relatively little is known about the rate of snow sublimation.

10.2.4. Sediment Flow, Surface Transport, and Denudation

Sediment movement in periglacial environments occurs through a combination of solution, suspension, and bedload transport. There are numerous sources of sediment for such transport. Weathering (see Chapter 4) produces fine rock particles, and mass-wasting processes (see Chapter 9) move material towards the stream channels. Moreover, many areas possess an abundance of loose and unconsolidated sediment inherited from previous glacial periods.

From a geomorphic point of view, it is of interest to know when, and under what conditions, the majority of sediment is transported. Unfortunately, few long-term studies monitor both stream discharge and sediment load. Table 10.4 summarizes some of the available data for Arctic Canada. Bedload transport is especially difficult to determine. One solution used (Church, 1972) is to derive solution and suspended-sediment values (see Table 10.4) by traditional field methods but to estimate bedload as values of "potential sediment transport" at full supply. Accordingly, calculated values may be overestimates. Moreover, because the Baffin Island study (see Table 10.4) was undertaken in a proglacial environment where current sediment yield is derived from unconsolidated glacial sediments, sediment yield bears no relationship to present rates of sediment production (Church, 1972, p. 63). In spite of this caution, the limited data available suggest that a dominance of bedload sediment transport is probably characteristic of many periglacial rivers, even in areas unaffected by recent glaciation. For example, over 80% of the sediment transfer in the Colville River of northern Alaska was of either a bedload or suspended nature (Arnborg et al., 1967), and, in a small permafrost drainage basin in northern Yukon Territory, a bedload transport approximately three times that of suspended load transport (444 t/km^2 versus 153 t/km^2) was reported for a 12-day period during peak discharge (Priesnitz and Schunke, 2002). Essentially similar conclusions are reached, by default, in several ongoing weathering and sediment-budget studies in northern Swedish Lapland (Beyrich et al., 2003, 2004a, b; Darmody et al., 2001; Thorn et al., 2001). Although chemical weathering and denudation is certainly important locally, as at Kärkevagge (Rapp, 1960a; see Chapter 4), more typical results from elsewhere in that region indicate that solute concentrations and chemical denudation amounts are generally low (Beyrich et al., 2003, p. 394). Moreover, Table 10.4 indicates that, for most catchments, suspended-sediment yields are generally higher than solute yields.

Further insight into the significance of fluvial processes in fashioning the periglacial landscape is provided by data in Table 10.5. A wide range of sediment yields are found in arctic regions. For example, when compared with a suggested North American mean yield of 130 tones km^{-2} year^{-1} (Gregory and Walling, 1973), these data sets suggest that periglacial yields are not significantly different from those of other environments (Table 10.5A). Furthermore, much solute transport is of the same order of magnitude as suspended sediment transport (Table 10.5B), and other data suggest that the dissolved organic carbon transport for major northern rivers is similar to that of other major river systems of the world (Table 10.5C).

Computations involving the total amount of material transported from the Baffin Island watersheds (see Table 10.4) indicate average ground surface lowering of the order of 200–450 mm/1000 years. However, as noted earlier, it is probable that normal sediment

Table 10.4. Values of sediment concentration and yield in rivers of the Canadian Arctic archipelago.

Basin location and year of study	Basin area (km^2)	Runoff (mm)	Maximum suspended sediment concentration (mg l^{-1})	Maximum solute concentration (mg l^{-1})	Suspended sediment yield (t km^{-2}yr^1)	Solute yield (t km^{-2}yr^1)	Bed load yield (t km^{-2}yr^1)	Reference
Hot Weather, Ellesmere Is., 1990–91	155	38 (1990) 13 (1991)	1870	502	17.3–51.9 1.7–1.9	4.7–4.4 1.8–1.8	n.d.	Lewkowicz and Wolfe (1994)
Mecham River, Cornwallis Is., 1970–71	95	297	1300	134	12.7	21.1	1.1–4.7	McCann and Cogley (1973); Cogley (1975)
Snowbird Creek, Bathurst Is., 1976	61	156	1300	42	25.6	4.4	2.5	Wedel et al. (1977)
Lewis River, Baffin Is., 1963–65	205	818 (1963) 294 (1964) 485 (1965)	1741	57 (average of 4 samples)	232.5 68.1 126.9	3.2 2.1 2.4	1049 236 482	Church (1972)
Upper South River, Baffin Is., 1967–68	90	620 (1967) 186 (1968)	800 (approx.)	8 (average of 7 samples)	48.8 0.8	4.0 1.9	1114 29	Church (1972)

n.d.: Not determined.
Source: Lewkowicz and Wolfe (1994).

Table 10.5. The magnitude of sediment transport in periglacial environments. (A) Some selected values of total sediment yield from periglacial catchments. (B) Typical volumes of suspended and solute sediment transport in periglacial catchments. (C) Dissolved organic carbon transport for major periglacial and non-periglacial rivers of the world.

A. Total sediment yield: (tonnes $km^{-2}a^{-1}$)

Sweden	0.8–4.7
Yenesei and Ob rivers, Russia	5.0–7.0
Colville River, Alaska	80–85
Baffin rivers, Canada	29–114

B. Suspended sediment transport (tonnes $km^{-2}a^{-1}$)

(i) Suspended transport:

Mackenzie/Yukon (catchments <10000 km^{-2})	0.2–11.5 t/km²/year
Mackenzie/Yukon (catchments >10000 km^{-2})	36–126 t/km²/year
Yukon coast	5–300 mg/1
Finnish Lapland	Typically 5–50 mg/1
Hoffellsjökull, Iceland	Mean 1337 mg/1 (glacier-fed);
Ellesmere Island	776 mg/1 (non-glacial)
	17.3–51.9 t/km²/year

(ii) Solute transport:	Total Dissolved Solids (mg/1)
Baffin Island	5–50
Fosheim, Ellesmere Island	1.8–4.7
Swedish Lapland	CaCO 25–55
Queen Elizabeth Islands	30–50 increasing to 70–100
Devon Island	CaCO peaked at 102
Banks Island slope plots	44–252
Hoffellsjökull, Iceland	Mean 271 (glacier-fed); 52 (non-glacial)

C. Dissolved organic carbon:

River	Discharge (km³/year)	Area (10³km²)	Runoff (mm/year)	Concentration Range (mg/1)	(t/km²/year)
Periglacial:					
Yukon	210	840	250		3.0
Mackenzie	249	1810	138	3–6	0.7
Lena	533	2430	219		2.1
Ob	419	2550	164		1.5
Yenesei	562	2580	218		1.6
Non-periglacial:					
Amazon	5520	6300	876	3–5	3.2
Orinoco	1135	950	996	2–5	6.1
Mississippi	439	3267	154	2–8	1.1
Ganges	366	975	375	1–9	1.7
Niger	152	1125	171	2–6	0.5
Yangse-Kiang	883	1950	453	5–23	6.1

Sources: Clark (1988), Goudie (1995, p. 60), Lewkowicz and Wolfe (1994).

production, and therefore average ground lowering, is at least one order of magnitude lower than the present yield rate. Thus, a rate of ~50 mm/1000 years, as calculated for the Colville River in northern Alaska (Arnborg et al., 1967), is probably more typical of most periglacial environments.

10.2.5. Fluvio-Thermal Erosion

The thermal effects of running water have already been commented upon (see Chapter 8). In addition to normal mechanical effects, running water possesses the ability to thaw permafrost. The most obvious result is the existence of taliks, or unfrozen zones, beneath the channels of all rivers and water bodies that do not freeze to their bottoms in winter (see Chapters 5 and 6). Where large river channels are incised within ice-rich and/or relatively unconsolidated sediments, lateral erosion can form thermo-erosional niches often several tens of meters deep (Figure 10.4A). These sometimes cause bank collapse, often in large blocks delineated by ice wedges (Figure 10.4B). This process, first described from the Colville River Delta (Walker and Arnborg, 1966), occurs widely along the banks of major rivers, such as the Yukon River in western Alaska and the Lena River in central Siberia, where they traverse ice-rich alluvial lowlands. Coastlines developed in ice-rich and unconsolidated sediments are also subject to rapid erosion and retreat by this process (see below). As regards channel dynamics, fluvio-thermal erosion is important for at least three reasons: (a) the collapse of river banks provides material for bedload transport (see above), (b) it promotes the development of wide, flat-bottomed channels (see below), and (c) it helps explain the efficiency of lateral stream migration and the formation of valley asymmetry in permafrost areas (see below).

10.2.6. Channel Morphology

River channels may be single or multiple, straight or meandering, and large or small. In periglacial environments, all types of channel pattern exist but undoubtedly the most common is the multiple, or braided, channel (Figure 10.5).

The dominance of braided channels is the result of at least three factors. First, braiding requires appreciable sediment load. Second, braiding is related to rapid and large variations in discharge. Third, bank erodibility is clearly a factor since excessive lateral erosion will not only lead to wide channels, shoaling, and the development of multiple channels, but also to the entrainment of large quantities of debris. All these requirements are present, in varying degrees, in periglacial environments underlain by permafrost. This is because the discharge of both nival and proglacial streams is subject to rapid and extreme fluctuation and the presence of loose glacigenic sediment supplies material for transport.

As a result, the floor of many stream channels in the high latitudes is covered in alluvial sediments ranging from coarse gravel to medium sand. The banks are often abrupt, giving a shallow box-like profile to the channel (Figure 10.6). At times of peak flow, the entire floor becomes covered with a layer of turbulent water, in which bedload transport dominates. As discharge decreases and competence drops, coarser material is deposited, and the stream assumes a new braided pattern. Thus, given time and repeated adjustments to the braided channel pattern, the channel floor is constantly reworked.

The importance of abundant bedload sediment in producing braided stream channels is illustrated by the absence of well-developed braided stream channels in those areas where bank erodibility is restricted and where debris suitable for transport is limited. For example, it is instructive to consider the 120 km-long Thomsen River on Banks Island, the largest river in the Canadian arctic islands. For the majority of its middle length, the channel is single, somewhat sinuous, with only occasional braiding. As such, it contrasts

(A)

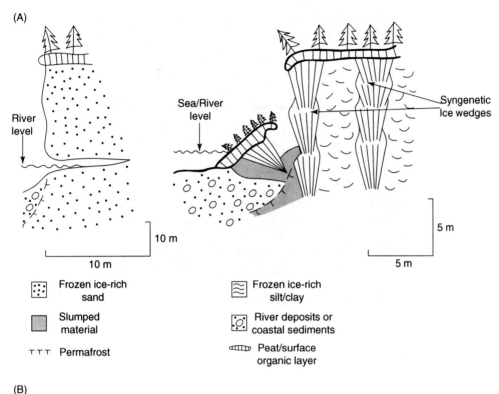

(B)

Figure 10.4. Fluvio-thermal erosion. (A) Forms of lateral and thermal riverbank and coastal erosion. (B) Coastal erosion by thermo-erosional undercutting and block collapse along ice wedges aligned parallel to the coast, Maitland Bluffs, Mackenzie Delta region, Canada.

with the other north- and west-flowing rivers on Banks Island, which are all braided. The reason is that the middle reaches of the Thomsen River are incised within relatively consolidated shale (Christopher Formation) while west-flowing drainage dissects either glacial materials or unconsolidated sand and gravel (Beaufort Formation). Thus, the availability

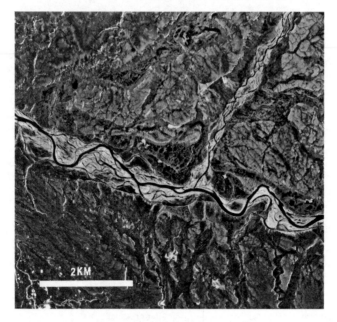

Figure 10.5. Vertical air photo of part of the Bernard River, Central Banks Island, Arctic Canada. The photograph is centered upon latitude 73°45′ north, longitude 122°15′ west. Note the meandering and braided main channel system, the braided system entering from the north, and the low fluvial terraces. Air photograph A 17564-48, National Air Photo Library, Natural Resources Canada, Ottawa.

Figure 10.6. Oblique air view from ~150 m elevation of small box-shaped valley with braided stream system incised within fluvial terrace. Note the well-developed thermal-contraction crack polygons. Prince Patrick Island, Arctic Canada.

of potential bedload is significantly less than in the west-flowing streams and, as a consequence, braiding is less apparent. On western Banks Island, channels can occasionally be observed to change from braided to meandering for limited distances. This is related to local exposures of sand (Eureka Sound Formation), and the relative absence of coarse bedload results in the stream becoming overcompetent. To compensate and equalize energy, it adapts the meandering pattern.

At the macro-scale, the large northern rivers, such as the Mackenzie, Lena, and others, usually possess single and well-defined channels. Despite the fact that these rivers freeze over during the winter months, flow beneath the ice cover is maintained throughout the year. The thermal effect of the water promotes the development of sub-channel taliks, and an abrupt shelving of the permafrost table. These factors favor steeply-inclined channel sides and deep, well-defined channels. Multiple channels only develop either where tributaries join or where the rivers exit to large water bodies where high sediment load results in large deltas (see below).

10.2.7. Valley Asymmetry

The presence of asymmetrical valleys in areas currently underlain by permafrost was first reported by W. B. Shostakovitch (1927) from Siberia. Today, there is an abundant literature that describes this phenomenon. To judge from the literature (Table 10.6), the north-facing slope is usually steeper in the high latitudes. However, steeper slopes of systematic orientation have also been reported to face south, west, and east.

The most probable explanation for the "normal" north-facing asymmetry involves greater mass-wasting activity on south-facing slopes and asymmetric lateral stream corrasion (Currey, 1964; Gravis, 1969). This is because south-facing slopes receive greater solar radiation and, as a consequence, usually have thicker active layers. Asymmetrical valleys raise problematic questions as to which of the slopes has either declined or steepened in angle and necessitates assumptions as to the manner of slope evolution. There is

Table 10.6. Some characteristics of valley asymmetry in northern regions.

Area and author	Locality	Valley alignment	Orientation of steeper slope
East Greenland:			
Poser (1948)	Wollston-Vorland	E-W	N
Malaurie (1952)	Disko	E-W	N
West Spitsbergen[1]:			
Dege	Andreeland	E-W	S
Dege	Conwayland	N-S	W
Klimaszewski	Kaffioya – Ebene	E-W	S
Klimaszewski	Brogger – Halbinsel	N-S	E
Siberia:			
Shostakovitch (1927)	Yakutia	E-W	N
Presniakow[2]	Yakutia	E-W	N
Gravis (1969)	Yakutia	E-W	N
Northern Canada:			
Bronhofer[2]	Southampton Island	E-W	N
French (1971b)	Banks Island	NW-SE	SW
Kennedy and Melton[2]	Caribou Hills, NWT	E-W	N, S
Alaska:			
Hopkins and Taber[2]	Central Alaska	E-W	N
Currey (1964)	Northwest Alaska	E-W	N

1. Quoted in Karrasch, 1970.
2. Quoted in French, 1976a.

Figure 10.7. Oblique air view of asymmetrical valleys on the Beaufort Plain, northwest Banks Island, Arctic Canada.

no simple, universal, explanation because asymmetrical valleys also form in non-permafrost regions (Hack and Goodlett, 1960) and many in mid-latitudes are interpreted as relict features related to Pleistocene periglacial conditions (see Chapter 13). In many cases, regional and site-specific conditions are also invoked to explain the asymmetry. For example, the asymmetrical valleys which exist on the Beaufort Plain of northwest Banks Island (Figure 10.7) are interpreted in the context of wind action, mass wasting, and fluvial incision. They are a quasi-equilibrium slope form closely related to the climatic and geomorphic environment of that area (see Chapter 2).

10.3. EOLIAN PROCESSES AND SEDIMENTS

In this section, the role of wind in periglacial environments is briefly outlined. More detailed summaries of eolian processes in general, usually in the context of hot deserts, can be found elsewhere (e.g. Cooke et al., 1993) and a more comprehensive account of cold-climate wind action can be found in M. Seppälä (2004).

Wind clearly plays an important geomorphological role in the high latitudes of the world. This is especially true for the Antarctic continent (Campbell and Claridge, 1987, pp. 110–111; Malin, 1987; Spate et al., 1995). Numerous data sets and anecdotal information illustrate the strength of wind in the ice-free peripheral areas that surround that continent (see Chapter 1 and Table 3.1). For example, near Terra Nova Bay in Northern Victoria Land, a number of automatic weather stations are maintained by the Italian Antarctic Program (PNRA) (Table 10.7). On Inexpressible Island, the locality where Scott's historic "Northern Party" over-wintered during 1911–1912 (Priestley, 1914), records indicate near-continuous north-northwest winds flowing off the Priestley and Reeves Glaciers that exceed 35 km (>20 knots) for over half the year. Incredibly, winds in excess of 70 km (>41 knots) are recorded for nearly 20% of the time. There is no question that large pebbles, in addition to silt and fine sand particles, can be transported considerable distances.

It should not be assumed that similar winds characterize all periglacial environments. In many areas, wind is of relatively minor importance, giving rise to only small-scale and

Table 10.7. Wind speed and directional data, recorded as a percentage of total time, for the coastal areas of the Northern Foothills, Southern Victoria Land, Antarctica, by the PNRA.

	Wind direction					
	NNW	NW	WNW	W	WSW	SW
Inexpressible Island	–	14.8	53.2	11.5	2.5	2.7
Terra Nova station	3.3	4.4	17.1	18.0	10.5	8.9
Priestley Glacier	49.3	31.1	2.3	–	–	–

	Wind speed (knots)*				
	1–6	7–16	17–27	28–40	>41
Inexpressible Island	12.7	18.3	23.2	24.9	18.6
Terra Nova station	50.4	19.0	14.1	8.9	4.4
Priestley Glacier	9.4	15.2	23.2	29.8	21.5

Source: Baroni (1996).
*1 knot = 1.85 km/h.

localized weathering effects, niveo-eolian deposits, and certain patterned ground phenomena (Bird, 1967, pp. 237–241; Pissart, 1966a; Washburn, 1969). In some areas, large Late-Pleistocene sand deposits occur and dune formation is a locally-dynamic process (Dijkmans and Koster, 1990; Dijkmans et al., 1986; Hamilton et al., 1988). Likewise, on the Qinghai-Xizang (Tibet) Plateau, dune formation is associated with areas of sand weathered from weakly-lithified sandstone bedrock (Wang and French, 1995c). In many tundra regions, as on central Banks Island, Arctic Canada, eolian process are active on alluvial and outwash plains adjacent to the large rivers (Pissart et al., 1977) and wherever Mesozoic- and Tertiary-age sandstone and shale bedrock is subject to deflation. Numerous other examples could be cited in which wind action has an important direct, but local, impact.

The indirect effects of wind are more subtle, but possibly more important. For example, wind plays a primary role in snow redistribution, and influences, therefore, slopewash and runoff processes. Wind also transports sediment, often in combination with snow, to form niveo-eolian deposits, and may entrain silt and fine-sand size particles and redeposit them over large areas as a thin veneer of what is loosely termed cover-sand. Wind also affects evaporation and latent heat loss from exposed slopes, which, in turn, influences the depth of the active layer and the magnitude of mass-wasting processes. Asymmetrical valleys (see above) and the oriented nature of many thaw lakes (see Chapter 8) are further examples of the indirect effect of the wind. Finally, wind is important to the movement of sea ice and in wave generation (see below). It is clear, therefore, that wind operates in a number of indirect ways to influence landforms.

10.3.1. Wind Abrasion

If wind is to carry out significant erosion, there must be a source of abrasive material suitable for wind transportation. Silt and fine-sand particles, picked up from exposed ground surfaces, are one source. Thus, the vegetation-free polar deserts of high latitudes are especially suited to wind abrasion. However, the ease of detachment of surface particles may quickly become limited by the formation of a lag composed of coarser particles.

It may also be hindered by the presence of a snow cover, or a salt crust or hard pan at the surface. For these reasons, wind abrasion is probably least effective in the summer months.

The majority of wind abrasion in high latitudes probably occurs during the 6–10 months of winter by wind-driven snow particles. It is well known that the hardness of ice is a function of temperature. Moh hardness values can reach the equivalent of those of orthoclase feldspar (Moh hardness of 6) in the extreme Antarctic winter (Figure 10.8A). Also, cold air is denser than warm air, and thus the corresponding threshold velocity that is needed to take a grain into saltation is lower in cold environments than in hot environments (Figure 10.8B). Laboratory experiments support the hypothesis that many wind abrasional features are produced in response to collisions by suspended dust particles rather than by saltating sand grains (Whitney and Dietrich, 1973). Field experiments using asbestos boards exposed to snow-laden katabatic winds in the Sor Rondane Mountains of Antarctica indicate maximum erosion takes place at a distance of ~30–40 cm above the ground surface (Matsuoka et al., 1996). It is also during winter that winds are strongest and most constant because well-developed high-pressure systems become established over the polar landmasses. Widespread wind erosion on the Qinghai-Xizang (Tibet) Plateau, however, is attributed to strong diurnal winds associated with night-time cooling of uplands. As such, it is a year-round process.

Wind abrasion is a form of natural sand-blasting. Its effectiveness is a function of hardness of the material being abraded, hardness of the abrading agent, and wind velocity. The most common wind-erosional features are ventifacts. These are smoothed, flattened, fluted, faceted, or grooved pebbles or boulders that have suffered modification from exposure to strong wind-transported abrasive material (Figure 10.9A).

In the high northern latitudes, ventifacts have been reported from northeast Greenland (Fristrup, 1953) and Siberia (Sverdrup, 1938) but are relatively rare in the lowlands of the Canadian Arctic islands (Bird, 1967; Pissart, 1966a), presumably because wind velocities are insufficient for entrainment of abrasive material. Long-term observations made upon a number of ventifacts from near Paulatuk, a small coastal community in Western Arctic Canada that is subject to year-round katabatic winds from adjacent uplands, suggest that

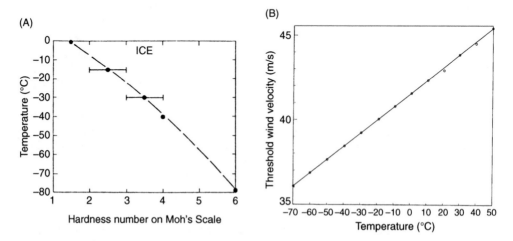

Figure 10.8. Snow as an abrasive agent. (A) Hardness of ice as a function of temperature; values may reach the equivalent of those for orthoclase feldspar (hardness 6) in the extreme Antarctic winter. More typically, at −40°C ice has a hardness of 4, similar to fluorite. (B) Threshold wind velocity required to lift a 2 mm granule to a height of 2 m at various temperatures. From Selby et al. (1974). Reproduced by permission of The Royal Society of New Zealand.

(A)

(B)

(C)

Figure 10.9. Wind-action effects. (A) Ventifact (wind-abraded rock) formed in fine-grained granite lying on coarse-grained syeno-granite bedrock in vicinity of Terra Nova Bay, Southern Victoria Land, Antarctica. Hammer alignment indicates probable wind-eroding direction. (B) Wind-eroded (deflation) hollow, or "blow-out," adjacent to Qinghai-Xizang (Tibet) Highway, China, exposes a small-diameter oil pipeline in July, 1990. (C) Niveo-aeolian sediments form as a snow bank ablates in early summer, Thomsen River lowlands, Central Banks Island, Arctic Canada.

ventifact formation is slow (Mackay and Burn, 2005). At Paulatuk, the ventifacts are glacial erratics, of varying lithologies. Despite 50 years of observation, little or no change in the form of the ventifacts could be seen.

The greatest variety and frequency of occurrence of wind-erosional features occurs in the ice-free areas of Antarctica (Lindsay, 1973; Nichols, 1966, pp. 35–36; Sekyra, 1969, p. 282; Selby, 1977). This is the result of the extremely strong and persistent katabatic winds flowing outwards from the vast continental interior. In the ice-free areas of Antarctica, some tafoni shows a preferred orientation that may relate to wind direction (Cailleux and Calkin, 1963). Elsewhere, observations suggest that the rate of ventifact formation is relatively fast because ventifacts can be found not only on surfaces of great antiquity but also on raised beaches of Holocene age. Any generalization is difficult because lithology is a major determinant of abrasion rate. However, crude approximations derived from the shape of ventifacts lying on Holocene-age raised beaches in the Terra Nova Bay area (French and Guglielmin, 1999, p. 343) gave abrasion (erosion) rates of 0.003 and 0.001 mm/a, respectively. These are an order of magnitude less than the rate of 0.015–0.022 mm/a estimated for the Larsemann and Vestfold Hills region of eastern Antarctica (Spate et al., 1995). When considered collectively, these limited data explain why J. R. Mackay was unable to detect significant abrasion over his 50-year period of observation at Paulatuk.

Case hardening and so-called "desert varnish" are problematic phenomena that may be wind-related and which are especially well developed in the ice-free areas of Antarctica. These terms are used to describe a poorly-understood process by which the exterior of a rock is made more resistant to weathering, probably by the evaporation of a mineral-bearing solution, leaving a thin cementation layer (Campbell and Claridge, 1987, pp. 124–129; Glasby et al., 1981) (see Chapter 4). Some argue that a biogeochemical origin should be considered (Dorn and Oberlander, 1982; Dorn et al., 1992). Even more so than ventifact formation, the development of rock varnish is thought to be extremely slow. Referring specifically to Antarctica, Campbell and Claridge (1987, pp. 127–129) state that "surfaces on which pitted and stained rocks are found are always old, of the order of a million years or more."

10.3.2. Wind Deflation

Deflation, the second aspect of wind activity, is the winnowing out of fine particles by wind and their transportation. Because vegetation is a major controlling factor, deflation reaches its greatest intensity on unvegetated surfaces and in the arid polar deserts.

An obvious indicator of deflation is the presence of a lag gravel or desert pavement on the ground surface. This reflects the removal, by wind, of finer particles. Another is the presence of shallow depressions or blow-outs, ranging from a few centimeters in width and depth to troughs many meters wide and deep (Figure 10.9B).

The amount of sediment removed by deflation depends largely upon the strength of the wind. For example, strong winds during the winter of 1990–91 on the Fosheim Peninsula of Ellesmere Island, in the Canadian High Arctic, resulted in an estimated soil loss of $\sim$4–5 kg/m^2 (4–5 mm), an amount equivalent to more than 20 years of denudation by wash and other processes (Lewkowicz, 1998). The maximum size (45 mm long) and weight (25 g) of particles transported during winter demonstrate that eolian transportation in the Canadian Arctic, like the Antarctic, is not confined to sand- and silt-sized materials. Elsewhere, M. Seppälä (1974) measured sand transport of 0.15 g/cm/hr during a 4-month period in a blow-out approximately 75 m wide. This translates into 3.2 tons of sand moving through this "gateway." Vegetation is the obvious inhibitor of deflation activity. For example, A. Pissart et al. (1977) describe deflation upon a sandy glacial outwash surface on southern Banks Island, Canada, where clumps of willow (*salix*) have led to the preservation of sandy mounds, 1.0–3.0 m high, separated by deflation blow-outs or troughs.

As with wind abrasion, the majority of deflation activity probably occurs during the winter when wind speeds are highest and when, typically, the upper few centimeters of frozen ground are desiccated and relatively friable. However, deflation may also occur in summer when dust storms, brought on by surface heating and instability of sandy areas during periods of intense solar radiation, may rise several hundreds of meters into the air blanketing the surrounding terrain with a thin cover of fine sand particles. Braided stream channels and outwash plains at low flow are particularly suited to this eolian activity.

10.3.3. Niveo-Eolian Sediments

The combination of wind-transported sediment and snow has long been noted (Fristrup, 1952; Pissart, 1966b). The resulting sediments, predominantly laminar in nature, are commonly termed "niveo-eolian" (Cailleux, 1974, 1978; Koster and Dijkmans, 1988). They occur not only in high latitudes but also in the snow environments of mid-latitudes (Ballantyne and Whittington, 1987; Jahn, 1972; Rochette and Cailleux, 1971).

Evidence for niveo-eolian deposition is frequently observed during spring snowmelt, when it is not uncommon for the snow surface to assume a gray, almost black color, and for pitted, small-scale thermokarst-like relief to form (Figure 10.9C). The sediment veneer, sometimes 1–2 cm thick, lowers the snow albedo and leads to earlier melt (Woo et al., 1991). Snow banks may also possess distinct sediment layers, each reflecting an event of wind deposition. Sediment concentrations in snow banks typically range from zero to as much as 0.2–2.0 g/l (Czeppe, 1965; Lewkowicz and Young, 1991, pp. 201–206).

As snow progressively ablates, wind-blown sediment is reworked and locally redeposited. The term "denivation" is sometimes used to refer to the sedimentological and morphological disturbances that occur. Because the sediments are predominantly silty in nature and susceptible to frost heave and cryoturbation activity, they are sometimes thought to promote the formation of slope hummocks (Lewkowicz and Gudjonsson, 1992). Niveo-eolian activity is closely related to slopewash activity (see Chapter 9). Facies analyses of niveo-eolian sediments are relatively few, and the processes by which these wind-derived sediments are reworked by snowmelt are still poorly understood. Experimental studies (Dijkmans and Mucher, 1989) suggest that the intercalation of snow during sedimentation does not, in itself, induce a laminated structure. Further investigation is required into the interactions between permafrost, moisture (snow) and eolian transport (sediment) in present periglacial environments.

10.3.4. Loess-Like Silt

Sequences of either buff or gray-colored silt-sized material are globally-widespread sediments attributed to wind transport and deposition. Not all are related to periglacial conditions since much is found in and around the margins of hot deserts of the world. Presumably, in periglacial environments, the silt is entrained from unvegetated floodplains, braided channels, glacial outwash plains, till plains, and lake shores.

Wind-blown silt deposits of cold-climate origin have been described from Alaska (Péwé, 1955, 1975), where they are called "upland silt," from central Siberia (Péwé and Journaux, 1983), and from Tibet (Péwé et al., 1995). They are largely Pleistocene in age (see Chapter 11) and analogous to the "loess" of mid-latitudes (see Chapter 13). In detail, wind-blown silts are well-sorted, homogeneous, and unstratified. Loosely coherent grains of between 0.01 mm and 0.05 mm in diameter often exceed 50–60% of the deposit. The typical grain-size distribution of wind-blown silt is illustrated in Figure 10.10.

Silt entrainment requires a dry soil surface. This probably explains why loess-like silt is rarely encountered in damp oceanic periglacial environments. In cold and arid environments, however, suitable conditions for entrainment from the ground surface occur during summer by evaporation and during the cold winters through sublimation. Both processes promote desiccation of the near-surface materials.

In Arctic North America, wind-blown silt of Holocene-age mantles upland surfaces in both central Alaska and parts of Yukon Territory. There is evidence of deposition from the Middle Quaternary onwards through to present day. Tephra (volcanic ash) layers and relict ice wedges indicate these sediments have been perennially frozen (i.e. permafrost) for several hundred thousand years. Many of the valley bottoms contain re-transported silt mixed with organic debris that is locally termed "muck." These predominantly Pleistocene-age deposits are discussed more fully in Chapter 11.

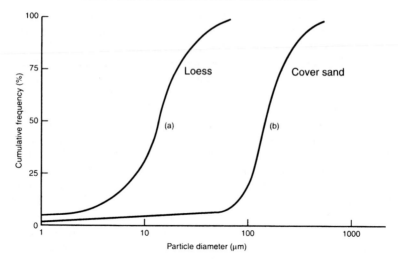

Figure 10.10.   Typical grain-size distribution curves for loess (a) and cover sand (b).

Figure 10.11.   A barchan (crescentic) sand dune on the Qinghai-Xizang (Tibet) Plateau south of Wudoaliang, viewed from the Trans-Plateau Highway in June 1987. By 1991, the dune had migrated a distance of ~350 m and crossed the highway.

10.3.5. Sand Dunes and Sand Sheets

Sand dunes are usually associated with hot dry environments but they also form in cold dry environments if suitable source material is present (Figure 10.11). Cold-climate dunes, like their hot counterparts, are transverse, crescentic (with horns pointing in the direction of movement), or parabolic (with horns pointing away from the direction of movement) in morphology. Occasionally, linear forms are superimposed on dune fields consisting mainly of transverse types. Sand bodies lacking a dune relief, i.e. without slip faces, are referred to as sand sheets. Usually, the sand particles are similar in grain size to those of hot deserts, commonly rounded, and possess subdued micro-relief.

Extensive sand dunes and sand sheets occur in parts of northern Alaska, the Canadian Arctic, Scandinavia, and the Qinghai-Xizang (Tibet) Plateau (Table 10.8). Tension cracks,

Table 10.8. Some major cold-climate dune fields in Arctic Canada, Alaska, northern Scandinavia and Tibet.

Location	Aeolian Forms	Age	Wind Direction	Comments	Source
Arctic Canada:					
Yukon Territory	Sand sheets		S-SW	Discontinuous, active	Nickling (1978)
Banks Island	Sand sheets	Mid-Holocene	SE	Continuous, active, niveo- and fluvio-aeolian	Pissart et al. (1977); Good and Bryant (1985)
Southeast Baffin Island	Dunes, sand sheets		N-ESE	Scattered, partly active	McKenna-Neuman and Gilbert (1986)
Antarctica:					
Victoria Valley	Transverse dunes; wholeback dunes		–	Active	Selby et al. (1974)
Alaska:					
Big Delta: Yukon Flats	Dunes: parabolic, sand sheets	Illinoian (?) – Wisconsinan	N	Scattered	Péwé (1975)
Tanana River Valley	Dunes: parabolic, longitudinal	Illinoian (?) – Wisconsinan	NE	Scattered, extent > 10000 km^2 scattered	
Kuskokwim River Valley	Dunes: parabolic	Illinoian (?) – Wisconsinan		Scattered	
Koyukuk Valley	Dunes: parabolic, transverse, longitudinal sand sheets	Late Wisconsinan – Holocene	NE	Scattered, partly active discontinuous	Koster (1988)
Kobuk Valley	Dunes: parabolic, transverse, longitudinal, barchanoid sand sheets	Late Illinoian (?) – Late Wisconsinan – Recent	NE-SE	Scattered, partly active megadunes up to 50 m high	Péwé (1975)
Arctic Coastal Plain	Dunes: parabolic, longitudinal	Late Wisconsinan – Recent	(SW-W)-NE-E	Scattered, partly stabilized	Carter (1981)
Scandinavia:					
Finnish Lapland	Dunes: parabolic	Late Wisconsinan – Holocene	NW	Partly stable; active	Seppälä, (1971); (1995b)
Northern Sweden	Dunes: parabolic	Late Wisconsinan – Holocene	NW	Partly stable	Seppälä, (1972a)
Qinghai Xizang (Tibet) Plateau:	Dunes: barchan	Recent	W	Active	Wang and French (1995c)

compressional features and irregular layering in the sand dunes reflect the melt ("deniva-tion") of sand-covered snow lenses on the slip faces of many dunes. Their cold-climate origin is deduced from the inclusion of plant, pollen, or faunal (insect) remains indicative of cold environments and the presence of wedge structures indicative of either permafrost or seasonal frost. Dune migration can be rapid; for example, on the Tibet Plateau a cres-centic (barchan) dune was observed to move approximately 350 m in four years (see Figure 10.11) (Wang and French, 1995c, pp. 266–167).

Sandstone bedrock often provides a local source of sediment supply for sand dunes and sand sheets. Also, coarse-grained glacigenic deposits (eskers, kames, glacio-fluvial sedi-ments) are the source for many dunes (McKenna-Neuman and Gilbert, 1986; Seppälä, 1971, 1972a; Van Vliet Lanoë et al., 1993). Late summer and winter transport is dominant, especially on surfaces associated with proglacial discharge regimes. By contrast, on allu-vial flood plains of rivers with nival discharges, sediment supplied at peak discharge in the spring provides material for eolian transport at low discharge throughout the summer (Fernald, 1964; Good and Bryant, 1985).

10.4. COASTAL PROCESSES AND LANDFORMS

A number of texts describe beach processes, coastal sedimentation, coastal morphology, and the applied aspects of the world's shorelines (Bird, 2000; Komar, 1976). Several geo-morphological summaries of cold-climate coasts are also available (John and Sugden, 1975; Viles and Spencer, 1995, pp. 254–288; Walker, 2005a, b).

The extent of coastlines currently experiencing periglacial conditions is considerable. In the northern hemisphere, the Arctic Ocean is almost completely landlocked and exten-sive coastlines occur in Alaska, Northern Canada, Greenland, and along the whole of the Russian mainland from the Kola Peninsula in the west to the Chuckchi Sea in the east. In addition, there are the various island groups such as the Canadian Arctic archipelago, Svalbard, Franz Josef Land, Novaya Zemblya, the North Land, the New Siberian Islands, and Wrangel Island. In Antarctica, although the majority of the coastline is composed of ice shelves and glaciers, the Northern Victoria Land coastline and the Antarctic Peninsula and South Shetland Islands are ice-free for several months of the year.

On the large scale, the geomorphology of cold-climate coasts is influenced by factors such as geological structure, tectonic activity, and Quaternary sea-level fluctuation. On the medium scale, cold-climate coasts consist of a suite of depositional and erosional landforms such as beaches, spits, inter-tidal flats, salt marshes, dunes, cliffs, and shore platforms. These occur in both periglacial and non-periglacial environments and reflect the wind and wave regimes and the weathering and sediment-supply characteristics of the climate. On the small scale, cold-climate coasts are dominated by the presence of ice. This results in a number of distinctive characteristics that can be summarized under four head-ings: the coastal–sea ice interface, the influence of sea ice upon wave generation, the effects of ice upon the beach, and the influence of permafrost and ground ice upon coastal development.

10.4.1. The Coastal–Sea-Ice Interface

Sea ice forms during winter from a suspension of small ice crystals ("frazil ice") which grow and coalesce to form a thin skin ("nila"). This evolves into a field of "pancake ice," small ice bodies about 3 m in diameter and up to 50 cm thick, which then coalesce to form the sea-ice cover. The maximum thickness of annual (single-year) sea ice varies between 30 cm and 200 cm. If sea ice were not to melt completely during the following summer, as is the case for over 50% of the Arctic Ocean sea ice, multi-year ice is created. This is thicker and harder than single-year ice and may exceed 2.5 m in thickness.

Sea ice also includes the large ice masses that enter the polar oceans from ice shelf break-up (where ice shelves are floating ice sheets) or from direct calving of tidewater glaciers. In the northern hemisphere, Baffin Bay, parts of Newfoundland and eastern Canada, and the northern North Atlantic are most affected by icebergs. In the Antarctic, most single-year ice melts as it floats northwards towards the Antarctic Convergence and the Southern Oceans but multi-year ice forms in the Weddel, Ross, and Bellinghausen Seas.

Sea ice dominates the surface of both the Arctic Ocean and extensive areas of the Southern Oceans that surround the Antarctica continent. In Canada alone, ~90% of its Arctic coastline of over 150000 km is affected by sea ice. The winter extent in the northern hemisphere is approximately $15 \times 10^{-6} \text{km}^2$ but reduces to approximately $8 \times 10^{-6} \text{m}^2$ in the summer (LeDrew et al., 1992). The extent of Antarctic sea ice is even larger: $20 \times 10^{-6} \text{km}^2$ in winter reducing to $3 \times 10^{-6} \text{km}^2$ in summer. Superimposed upon this significant seasonal variability is an inter-annual variability that is now complicated by global warming trends. For example, in the last 30 years, the thickness of the Arctic sea ice has thinned significantly (some estimates indicate a 30% loss) and in Antarctica large parts of the Larsen and Ross Sea ice shelves have recently broken away.

10.4.2. Sea Ice, Wave Generation, and Sediment Transport

Because of sea ice, wave action on the beach may be restricted to as little as 8–10 weeks a year in certain sheltered parts of the Canadian arctic (Taylor and McCann, 1976). In other areas, such as the southern Beaufort Sea and the northern Siberian coast, open-water conditions may develop for several months of the year (Are, 1998; Hequette and Barnes, 1990; Hume and Schalk, 1964, 1967). In the North Atlantic, warm Gulf Stream water keeps parts of Svalbard virtually ice-free for most of the year. Thus, ice cover, or the nearby presence of the permanent pack, is an important control over the distance of fetch and the time period over which wave action can operate. The effectiveness of wave action may be further restricted by the presence of a narrow strip of land-fast ice (the "ice foot"), which remains frozen to the shore and is unaffected by tidal movements. Thus, Arctic beaches are essentially low-energy environments in which normal wave action and coastal processes are limited. However, shore-ice action is especially important in the cold meso- and macro-tidal environments of the mid-latitudes, where it disturbs the intertidal zone by pushing and disrupting sedimentation and vegetation growth (Allard and Tremblay, 1983; Dionne, 1975, 1989; Jahn, 1977).

The magnitude and frequency of periods of storm events assume great importance in most assessments of coastal conditions. The majority of sediment transport and beach reworking occurs during the few major events which have a frequency of occurrence of two or three times a year. This is also true for periglacial environments (Harper et al., 1988; McCann, 1972; Owens and McCann, 1970; Reimnitz and Maurer, 1979). For example, observations along the northern Alaskan coast near Point Barrow (Hume and Schalk, 1967; Hume et al., 1972) indicate that coastal cliff retreat rates between 1948 and 1968 were related to the annual frequency of westerly-wind storms. It was concluded that a 50% decrease in storms during the open-water season resulted in a decrease in cliff retreat from approximately 4 m/year to 1 m/year.

Standard frequency and magnitude concepts can be used in the study of cold-climate coasts. Periods of maximum wave generation are likely to occur during those storm events which coincide with the existence of maximum ice-free fetch. For example, in the Western at Sachs Harbour during the period 1971–1977, no storms acted over a fetch greater than 100 km and the long-term probability of such an event was less then 0.1%. Of the 47 storm events with ice-free fetch in the same period, less than 40% had predicted significant wave heights greater than 2.0 m (Harry et al., 1983). Storm surges and storm-generated sediment transport in the Beaufort Sea area are also described (Forbes, 1989; Hequette and Hill, 1993). These, and other such studies, demonstrate not only that sea ice is an important restraint upon wave generation and coastal erosion, but also that traditional methods of coastal analysis are useful, even in ice-infested waters.

There are problems inherent in isolating and defining the geomorphic effectiveness of single, major events. For example, beach borrow associated with construction activity may increase the effectiveness of a storm event, while coastal engineering structures will reduce such events (Hume and Schalk, 1964; Walker, 1991).

As the direct result of the limited wave action experienced by many Arctic beaches, beach processes and the transport and redistribution of material are thought to operate at slow rates. In general, beaches are poorly developed and narrow, often composed primarily of coarse sand and cobbles. Beach sediments are poorly sorted and possess low roundness values compared to other beach environments (Table 10.9). However, there is abundant evidence that considerable long-shore transport of material may take place because many coastlines show dynamic growth of complex depositional features such as spits and offshore bars (Hequette and Ruz, 1991; Ruz et al., 1992). In the Western Arctic, two of the largest are those at Point Barrow, Alaska, and Cape Kellett, southwest Banks Island. The latter extends for a total distance in excess of 12 km. In the vicinity of Sachs Harbour, M. J. Clark et al. (1984) concluded that annual sediment gain over a 29-year period was ~57 000 m³ and that a 1.6 km long spit at the mouth of the harbor could have grown in a period estimated at between 70 and 230 years (Figure 10.12). Along the northern coast of Alaska, Hume and Schalk (1967) report approximately 10 000 m³ of annual net sediment transport is typical.

Table 10.9. Some roundness values for beach materials in various periglacial and non-periglacial environments.

Locality	Beach Environment	Rock Type	Cailleux roundness values
Periglacial:			
Devon Island, NWT, Canada	Sheltered	Limestone	25–267
Hall Beach, NWT, Canada	Sheltered	Limestone	216
Jacobshaven, west Greenland	Sheltered	Quartz	90–105
Kuggsa Dessa, west Greenland	Exposed	Quartz	135–160
Godthaab, west Greenland	Exposed	Quartz	270–388
Non-periglacial:			
Western Mediterranean (various sites)	Enclosed sea	Limestone	355
Lake Ontario, Canada	Enclosed sea	Gneiss	388
Finistère, northwest France	Exposed	Quartz	250–270 400–460

Source: McCann and Owens (1969).

(a)

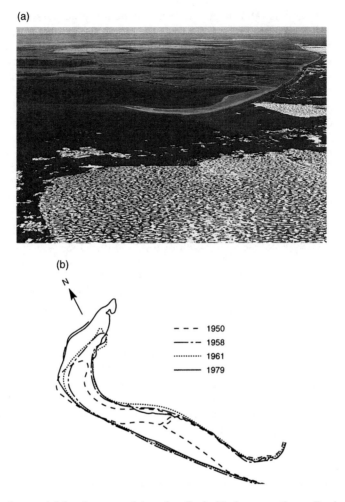

(b)

Figure 10.12. Sequential development of the spit at Sachs Harbour, southwest Banks Island, NWT, Canada, 1950–1979. (a) Oblique air view of the spit in July 1972. Note the pack ice offshore. (b) Change inferred from air photo coverage, 1950, 1958, 1961, and ground survey, 1979. From Clark et al. (1984), Harry et al. (1983).

10.4.3. Ice on the Beach

The ice foot plays an important role in limiting beach reworking. Ice forms in the inter-tidal zone during autumn when swash or spray from breaking waves freezes on contact with the beach. Once a layer of ice has formed, wave-induced movement of beach material ceases. Likewise, in spring, ice may remain frozen to the beach long after the sea ice breaks up and prevents the direct action of waves for several weeks. Multi-year fast ice may also develop (Reimnitz et al., 1995). The development and persistence of beach-fast ice is helped if permafrost is present in the inter-tidal sediments (Allard et al., 1992; Dyke, 1991; Dyke and Wolfe, 1993; Hansell et al., 1983). The presence of beach-fast ice also means that conventional freeze-up and break-up dates do not necessarily apply to the Arctic beach zone.

The size and extent of the ice foot varies from year to year depending primarily upon the sea conditions the previous autumn; moderate to strong wave action is an optimum for ice-foot accretion. The tidal range and beach slope are other factors which influence ice-foot width and thickness. The greater the tidal range, the greater is the zone of spray and swash accretion, while the steeper the beach profile, the smaller is the width and thickness of the ice foot. Usually, the ice foot ablates during the early part of the open-water season. Initially, a shore lead develops some 5–15 m off-shore to isolate the sea ice proper from the ice foot. Then the ice foot is breached by water draining seawards from melting snow in the backshore zone.

The effectiveness of storm events can be influenced by either the presence of unusually long-lasting beach-fast ice, which would inhibit erosion, or its removal by a prior storm event, which would accentuate erosion.

The more direct effects of ice on the beach occur when sea ice is forced, by strong onshore winds, to impinge upon the shoreline. Typically, tabular ice bodies push and scour beach material to form irregular ridges and scours in and above the high-tide zone (Barnes, 1982; Dionne, 1975, 1989; Hume and Schalk, 1964; Owens and McCann, 1970; Taylor and McCann, 1976). Where ice becomes buried by beach material, its eventual melt produces pits, depressions, and irregular topography. In general, these beach forms are short-lived because the shorelines are systematically reworked every year.

Boulder barricades are also typical of many cold-climate coastlines. These are elongate rows of closely packed boulders that flank the coastline and which are usually separated from the shore by a low-gradient nearshore zone (Rosen, 2005). They are the result of ice movement and the pushing of boulders during late winter and spring (Lauriol and Gray, 1980). They are especially well developed along the coastline of eastern Canada (Dionne, 1994, 2002), where they consist of glacial erratic boulders that rest on rock-cut shore platforms. Ice-push ridges also occur around the shores of inland-water bodies that are large enough to permit wind-induced movement of the ice cover. Smaller ice-push ridges can also form around large thaw lakes. Because of the absence of a tidal range and no well-developed ice foot, all these lacustrine push-ridges form at the same elevation each year and are relatively regular in plan and form.

10.4.4. The Influence of Permafrost and Ground Ice

In areas where the coastline is developed in permafrost terrain composed of unconsolidated and ice-rich sediments, rapid bluff retreat may occur. Fine-grained sediments are usually associated with high ice contents, and storm waves result in either the exposure of ice bodies to rapid melt or the undercutting and formation of thermo-erosional niches at high-tide level. The former induce retrograde thaw-slumping (see Chapter 8) while the latter lead to dramatic block collapse, often controlled by ice wedges which act as lines of weakness (see Figure 10.4B). In both cases, fine sediment is released, carried away, and subsequently deposited elsewhere along the coast. The coastal processes of melt and erosion are sometimes termed thermo-abrasion (Are, 1983, 1998). This is common along the coasts of the permafrost lowlands of Siberian and arctic North America and along the banks of major rivers (see earlier). It is especially effective wherever coastal bluffs are developed within ice-rich and unconsolidated sediments (Dallimore et al., 1996a; Hume et al., 1972; Mackay, 1963, 1986a; Shur et al., 2002).

In northern Alaska and the western Canadian Arctic, average rates of bluff retreat of between 2 and 4 m/year are typical (Harry et al., 1983; Hume et al., 1972; Mackay, 1986a) but where beach gravels are present, retreat is less. Along the Yukon coast and in the Mackenzie District of Canada, a similar pattern of rapid coastal retreat can also be observed (Mackay, 1963; McDonald and Lewis, 1973; Ruz et al., 1992, table 1). Measurements at the Yukon–Alaska border indicate the coast there has retreated 43 m since 1912, while the southeast side of Herschel Island has probably undergone 1–2 km of retreat in postglacial times. In the vicinity of Tuktoyaktuk, coastal retreat of over 150 m has occurred since 1935 (Mackay, 1986a). In one locality this has caused the draining of a lake and the growth of a number of pingos in the drained-lake bottom (see Chapter 6). Similar rates of retreat characterize the coastal lowlands of Siberia, where retreat rates of between 4 m and 6 m per year are not uncommon (Are, 1972). Barr (1976) cites historical evidence for the complete disappearance of certain islands in the Laptev Sea and attributes this to rapid coastal retreat. Some erosion adjacent to settlements may be due to beach borrow to maintain townsite and infrastructure facilities (Walker, 1991).

The rapid coastal retreat observed along the Beaufort Sea and northern Siberian coasts is clearly exceptional, being the result more of thermo-abrasion than true coastal erosion. For the most part, wave action and coastal processes play a subsidiary role in transport of thawed sediments away from "source" areas towards "sinks" or zones of deposition. While these coastal landforms are not dissimilar to those in non-permafrost environments, it is their rate of formation and the speed of sediment production that is distinctive. The sequence of coastal evolution characteristic of ice-rich thermokarst lowlands is summarized in Figure 10.13. Rapid coastal retreat that results from thermo-abrasion leads to the formation of headlands and bays as thermokarst lakes and depressions are breached. Abundant sediment is transported by currents and littoral processes to form spits and tombolos. With time, coarser sediment begins to armor the beach and offshore barriers and barrier islands are formed.

Finally, it must be emphasized that extensive coastlines in the central and eastern Arctic, Greenland, Svalbard, Northern Victoria Land, and the Antarctica Peninsula are developed in coherent bedrock in which ice amounts are low and where erosion rates reflect the more normal relations between rock resistance and wave action. The absence of fine-grained sediments limits depositional features to coarse-grained pebble beaches often related to storms and extreme events. Shorelines are dominated by recent, glacially-induced, isostatic uplift rather than submergence. Abrupt coastal cliffs, shore platforms and raised beaches are typical coastal features.

10.4.5. Cold-Climate Deltas

Deltas form wherever rivers carrying high sediment loads enter the marine environment. Their formation in cold-climate environments is informative on several accounts. First, they represent sedimentation under cold-climate conditions. As such, they provide an ideal setting for the growth of syngenetic permafrost bodies. Second, they illustrate the interactions of river and sea ice and of fresh water mixing with saline water. Third, deltas supply large volumes of sediment to the coastal system, which, in turn, promotes the formation of depositional features.

The large deltas that concern periglacial geomorphology include those of the Mackenzie, Colville, and Yukon rivers in North America and the Ob, Yenisei, Lena, Indigirka, and Kolyma in Siberia. They are well studied primarily because of their abundance of ice wedges, pingos, and other permafrost aggradational landforms and for the thermokarst features that reflect the ice-rich silty nature of the underlying sediments. All these features have been discussed earlier in Chapters 5, 6, 7, and 8. A number of studies document the geomorphology of several of these large cold-climate deltas (Korotayev, 1986; Mackay, 1963; Walker, 1974, 1978, 1983, 1998).

More relevant here are the many smaller meso-scale deltas that occur in periglacial coastal environments. They have been little studied but one such delta is that of the

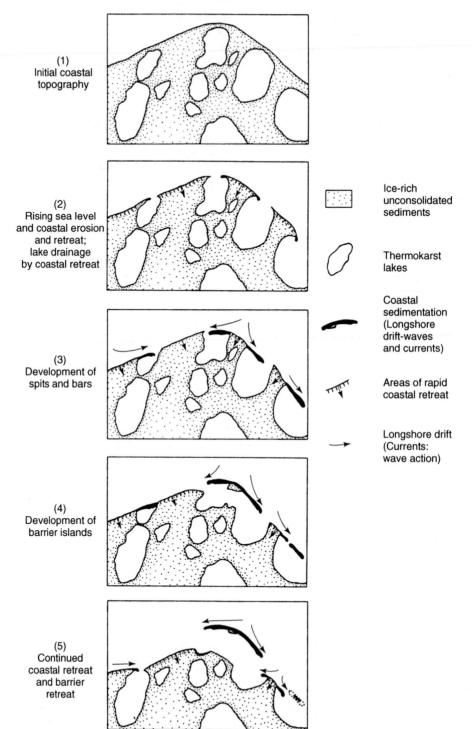

(1)
Initial coastal
topography

(2)
Rising sea level
and coastal erosion
and retreat;
lake drainage
by coastal retreat

Ice-rich
unconsolidated
sediments

Thermokarst
lakes

(3)
Development of
spits and bars

Coastal
sedimentation
(Longshore
drift-waves
and currents)

Areas of rapid
coastal retreat

(4)
Development of
barrier islands

Longshore drift
(Currents:
wave action)

(5)
Continued
coastal retreat
and barrier
retreat

Figure 10.13.   Typical coastal evolution in lowland thermokarst terrain adjacent to the Beaufort Sea that is subject to a slowly transgressing sea level, according to Ruz et al. (1992).

modern Horton River, in the western Canadian Arctic (Mackay and Slaymaker, 1989). The delta is distinctive on account of its relative absence of vegetation and the numerous scour troughs and other disturbances at the surface. The major process affecting the surface morphology is the annual river break-up. The low gradient, numerous channels, and the presence of sea ice early in the summer means that ice jams commonly form. The river ice bulldozes, scrapes, and flutes the delta surface. A further characteristic is that the surface lacks typical levees because much of the delta floods because of the high water levels induced by the ice dams. It is probable that similar topography and pattern of geomorphic change characterize other meso-scale deltas in cold-climates.

ADVANCED READING

John, B. S., Sugden, D. E. (1975). Coastal geomorphology of high latitudes. *Progress in Geography*, **7**, 53–132.

Prowse, T. D., Ommanney, C. S. L., eds. (1990a). Northern hydrology. Canadian perspectives. National Hydrology Research Institute, Environment Canada, science report no. 1, 208 pp.

Prowse, T. D., Ommanney, C. S. L., eds. (1990b). Northern hydrology. Selected perspectives. National Hydrology Research Institute, Environment Canada, science report no. 6, 532 pp.

Seppälä, M. (2004). *Wind as a Geomorphic Agent in Cold Climates*. Cambridge University Press, Cambridge, 358 pp.

DISCUSSION TOPICS

1. What is the importance of fluvial processes in fashioning periglacial landscapes?

2. How similar are the channel and valley forms of periglacial and non-periglacial environments?

3. What is the importance of wind in periglacial landscape evolution?

4. How unusual are the coastal processes and landforms of cold environments?

PART III

Quaternary and Late-Pleistocene Periglacial Environments

11 Quaternary Periglacial Conditions

Many of the high latitudes areas of North America and Eurasia that currently experience periglacial climates also experienced episodes of cold, non-glacial conditions during the Quaternary. At the same time, extensive areas in the mid-latitudes of the northern hemisphere that lay south of the continental ice sheets were also subject to periods of intense cold. This chapter examines the evidence for Pleistocene periglacial conditions in the high latitudes and the roles played by past permafrost and frost-action processes in fashioning the mid-latitude landscapes.

11.1. INTRODUCTION

The Quaternary is the chronostratigraphic unit that refers to the last 2–3 million years. At its lower boundary, it merges into the Late Pliocene. Some regard the Quaternary as beginning as early as ~3.0 Ma, others take a paleomagnetic reversal at 2.6 Ma (Gauss) as the cut-off.

Traditional, the Quaternary is divided into two units: first, the Pleistocene, subdivided into Early, Middle, and Late Pleistocene, and, second, the Holocene, the last 10 000 years, that extends into today. Colloquially, the Quaternary is sometimes referred to as "The Ice Ages" because the fluctuating climates of the time enabled the growth and decay of continental ice sheets on many occasions.

The onset of growth of the Antarctic ice sheet began much earlier than the Quaternary, possibly during the Late Miocene, approximately 10–13 million years ago. The reason for this is still unclear. A contributing factor may have been related to the migration of the Antarctic Plate towards its present polar position in Late Tertiary times, thereby allowing a strong zonal westerly-wind circulation to develop across the entire southern latitudes, effectively isolating the Antarctic continent.

11.2. THE TIME SCALE AND CLIMATIC FLUCTUATIONS

Changing oxygen-isotope ratios measured in the carbonate shells and skeletons of small benthic and planktonic organisms (foraminifera) buried within deep-sea sediments provide the best record of Quaternary climatic fluctuations. This is because the temperature of the deep ocean has probably remained unchanged throughout the entire Pleistocene and these bottom-living organisms incorporate oxygen δO^{16} and δO^{18} in direct proportion to that contained in seawater. The composition of the latter can change and reflects the fact that the heavier isotope, δO^{18}, is preferentially enriched in the liquid phase and depleted in the vapor phase. These processes are dependent on temperature. Consequently, on ice sheets and glaciers today, the snow in winter can be distinguished from isotopically-heavier summer layers. In the context of the Quaternary, a large proportion of global water must have been locked up in ice sheets as global sea level fell. As a result, deep ocean seawater was gradually depleted of δO^{16}. Thus, variations in the $\delta O^{16} : O^{18}$ ratio can be used to indirectly estimate the extent of glaciation during time of deposition of the shells.

Comparison of deep-sea cores demonstrates that oxygen-isotope curves can be correlated worldwide. A number of so-called "oxygen-isotope stages" (OIS) are now recognized; those for the last 1.6 million years are presented in Figure 11.1. The corresponding magnetostratigraphy is also shown in Figure 11.1. The latter is based upon the fact that, when lava cools, any newly-formed crystals of magnetic minerals are aligned with the Earth's magnetic field. At different times in Earth history, the magnetic polarity has changed. The recent (normal) Brunhes Epoch began about 780000 years ago and followed upon the reversely-polarized Matuyama Epoch, which began about 2.6 million year ago. The Matuyama was interrupted by short periods of normal polarity (Jaramillo and Olduvia Events). These various events (Figure 11.2) now assist in dating many Quaternary-age cold-climate sediments.

Other lines of approach used in paleo-environmental reconstruction include palynology, where maximum July air temperatures are inferred from plant pollens found in terrestrial stratigraphic cores, the analysis of tree rings (dendrochronology), where precipitation and temperature are inferred from the width of annual growth bands, and the identification of remains of insects, such as beetles (coleoptera), that have narrow temperature tolerance limits or faunal remains (bone) that are capable of amino-acid dating.

On the time scale of the Quaternary (i.e. the last ~3 Ma), periglacial conditions roughly coincided with the various advances and retreats of continental ice sheets. In broad outline, the various glacial and interglacial stages are well known for both North America and Europe. For example, in Central and Western Europe, the stratigraphic position of numerous frost-fissure casts, involutions, and various cold-climate mass-wasting deposits suggests a number of periods during the Middle and Late Pleistocene when either permafrost or cold-climate (frost action) conditions were present (Figure 11.2). The Early Pleistocene is least well known.

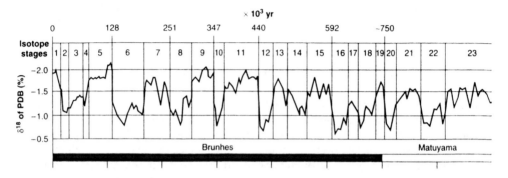
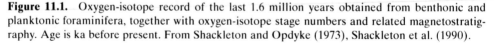

Figure 11.1. Oxygen-isotope record of the last 1.6 million years obtained from benthonic and planktonic foraminifera, together with oxygen-isotope stage numbers and related magnetostratigraphy. Age is ka before present. From Shackleton and Opdyke (1973), Shackleton et al. (1990).

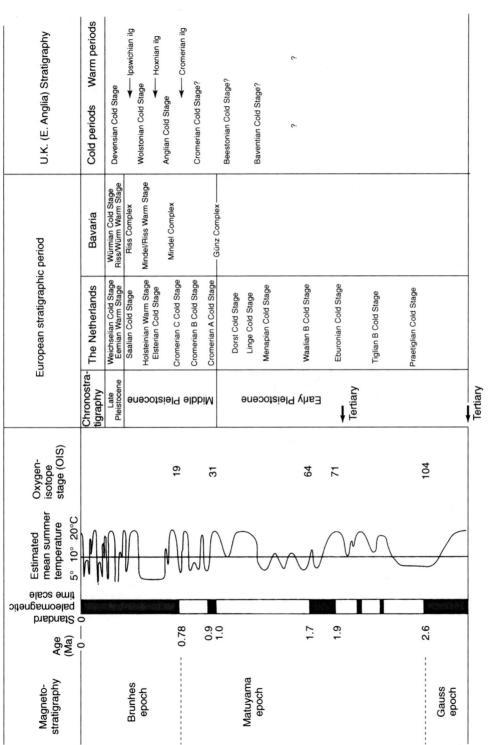

Figure 11.2. Generalized mean summer temperature for the Quaternary together with the paleomagnetic time scale, associated oxygen-isotope stages, and the European stratigraphic record with the occurrence of major cold-climate events and their nomenclature. Both possible Quaternary/Tertiary boundaries are indicated by arrows. Modified from West (1977), Ehlers (1996, table 12).

From a practical point of view, we need only be concerned with the more recent cold-climate fluctuations. This is because most periglacial landforms and deposits that formed in earlier cold periods are unlikely to have survived to the present day without having been modified beyond recognition by either later cold stages or by intervening (interglacial) periods of more temperate climate.

The record of the last 100000–150000 years is also illustrated by the oxygen-isotope curves obtained from the Byrd Station (Antarctica) and Camp Century (Greenland) ice cores (Figure 11.3). They indicate that, in both Europe and North America, the penultimate cold stage (Illinoian/Riss; OIS-6) occurred some time prior to 150000 years BP. The last interglacial, or warm, period (Sangamon/Eemien; OIS-5e) is generally believed to have lasted until ~120000 years ago, after which the climate deteriorated during OIS-4 through OIS-2. The final cold stage lasted until approximately 10000 years ago. Not surprisingly, the last cold period (Wisconsinan/Weichselian) is best known, and several

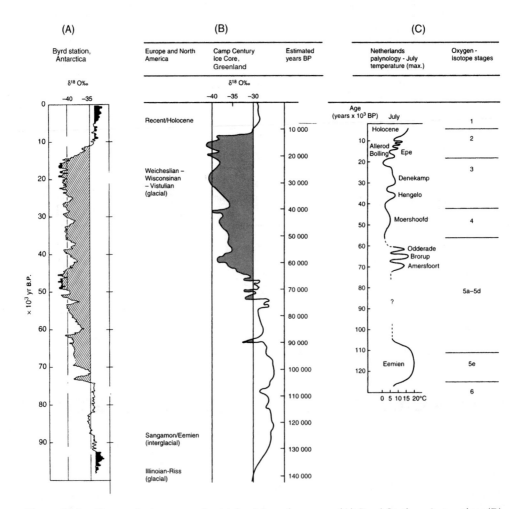

Figure 11.3. Oxygen-isotope records obtained from ice cores. (A) Byrd Station, Antarctica. (B) Camp Century, Greenland, together with time scale. (C) Generalized Late-Quaternary isotope stratigraphy, and July temperatures inferred from palynology. Colder temperatures are indicated by decreasing O[18] values (shaded areas). From Johnsen et al. (1972), West (1977, figure 10.10), Vandenberghe and Pissart (1993).

climate fluctuations within this time period are also recognized. For example, Figure 11.3 also shows the generally accepted European stratigraphic record for The Netherlands, Central Germany, and the UK (East Anglia), together with inferred July air temperatures compiled from palynological data for the last 120000 years. To add complexity, however, the North American chronostratigraphy is equally detailed. For example, Figure 11.4 attempts to correlate the Pliocene and Pleistocene cold-climate events in the western Cordillera with those of the continental glaciations of North America.

During the last glacial stage, environmental conditions fluctuated considerably, both in time and space, in tandem with the growth, advance, and retreat of the ice sheets. Smaller-scale oscillations of more temperate, or interstadial, climatic conditions can be recognized. Thus, Late-Pleistocene periglacial conditions varied greatly in both intensity and duration. In northwest Europe, for example, fluctuations during the last 60000–70000 years (the Weichselian glacial period) resulted in arctic and sub-arctic conditions for only about 25000–30000 years. There is evidence to believe that broadly similar climatic fluctuations occurred during the Wisconsinan in North America (see Figure 11.4).

It seems reasonable to assume that periglacial conditions developed around the ice margins at the height of the cold stages, and that interstadials saw fluctuations between periglacial and non-periglacial conditions. The latest transition into the Holocene illustrates the oscillating nature of cold-climate conditions. For example, in Europe, the late and post-glacial stages can be divided into pollen zones, indicating the dominance of different vegetation sequences (Table 11.1). Fluctuations in the arboreal and non-arboreal pollen ratio indicate that during deglaciation, two periods of relatively warmer conditions occurred, the Bolling and Allerod interstadials, at which time tundra vegetation was replaced temporarily by shrub-tundra (dwarf birch) and birch forest. Then, in postglacial times, there was a peak of warmth and dryness between 5000 and 3000 years BP. In Eastern North America, the vegetation sequence is more general (Table 11.1) and suggests that the tundra belt was narrower and the climate more humid than in Europe. It is highly likely that the climatic amelioration in eastern North America was accentuated by large proglacial water bodies, such as those of the proto-Great Lakes and the Champlain Sea. As a result, the tundra belt was quickly replaced by spruce-fir, and then by pine.

More detailed descriptions of the Quaternary climatic fluctuations are beyond the scope of this book. An abundant literature is available, and the reader is referred to recent texts by Bennett and Glasser (1996), Bell and Walker (1992), Ehlers (1996), and Bradley (1999).

11.3. GLOBAL (EUSTATIC) CONSIDERATIONS

Any discussion of Pleistocene cold-climate conditions must be placed within the broader context of the other global changes that occurred during the Quaternary. Two are of special relevance to periglacial environments.

11.3.1. Sea-Level Changes

Today, if the Antarctic ice sheets were to melt completely, global sea level would rise by ~60 meters. Thus, fluctuations undoubtedly occurred during the Quaternary in response to the growth and decay of the continental ice sheets. Thus, many terrestrial lowlands that were subject to Late Pleistocene cold-climate conditions are now being either submerged or eroded by the latest (Holocene) transgression (see Chapter 10). Based on borehole data from the North Sea, global sea level was more than 50m below present level during the

Northwestern North America   **Continental USA**

AGE			CORDILLERAN GLACIATIONS (Northern Canada, east-central Alaska)	CONTINENTAL GLACIATIONS (Banks Island)	
Geomagnetic time scale and magnetic polarity	Geologic time scale	Marine Isotope Stage			
Late Brunhes < 0.25 Ma	LATE PLEISTOCENE	2	Salcha; Eagle	Laurentide Ice Sheet	Wisconsinan
		8 / 10 / 12	McConnell Gayna River		
Early Brunhes 0.78 - 0.25 Ma (N)	MIDDLE PLEISTOCENE		Mount Harper; Black Hills	Thomsen and 2 unnamed glaciations	Illinoian
			7th and 8th glaciations (no formal names)		
		19			
Latest Matuyama 0.99 - 0.78 Ma (R)	EARLY PLEISTOCENE		Sixth glaciation (no formal name) / unnamed event / unnamed event	Banks Island Glaciation (s) ?	Pre-Illinoian
Late Matuyama (Jaramillo) 1.07 - 0.99 Ma (N)		30	Fifth glaciation (no formal name) / unnamed event / unnamed event		
Late Matuyama 1.77 - 1.07 Ma (R)			Forth glaciation (no formal name) / unnamed event		
Early Matuyama (Olduvai) 1.95 - 1.77 Ma (N)	PLIOCENE	70 / 72	Darling Creek Glaciation > 1.6 Ma Oldest Drift > 1.6 Ma / Third Glaciation (no formal name)	unnamed ?	
Early Matuyama 2.58 - 1.95 Ma (R)		100	Second Glaciation (no formal name)	unnamed ?	
Gauss 2.9 - 2.58 Ma (N)			First Cordilleran Ice Sheet	unnamed ?	

Figure 11.4. Simplified Quaternary magnetostratigraphy and chronostratigraphic correlations between northwestern North America (northern Canada, east-central Alaska, Banks Island), and the northern continental United States. From Duk-Rodkin et al. (2004). Reprinted from Ehlers et al, 2004, with permission from Elsevier.

Table 11.1. Generalized sequence of Late-glacial (Wisconsinan/Devensian) and Post-glacial (Flandrian/Holocene) vegetation successions in the British Isles, New England, and southern Québec-St Lawrence Lowlands. Shaded area indicate the probable time periods when either discontinuous permafrost or deep seasonal frost was present.

Stage	Age (BP)	Period	British Isles (West, 1977) Pollen zone	British Isles (West, 1977) General characteristics of Vegetation (dominants)	New England (Peteet et al., 1994)	Southwest Québec (Richard, 1994)
Postglacial (Holocene)		Sub-Atlantic	VIII-modern	Afforestation	North American mixed forest (boreal-broadleaf)	
			VIII	Alder, oak, birch, beech, ash		
			VIIb	Alder, oak, lime-deforestation		
				Elm decline		
	5500	Atlantic	VI	Alder, oak, elm, lime		
	6500	Boreal	VI	Hazel, pine	Mixed white pine oak forest	
				(a) Oak, elm, lime	White pine forest	
				(b) Oak, elm		
				(c) Hazel, elm		Maple; pine
	9000	Pre-Boreal	V	Hazel, pine, birch		
	10000		IV	Birch, pine	Boreal forest	
Late Glacial (Wisconsinan)		Younger Dryas	III	Shrub tundra (dwarf birch)	? ?	Birch, pine
		Allerod	II	Birch with shrubs-tundra	Woodland (Picea, Betula)	Poplar spruce
	11000	Older Dryas	Ic			Dwarf birch; Herb tundra
		Bolling	Ib	Shrub tundra/tundra	Tundra	
	12000	Early Dryas	Ia		ICE	ICE Champlain Sea

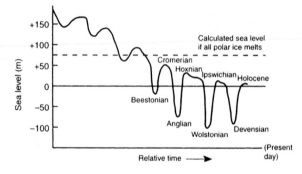

Figure 11.5. Generalized sequence of Quaternary sea-level changes. From Pethick (1984, pp. 220–221).

last Weichselian glacial maximum approximately 18 ka ago (Laban and van der Meer, 2004, p. 257).

Most discussion of Pleistocene-age shorelines usually centers upon glacio-isostatic rebound and raised beaches (Sissons, 1981; Synge, 1981; West, 1977, pp. 151–179). Moreover, if one accepts the "staircase" theory of Quaternary sea-level change (Figure 11.5) and ignores isostatic and tectonic uplift, it follows that shorelines and/or beach deposits directly related to the low sea levels of the Pleistocene are rare, preserved only in the stratigraphic record. For example, the well-known Pleistocene raised beaches that are found along the coasts of southern England and generally ascribed to marine oxygen-isotope stages 9, 7, and 5e (Bates et al., 2003; Davies and Keen, 1985), are covered by solifluction or "head" deposits (see Chapter 12). Their present elevation is the result of the slow uplift of the coastal zone caused by isostatic response to sediment unloading during accelerated phases of erosion in the cold stages (Bates et al., 2003).

In theory, on stable coastlines constructed in frost-susceptible bedrock, the combination of intense frost action and marine action should lead to well-developed coastal features such as cliffs, shore platforms, and beaches. For example, the coastline of the maritime Antarctica reflects the combination of recent deglaciation, isostatic uplift, and current marine action. However, the formation of similar shorelines during the cold stages of the Pleistocene is problematic, and the extent, nature, and duration of Pleistocene sea ice are highly speculative.

11.3.2. Uplift of Qinghai-Xizang (Tibet) Plateau

The Tibet Plateau is a vast upland that ranges in elevation from approximately 4000 meters a.s.l. to over 5000 meters a.s.l. Uplift commenced at the end of the Pliocene as the Indian and Eurasian plates collided. During the Quaternary, it is estimated that at least 3500 m of uplift has occurred. Uplift was especially rapid during the Late Pleistocene with rates of more than 10 mm/year (i.e. 1500–2000 m over the last 130 ka). Differential movement (faulting, geothermal springs) accompanied uplift. Today, the Kunlun and Himalayan mountain chains, to the north and south respectively, are youthful landscapes characterized by slope instability and rapid stream incision.

Just as the separation of the Antarctic Plate during Late Tertiary times helped create the strong zonal wind belt in the southern hemisphere mid-latitudes, so uplift of the Tibet Plateau affected the zonal wind systems of the northern hemisphere during the Quaternary. Penetration of monsoon wind into central Asia was prevented and the central

Asian climate became drier and more continental. Outward-flowing winds from the high-pressure systems that inevitably developed in winter transported dust particles (loess) southwards towards the lowlands of interior China, and northwards towards the Eurasia plains of Mongolia, Kazakhstan, and central and southern Siberia. The zonal westerly winds were also diverted to the north and south of the Tibet uplift while at the same time, the expansion of the northern hemisphere ice sheets led to an increase in the latitudinal pressure gradients and to generally stronger winds in the mid-latitudes.

11.4. PLEISTOCENE PERIGLACIAL ENVIRONMENTS OF HIGH LATITUDES

Intuitively, it seems reasonable to assume that the high latitudes experienced climatic conditions during the Pleistocene that were similar to those of today. However, there were two major differences. First, the ice-free areas of the high latitudes would have been in relatively close proximity to the continental ice sheets. As a consequence, strong gravity-induced winds, similar to those experienced today in the ice-free areas that surround the Antarctic continent, would have been common. Evidence for this is provided by the wide-spread occurrence of thick deposits of wind-blown silt in north-western North America, Western Europe, and the lowlands of central Russia. Second, because of lowered global sea level, many areas of high latitude would have been in more continental locations than they are today.

In essence, the high-latitude Pleistocene periglacial environments would have been dominated not only by a seasonal temperature regime, as today, but also by (1) greater continentality, (2) greater aridity, and (3) more intense wind action.

11.4.1. Extent of Past Glaciations

The extent of Quaternary glaciations determined, to a large degree, the extent of the Pleistocene periglacial environments of high latitude. Here, discussion is restricted to the northern hemisphere because it is believed the Antarctic ice sheet has remained relatively stable throughout the Quaternary.

In the northern hemisphere, continental-scale ice sheets developed over Eurasia and North America on several occasions (Duk-Rodkin et al., 2004; Rozenbaum and Shpolyanskaya, 1998a, b; Westgate et al., 2001). There is debate as to the extent of the Early and Middle Pleistocene ice sheets, and several areas that are thought to have been "never-glaciated" may have been covered by a thin veneer of cold-based ice rather than being "ice-free." This is probably the case for parts of the western North American Arctic (Vincent, 1983, 1989), where the relative lack of cold-climate faunal and plant remains that date to the time of the last glacial period suggests some areas may have been snow- or ice-covered. Likewise, the "unglaciated" upland terrain of the southern Mackenzie Mountains, NWT, which is thought not to have experienced glaciation for at least the past 350 ka (Ford, 1996), may also have been covered by stationary, cold-based snow bodies.

The nature and extent of the Late-Pleistocene glaciations of northern Siberia and northern North America and Greenland are much better known (Dyke and Prest, 1987; Fulton, 1989; Prest, 1984; Romanovskii and Hubberten, 2001; Romanovskii et al., 2000). Figure 11.6 is an international world map of glacier-ice extent during the last glacial maximum (ca 18000 ka) that shows the maximum extent of ice sheets and areas underlain by permafrost. The extent of ice-free land takes into account the lower global sea level of the time. Figure 11.6 shows the maximalist hypothesis of the extension of the Baltic-West

Siberian ice sheet. In North America, the Cordilleran, Laurentide, Queen Elizabeth Islands, and Greenland ice-sheets are recognized.

The largest ice-free areas existed in the western arctic of North America and in eastern Eurasia. These were joined during the Pleistocene to form an extensive land area known as Beringia. Remnants of Beringia occur today in central and eastern Siberia, central Alaska, and northwestern Arctic Canada. Much of the Eurasian portion of Beringia now lies submerged beneath the Laptev and East Siberian Seas. Not shown in Figure 11.6 is the sea-ice cover that, presumably, filled the Arctic Basin. One must envisage, therefore, that extensive areas of Beringia consisted of high-latitude tundra and steppe-tundra terrain that extended northwards to an ice-covered polar ocean. The southern parts of Beringia would have either bordered the northern Pacific Ocean or merged into the interior steppe-lands of central Asia (Tibet, Kazakhstan, Mongolia, and northeast China). In North America, a narrow ice-free area of Beringia extended southwards in Central Yukon Territory and other ice-free areas probably existed in the northern Mackenzie District and on western Banks Island.

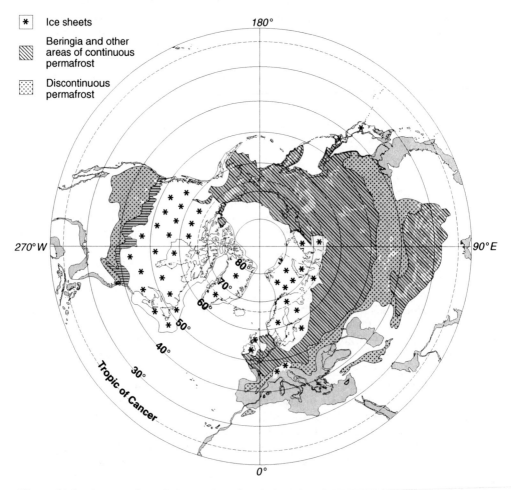

Figure 11.6. An overview of the northern hemisphere ice sheets and extent of the periglacial domain at the height of the last glacial maximum (~18 ka). From CCGM/CGMW-ANDRA (1999) and other sources mentioned in the text.

11.4.2. Relict Permafrost

Ice sheets may destroy, inhibit, preserve, or stimulate the growth of permafrost beneath the ice. This depends upon the thickness of the ice body, its cold- or warm-based nature, and its duration. Likewise, the cold-climate conditions that induced the ice sheets would cause the formation of permafrost in adjacent ice-free terrain. In general, one can envisage thick and cold permafrost in never-glaciated terrain and thinner or non-existent permafrost beneath ice sheets. If permafrost that formed during one cold period is preserved into the following cold period and then overridden by ice, complex permafrost bodies may result. For example, Figure 11.7 provides schematic illustrations of the sorts of relationships that might exist between permafrost thickness and the nature and extent

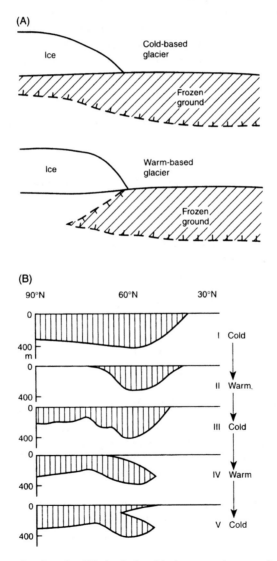

Figure 11.7. Diagram showing simplified relationship between the growth and decay of permafrost. (A) Cold-/warm-based ice sheets; (B) climatic fluctuations. From Romanovskii (1993).

of Pleistocene ice bodies and how permafrost bodies might reflect different climatic fluctuations.

There is abundant evidence to indicate that, in high latitudes, ancient permafrost survived the impact of warmer-than-modern periods (interglacials). For example, J. R. Mackay et al. (1972) were some of the first in North America to show that ground-ice bodies predated the last glaciation in the Western Arctic because the ice was beyond the limit of radiocarbon dating. In more recent years, many bodies of massive ice and icy sediment, both in the western North American Arctic and in western Siberia, have been interpreted as deformed buried glacier ice (see Chapter 7). Thaw unconformities and other ancient ice bodies provide additional evidence (Mackay and Matthews, 1983; Froese et al., 2000). For example, in central Yukon, relict ice wedges, at two sites separated by ~100 km, are overlain by a thaw unconformity which includes tephra beds dated at 740 000 +/− 60 000 years (Froese et al., 2006). Permafrost is thought to have persisted in central Yukon through at least six interglacials, including oxygen-isotope stages 11 and 5e, both considered to be longer and warmer than the present interglacial (the Holocene).

Insight into the actual paleo-permafrost environments can be inferred from various sources. These include speleotherms in caves, beetles (coleoptera) and snails (non-marine mollusca), pollen, and plant and faunal remains. There is a wide-ranging literature (Hopkins et al., 1982; Péwé et al., 1997; Zazula et al., 2003). Rodent middens, from ice-rich loess deposits in central Yukon, provide especially useful information (Zazula et al., 2005). They contain a diverse assemblage of graminoids, forbs, and mosses that suggests a range of floristic and animal habitats in Beringia, including steppe-like tundra on well-drained soils, wet tundra meadows on lowlands, and hydric habitats in valley bottoms. Inevitably, these northern ice-free areas acted as refugia for Pleistocene plants and animals at the height of the glacial periods. Some of the most explicit evidence is provided by the frozen carcasses of woolly mammoths and other mammals that are found preserved in the permafrost of northern Siberia, Alaska, and Canada (Figure 11.8). The fact that these large mammals were able to survive in relative abundance testifies to the productivity of the Pleistocene tundra ecosystem. Rodent middens are also excellent indicators of the cold-climate environments in which the animals lived. For example, the remains of arctic ground squirrel burrows and nesting sites in west-central Yukon Territory were dated at ~25 000 years BP (Zazula et al., 2005). If one assumes that the animals were unable to burrow into underlying permafrost, the maximum depth of the burrow reflects the active-layer depth. These are higher than those currently recorded, suggesting a more continental climate (higher summer temperatures) than today.

11.4.3. Syngenetic Permafrost Growth

Long-continued cold-climate sedimentation leads to syngenetic permafrost growth. Sediment aggradation occurs for a variety of reasons; these include eolian deposition on upland surfaces and in lee-slope positions, mass-wasting and sediment redeposition on lower slopes, and fluvial deposition on flood plains and alluvial surfaces in valley bottoms.

Pleistocene-age syngenetic permafrost is typically formed within fine-grained unconsolidated sediments. It is often ice-rich and associated with layered and lenticular cryostructures (see Chapter 7). These are regarded as typical of primary (unmodified) syngenetic permafrost (Bray et al., 2006; Shur et al., 2004) and reflect the upward growth of permafrost in response to an aggrading ground surface. Syngenetic permafrost also commonly contains locally-modified, or secondary, deposits possessing cryostructures

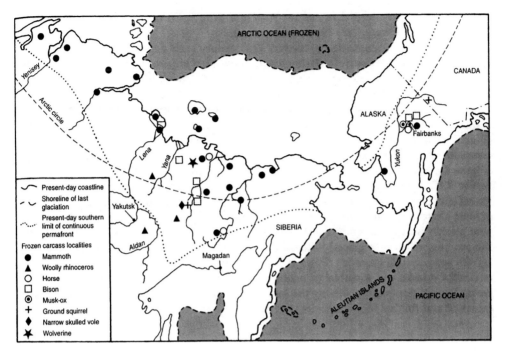

Figure 11.8. Distribution of frozen Pleistocene mammalian carcasses in Siberia, Alaska and northwestern Canada. Compiled from Sutcliffe (1985), Guthrie (1990), and others.

indicative of epigenetic freezing (see Figure 8.7). Often ice wedges act as lines of preferential thaw-erosion. These secondary or modified deposits are described as pseudomorphs. Cryostructures, and gravel, sand, silt, and ice pseudomorphs have all been discussed in Chapter 8.

Syngenetic permafrost growth is an important concept when understanding the nature of Pleistocene loess and "muck" deposits.

11.4.4. Loess Deposition

Wind action played a dominant role in fashioning the Pleistocene periglacial environments of high northern latitudes. The evidence lies in the form of extensive and thick deposits of loess-like silt (dust) that mantles extensive areas of central Alaska, south-central Yakutia, the Qinghai-Xizang Plateau, and other areas (Péwé and Journaux, 1983; Péwé et al., 1995). These wind-blown dust layers, generally termed loess (see Chapter 10), commonly contain ash layers from volcanic eruptions, ancient tree trunks and other plant and animal remains. They are well sorted; the typical grain size is shown in Figure 10.10.

Loess deposition in the high latitudes was favored by well-developed high-pressure systems that formed over the continental ice sheets and the gravity-induced (katabatic) winds that flowed outwards as a consequence. Other factors included the abundance of fine-grained material available for wind transport on the outwash plains and braided channel systems that were adjacent to the ice margin. For example, in central Alaska, much of the loess was derived from the alluvial floodplains of the Tanana and Yukon Rivers. In Siberia, the origin of the silt is more complex since much was probably derived from the gobi (cold) deserts of central Asia following uplift of the Tibet Plateau during late Tertiary and Quaternary times (see earlier).

One of the best documented records of Quaternary loess deposition comes from Gold Hill, near Fairbanks, Central Alaska, where large sections are regularly exposed during placer gold-mining operations. Originally, the loess that mantled the upland surfaces was termed "upland silt" (Péwé, 1955). Following initial deposition, the loess was re-transported by mass-wasting processes towards lower elevations. Typically, the thickest deposits of these loess-like materials now occur on valley-side slopes and in valley bottoms (Figure 11.9A) (Péwé, 1975). At Gold Hill, their magnetostratigraphy provides a near-complete record of deposition from approximately 3 Ma to the present (Preece et al., 1999) (Figure 11.9B). The silty sediments are characterized by large syngenetic ice wedges (see Figure 7.16A) and by numerous faunal and organic remains. Several major periods of global climate warming are recorded, with times of permafrost thawing and great erosion of loess alternating with periods of loess deposition and permafrost aggradation. A prominent marker horizon is the Eva Forest Bed, a layer of now-frozen organic material that is the remnant of an interglacial boreal forest that flourished approximately 125 000 years ago (oxygen-isotope stage 5e) (Péwé et al., 1997). It is clear that the last interglacial period in central Alaska was a period of major erosion of loess and deep and rapid thawing of permafrost. Subsequently, during the last 100 000 years, the treeless steppe environment returned and thawed loess deposits were refrozen.

11.4.5. Mass Wasting and "Muck" Deposits

The organic-rich and perennially-frozen silty sediments derived from reworked loess were first encountered by placer miners in the Kolyma, Klondike, and Alaska (Yukon) regions at the end of the 19th Century. They were colloquially termed "muck" deposits, a term now widely accepted in the English-language scientific literature. In central Siberia, similar materials are termed "Yedoma Suite."

Muck deposits are highly variable in composition yet, because of their ice-rich and organic-rich nature, provide useful cryostratigraphic and paleo-environmental information. For example, a typical cryostratigraphic section of Klondike muck deposits shows silty material, containing large ice wedges and numerous organic and faunal remains, overlying Late Tertiary/Early Quaternary gold-bearing creek gravels (Figure 11.10). Although the Early Quaternary record is often incomplete, several periods of thaw and/or erosion are usually indicated in the Late-Pleistocene sediments (Fraser and Burn, 1997; Kotler and Burn, 2000). They are also clearly cold-climate in origin because ice wedges are preserved in the basal unit while the main unit contains ice that has an isotopic composition indicative of full-glacial conditions (δO^{18} of -32 to $-29‰$; δD of -234 to $-257‰$). Evidence that the permafrost is syngenetic in nature is provided by aggradational ice (see Chapters 5 and 7). Finally, thermokarst-cave ice ("pool ice") indicates fluvio-thermal erosion and localized secondary modification (see Chapter 8). In some places, ground-water intrusion, probably along the base of the active layer, has resulted in massive icy beds.

These typical characteristics of muck deposits provide useful information for paleo-environmental reconstruction. They can also be reconciled with our understanding of permafrost conditions today (see Part II). Fuller accounts of the numerous faunal and organic remains that are contained within these perennially-frozen muck deposits can be found in Guthrie (1990, 2001), Harington (2003), Schweger (1997), Zazula et al. (2003), and others.

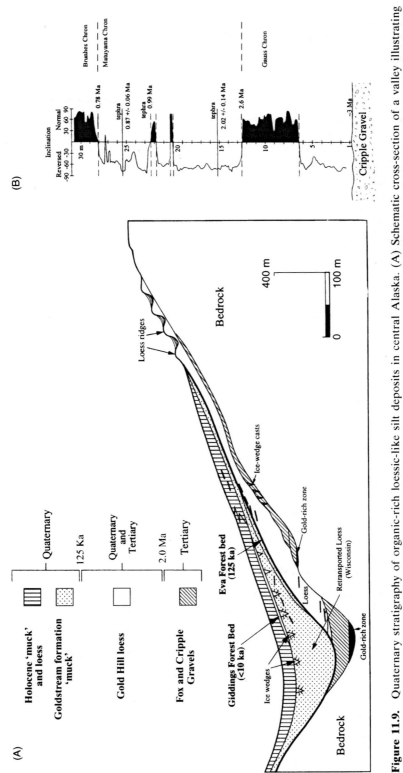
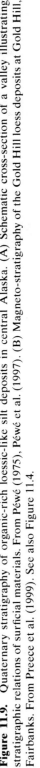

Figure 11.9. Quaternary stratigraphy of organic-rich loessic-like silt deposits in central Alaska. (A) Schematic cross-section of a valley illustrating stratigraphic relations of surficial materials. From Péwé (1975), Péwé et al. (1997). (B) Magneto-stratigraphy of the Gold Hill loess deposits at Gold Hill, Fairbanks. From Preece et al. (1999). See also Figure 11.4.

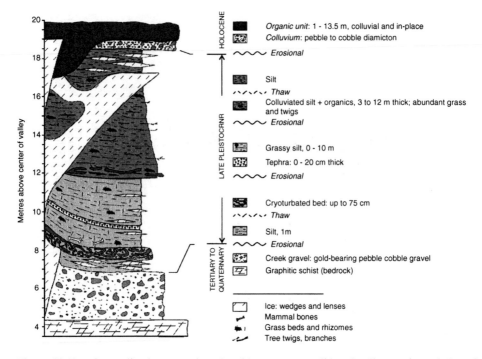

Figure 11.10. Generalized cryostratigraphy of frozen unconsolidated sediments ("muck deposits") in the Klondike, Yukon Territory, Canada. From Fraser and Burn (1997), Kotler and Burn (2000).

11.5. PLEISTOCENE PERIGLACIAL ENVIRONMENTS OF MID-LATITUDES

This section discusses the nature and extent of the mid-latitude periglacial environments of the Pleistocene. There is lack of a satisfactory modern analogue (French, 2000). Other accounts are also available (Dawson, 1992, pp. 109–125; West, 1977, pp. 71–110) and more advanced reading is indicated at the end of this chapter.

11.5.1. General Considerations

The mid-latitude periglacial environments of the Pleistocene were different from those of high latitude. The major difference related to solar radiation. In mid-latitudes, there is no equivalent of the "arctic night" or "arctic day." Therefore, the annual ground-thermal regime would have been dominated by diurnal freeze–thaw cycles of shallow depth and short duration. Permafrost may, or may not, have formed. Given the number of freeze–thaw cycles that occur in both high-latitude and alpine mid-latitude environments today (see Chapter 4), an average annual frequency of between 50 and 100 cycles might be a realistic assumption for the Pleistocene mid-latitudes. This would have contributed to mechanical rock disintegration and to the mass-wasting of any frost-shattered debris. Second, the effects of orientation, with respect to solar-insolation, were more marked than in high latitudes. The effect of orientation is complex, however, because climate fluctuations promote a space–time change. For example, with the onset of cooling following upon either an interglacial or interstadial period, north- and east-facing slopes would have experienced freeze–thaw conditions first. But, during the height of a cold period, south- and west-facing slopes would probably have experienced greater frequencies of freeze–thaw because north- and east-facing slopes would have remained frozen for a longer period of time. Therefore, depending upon the severity of the temperature drop, frost-action processes might be favored either on (i) south- and west-facing slopes (being "warmer") or (ii) north- and east-facing slopes (being "colder"). This has obvious implications for the effectiveness of mass-wasting processes and slope evolution.

Related to solar radiation inputs is the question of permafrost. In high latitudes the long Arctic "night" enables the intense cooling of the ground surface that favors permafrost formation. However, the relatively short duration of the mid-latitude winter and the fact that, even in winter, temperatures would have followed a diurnal pattern, means that the extremes of winter temperatures, currently experienced by high latitudes today, may not have existed. This would have been especially true for oceanic environments, less so for continental environments. For example, it has been suggested that temperature depressions as great as 10–12 °C might have existed in the central European lowlands. Permafrost would certainly have formed. However, in more oceanic locations of western England and southwest Ireland, it is unreasonable to envisage the formation of thick permafrost. Snow cover would also have been an important factor influencing the ground-thermal regime. In all likelihood, permafrost was a transitory phenomenon, forming only during the coldest periods of the cold-climate oscillations. Deep seasonal frost, rather than permafrost, was probably more typical for much of the cold periods.

Another major difference between the mid-latitudes of the Pleistocene and the high latitudes today arises from the fact that global atmospheric circulation patterns were different during glacial times when large ice sheets formed in the mid- and high-latitudes. In all probability, there was an increased intensity of the climatic gradients equator-wards away from the ice margins. The mid-latitude westerlies would not only have been displaced towards the equator but they would also have been strong (see earlier). At the same time, anticyclonic (high pressure) conditions would have developed over the ice sheets and over ice-free continental lowlands. In Central Europe, for example, the development of strong high-pressure systems, especially in winter, would have led to the "blocking" of traveling disturbances in the westerly winds. Disturbances would have been diverted either north-wards towards Iceland, Svalbard, and northern Scandinavia or southwards towards the Mediterranean. Under these conditions, central and eastern Europe would have experienced below-average precipitation while western France, northwestern Spain, southwest England, and southern Ireland would have experienced above-average amounts. Thus, one might expect only 40% of today's amount to have fallen in parts of Central and Eastern Europe. In Poland, for example, where current precipitation amounts are approximately 400–500 mm per annum, this would mean less than 250 mm per annum, an amount comparable to many High-Arctic regions today. It must also be remembered that the Atlantic Ocean would have been several hundred kilometers further to the west, the North Sea would not have existed, and the eastern half of England would have been connected to the European continent.

Similar paleo-geographic considerations are relevant in other areas of the mid-latitudes. For example, in northeastern China, a complicating factor was the continuing uplift of the Tibetan Plateau throughout the Quaternary. This caused major realignment of the monsoon winds.

There are many reasons to suggest that wind action was important in mid-latitudes during the Pleistocene. First, as mentioned earlier, wind and pressure gradients would have been high due to the contraction of the various zonal climate belts south of the high-pressure systems that would have formed over the continental ice sheets. Second, these anticyclonic conditions would have promoted strong gravity-induced (katabatic) winds. Third, an abundant supply of fine material suitable for abrasion and eolian transport was readily available on the outwash plains located at the margins of the ice sheets. Fourth, the general decrease in precipitation resulting from the colder conditions of glacial times and the anticyclonic blocking of traveling disturbances in the zonal westerly winds would have favored aridity and deflation.

The nature of mid-latitude Pleistocene rivers also needs brief mention. In all probability, fluvial activity was greater than conventionally imagined. Because of the absence of the Arctic "night," running water would have been a year-round phenomenon, and certainly more than just the two- or three-month activity that is typical of many high latitudes today (see Chapter 10). The spring flood (nival freshet) would not have been so marked because some winter snowfall would have entered the drainage system during the winter period. Finally, if permafrost were either discontinuous or absent, there would have been more infiltration into the ground and less direct runoff. The large Pleistocene drainage systems of the cold periods are discussed more fully in Chapter 12.

11.5.2. Mammals and Ecosystems

The most reliable criterion for identifying the extent of the mid-latitude periglacial environments of the Pleistocene is the distribution of plants, and the previous extent of the tundra, steppe tundra, and forest-tundra ecozones. Palynological studies, together with those involving steppe and tundra faunal remains, help determine environmental conditions.

For our purposes, faunal and insect remains, pollen taxa, and even the drawings of ancient cave men (Sutcliffe, 1985) provide the most convincing and dramatic evidence. Certain insects, such as beetles (coleoptera) and land snails (non-marine mollusca) survive only under limiting temperature and moisture ranges. Many species are only found today in sub-arctic and arctic environments. Both beetles and land snails have been used successfully in paleo-reconstruction (Keen, 1985; Kerney, 1963; Morgan and Morgan, 1980; Morgan et al., 1983; Sher et al., 1979). Because they live at or near the ground surface, they provide good indications of ground temperatures. The same is true for pollen with regards to July air temperatures (Kolstrup, 1980). The larger mammals, with their greater mobility, have less rigid environmental controls, but even here the evidence is clear. For example, caves in the Dordogne region of southern France were occupied by early man ~20–30 ka ago at the height of the last glacial maximum (Delluc et al., 2001). Crude wall drawings depict a variety of now-extinct animals, including woolly mammoths, bears, woolly rhinoceros, and reindeer (Figure 11.11). Likewise, in the mid-latitude prairies of North America, the Middle- and Late-Quaternary stratigraphic records reveal mammal bones from a variety of animals that, now, are either extinct or live in sub-arctic and arctic environments (Stalker, 1984; Stalker and Churcher, 1982).

11.5.3. Perennial or Seasonal Frost?

In the context of mid-latitudes, it is important to distinguish, without ambiguity, between perennially-frozen and seasonally-frozen ground. This is because the warm limits of the mid-latitude periglacial domain merged with the non-periglacial environments of lower latitudes. It is probable that deep seasonal frost, rather than permafrost, characterized major areas of the mid-latitudes during much of the Pleistocene.

(A)

(B)

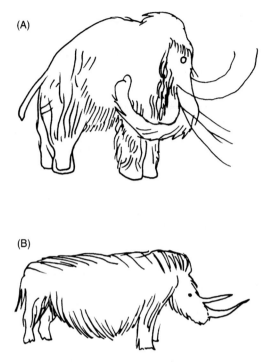

Figure 11.11. Cave drawings from the Dordogne region, southern France, indicate the cold-climate fauna, now extinct, that once lived in the mid-lattitude Pleistocene periglacial zone. (A) Woolly mammoth, Cave of Les Combarelles; (B) woolly rhinoceros, Cave of Font de Gaume. Reproduced in Sutcliffe (1985), Archambeau and Archambeau (1989), and Delluc et al. (2001).

From a geomorphological viewpoint, this distinction is also useful. For example, one must distinguish between seasonal-frost cracks (ground veins or "soil wedges" in the Russian literature), and perennial-frost cracks (see Chapters 6 and 12). The former are limited to deep seasonal frost, the latter penetrate permafrost. A second example relates to frost mounds. It is necessary to distinguish between seasonal- and perennial-frost mounds because only the latter are unambiguous indicators of permafrost (see Chapter 6).

In terms of cryostratigraphy, it is useful if one can identify the boundary between the base of former seasonal-frost penetration and the top of former permafrost. In the currently non-frozen sediments of the mid-latitudes, this is difficult although it has been attempted in a number of ways. For example, G. C. Maarleveld (1976) used the depth to which involutions occur to infer the depth of the active layer but thermokarst involutions must be differentiated from active-layer cryoturbation features, as outlined in Chapter 6. Another method is to identify the boundary based on micromorphological characteristics of the sediments (Harris, 1985; Van Vliet-Lanoë, 1982, 1985). However, specific differences may reflect the lithological and mineralogical (weathering) complexes involved. The ancient permafrost table can also be inferred from pedological studies that involve the recognition of fragipans and other weathering characteristics thought associated with the former existence of permafrost (Nikiforoff, 1955; Vogt et al., 1995) (see Chapter 4). The most reliable method probably involves the recognition of mineralogical changes above and below an inferred thaw unconformity (Burn et al., 1986; Xing et al., 1980).

11.5.4. Problems of Paleo-Environmental Reconstruction

There are several reliable indicators of past permafrost. These are (i) sedimentary structures that indicate thermal-contraction cracking (i.e. ice-wedge pseudomorphs, sand-wedge casts, and composite-wedge casts), (ii) pingo remnants, and (iii) thermokarst phenomena. However, the actual air and ground temperatures required for thermal-contraction cracking, and for the formation of permafrost, are known only in general terms (see Chapters 5 and 6). For example, thermal-contraction cracking can occur when mean annual air temperatures are significantly warmer than $-6\,°C$ and frost cracking has been observed in deep seasonal frost in addition to permafrost. J. B. Murton and E. Kolstrup (2003) summarize the now well-known arguments against the uncritical use of frost fissures as paleotemperature indicators. Thermokarst features (see Chapter 8) clearly indicate the previous existence of permafrost, but the "self-destroying" nature of much thermokarst activity limits its usefulness. Even when the former occurrence of permafrost has been established, there is still no simple relationship between permafrost and air temperatures. Permafrost may also be relict and inherited from an earlier cold period. If it is discontinuous, site-specific lithological, moisture, snow cover, and vegetation conditions may be involved.

Without the actual presence of either seasonally- or perennially-frozen ground, there are few reliable indicators of past frost action. Attention in the past often concentrated upon the identification of "relict" features of assumed frost-action significance. For example, H. T. U. Smith (1949) listed 13 different features as being of climatic relevance. Unfortunately, many, such as landslides, superficial folds, and asymmetric valleys, also occur under non-periglacial conditions. Others, such as blockfields and patterned ground, are ambiguous because their relationship with air and ground temperature is unclear.

Several problems hinder attempts to calculate possible temperature depressions during the Pleistocene, and hence, to provide quantitative parameters to the cold-climate conditions. Usually, temperature depressions are calculated on the basis of modern lapse rates and present snowlines in alpine regions. For example, assuming a lapse rate of $0.5\,°C/100\,m$, a mean annual temperature at the snowline of $0\,°C$, and an elevation difference between present snowline and an inferred Pleistocene snowline of approximately $1000\,m$, it was calculated that Europe probably experienced a temperature depression of between 4 and $6\,°C$ during the last glacial period (Wright, 1961, pp. 966–970). This sort of analysis masks the considerable variability of modern lapse rates which occur in mountains. Moreover, there is no guarantee that lapse rates were similar during the Pleistocene to those of today and, furthermore, that alpine conditions are representative of the lowland zone lying between the Alps and the continental ice sheets. In fact, in continental lowlands of mid-latitudes, strong temperature inversions up to $1500\,m$ in elevation were probably common. As a result, it has been suggested that mean annual air temperatures $10–12\,°C$ lower than today were probably more realistic for the European lowlands during the last cold stage (Shotton, 1960).

Estimates of any possible drop in elevation of the Pleistocene snowline are also difficult because (1) the Pleistocene snowline is usually identified upon morphological evidence, such as cirque heights or nivation hollows, which themselves may be of considerable amplitude or variability with respect to elevation, (2) the modern snowline is not necessarily at $0\,°C$, as is often assumed, but generally lower, and (3) in certain areas, such as southern Africa, a hypothetical snowline has to be assumed.

11.5.5. Extent in the Northern Hemisphere

One of the earliest attempts at periglacial morpho-climatic reconstruction involved the identification of Late-Pleistocene ecozones in Europe (Büdel, 1951). Three major zones were recognized: (i) a frost-rubble, or tundra, zone ("frostschuttzone") lay to the immediately south of the continental ice sheets, (ii) a forest-tundra zone, generally more restricted in extent, lay south of the treeline and the frost-rubble zone, and (iii) a steppe zone of open parkland vegetation lay to the east of the treeline and extended as far as the Ural Mountains in the east and as far south as the northern shore of the Black Sea. Collectively, these three ecozones were thought to represent the extent of the periglacial domain at that time. Under this reconstruction, much, if not most, of Europe outside of the continental ice sheets was affected by periglacial conditions at the height of the last glacial advance. Only the southern parts of the Iberian Peninsula and coastal areas bordering the Mediterranean were thought to have escaped.

A more specific indicator is the former extent of Pleistocene permafrost. The southern limit can be inferred from the recognition of frost-fissure pseudomorphs and casts, and other structures indicating the former existence of ground ice (Figure 11.12). It is clear that permafrost was present in a zone across eastern, central, and northwestern Europe. A wide periglacial belt, approximately 500 km wide in central and eastern Europe, extended westwards into eastern and central England. The difficult, still unresolved, problems relate to (a) the severity of the periglacial climate, especially the mean January

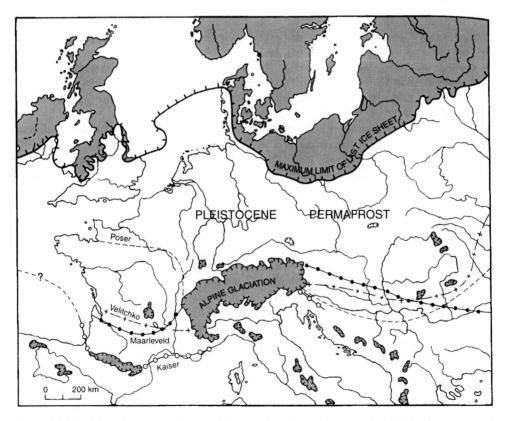

Figure 11.12. Map of western and central Europe showing southern permafrost limits as proposed by Poser (1948), Kaiser (1960), Maarleveld (1976), and Velichko (1982). Modified from Vandenberghe and Pissart (1993). Numerous other studies, including Dylik (1956), Johnsson (1959), Morgan (1971), Gozdzik (1973), Karte (1987), Kolstrup (1987), Svensson (1988a) and Eissmann (1994) document inferred Pleistocene-age permafrost structures in Europe.

air and ground temperatures, (b) the thickness and distribution of the permafrost and whether discontinuous permafrost or deep seasonal frost was widespread, and (c) the temporal duration of these conditions.

In North America, an early attempt to map the various ecozones which existed in the mid-latitudes at the height of the last glaciation was made by D. Brunschweiller (1962). A more recent synthesis (Figure 11.13) now permits a more definite reconstruction but several considerations suggest this map overemphasizes the extent of tundra and steppe. First, the pollen record for many parts of the United States and southern Canada does not suggest a widespread frost-rubble tundra zone, and, instead, indicates that forest zones intermingled with loess-steppe zones. Second, the maximum southern extent of the last glacial ice sheet (the Late Wisconsinan) was further south than that in Europe; as a consequence, the ice-marginal zone of severe periglacial conditions was probably more restricted due to a tighter alignment of the displaced climatic zones. Third, retreat of the Wisconsinan ice was accompanied by formation of extensive proglacial lakes, brought about by isostatic depression and either the ponding of meltwater or the influx of marine waters in the St Lawrence Lowlands These water bodies exerted modifying influences upon regional climate and limited the land areas exposed to cold sub-aerial conditions. By the time they disappeared, either hundreds or several thousands of years later, climate had ameliorated sufficiently to allow forest growth.

The most convincing evidence for Late-Pleistocene permafrost in the continental United States comes from frost-fissure pseudomorphs found in north-central Wisconsin, the interior and western plains (Iowa, Nebraska, Wyoming), and southern New Jersey. In the Appalachians, large amounts of shattered bedrock (blockfields) suggest intense frost action. As in Europe, one must envisage a time-transgressive shift of the periglacial zone northwards as the Laurentide and Cordilleran ice sheets progressively withdrew.

In Russia, several detailed maps of the extent of Late-Pleistocene periglacial conditions have been produced (e.g. Kondratjeva et al., 1993; Popov, 1961; Popov et al. 1985, 1990; Rozenbaum and Shpolyanskaya, 1998a, b; Velichko, 1975, 1982). There is general agreement as to the changes in permafrost extent during the Late Quaternary. For example, during the Zyriansk (50000–60000 years ago) and Sartansk (15000–27000 years ago) cold stages, permafrost reached its maximum southern extent (48–49°N) on the Russian Plain. However, much of the permafrost in Siberia and the Russian Plain is relict. In formerly-glaciated areas, such as western Siberia, relict Early- and middle-Pleistocene-age permafrost is separated at depth from Late-Pleistocene permafrost by an unfrozen zone. By contrast, in unglaciated areas, the permafrost is continuous and thick (see Figure 11.7). In general, one must envisage thick and cold permafrost in never-glaciated terrain and thin or no permafrost beneath ice sheets.

Other mid-latitude regions of the world were also affected by Late-Pleistocene periglacial conditions. In northern China, for example, the southern limit of Late-Pleistocene permafrost has been mapped as far south as 40°N, based primarily upon ice-wedge pseudomorphs and sand-wedge casts (Figure 11.14.) In Mongolia, sand-wedge casts have been recognized in weathered Pliocene sandstone and gravelly layers that date from between ~20 and 26ka (Vandenberghe et al., 2004), and others have been reported from Tibet (Porter et al., 2001). In the plains of central and northern Kazakhstan, several episodes of Middle- and Late-Pleistocene permafrost are inferred on the basis of numerous sand-wedge casts and cryogenic horizons (Aubekerov and Gorbunov, 1999). Finally, in Western Transbaikalia (east of Lake Baikal, latitude 51–57°N), cryogenic soil structures thought indicative of permafrost (Vogt et al., 1995) occur in sediments that, in age, span the Bruhnes-Matuyama boundary.

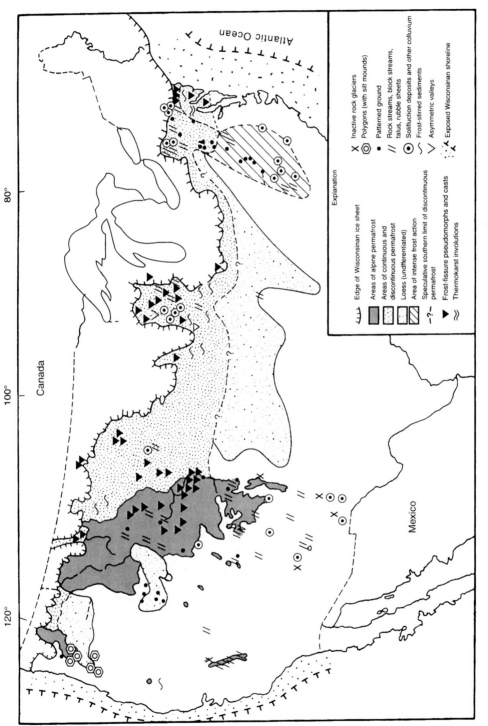

Figure 11.13. Reconstruction of the maximum extent of Late-Pleistocene (Wisconsinan) periglacial conditions in the USA south of the maximum ice-sheet limits. Based upon numerous sources, including Schafer (1949), Wayne (1967, 1991), Black (1964), Walters (1978, 1994); Péwé (1983a), Johnson (1990), Clarke and Ciolkosz (1988), Braun (1989), and French et al. (2003, 2005).

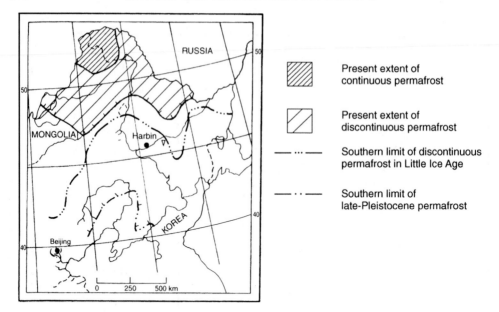

Figure 11.14. Changes in permafrost distribution since the Late Pleistocene in northern China. Compiled from data in Brown (2001), Vandenberghe et al (2004), Qiu and Cheng (1995).

11.5.6. Extent in the Southern Hemisphere

In the southern hemisphere, the majority of Pleistocene frost-action phenomena have been described from the southern parts of South America and Southern Africa. There are also reports from Tasmania and New Zealand. Because of its wide-ranging nature, it is difficult to evaluate the significance of this information. For example, most of the frost-action phenomena described from Tasmania and New Zealand (tors, talus, blockstreams, hummocks, bedded scree) (Derbyshire, 1973; Soons and Price, 1990) appear not to require permafrost. Only in southern Patagonia and Tierra del Fuego, Argentina, have unambiguous features of past permafrost and intense frost action been identified (Galloway, 1985; Grosso and Corte, 1989, 1991; Rabassa et al., 2004; Vogt and del Valle, 1994). There, a cold lowland steppe-like ecozone developed to the east of the ice-covered Fuegian mountains. On the Falklands Plateau, no evidence for permafrost has yet been identified in the Falkland Islands. It seems reasonable to conclude that the southern hemisphere mid-latitudes did not experience the same magnitude of temperature depression as the northern hemisphere. Derbyshire (1973) suggests a temperature depression of only 6.5°C for the last cold period in Tasmania. Presumably, this was because the presence of the Southern Oceans and the strong zonal westerly winds isolated the southern continents from the effects of the huge Antarctic ice mass to the south.

11.6. CONCLUSIONS

Estimates as to the global extent of the Late-Pleistocene periglacial domain are difficult. Periglacial environments varied in time and space; some were characterized by both intense frost action and permafrost; others, by intense frost action only. A conservative guess would be that as much as 20% of the Earth's surface experienced cold-climate, non-glacial conditions during the Pleistocene.

ADVANCED READING

Bell, M., Walker, M. J. C. (1992). *Late-Quaternary Environmental Change*. Addison Wesley Longman, Harlow, 273 pp.

Bradley, R. S. (1999). *Palaeoclimatology. Reconstructing Climates of the Quaternary*, 2nd edn. Harcourt/Academic Press, Boston, MA, 613 pp.

Guthrie, R. D. (1990). *Frozen Fauna of the Mammoth Steppe: The Story of Blue Babe*. University of Chicago Press, Chicago, IL, 323 pp.

Péwé, T. L., Berger, G. W., Westgate, J. A., Brown, P. M., Leavitt, S. W. (1997). Eva Interglaciation Forest Bed, unglaciated east-central Alaska: global warming 125 000 years ago. *Geological Society of America*, special paper 319, 54 pp.

Rozenbaum, G. E., Shpolyanskaya, N. A. (1998). Late-Cenozoic permafrost history of the Russian Arctic. *Permafrost and Periglacial Processes*, **9**, 247–273.

Velichko, A. A. (1982). *Atlas Monograph: Palaeogeography of Europe for the Last 100 000 Years*. Nauka, Moscow, 155 pp. (in Russian).

DISCUSSION TOPICS

1. How extensive, in time and space, were the Pleistocene periglacial environments in mid-latitudes?

2. What are the problems of paleo-reconstruction of Pleistocene cold climates?

12 Evidence for Past Permafrost

The evidence from which the previous existence of perennially-frozen ground can be deduced is reasonably clear and unambiguous. It can be categorized as that which indicates either (i) permafrost aggradation or (ii) permafrost degradation. The distinction between permafrost and deep seasonal-frost can be obscure. There is widespread evidence for Late-Pleistocene thermokarst activity in the mid-latitudes.

12.1. INTRODUCTION

The periglacial mid-latitude environments of the Pleistocene were characterized by the presence of either seasonal or perennial frost, intense frost action, and the enhanced activity of azonal processes such as snow, wind, and running water. This chapter reviews the evidence that indicates the former existence of permafrost. It warrants treatment as a separate chapter because the permafrost conditions of high latitudes are a central component to present-day periglacial environments.

12.2. PAST PERMAFROST AGGRADATION

12.2.1. The Paleo-Permafrost Table

If permafrost previously existed, it must have formed in association with a ground-surface layer that was subject to seasonal freezing and thawing. This layer is termed the active layer. In regions where permafrost is forming today, the base of the active layer constitutes the permafrost table. Because of annual variations in the maximum depth of thaw, the base of the active layer can fluctuate slightly. This interface, between the active layer and the underlying permafrost, is referred to as the transient zone. Both the active layer and the transient zone have been discussed earlier in Chapter 5.

If the top of permafrost (or the base of the active layer), so defined, can be identified in unfrozen sediments in non-permafrost regions, this is highly significant in terms of paleo-climatic reconstruction. This is because it provides a vital distinction between seasonally- and perennially-frozen ground conditions. For example, if such a layer can be identified, its thickness can be used, through application of the Stefan equation (see Chapter 5), to approximate the mean ground surface temperature (MAGST) (see Figure 3.3). If one then makes assumptions as to the magnitude of the surface offset (see Chapter 3), an approximation of the mean annual air temperature (MAAT) can be inferred.

A complicating factor is the presence of relict permafrost. Where this occurs, the base of the seasonally-frozen layer may not be the top of permafrost because the latter may be at depth and separated from the base of seasonal freezing by an unfrozen layer, termed a residual thaw layer. In this case, the top of permafrost reflects a paleo-thaw unconformity rather than a paleo-active layer. While the paleo-active layer has paleo-climatic significance, and a paleo-thaw layer may have regional climatic significance, a paleo-thaw unconformity may also be of only local significance. For example, it might reflect the previous existence of a localized water body or thaw lake. Care is required, therefore, in the identification of the paleo-permafrost table.

Unfortunately, as explained in Chapter 11, it is also not easy to recognize the paleo-permafrost table in now-unfrozen sediments. Moreover, it is frequently ignored when attempts are made to identify past permafrost features and structures.

One of the more reliable methods involves recognition of the mineralogical and weathering differences that might occur above and below a paleo-permafrost table. Both would reflect changes associated with the moisture migration that occurs within frozen ground in response to the temperature gradient (see Chapter 5, Figure 5.13). It can be demonstrated that, in areas of permafrost today, mineralogical differences may occur at thaw unconformities. For example, using X-ray diffraction, trends in the peak ratios of mica to chlorite (10Å:7.1Å) and mica to quartz (10Å:4.6Å) are sufficiently distinctive as to assist in the recognition of relict (early Holocene) active layers (i.e. paleo-thaw unconformities) on the Tibet Plateau, China, at both Fenghuo Shan and Wudoaling (Xing et al., 1980), and in northwestern Arctic Canada at Mayo (Burn et al., 1986) and Garry Island (Burn, 1997). However, if the active layer were developed in soil and rock composed of minerals that are less susceptible to weathering, if permafrost episodes were short-lived, or if the ground surface were not stable, mineralogical changes might be less discernable.

If a permafrost table exists for an extended period of time beneath a non-aggrading ground surface in an arid environment, some field observations suggest that an indurated layer, or hardpan, may form. However, the nature of the cryo-pedological transformations that are involved is still unclear. For example, in the western Transbaikalia region of southern Siberia, where permafrost is not present today, a hard calcrete layer lies beneath a silt-dust surface horizon that contains faunal remains that indicate cold dry steppe-like conditions about 750000–800000 years ago (Bruhnes-Matayama boundary) (Vogt et al., 1995). The calcrete layer is broken and displaced and the overlying sand and calcareous dust contains vertical and oblique cracks, small placations, and folds. The calcrete is interpreted as having formed at the permafrost table and its dislocation occurred when underlying sandy layers thawed. In Eastern North America, similar dislocated hardpan horizons can be observed in Maryland, Delaware, and southern New Jersey (Nikiforoff, 1955). In the Pine Barrens of New Jersey (see Chapter 2), the hardpan preserves sand-wedge casts that are associated with an old land surface that relates to the penultimate cold stage (OIS-6) (French et al., 2003).

If an ice-rich layer forms near the top of permafrost and at the base of the active layer ("aggradational" ice), as can be observed in permafrost environments today (see Chapter 7), this may be reflected in soil micro-morphology upon thaw. For example, ice segregation is favored in fine-grained sediments and a platey microstructure may result when the ice lenses thaw. In coherent bedrock, frost action in the active layer and the formation of ground ice in the near-surface bedrock can lead to brecciation (Murton, 1996a). However, unless other evidence for previous permafrost is present, brecciation and platey soil microstructures can also be the result of seasonal freezing rather than permafrost.

Relatively few studies have convincingly demonstrated the existence of a paleo-permafrost table in now-unfrozen sediments. Given the widespread occurrence of other indicators of past permafrost that are described below, the paleo-permafrost table must also occur widely in the stratigraphic record.

12.2.2. Frost-Fissure Pseudomorphs and Casts

Wedge-shaped sedimentary structures, interpreted as the result of thermal-contraction cracking of perennially-frozen ground, are convincing evidence for the previous existence of permafrost. These structures are frequently reported from mid-latitudes. Although there is general agreement as to their paleo-environmental significance, their specific relevance to air temperature is far less clear. As discussed earlier in Chapter 6, cracking appears to be controlled not only by ground temperature but also by site-specific conditions such as lithology and associated thermal conductivity, antecedent conditions, and the duration and thickness of the snow cover. The fact that certain structures may result from seasonal frost (soil or "ground wedges") further complicates the situation (see Chapter 13). Finally, one must distinguish thermal-contraction cracks from dilation cracks, desiccation cracks, sand dykes, and various other soft-sediment deformations (Burbidge et al., 1988; Butrym et al., 1964). Pleistocene frost fissures may also be epigenetic, syngenetic, or anti-syngenetic in nature (see Chapter 7).

When discussing frost fissures in non-frozen sediments as opposed to those present in perennially-frozen sediments, slightly different terminology must be used in order to take account of the changes that occur as permafrost degrades (Harry and Gozdzik, 1988; Kudryavtsev, 1978; Murton and French, 1993b). This is essential for correct interpretation. For example, following N. N. Romanovskii (in Dylikowa et al., 1978), one can recognize both primary and secondary wedges. Primary wedges can be filled with ice, ice and mineral soil, eolian sand and ice, and eolian sand. They can also be divided into those that form in the seasonally-frozen ground (initially-ground wedges) and those that penetrate permafrost (ice wedges, sand-ice wedges, and primary-sand wedges). Secondary wedges result when frozen ground, either seasonal or perennial, thaws. The resulting structures are ice-wedge pseudomorphs and sand-wedge casts. The presence of Pleistocene permafrost bodies further complicates the Romanovskii scheme because one must also recognize not only active and inactive wedges but also ancient wedges.

The nature of the wedge depends upon the degree of thaw-deformation that the structure has experienced. This means the resulting form can be categorized as being either a cast (i.e. bears some resemblance to the original form) or a pseudomorph (i.e. bears little resemblance to the original form). The degree of deformation largely depends upon a number of factors: the nature of the enclosing sediments, the amount of ice originally contained within the fissure, the rapidity of thaw, and the degree to which water or other processes have eroded the fissure.

Thaw-modification results in the selective preservation of pseudomorphs and casts. Most ice-wedge pseudomorphs and sand-wedge casts are found in sand and gravel. This is because these coarse-grained sediments are usually ice-poor, while few pseudomorphs or casts are recognizable in silt and clay, which are usually ice-rich. Thus, sand wedges are also more likely to be preserved as casts than are ice wedges to be preserved as pseudomorphs. Furthermore, whereas ice wedges preferentially develop in ice-rich, fine-grained sediments (thaw-sensitive), their pseudomorphs are selectively preserved in ice-poor, coarse-grained sediments (thaw-stable).

The fill of a frost-fissure cast or pseudomorph is either primary (i.e. the initial sand/mineral soil remains, as in a sand-wedge cast) or secondary (i.e. the initial icy fill thaws and the void left is filled with different material, as in an ice-wedge pseudomorph). Where fill is both primary and secondary, in varying amounts, the cast or pseudomorph is deemed "composite" in nature. The range of structures that ice-wedge pseudomorphs can assume is large (Figure 12.1).

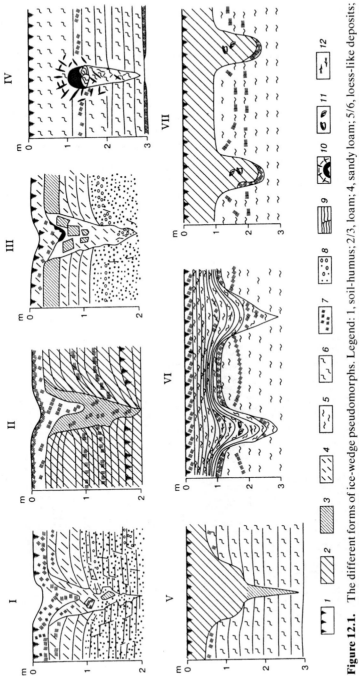

Figure 12.1. The different forms of ice-wedge pseudomorphs. Legend: 1, soil-humus; 2/3, loam; 4, sandy loam; 5/6, loess-like deposits; 7, peat; 8, sand and gravel; 9, stratification and small faults; 10, voids; 11, freshwater mollusca shells; 12, plant residues (transported). From Romanovskii (1973, figure 10), Melnikov and Spesivtsev (2000).

Because Pleistocene frost fissures are important in terms of their paleo-environmental significance, each type is briefly discussed.

Ice-wedge pseudomorphs (Figure 12.1) are structures of secondary fill. They form when the ice wedge slowly melts, usually as the permafrost degrades but sometimes by running water and thaw erosion acting preferentially along the wedge (see Chapter 8). As the ice melts, sediment collapses into the void. In arctic regions today, a range of processes are involved, including thermal erosion, collapse, subsidence, refreezing, loading, buoyancy, spreading, folding, and shearing (Murton and French, 1993b). A sketch of an immature pseudomorph forming above a partially-thawed ice wedge is illustrated in Figure 8.6. The bent, u-shaped involution (60–70 cm deep and ≤20 cm wide) above the ice wedge is characteristic of the thaw-transformation stage (Harry and Gozdzik, 1988). A fully developed pseudomorph (Figure 12.2) is characterized by the penetration of material into the fissure from above and from the sides, the downward inflection of layers if the enclosing sediments are stratified, and various systems of miniature faults.

Sand-wedge casts contain a fill of relatively well-sorted eolian fine sand that is quite different from the enclosing sediment. They often possess a more truly wedged shape with a relatively clear boundary between the wedge and enclosing material. This is because

(A)   (B)

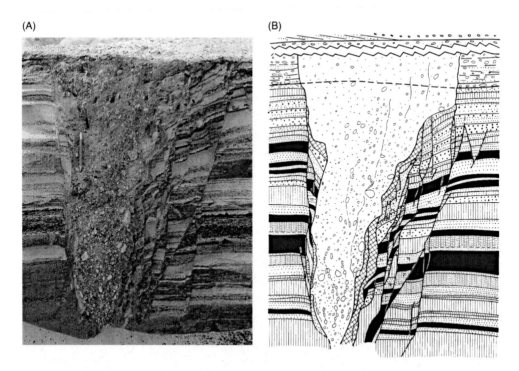

Figure 12.2. Saalian-age ice-wedge pseudomorph with well-developed en echelon downthrows exposed in the Dilitzsch-SW brown coal (lignite) mine near Lissa, eastern Germany. (A) Photograph; (B) sketch. Pencil in A is 15 cm long. Photograph, sketch, and caption information are supplied courtesy of Professor L. Eissmann.

Figure 12.3. Late-Pleistocene sand-wedge structure, ~2.7 m deep and 5–20 cm wide, containing well-sorted and wind-abraded fine sand dated by OSL at >65 ka, formed within indurated Late-Tertiary gravel (Bridgeton Formation), Cumberland County, Southern New Jersey, Eastern USA.

thaw of the non-icy infill material does not allow secondary material and the cast retains much of its original sedimentary structure. A photograph of a sand-wedge cast developed in sand and gravel is illustrated in Figure 12.3. By comparison, a composite wedge consists of a mixture of both primary and secondary infill (Figure 12.4), and shows an inner wedge of primary fill and an outer wedge showing signs of secondary fill. In this case, the fissure was initially filled with a mixture of sand and ice but the ice amount was so small that the foliated structure was not destroyed upon thaw.

Sand-wedge casts are sometimes thought to reflect colder conditions than ice-wedge pseudomorphs (Murton and French, 1993a). This is not the case because field experience demonstrates that all types of fissures can form in the same geographical area. Local site characteristics, exposure to wind, the availability of wind-blown sediment, and drainage conditions all influence the nature of the fill material. It is not surprising, therefore, that observations in western Jutland, Denmark, indicate that ice-wedge pseudomorphs, sand-wedge casts, and composite-wedge casts all exist in the same area (Kolstrup, 1987).

Thermal-contraction cracking leads to the development of large-scale polygonal patterns at the surface. These are sometimes recognized on air photographs by differences in vegetation or crop ripening (Figure 12.5). However, only the most recent features are usually identified, and older polygon nets may be buried by more recent sediments. Based upon the distribution of polygons and frost-fissure casts and pseudomorphs in Western

Figure 12.4. Late-Pleistocene composite wedge cast formed within loess and lying beneath a fossil soil horizon, Miechow Plateau, southwestern Poland. The upper, inner (sand) wedge is formed by primary infilling; the lower (pseudomorph) wedge shows evidence of secondary infilling.

Figure 12.5. Oblique air photograph showing traces of a random orthogonal thermal-contraction crack network, as revealed in crop marking, near Vessige, south-western Sweden. Photograph and caption information are supplied courtesy of Professor H. Svensson.

and Central Europe, it is possible to recognize a broad zone, ~200–300 km wide, where permafrost probably existed at different times during the Pleistocene (see Figure 11.12). To the east, the zone broadened to over 500 km in the Russian Plain. Distribution of the same phenomena in the mid-latitudes of North America suggests a similar, but much narrower, permafrost zone existed along the maximum southern limit of the ice (see Figures 11.6; 11.13). The uneven quality of the information means that we are unable to determine the variability in extent of permafrost in the different cold periods; Figures 11.12 and 11.13 reflect mainly the extent of Late-Pleistocene permafrost. On both continents, the presence of casts and pseudomorphs in areas within the maximum limit of last glaciation means that the Late-Pleistocene periglacial zones followed the retreating ice. In Central Asia and northeastern China, sand-wedge casts are the predominant line of evidence that is used to map the former extent of Pleistocene permafrost (see Figure 11.14).

12.2.3. Frost-Mound Remnants

Pingo remnants provide incontrovertible proof of the former existence of permafrost. However, as shown in Chapter 6, a range of frost mounds exist (see Figure 6.11), some of which merely require cold-climate conditions, not permafrost, for their formation. Therefore, the identification of pingo remnants needs careful consideration. Moreover, it must be stressed that pingos are not common features of the periglacial landscape; they are abundant in just a few geographic areas and, for the most part, are relatively small and isolated features.

In present permafrost environments, collapsed pingos possess a raised rim or rampart, formed by mass wasting down their sides, that encloses a central depression where the ice core has melted (see Figure 6.13). The rampart is significant because this distinguishes a depression caused by the previous growth of a frost mound from one of simple thermokarst (thaw lake) origin. If modern collapsed pingos are a guide, the ramparts vary from as little as 0.5 m to over 5.0 m in height, and the diameter of the depressions within the ramparts may be as great as 200–300 m. Some pingos are known to exist for several thousands of years. Finally, it must be remembered that pingos are hydrologic phenomena, and there are two major types, hydraulic and hydrostatic (see Chapter 6).

It is important to transfer this knowledge selectively to the possible recognition of former pingos in the now-unfrozen mid-latitude landscapes. Presumably, the same hydrologic conditions that favor pingo growth in present-day permafrost environments must also have existed if pingos were to have formed. It should also be stressed that pingos are aggradational permafrost landforms. This is important for two reasons. First, they are progressively destroyed by the thaw of the ice core during any subsequent permafrost degradation. Such localized thaw-modification, accentuated by on-site groundwater flow if the feature is of hydraulic origin, makes pingo – remnant recognition, on both morphological and sedimentological grounds, extremely difficult. Second, it follows that Pleistocene pingo remnants should be found at locations that either possessed the appropriate hydrological conditions (e.g. groundwater discharge, if hydraulic in nature) or allowed the localized freezing of a talik (e.g. drained lake, abandoned river channel, if hydrostatic in nature). Such specific requirements should allow one to eliminate a pingo hypothesis in the early stages of any Pleistocene paleo-reconstruction investigation. There is also the necessity of eliminating non-permafrost explanations. For example, seasonal-frost mounds do not necessarily require permafrost for their formation; deep seasonal frost at a site of groundwater discharge together with an underlying impermeable layer is an equally suitable environment for their occurrence (see Chapter 6). In previously-glaciated terrain, a kettle origin (i.e. a partial melt-out or ablation feature) must also be considered.

Numerous shallow ridges and ramparts that sometimes enclose indistinct depressions occur widely in the lowlands of western Europe and eastern North America (Figure 12.6). Frequently, they have been interpreted as the remnants of Pleistocene frost mounds, especially pingos (de Gans, 1988; de Gans and Sohl, 1981; de Groot et al., 1987; Flemal, 1976; Marsh, 1987; Pissart, 1963; Weigand, 1965). While a frost-mound interpretation is probably correct in general terms, it is likely that few are remnants of either hydrostatic- or hydraulic-system pingos. Most are inconsistent with the hydrologic conditions necessary for pingo growth. Instead, the ramparts are often irregular in plan and "mutually interfering," giving rise to a complex alignment and distribution. Some are semi-circular and open in an upslope direction while others are elongate in the direction of slope. These morphological characteristics are more typical of seasonal-frost mounds in which new generations are repeatedly born at the same site by the continual movement of groundwater to the surface. Many occur in typical open-system localities, such as on lower valley-side slopes or at spring-line locations.

Some depressions are interpreted as the remnants of palsas or mineral palsas ("lithalsas"). However, palsas usually reflect icy permafrost bodies lying beneath localized peaty sediments (see Chapter 6). Their degradation does not take place in the same manner as a pingo (where thaw is induced along dilation cracks followed by collapse of the summit) but through slumping around the base of the feature. Thus, no ramparts develop and, in the absence of a large ice core, there is no central depression. In this context, features that have attracted considerable attention over the years are the circular, rampart-surrounded depressions known as "viviers" (fish ponds) in the Hautes Fagnes uplands of Belgium. First described and interpreted by A. Pissart (1956) as pingo remnants, they were subsequently reinterpreted as remnants of mineral palsas and, most recently, as lithalsas. According to Pissart (2000, 2002), they are analogous to small frost mounds that occur today in areas of discontinuous permafrost where mineral soil overlies impermeable bedrock. Several of these frost-mound remnants have been investigated by stratigraphic and palynologic studies of the material within the central depression. Pollen analyses suggest that they formed during the Late Dryas (10 500–11 000 years BP) (Mullenders and Gullentops, 1969). In the UK, apparently similar features formed and quickly degraded during and at the end of Zone III of the Late Glacial (Sparks et al., 1972; Watson, 1977).

Figure 12.6. Oblique air view of part of Walton Common, Norfolk, eastern England, showing ramparts and enclosed depressions associated with the formation of Late-Pleistocene frost mounds, probably seasonal in nature. Photograph is reproduced by permission of the University of Cambridge Photo Collection.

Many of the frost-mound remnants that occur in mid-latitudes are, almost certainly, a legacy from the most recent period of cold-climate conditions. The significance of any sort of open-system interpretation lies in the fact that this does not require continuous permafrost. Instead, their irregular distribution probably reflects the limiting hydrologic conditions necessary for growth. The discharge of water flowing beneath or within permafrost must be relatively small in amount and close to 0 °C in temperature, otherwise either a perennial spring will result if the temperature is considerably in excess of 0 °C, or the conduit will be sealed and the mound will cease to develop if the temperature drops significantly below −2 to −3 °C. The possibility that icings and related features formed at the site of these frost-mound features is also likely yet direct morphological and stratigraphic evidence for such occurrences is difficult to find.

Frost-mound remnants have not been reported with the same frequency from the mid-latitudes of North America and elsewhere. The reason for this is not clear. The depression infill from one so-called "pingo" locality in central Pennsylvania (Marsh, 1987) also indicates a late-glacial age for sediment deposition. Subsequent investigations concluded that this depression was a "ground-ice scar" (seasonal-frost mound?) that formed between two wind-aligned ridges at a groundwater seepage zone (Clark et al., 1992; Marsh, 1998, 1999, pp. 24–29). Elsewhere, the DeKalb Mounds of north-central Illinois (Flemal et al., 1973) and the "prairie mounds" of southern Alberta (Bik, 1969) were initially interpreted as remnants of large Pleistocene pingo "fields". However, it seems best to regards these features as ablation-till phenomena (Stalker, 1960).

12.3. PAST PERMAFROST DEGRADATION

The most widespread evidence for the former existence of permafrost comes from the mass-wasting and landscape modification that must have accompanied the thaw and eventual disappearance of permafrost. This modification would have been especially drastic in areas where the permafrost was ice-rich and had developed in soft bedrock or unconsolidated materials. In fact, one might argue that the identification of past thermokarst activity should constitute the vast majority of Pleistocene periglacial paleo-environmental studies.

Unfortunately, the "self-destroying" nature of much thermokarst activity hinders such study. By definition, thermokarst sediments are heterogeneous diamicts. They often defy description. Landforms that result from thermokarst activity assume a variety of shapes and sizes, and many have experienced repeated modification. There are few diagnostic sedimentary structures associated with thermokarst-derived sediments, a notable exception being thermokarst lake-basin sedimentation (Murton, 1996b). Finally, once permafrost has disappeared, non-periglacial processes act to further modify, or degrade, whatever form the evidence takes.

12.3.1. Thermokarst Depressions

Pits, ponds, and shallow depressions that lack surrounding ramparts are sometimes given a thermokarst interpretation. Present-day analogues are, presumably, thaw lakes and depressions (see Chapter 8). In the northwest European lowlands, inferred thermokarst depressions were first referred to as "mares," "mardelles," or "solle" (Cailleux, 1956, 1957; Troll, 1962). In the Beauce and Brie areas of northern France their density reaches 35/km². A. Pissart (1958, 1960) was the first to explicitly interpret these depressions as thermokarstic in origin. However, some are certainly pits dug by man (Prince, 1961). The possibility has also been raised that the Breckland Meres of eastern England may be of a thaw-lake origin (Sparks et al., 1972). In North America, numerous enclosed depressions ("spungs") in southern New Jersey were first interpreted as periglacial "frost-thaw" basins (Wolfe, 1953), then as "pingo scars" (Bonfiglio and Cresson, 1982), and more recently as cold-climate wind-deflation hollows (French and Demitroff, 2001). Generally speaking, the separation of thermokarst depressions from the multitude of enclosed depressions of all sizes and shapes of other origins seems almost impossible.

12.3.2. Paleo-Thaw Layers

Past thermokarst activity may be recognised in the stratigraphic record by the existence of a paleo-thaw layer (i.e. a horizon corresponding to the depth to which previous thawing of permafrost had proceeded). In present permafrost regions, paleo-thaw layers often correspond to secondary thaw unconformities. These have been described in Chapter 6, in the context of present permafrost environments, in Chapter 11, in the context of relict permafrost, and earlier in this chapter in the context of the paleo-permafrost table.

Paleo-thaw layers have been described from perennially-frozen sediments in several areas of the western North American Arctic. Involuted structures ("thermokarst involutions") are a diagnostic feature of these layers in addition to isotopic differences in the ground ice that is present both above and below the actual unconformity. If similar to a paleo-permafrost table (see earlier), mineralogical and other weathering differences should also be present in and above a paleo-thaw layer.

Late-Pleistocene paleo-thaw layers have been inferred from studies at a number of localities in the lowlands of western and central Europe (Eissmann, 1978; Maarleveld, 1981; Vandenberghe, 1983; Vandenberghe and Broek, 1982; Vandenberghe and Krook, 1982). Such horizons reflect the deepening thaw layers that would have existed during partial permafrost degradation. Typically, the sediments lying above the paleo-thaw layer are deformed and chaotic. These soft-sediment deformations can be attributed to degrading ice-rich permafrost at depth.

12.3.3. Thermokarst Involutions and Sediment-Filled Pots

Thermokarst involutions and various other thermokarst structures and sediments that can be observed in present permafrost environments have been discussed in Chapter 8. Here, attention focuses upon the different types of disturbed soil horizons ("cryoturbations") that are attributed to Pleistocene thermokarst. These structures occur widely in the near-surface sediments of many mid-latitude areas.

Amorphous deformations appear to be the most widespread (Figure 12.7). In some instances, the deformed structures extend for several tens of meters in horizontal extent and affect sediments to depths as great as 3–4m below the ground surface. Because of this, they cannot be interpreted as the product of traditional cryoturbation that occurs during the repeated freezing and thawing of a seasonally-frozen, near-surface layer (see Chapter 6). Even in continental Siberia, where summer temperatures exceed 30°C, the maximum depth of thaw in unconsolidated sediments rarely exceeds 2.0m. Instead, it is much more likely that, during permafrost degradation, density-controlled mass displacements in water-saturated sediments would have caused underlying sand to ascend and overlying gravel to descend. Smaller deformation structures of the "bird foot" or "drop soil" type (see Figure 12.8B) are also caused by loading

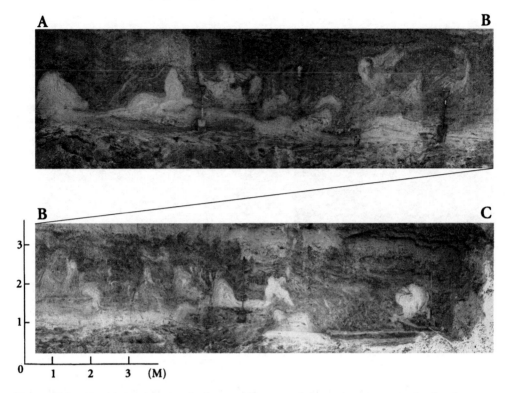

Figure 12.7. Composite photograph of post-sedimentary deformation structures that involve overlying sandy gravel and underlying fine sand, Newtonville, Atlantic County, Southern New Jersey, eastern USA. The flame structures indicate underlying sand has risen while overlying gravel has descended. Several phases of activity are indicated. See French et al. (2005).

and density differences in water-saturated sediments, probably during the degradation of underlying permafrost.

Irregular tongues of material that penetrate underlying sediment may also be explained as the preferentially thaw-modified remnants of ancient sand wedges and ice wedges. These structures occur in proximity to an inferred ground surface. They defy simple generalization because they are irregular kettle-like structures that consist of chaotic assemblages of silt, sand, and gravel in which evidence of running water, erosion, slumping, and mass displacement may all be present. As such, the sediments are quite different to those of thermokarst lake-basin sedimentation (see Chapter 8). These collapse structures are termed "sediment-filled pots" (Conant et al., 1976; French et al., 2005). Typically, they may be 2–4 m in both vertical and horizontal dimensions (Figure 12.8A), and occasionally are penetrated by wedge structures (see Figure 12.11B). Sediment-filled pots presumably form at the intersection of two or more wedges by a combination of thermal erosion and the mixing, slumping, and redeposition of material from both the wedges and the enclosing sediments. The modern permafrost analogue is the "thaw sink" that occurs in tundra lowlands in thin and/or discontinuous permafrost (Hopkins, 1949).

Sediment-filled pots, ice-wedge pseudomorphs, and degraded sand wedges are regarded as entirely consistent with the nature of thermokarst activity and sedimentation. As described in Chapter 8, fluvio-thermal erosion operates preferentially along ice wedges and other lines of weakness in present permafrost environments. Similar processes would

(A)   (B)

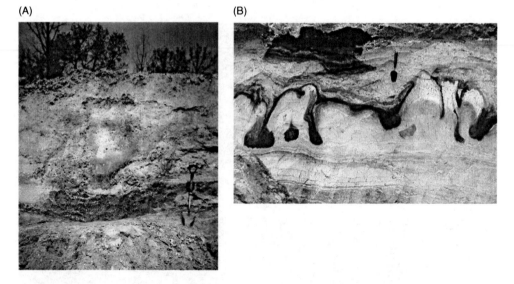

Figure 12.8. Late-Pleistocene thermokarst structures. (A) Typical "sediment-filled pot" containing heterogeneous locally-derived materials developed in Tertiary-age gravel, Atlantic County, Southern New Jersey, eastern USA. (B) Bird-foot involutions in fluvial silt and sand with organic layers, caused by loading and density differences in water-saturated sediments during degradation of Pleistocene-age permafrost northern Belgium.

have occurred during the thaw-degradation of near-surface permafrost bodies in mid-latitudes during the Pleistocene.

12.3.4. Large-Scale Soft-Sediment Deformations

Much Pleistocene-age permafrost would have formed in fine-grained consolidated bedrock such as clay and shale, and in unconsolidated sand, silt, and gravel. All these sediments, when frozen, would have probably been ice-rich because this is certainly the case for most of these lithologies in modern-day permafrost terrains (see Part II and especially Chapter 7). It is also likely that many Pleistocene permafrost bodies were several tens of meters thick. Therefore, as these bodies progressively thawed at the end of each cold stage, any ice layers or massive icy bodies at depth within the permafrost would have melted. Excess water would have resulted in super-saturation, high pore-water pressures, and deformation through either gravitation (i.e. collapse), loading (i.e. upturning or injection), water-escape (i.e. high pore-water pressures and low inter-particle cohesion), or a combination of the above.

Undoubtedly, the most impressive cryogenic-induced deformations that have been observed anywhere in the mid-latitudes are exposed in the thick sequences of unconsolidated deposits in the opencast lignite workings of central Germany. They reveal large, numerous, and complex plastic-deformation structures that indicate a near-continuous record of changing permafrost conditions throughout the Quaternary (Eissmann, 1978, 1994, 2002). Figure 12.9 schematically illustrates the range of structures. It should be noted that they are not all drawn at the same scale. Many (categories 1–3, Figure 12.9) are glacigenic in nature and relate to ice movements in the North German Plain. These are beyond the scope of this book. Others (category 5, Figure 12.9) relate to ice-wedge

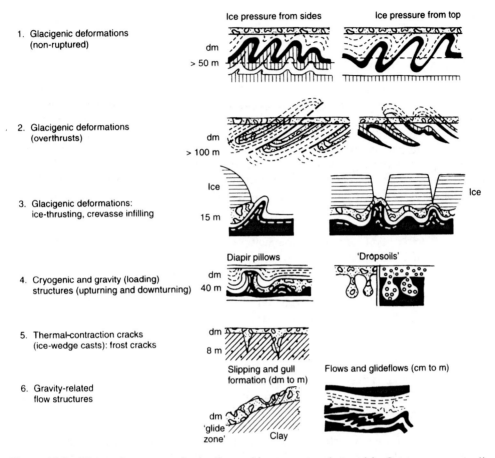

1. Glacigenic deformations (non-ruptured)
   dm
   > 50 m

   Ice pressure from sides

   Ice pressure from top

2. Glacigenic deformations (overthrusts)
   dm
   > 100 m

3. Glacigenic deformations: ice-thrusting, crevasse infilling
   15 m

   Ice

   Ice

4. Cryogenic and gravity (loading) structures (upturning and downturning)
   dm
   40 m

   Diapir pillows

   'Dropsoils'

5. Thermal-contraction cracks (ice-wedge casts): frost cracks
   dm
   8 m

6. Gravity-related flow structures

   Slipping and gull formation (dm to m)

   Flows and glideflows (cm to m)

   dm
   'glide zone'   Clay

Figure 12.9. The various types of non-diastrophic structures observed in Quaternary-age sediments in eastern Germany. Some are cryogenic (3 and 4), some are glacigenic (1, 2, 3), and one (6) has no specific cold-climate significance. Note that they are not all drawn at the same scale. From Eissmann (1994). Reproduced by permission of the Mauritianum Altenburg.

pseudomorphs that have already been discussed. Here, attention focuses upon the large-scale deformations (category 4, Figure 12.9) that are permafrost-related. They take the form of large diapiric upturning and drop-soil structures within lignite beds. Many are tens of meters in height. Almost certainly, these massive deformations formed when icy layers, developed preferentially within and beneath the lignite beds, melted as permafrost degraded. This cryostratigraphic (i.e. ground ice) situation would have developed during previous permafrost aggradation because the low thermal conductivity of organic material such as lignite (see Table 5.1) slowed the downward growth of permafrost and allowed groundwater to migrate upwards by cryosuction towards the near-stationary freezing front.

In central Germany, it appears that permafrost aggradation, and the growth of ground-ice bodies within and beneath the lignite beds, occurred on several occasions during the Quaternary. This is because different sets of ice-wedge pseudomorphs are preserved within overlying gravels of Elsterian (Kansan or Nebraskan), Saalian (Illinoian), and Weichselian (Wisconsinan) age. Thus, there is clear evidence that at least three major periods of permafrost aggradation and degradation occurred during the Middle and Late

Pleistocene. At those times, this part of central Europe lay to the south of the main Baltic–West Siberian ice-sheet complexes.

12.3.5. Non-Diastrophic Structures in Bedrock

At a similar scale to the diapiric lignite deformations described above are non-diastrophic structures that occur in relatively consolidated bedrock. These were first given Pleistocene, or periglacial, significance by bedrock geologists in the U.K. (Hollingworth et al., 1944; Kellaway, 1972). The structures take the form of gently-dipping strata adjacent to valleys (cambering), up-arching of strata in valley bottoms (valley bulging), and the formation of expanded joints (gulls) in coherent bedrock which appears to have preferentially moved across plastically-deforming softer sediment beneath (Figure 12.10). The significance of these non-diastrophic structures is frequently overlooked by geomorphologists involved in Pleistocene periglacial reconstruction.

The nature of permafrost creep and its dependence upon temperature has been discussed in Chapter 9. This information supports the suggestion that deep-seated creep-deformation in a plastic but still frozen state, as might exist when permafrost temperatures approach 0 °C, was the cause of these non-diastrophic structures.

Two field examples illustrate the significance of the thaw-rheology of perennially-frozen ground. First, cambering and valley-bulging phenomena are well developed in the Jurassic and Cretaceous bedrocks of central and southern England (Dury, 1959, pp. 179–181; Hawkins and Privett, 1981). These strata show a succession of thick, often over-consolidated, argillaceous rocks separated by more resistant arenaceous and carbonate rocks. An attractive explanation for these phenomena is that they developed as permafrost degraded at various times during the Pleistocene (Horswill and Horton, 1976; Whiteman and Kemp, 1990). Permafrost creep would have been accentuated in the more ice-rich clay strata because increasing amounts of unfrozen water would become available to lubricate movement as ground temperature approached 0 °C (see Figure 4.2). The second example relates to the occurrence of disrupted and broken blocks of bog ironstone that are found in Tertiary-age gravels in the Pine Barrens of Southern New Jersey, eastern USA. By its very nature, bog ironstone forms in horizontal layers at the level of the regional water table. During the Quaternary in New Jersey, it formed continuously and in accordance with fluctuations of the groundwater table. Today, where ironstone beds are encountered near the surface, they are invariably inclined yet at depth are inevitably horizontal. An explanation is that this heavy rock "foundered," or sank, in the water-saturated conditions that would have temporarily existed during the thaw of icy permafrost beds below (French et al., 2003, 2005).

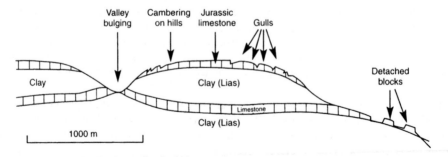

Figure 12.10. Types of large non-diastrophic structures associated with Pleistocene permafrost in central England described by Hollingworth et al. (1944) and Dury (1959).

12.3.6. Discussion

Thermokarst continues to be neglected in many Pleistocene periglacial investigations. This was first highlighted over 40 years ago by J. Dylik (1964b) and remains true today. The reality is that evidence for the degradation (thaw) of permafrost should be far more widespread and much better preserved than evidence for the earlier growth of permafrost. However, the literature indicates an over-emphasis upon the recognition of features and structures associated with aggrading permafrost (pingo remnants, frost fissures) and a relative neglect of thermokarst sediments and structures.

A classic illustration is provided by the Pleistocene freeze–thaw phenomena (cryoturbations) that are discussed in Chapter 13. Some were first interpreted within the context of either seasonal freezing and thawing, or a deep active layer, rather than in the context of degrading permafrost (Edelman and Tavernier, 1940; Sharp, 1942b; Wolfe, 1953). Yet interpretations such as these need careful re-evaluation in light of modern understanding of thermokarst processes. For example, the "frost kettles" or "barrel-shaped macropolygons" that are described from central Europe (Pecsi, 1964, p. 287; Sekyra, 1956, p. 358) are better interpreted as "sediment pots" (Figure 12.11). Likewise, many of the deformed structures and large-scale patterned ground (stripes) that characterize the chalky landscapes of southern and southeastern England (Te Punga, 1957; Watts et al., 1966; Williams, 1964) are now interpreted in the context of active-layer deepening through ice-rich permafrost that occurred during a shoft-lived climatic oscillation approximately 21 ka ago (Murton et al., 1995, 2003). For example, on Thanet, Kent, pebbly silt was involuted into underlying chalk diamicton during an episode of soft-sediment deformation, and sandy stripes were subsequently formed by cell-like circulation within the deepened active layer (Murton et al., 2003). Reference is made to the inferred movements associated with earth hummocks that appear characteristics of the modern active layer in present permafrost environments (see Chapter 6).

A final example comes from several stratigraphic and geotechnical investigations in southern England that indicate the previous instability of slopes developed in lithologies that, under current conditions, would be regarded as stable (see Chapter 9, Table 9.7). For example, the degraded form of a slope developed in London Clay beneath an old cliff-line at Hadley, Essex, can best be understood in terms of thaw-consolidation theory rather than by conventional geotechnical analyses (Hutchinson and Gostelow, 1976). It must be assumed that many slopes in many mid-latitude areas reflect, to varying degrees, the imprint of Pleistocene thermokarst activity.

12.4. SUMMARY

The evidence for Pleistocene thermokarst in the mid-latitudes is widespread but the complex nature of thermokarst activity, in both bedrock and unconsolidated sediments, makes that evidence difficult to recognize. The most reliable information for the former existence of permafrost comes from various casts and pseudomorphs that reflect thermal-contraction cracking, a process that clearly demands permafrost for its operation. Even here, the complexity of form makes recognition difficult. Likewise, evidence for the aggradation of permafrost, such as the paleo-permafrost table, and various types of frost mounds, is also difficult because much will have been destroyed by subsequent thermokarst activity. Finally, at the warm limits of the distribution of past permafrost, the distinction between perennial and seasonal frost, and continuous and discontinuous permafrost conditions, may be virtually impossible to determine.

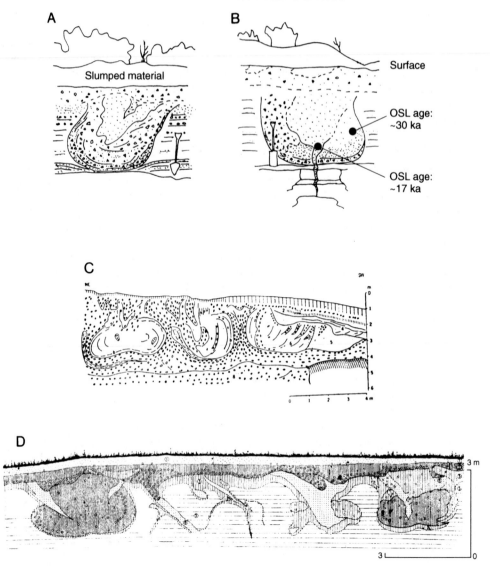

Figure 12.11. Field sketches of Late-Pleistocene thermokarst structures. (A) and (B) are from the Pine Barrens, Southern New Jersey, Eastern USA. (A) Sketch of the sediment-pot illustrated in Figure 12.8A. (B) OSL dating of a sediment-pot penetrated by a sand-wedge cast indicates a mid-Wisconsinan period of thermokarst activity followed by a late-Wisconsinan period of permafrost aggradation (frost-fissure formation). From French et al. (2005). (C) Sedimentary structures described as "barrel-shaped macropolygons with river gravels," Hungary. From Pecsi (1964). (D) Pleistocene "ice wedges and kettles" in Cretaceous marl, Czechoslovakia. From Sekyra (1956). Reproduced by permission of John Wiley & Sons Ltd.

ADVANCED READING

Dylikowa, A., Gozdzik, J. S., Jahn, A. (1978). Methodology of field studies: fossil frost- and ice wedges. *Biuletyn Peryglacjalny*, **27**, 171–178.

Eissmann, L. (2002). Quaternary geology of eastern Germany (Saxony, Saxon-Anhalt, South Brandenburg, Thuringia), type area of the Elsterian and Saalian stages in Europe. *Quaternary Science Reviews*, **21**, 1275–1346.

Murton, J. B., Bateman, M. D., Baker, C. A., Knox, R., Whiteman, C. A. (2003). The Devensian periglacial record on Thanet, Kent, UK. *Permafrost and Periglacial Processes*, **14**, 217–246.

Romanovskii, N. N. (1973). Regularities in formation of frost-fissures and development of frost-fissure polygons. *Biuletyn Peryglacjalny*, **23**, 237–277.

DISCUSSION TOPICS

1. How conclusive is the evidence for the former existence of perennially-frozen ground in mid-latitude regions during the Pleistocene?

2. Why is the distinction between seasonal and perennial frost important in Pleistocene periglacial reconstruction?

3. What were the physical and structural changes that occurred when Pleistocene permafrost degraded?

13 Periglacial Landscape Modification

During the cold periods of the Pleistocene azonal processes undertook significant landscape modification in the ice-free mid-latitudes.

Frost action in soils and on bedrock produced disturbed horizons and frost-derived debris. The latter was transported by mass-wasting processes towards lower elevations. Strong katabatic and zonal winds transported and redeposited fine sediment (loess) derived either from outwash plains adjacent to the ice margins or from the cold deserts of central Asia. In some areas, deflation activity and locally-derived dunes and sand sheets were important. Large rivers crossed the periglacial zone, and ice-marginal channels (pradolinas), sandur, and proglacial water bodies paralleled the ice margins.

Landscape modification was by fluvial incision, modification of valley-side slopes, snowmelt-induced erosion, and sediment transport on slopes. Mass wasting caused valley-bottom aggradation. Thermokarst processes, described in Chapter 12, further modified landscapes.

13.1. INTRODUCTION

Evidence for past frost action is often ambiguous because the frost-action effects of present climate must also be considered. Even when present frost action is clearly not involved, the "relict significance" of the phenomenon concerned depends upon an evaluation of the lithological susceptibility of the soil or rock in question, and site-specific conditions. Further problems arise because certain phenomena can be produced without invoking frost action. Finally, anthropogenic activities can modify and even obliterate structures. For example, it is easy to mistake a series of old plough furrows for a sequence of frost-disturbed structures and old post holes for ice-wedge pseudomorphs. For these reasons, the recognition of relict frost-action phenomena requires caution. Only where several different features occur either together or in close association can one assume a frost-dominated environment.

It is also relevant to ask how much of the present mid-latitude landscape reflects its periglacial legacy. Unfortunately, there is no easy answer to this question. Not only did the cold-climates vary in intensity and duration in different areas, but different rock types would have produced different landscape responses.

The Periglacial Environment, Third Edition Hugh M French
Copyright © 2007 John Wiley & Sons Ltd

13.2. INTENSE FROST ACTION

13.2.1. Soil Wedges

The occurrence of soil wedges ("ground wedges" or "initially-ground wedges") that result from seasonal-frost cracking has already been discussed (see Chapter 6). Their occurrence implies deep seasonal-frost penetration. The Russian literature recognizes a range of forms (Figure 13.1). Although seasonal-frost cracking has been reported from several non-permafrost regions (Svensson, 1983, 1988b; Washburn et al., 1963), there are few descriptions of relict soil wedges in mid-latitudes in the North American and European literature. Brief observations made by French et al. (2003, pp. 267–268) for Southern New Jersey, Eastern USA, suggest they can be distinguished from sand-wedge casts by their younger stratigraphic position and by their shape, size, and spacing. They are broad structures, usually less than 2.0 m deep and between 20 cm and 60 cm wide, spaced approximately 2.0 m to 3.0 m apart, and filled with locally-derived mineral soil of a primary nature. It is speculated that they formed under conditions of deep seasonal frost that occurred when permafrost degraded towards the end of the Late Pleistocene. Their recognition is difficult because of the effects of agricultural practices (plowing, post holes, service lines etc.).

It is likely that some of the wedge structures reported from elsewhere in the mid-latitudes (Chapter 11) may reflect deep seasonal frost rather than permafrost.

13.2.2. Frost-Disturbed Bedrock

Mechanically-weathered bedrock is often interpreted to reflect intense frost action. As discussed earlier in Chapter 4, frost action can imply several different mechanical weathering processes. For example, rock may be shattered by the downward infiltration of water, and its subsequent freezing, in cracks and interstices. Rock shattering may also result from

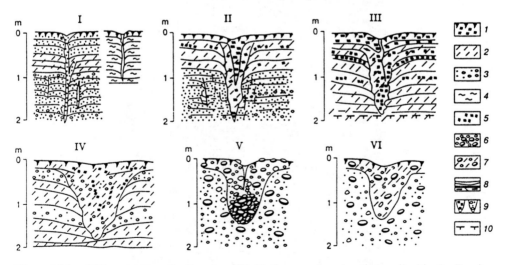

Figure 13.1. Different types of soil wedges ("initially-ground wedges") described in the Russian literature. Legend: 1, soil/humus; 3, sand and gravel; 4, loess-like deposits; 5, peat; 6/7, boulders and pebbles; 8, stratification and small faults; 9, ground wedges; 10, permafrost table. From Romanovskii (1973), Melnikov and Spesivtsev (2000).

thermal stresses set up in the rock itself. Finally, the growth of segregated ice lenses at depth, from water migrating towards the freezing plane, may force rock layers apart.

Ice-rich brecciated bedrock can be observed in many permafrost regions today (Büdel, 1977; see English translation, 1982, p. 82; French et al., 1986) (see Section 7.3.3). It is the result of the growth of segregated ice, often reticulate in structure (Mackay, 1974b; Murton and French, 1994). In sedimatary bedrock, expanded joints (see Figures 7.6C, D) may cause buckling of adjacent rock (see Section 4.5.1).

Brecciated bedrock, often to depths of 3–6 m, can be observed in a number of lithologies in mid-latitudes today (Figure 13.2). For example, in central and southern England, brecciation has been described from Portland Limestone, Mercer Mudstone, Upper Lias clay, and Chalk (Bradshaw and Smith, 1963; Horswill and Horton, 1976; Murton, 1996a). The Chalk is especially suited to brecciation by ice segregation because it is relatively soft, and highly porous and permeable (Williams, 1987). Brecciation takes the form of a near-surface mantle of loose bedrock in which vertical and horizontal joints are spaced a few centimeters apart (see Figure 7.6B). Typically, the lower contact with unbrecciated bedrock is gradual.

Brecciated bedrock has attracted only limited attention from periglacial geomorphologists but is probably far more widespread in occurrence than is currently reported. One

Figure 13.2.   Brecciated Chalk bedrock. The section shows, in ascending sequence, unweathered chalk, brecciated chalk, an involuted layer, and wind-blown sediment ("brickearth"). Isle of Thanet, Kent, southern England.

must note that this phenomenon fits well with our current understanding of permafrost and ground ice conditions in arctic regions today.

Slightly more ambiguous phenomena are the angular rock-rubble accumulations, known variously as "blockfields," "boulder fields," or "felsenmeer," that cover many of the uplands and low mountain ranges of the mid-latitudes (Derbyshire, 1973; Dylik, 1956; Clark et al., 1992; Kwon, 1978, 1979; Lewis, 1988; Pissart, 1953; Smith, 1962). They are usually developed upon hard, siliceous rocks such as sandstone, quartzite, and granite (Figure 13.3A). These are the "periglacial facies" of Lozinski (see Chapter 1). Their origin is not fully understood and the broader issue of cryogenic weathering must be considered in their interpretation (see Chapter 4). While frost-wedging along joints and bedding planes is applicable to sandstone bedrock, differential thermal expansion of silica minerals under cryogenic conditions may be more appropriate to explain rock

(A)

(B)

Figure 13.3. Relict bedrock features of supposed frost-action significance. (A) Blockfield developed on granite, Mount Wielki Szyszak, Karkonosze, Carpathian Mountains, southern Poland. Note that the "periglacial facies" that Lozinzki described were in sandstone and further to the east in the Gorgany Range. Photograph is supplied courtesy of R. Zurawek. (B) Angular hillslope tor, 10 m high, fashioned by Pleistocene frost action and formed in metadolomite, Dartmoor, North Devon, UK. Photograph is supplied courtesy of Professor D. Mottershead.

disintegration of quartzite and granite. There are also other considerations. First, it must be proven that these blocky accumulations are not forming under present environmental conditions. In the Appalachian Mountains of the eastern United States, both actively forming and relict blockfields exist (Hack and Goodlett, 1960, pp. 31–32; Rapp, 1967). Second, many of the kurums and boulder fields in northern environments, such as Transbaikal, northern Ural and the northern Yukon mountains (French, 1987; Perov, 1969; Romanovskii, 1985; Tyurin et al., 1982), appear largely inactive. One must question, therefore, their validity as modern analogues. Third, blockfields may form under glacial, as opposed to periglacial, conditions (Dahl, 1966; Ives, 1966). It has been suggested that blockfields, and entire periglacial landscapes, can be preserved beneath protective cold-based ice covers (Kleman and Borgström, 1990). It follows that blockfields, although certainly cold-climate phenomena, cannot be uncritically regarded as diagnostic of either intense frost action or permafrost conditions.

Equally ambiguous are the bedrock outcrops, known generally as tors, which exist on many hillslopes and upper summits of the mid-latitude landscapes (Figure 13.3B). They have attracted considerable attention (Caine, 1967; Czudek, 1964; Demek, 1964; Fahey, 1981; Jahn, 1962; Linton, 1955, 1964; Palmer and Nielson, 1962; Palmer and Radley, 1961; Te Punga, 1956; Wood, 1969) but their climatic significance is uncertain. If of periglacial (i.e. frost action) significance, tors reflect the remnants of frost-shattered bedrock that is now surrounded by a low-angled surface across which frost-shattered debris has been transported by mass wasting. The alternative interpretation is that tors result from differential deep weathering during Tertiary times and have subsequently experienced regolith removal by mass wasting during the Quaternary to expose intact bedrock.

Both interpretations have periglacial merit since they stress the importance of either frost action or mass wasting. It is significant that tors of both origins have both been described from within the same areas (Demek, 1964; Jahn, 1962; Linton, 1955; Palmer and Nielson, 1962). On the other hand, it is unwise to uncritically accept a simple frost-action explanation. This is because tors in high latitudes are just a further example of the problem of cryogenic weathering discussed in Chapter 4. Moreover, tors are also explained in terms of chemical weathering (Derbyshire, 1972) and wind abrasion (St-Onge, 1965). Therefore, while they may be regarded as indicative of previous cold-climate conditions, their exact significance is still imprecise. Closely related to tors are so-called "cryoplanation terraces." While the existence of these surfaces in northern latitudes is unquestionable (see Chapter 9), no study has yet demonstrated their active formation under cold-climate conditions. Some, initially interpreted as cryoplanation (altiplanation) forms, have been shown to be raised marine benches (Péwé et al., 1982, pp. 2533). Thus, it seems more appropriate to stress the role of cold-climate mass wasting rather than frost action in any interpretation of tors and associated bedrock surfaces.

13.2.3. Stratified Slope Deposits

Intense freeze–thaw action, together with slopewash activity, is generally believed to result in rhythmically-stratified slope-waste deposits (Figure 13.4). First described from the Charente region of western France, these deposits were termed "grèzes litées" or "éboulis ordonnés" (deWolf, 1988; Guillien, 1951; Malaurie and Guillien, 1953). Other stratified slope deposits were subsequently described from central Poland (Dylik, 1960, 1969a) and southern England (Kerney et al., 1964).

The sediments involved are essentially frost-shattered debris, the nature of which depends upon the lithology concerned. The deposits, which may attain a thickness in excess of 10 m, possess coarse bedding and a degree of sorting. Typically, they mantle

Figure 13.4. An exposure of stratified slope deposits ("grèzes litées"), near Sonneville, Charente region, southwest France.

lower valley-side slopes. In Charente, the stratified sediments consist of small angular limestone particles, up to 2–3 cm in diameter, together with inter-layered finer fractions. These materials, and their stratification, reflect frequent, probably diurnal, oscillations of freezing and thawing. The sorting, bedding, eluviation, and redeposition of finer particles are all attributed to slopewash following the melt of ice and snow. Bedded scree is a form of stratified-slope deposit found in upland areas (Boardman, 1978; Derbyshire, 1973; Karte, 1983).

The apparent absence of similar deposits from most high latitudes can probably be explained by the relative lack of repeated freeze–thaw oscillations in high latitudes (see Chapter 3). The closest present-day analogue is the stratified scree described from the sub-arctic region of Gaspésie, Eastern Canada (Hétu, 1995; Hétu et al., 1994). There, stratification is achieved by the sliding of icy block fragments over each other on the slope in winter together with niveo-eolian processes in spring. Another analogous mechanism, especially appropriate for high-mountain relief in low latitudes, is a combination of solifluction and mudflow/eluviation following snowmelt and active-layer thaw in spring (Francou, 1988, 1990).

In summary, stratified-slope deposits should be regarded as having some qualitative cold-climate significance. However, it must also be borne in mind that similar deposits may reflect sedimentological mechanisms, such as dry-grain flow and debris flow, which are unrelated to periglacial conditions (Hétu et al., 1995; Van Steijn et al., 1995).

13.2.4. Head and Solifluction Deposits

The slow mass-wasting processes that occur on slopes in cold climates today have been described in Chapter 9. Given the importance of freeze–thaw activity in the Pleistocene mid-latitudes, many unstratified and heterogeneous materials which mantle lower slopes and valley bottoms are regarded as the mid-latitude equivalents of the solifluction and "muck" deposits that formed in the high latitudes during the Pleistocene (see Chapter 11).

These deposits were first reported from southwest England, where they were termed "head" since they rested on the top of many coastal cliff sections (De la Beche, 1839). Later, C. Reid (1887) described the "coombe rock" of the Chalkland valleys as a form of "head" deposit, and J. Prestwich (1892) documented "rubble-drift" deposits elsewhere in southern England. Subsequently, the mapping of "head" became standard practice by both the Geological Survey and the Soil Survey of Great Britain (Avery, 1964; Dines et al., 1940). On the European Continent too, the significance of "pseudo-glacial" sediments had been appreciated at an early date and the recognition of frost-derived sediments in the lowlands of western and central Europe quickly became commonplace (Büdel, 1944; Edelman et al., 1936; Troll, 1944). In North America, the mapping of "surficial" materials quickly became standard practice, although much surficial material is clearly glacial rather than periglacial in origin. Today, "head" deposits are synonymous with solifluction deposits.

Head deposits vary considerable in their composition. Their cold-climate origin is based upon two characteristics. First, they consist of predominantly poorly-sorted, unstratified, angular debris of local derivation with larger elongate clasts often aligned downslope and tilted upwards. These are characteristics of modern solifluction deposits (see Chapter 9). Second, some head deposits contain faunal remains (mollusca, coleoptera) indicative of cold-climate conditions. Today, slopewash, soil creep, and eolian deposition account for much of the head that is forming in temperate mid-latitudes today and which is generally termed "colluvium."

As general indicators of cold-climate conditions, head deposits represent an important line of evidence. However, as with much of the evidence already discussed, their interpretation in terms of frost action is not always easy. A fundamental problem lies in our inability, even in present-day periglacial environments, to differentiate between a solifluction deposit that is primarily the result of frost creep and retrograde movement, and one primarily the result of gelifluction. The former implies seasonal frost and the latter implies permafrost. A second problem is that not all solifluction deposits contain readily identifiable cold-climate faunal remains. A third is that solifluction is sometimes difficult to distinguish from till. For example, where a slope deposit has been derived from glacial sediments, as in ablation till, it is impossible to distinguish till from solifluction because, to all intents and purposes, they are the same.

13.2.5. Frost-Disturbed Soils and Structures (Involutions)

Disturbed, distorted, and deformed structures occurring in unconsolidated Quaternary sediments have been described frequently from mid-latitudes. In the European literature, these structures are usually referred to as "cryoturbations" (Edelman et al., 1936), "brodelboden" (Gripp, 1926; Troll, 1944), or "pseudo-solle" (Maarleveld and van den Toorn, 1955). In the English-language literature, the term "involution" is preferred (Denny, 1936; Sharp, 1942b).

Use of the term "involution" requires clarification because it is possible to identify large structures that form during permafrost degradation and smaller ones that form within the active layer. Therefore, one must distinguish between (i) "thermokarst involutions," formed in water-saturated sediments by loading during thaw-degradation of an ice-rich permafrost body, and (ii) "periglacial involutions," formed by repeated frost action (cryoturbation) within a seasonally-frozen layer. Almost by definition, periglacial involutions are usually less than 1.0–2.0 m in vertical extent and are restricted to either a paleo-permafrost active layer or, less likely, to a paleo-zone of deep seasonal frost. It is the latter type of involution that is of concern in this section.

Most periglacial involutions are more regular in form and occurrence than thermokarst involutions. In spite of this, the diversity of forms is high (Figure 13.5A), ranging between

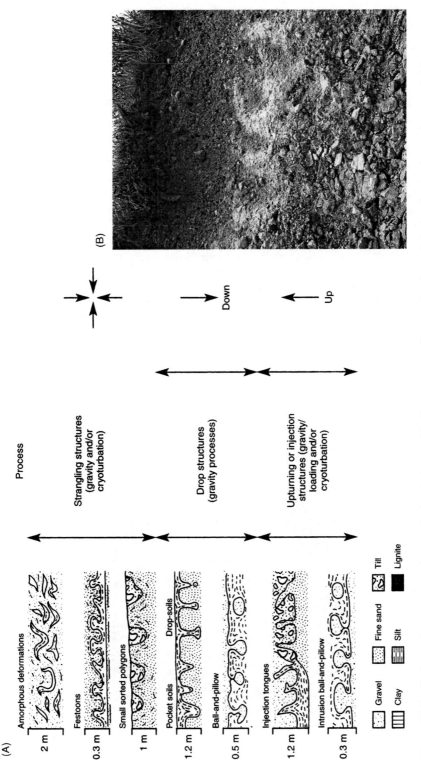

Figure 13.5. Deformation structures (cryoturbations) observed in Quaternary sediments that formed during the previous existence of either seasonally-frozen ground or in a former active layer overlying permafrost. (A) Range of structures observed in Quaternary sediments in central Germany. From Eissmann (1994). (B) Cryoturbations ("festoons") in near-surface sediments, Pays de Caux, Northern France.

amorphous deformations, drop soils, injection tongues, intrusion features, and festoons (Figure 13.5B). All these structures are probably related to the previous thawing and freezing of ground. However, a cell-like convective movement within the active layer, as inferred for permafrost regions today (see Chapter 6, pp. 144–152), does not adequately explain all the structures than can be observed, and the significance of any difference between seasonal- and perennial-frost conditions is still unclear. Therefore, these structures need careful interpretation.

13.3. INTENSE WIND ACTION

A seminal study by A. Cailleux (1942) analyzed the wind-abraded form of over 3000 sand samples collected from across Europe. It showed that sand grains that had undergone transportation and saltation by wind are rounded and possess a matt or "frosted" appearance (Figure 13.6). Furthermore, the percentage of frosted grains increases eastwards in a broad belt extending from the lowlands of central Europe into European Russia and reaches nearly 100% in areas immediately south of the glacial limits (Figure 13.7). The conclusion was that intense wind action was an important characteristic of the mid-latitudes during the cold periods of the Pleistocene.

Since then, other features and deposits have confirmed the intensity of Pleistocene wind action. These include wind-abraded rocks, eolian sand veneers, and wind-blown silt. All are described below.

13.3.1. Wind-Abraded Rocks

Polished, eroded, and faceted pebbles and rocks occur widely in the mid-latitude areas that surrounded the Pleistocene ice sheets. They are commonly termed ventifacts. They are attributed to the strong katabatic and zonal winds that would have characterized the ice-marginal areas. As such, they are well developed in areas such as central Poland (Dylik, 1956), southern Scandinavia (Svensson, 1983; Schlyter, 1995), northern

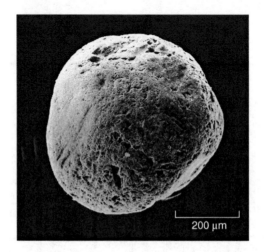

200 µm

Figure 13.6. Rounded matt ("frosted") sand grain caused by eolian action. The sample was obtained from Late-Pleistocene-age dunes, Belchatow, central Poland. Photograph is supplied courtesy of Dr J. Gozdzik.

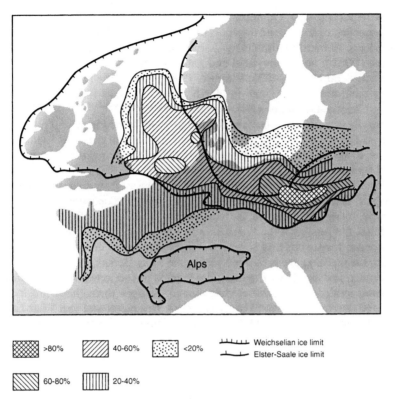

Figure 13.7. The distribution and frequency of occurrence of wind-modified sand grains (0.4–0.1 mm) in Europe according to A. Cailleux (1942).

Saskatchewan (Fisher, 1996), and north-central United States (Sharp, 1949). In most cases, ventifacts are found in association with sand, often buried by younger sediments. Sometimes, they form layers of abraded pebbles from between which finer particles have been winnowed out.

Although ventifacts allow paleo-environmental inferences to be made as to wind direction and intensity in certain areas, the majority can only be used as supporting evidence. This is because the grooves, fluting, and facets are rarely of constant orientation and most ventifacts have probably experienced movement over time. Even when large wind-polished boulders are examined, as in Denmark (Christiansen and Svensson, 1998), it is difficult to unambiguously identify the wind systems that were involved (Christiansen and Svensson, 1999; Vandenberghe et al., 1999).

13.3.2. Eolian Silt (Loess)

Well-sorted eolian silt is termed loess. Pleistocene-age loess is thought either to have originated from glacially-derived outwash surfaces at the margins of the retreating ice sheets or to have been transported from the deserts of Central Asia. Loess is largely a northern-hemisphere phenomenon and usually confined to the mid-latitudes. Loess-like silt of the higher latitudes (Péwé, 1955; Péwé and Journaux, 1983; Péwé et al., 1995) is discussed in Chapter 11.

Loess is important as a Quaternary paleo-environmental indicator because a near-complete sequence of loess deposition occurs on the Loess Plateau of Central China. By its nature, loess is highly suitable to thermo-luminescence (TL), optically-stimulated luminescence (OSL), and amino-acid and paleo-magnetic dating techniques (Forman et al., 1995; Oches and McCoy, 1995; Whintle, 1990). One of the most complete records of Quaternary environmental change in North America comes from the loess of Central Alaska (see Chapter 11, Figure 11.9).

It is generally accepted that most mid-latitude loess formed in glacial rather than inter-glacial times. The environments must have been cold and dry. This is indicated by three things. First, if wetted, loess experiences shrinkage and compaction, proving that it was not soaked by water or deposited by it. Second, much loess is calcareous and shows no signs of leaching by water (precipitation). Third, loess often contains faunal remains, especially land snails (mollusca) suited to cold steppe-like conditions.

In Europe, loess is thickest in the east, where it mantles both upland plateaus and lowland plains and reaches a thickness of over 30 m. In southwestern Poland, at least five Late-Pleistocene fossil soils and several horizons containing cryogenic structures (sand-wedge casts, involutions) are preserved within the loess (Jersak, 1973, 1977). In Belgium, northern France and southern England, the loess is not so extensive or as thick but still provides good evidence of cyclic environmental changes throughout much of the late Pleistocene (Balescu et al., 1988; Haesaerts, 1983; Haesaerts and van Vliet-Lanoë, 1981; Vandenberghe and Nugteren, 2001). In southern England, many surficial deposits that mantle the Chalk and other upland surfaces clearly possess a loess fraction (Bateman, 1998; Loveday, 1962). Loess is also reported from the head deposits of southwest England (Mottershead, 1977; Roberts, 1985).

The source area for the loess of Western Europe was probably the exposed floor of the North Sea when sea level dropped during the glacial periods (see Chapter 11). This is certainly the case for The Netherlands (Kasse, 1997). Typically, successive episodes of loess deposition in Western Europe have left a widespread veneer of silty sediment that has been reworked by water and incorporated into underlying materials through frost action and solifluction.

In North America, extensive loess deposits occur in the Missouri–Mississippi drainage basins of north-central United States (Smith, 1964). Here, the Late-Pleistocene loess depositional record is closely related to fluctuations in the Laurentide ice (Forman and Pierson, 2002). In all probability, the loess was derived from the large braided alluvial channels formed by rivers draining from the ice margins. This is because loess thickness progressively decreases with increasing distance from these channels. As in Europe, various episodes of loess deposition are recognized. Faunal remains usually suggest a forest/grassland cover rather than the steppe/tundra cover inferred for Europe.

The Loess Plateau of western China, where loess thicknesses can exceed 200–300 m, has attracted attention from both Chinese and western scientists. Proximity to the deserts of central Asia, combined with the progressive uplift of the Tibet Plateau during Quaternary times, may explain its occurrence, but a complete understanding of the origin of much Chinese loess is still lacking. The extensive literature on the loess of central China is not discussed here.

13.3.3. Cold-Climate Eolian Sand

Eolian sand is coarser than loess and usually in the size range 0.06–1.00 mm in diameter. The term "cover-sand" is used to describe these sediments when a cold-climate depositional environment is inferred and where the sand constitutes a thin surficial veneer. In Europe, cover-sand is extensive over large areas of the western and central European lowlands (Catt, 1977; Maarleveld, 1960; Nowaczyk, 1976). It is typically between 0.5 m and 5.0 m in thickness.

Cover-sand is related to a nearby sediment source. It shows a degree of alternating bedding involving layers of silty sand and sand that is interpreted as the result of deposition on an alternating wet and dry surface. The loam is reworked loess. Cryoturbation structures and occasional faunal remains in cover-sand further indicate a cold climate. Thus, cover-sand is regarded as being largely niveo-eolian in origin, having been modified by sheetwash processes (Koster, 1988; Schokker and Koster, 2004). A variant of cover-sand occurring in northwestern France is the silty sandy layers termed "limons-a-doublets" (Lautridou and Giresse, 1981). These layers are best explained by the migration of clay and iron within loess following initial loss of carbonate. In Eastern England, much of the cover-sand is thought to have been deposited during the Younger Dryas (12.5–11.4 ka) (Bateman, 1998).

Cold-climate eolian sand may also accumulate in thicker layers to form sand sheets and sand dunes. In both Europe and North America, they occur in areas located immediately beyond the Late-Pleistocene ice margins (Figure 13.8). According to Koster (1988), the sands were derived from flood plains, glacial outwash plains, till plains, and lake shores. Parabolic and transverse dunes are common, similar to those that form in warm semi-arid regions today. Facies analyses indicate a complex interaction between fluvial, lacustrine, and eolian processes in their formation (Ruegg, 1983; Schokker and Koster, 2004; Schwan, 1986). Under favorable conditions, wind direction can be inferred from dune morphology and orientation, the dip of the foreset beds, and the relation of dunes to known sediment-source areas (Koster, 1988; Maarleveld, 1960). In the southeastern Netherlands, the majority of the sediments were deposited during successive Middle- and Late-Pleistocene glacial periods (Kasse, 1997). The latest period of sand-sheet formation in Northwestern Europe occurred between ~14 and 12 ka, and coincided with a period of permafrost degradation and increased aridity.

Like the loess deposits described earlier, cold-climate eolian sand represents further evidence of the importance of wind in the mid-latitudes during the cold periods of the Pleistocene.

13.4. FLUVIAL ACTIVITY

The drainage of the immediate ice-marginal, or proglacial, zone surrounding the continental ice sheets was dominated by large water bodies. These were of two kinds. First, meltwater discharge was concentrated within wide shallow channels aligned approximately parallel to the ice margin. On the North European Plain, these channels are termed "pradolinas" or "ürstromtalers." They experienced episodic and repeated use, especially toward the end stages of each cold period. In parts of Canada and the Russian Plain, large freshwater lakes, often interconnected, were impounded against the retreating ice margin. Detailed accounts of these water bodies are contained in the many regional geology reports of various areas, and useful summaries, including maps, are provided by Velichko (1982) and Fulton (1989). Second, larger fluvial drainage systems extended away from the ice margins. Rather then being constrained within erosive channels, these rivers formed broad outwash plains characterized by braided-channel surfaces composed of coarse clastic sediment. Today, similar surfaces occur around the margins of ice sheets and glaciers, and are termed sandur (plural: "sandar"; Church, 1972; Krigstrom, 1962).

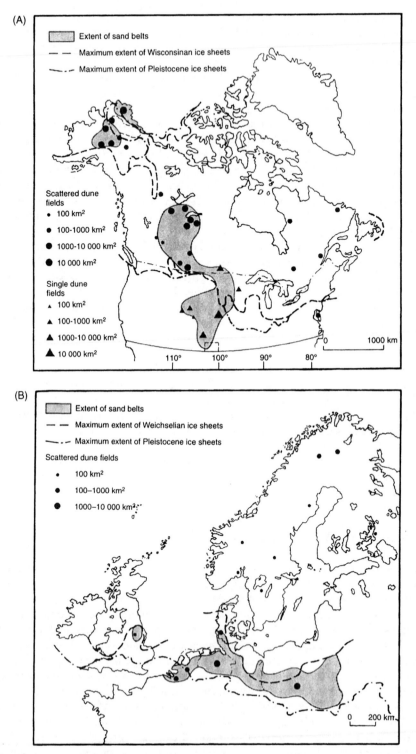

Figure 13.8. Maps showing the distribution of Pleistocene cold-climate dune fields and sand sheets. (A) Europe; (B) North America. Also indicated are the maximum extents of Pleistocene (Wisconsinan/Weichselian) ice sheets. From Koster (1988). Reproduced by permission of John Wiley & Sons Ltd.

Paleo-sandur surfaces are not easy to recognize in the mid-latitudes today because of vegetation growth and subsequent modification by land-use practices. Where ancient paleo-surfaces have been preserved, aerial photographs sometimes reveal palimpsests of these large paleo-channel systems. This is the case for the Pine Barrens of southern New Jersey, Eastern USA (Farrell et al., 1985).

The following discussion concentrates upon the fluvial modifications that occurred further from the retreating ice margins, that is, in the Pleistocene periglacial zone senso-stricto.

13.4.1. Major Rivers

Brief mention of the large-scale paleo-hydrology of mid-latitudes is justified because vestiges of early non-periglacial landscapes complicate recognition of cold-climate features. For example, in northwestern Arctic Canada and in Alaska, complex changes to the present-day Yukon and Mackenzie drainage systems were effected by late-Cretaceous tectonic activity, followed by the various Cordilleran and continental glaciations during the Quaternary (Duk-Rodkin et al., 2004; see Figure 2.2). Likewise, in Western Europe, the Pleistocene proto-Rhine drainage flowed across broad lowlands now occupied by the North Sea. The proto-Thames also drained eastwards towards the North Sea, probably joining with the proto-Rhine.

Evidence for these former fluvial systems can be found in flights of terraces and upland surfaces. In northwestern Canada and Alaska, high-level benches and pediment-like surfaces reflect these old landscape systems (Duk-Rodkin et al., 2004). In Europe, the best known terrace systems are those of the Thames, Rhine, and Maas catchments (Gibbard, 1988; van Huissteden, 1990; van Huissteden and Vandenberghe, 1988). In the North Sea basin itself, "cold" fluvial sediments, present in boreholes within the Dutch sector, provide proof of the major river systems that extended across this lowland when sea level was lower than today during the glacial periods (Laban and van der Meer, 2004).

There is a possibility that many so-called "cryoplanation" terraces, described in the Pleistocene periglacial literature, from the hard crystalline uplands of both Europe and northwestern North America, relate to these early landscape systems. As such, they may not have a cryogenic origin but are largely inherited features that were merely "trimmed" by cold-climate conditions (see Chapter 9). It is significant that there is little evidence for the current formation of cryoplanation terraces and cryopediments in today's high latitudes.

The cold-climate nature of many of the ancient fluvial deposits that are found in the mid-latitudes is inferred from the presence of frost-fissure pseudomorphs, involutions, and faunal remains contained within them. Sedimentological studies suggest a cyclic pattern of river incision and subsequent infilling (Vandenberghe, 1993) that corresponds to the start and end of each cold period (Figure 13.9A). In theory, fluvial erosion would have been favored at the beginning of the cold stages because slope and riverbank stability would have been maintained by the degrading vegetation cover relative to a decline in evapotranspiration and an increase in runoff. At the cold period maximum, sediment supply and discharge would probably have peaked, and aggrading, braided-channel systems would have been characteristic. Then, during the transition back to the warmer interstadial or interglacial conditions, there would have been incision due to the reduction in sediment load. Although somewhat speculative, this model is supported by stratigraphic observations. For example, Figure 13.9B shows a typical channel-fill sequence from a Maas river terrace in the Southern Netherlands. From the analyses of paleo-fluvial systems in both The Netherlands and the Central Poland lowlands, Vandenberghe et al. (1994) conclude that there were frequent transitions between braided and high-sinuosity meandering systems.

(A)

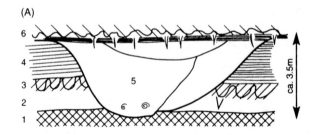

(B)

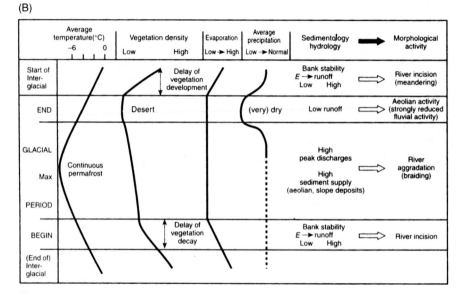

Figure 13.9. The Pleistocene periglacial fluvial cycle. (A) Typical infill sequence associated with Middle- and Late-Weichselian fluvial sedimentation, Maastricht, The Netherlands. Legend: 1, Eemian soil, 2, Early-Weichselian silt loam, 3, cryoturbation (frost-disturbed) layer, 4, Middle-Weichselian silt loam, 5, Late-Weichselian gully infilling, 6, Late-Weichselian frost-disturbed horizon and frost fissures. (B) The generalized cycle that occurs during a complete glacial (cold) period. From Vandenberghe (1993). Reproduced by permission of E. Schweizerbart, Borntraeger and Cramer Science Publishers.

13.4.2. Asymmetrical Valleys

The asymmetrical modification of valleys was also characteristic of the mid-latitude periglacial environments. While slope asymmetry can reflect either structural control or microclimatic differences on slopes of varying orientation, the latter must have been especially pronounced in the mid-latitudes during the Pleistocene.

In Western Europe, asymmetrical valleys are widespread, the steeper slope commonly facing west or southwest (Table 13.1). The cold-climate explanation involves differential insolation and freeze–thaw on the south- and west-facing slope, with snow- or

Table 13.1. Some characteristics of slope asymmetry attributed to Pleistocene periglacial conditions in the European mid-latitudes.

Location	Reference	Orientation of Steeper Slope	Processes Involved[1]					Mechanism Involved[1]	
			(1) Differential insolation and freeze-thaw	(2) Differential solifluction	(3) Wind and snow	(4) Wind and loess	(5) Lateral stream erosion	Decline of N- and E-facing slope	Steepening of S- and W-facing slope
United Kingdom:									
Chiltern Hills	Ollier and Thomasson (1957)	W/SW	X	x			x		X
Hertfordshire	Thomasson (1961)	W	X	x			x		X
Chalk, southern England	French (1972a, 1973)	W/SW	X	x	x		x		X
France:									
Gascony	Taillefer (1944)	W		x	X		x	x	X
Gascony	Faucher (1931)	W		x		X	x		x
North France	Gloriad and Tricart (1952)	W	x	X			x		x
Netherlands and Belgium:									
Haspengouw	Geukens (1947)	W	X	x	x		x	x	x
Veluwe	Edelman and Maarleveld (1949)	W/SW		x	X		x	x	x
Hesbaye	Grimberbieux (1955)	W	X		x		x		x
Germany:									
South Germany	Büdel (1944, 1953)	W/SW		x	X	X	x	x	x
Muschelkalk	Helbig (1965)	W		X	x			x	
Czechoslovakia:									
Bohemia	Czudek (1964)			X			x		
Poland:									
Łódź plateau	Klatkowa (1965)	W	X	x	x		x	x	x

[1]X dominant process; x secondary proces.

loess-induced mass-wasting on the opposing slope, combined with a migrating stream in the valley bottom (Edelman and Maarleveld, 1949; Geukens, 1947; Ollier and Thomasson, 1957). Under this interpretation, asymmetry formed as valley incision was occurring and the steeper slope is regarded as the "warmer" slope. If this asymmetry is compared to that observed in high latitudes today (see Chapter 10, Table 10.6), the main difference is that, in high latitudes, the steeper slope, irrespective of orientation, is usually the "colder" slope (Currey, 1964; French, 1971b). Presumably, this reflected the greater activity of diurnal freeze–thaw processes on "warmer" slopes in mid-latitudes leading to erosion and retreat of the slope. By contrast, in high latitudes, the so-called "warmer" slope gives rise to a thicker active layer, greater mass-wasting, and an overall lowering of slope.

13.4.3. Dells and Periglacial Valleys

Additional cold-climate fluvial modification took the form of shallow depressions or gullies that developed within existing catchments. Presumably, they were eroded by snow melt, either by slopewash, seepage, or concentrated channel flow, acting over seasonally or perennially-frozen (impermeable) substrate.

The Chalk uplands and plateaus of southern England, northern France, Belgium, and southwestern Germany illustrate this type of fluvial modification (Gloriad and Tricart, 1952; Grimberbieux, 1955; Helbig, 1965; Ollier and Thomasson, 1957). All are dissected by well-developed networks of dry valleys that must have formed when the Chalk was able to support a transitory surface drainage. By its very nature, the Chalk is highly permeable and porous. Today, many streams are ephemeral and many valleys are dry. Undoubtedly, many of the larger valleys are polycyclic because their long profiles show signs of successive rejuvenation. These valleys probably developed by normal fluvial action when the groundwater table was higher but, as it fell in response to progressively lower sea levels during the Quaternary (see Figure 11.5), the upper valley sections were left dry. However, the much smaller depressions and gullies that are discussed here were eroded when frozen subsoil would have temporarily reversed the desiccation sequence and promoted surface runoff (Bull, 1940; Reid, 1887).

The broad shallow depressions that exist on the Chalk outcrops are referred to as "dells." They are analogous to the "nivation hollows" that have been described from high latitudes (Ballantyne, 1978; Cook and Raiche, 1962). While the utility of the nivation concept is debatable in the context of the high latitudes (see Chapter 9), the importance of snow in the mid-latitudes should not be underestimated. The possibility exists that these shallow depressions are true periglacial valleys in the sense that they are totally the result of snowmelt erosion under cold-climate conditions. Periglacial dells are described from the Lódź Plateau of central Poland (Klatkowa, 1965), and relict nivation features in general are discussed by Ballantyne and Harris (1994, pp. 245–250).

Other truly periglacial valleys are also associated with the Chalk escarpments of southern England. These take the form of deeply incised gullies possessing steep slopes, flat bottoms, and abrupt headwalls (Brown, 1969; Kerney et al., 1964). They appear concentrated at groundwater-discharge localities. The valley infill is composed primarily of well-sorted and well-stratified calcareous mud, suggesting the valleys were eroded by a combination of intense mechanical weathering and snowmelt runoff, probably in the late glacial. Groundwater seepage, as indicated by springs today, would have been a major factor that intensified frost shattering.

In summary, there is abundant evidence to suggest that fluvial activity was important in the Pleistocene periglacial environments of mid-latitudes. Adjacent to the ice margins, meltwater discharges eroded spillway channels, deposited broad alluvial fans, and provided the source waters for major river systems. Further away from the ice margin, snowmelt and groundwater discharges effected the asymmetrical modification of valley-side slopes and the erosion of periglacial dells and valleys. Whether or not runoff was upon seasonally- or perennially-frozen ground is difficult to determine.

13.5. SLOPE MODIFICATION

Here, three examples of cold-climate slope modification are described from the lowlands of Europe. They illustrate the importance of mass-wasting processes, the infilling of valleys with both fluvial sediment and mass-wasted material, and an overall lowering of relief.

13.5.1. Mass Wasting on Slopes

Blockfields have already been discussed in the context of frost-weathered bedrock (see earlier). Where concentrations of blocks occur on lower slopes and in valley bottoms, they have clearly been moved to their present positions. These accumulations, termed "stone streams," "rock streams," or "coulées pierreuses," occur at a number of localities in western Europe and North America. They are interpreted to be the result of cold-climate mass wasting (Péwé, 1983b; Pissart, 1953; Smith, 1962).

Periglacial "rock-streams" that occur within the Chalk valleys of southern England (Small et al., 1970; Te Punga, 1957; Williams, 1968) serve to illustrate the role played by mass-wasting in slope evolution during the cold periods (Figure 13.10). Thaw-consolidation and solifluction would have been the mechanisms of transport.

The stone streams of the English Chalklands consist of large blocks of silicified sand and flint conglomerate ("sarsens") that presumably originated on the upland surfaces of the Chalk as a form of "duricrust" or silcrete in Late Tertiary times. During the Quaternary, these heavy blocks moved, first downslope and then downvalley, some as much as 4000 m on an average gradient of between 1° and 3°. Any explanation involving sarsen movement by early man is highly unlikely because many stones weigh several tons. It is likely that there were several episodes of movement during the Pleistocene. On each occasion, the valley became plugged with material and the stream was pushed laterally to undercut the opposing slope. It has yet to be established whether the mass wasting was associated with either seasonal or perennial frost. If it were the latter, the sarsen stones probably slid across the base of the active layer as a form of plowing block. If it were seasonal frost, the stones were probably rafted downslope within the waste mantle. It is likely that both mechanisms operated at different times.

In terms of landscape modification, rocks streams indicate denudation from upland surfaces and the transport and deposition of debris on lower slopes and in valley bottoms.

13.5.2. Valley-Bottom Aggradation

Mass wasting on slopes and the movement of material towards the valley bottom was often accompanied by the aggradation of fluvial sediment within the actual stream channel. It must be remembered that the rivers that flowed away from the ice margin carried high

Southwest   Northeast

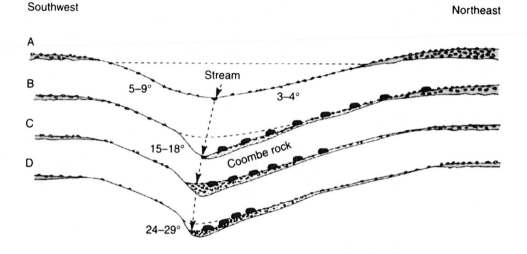

A Initial valley incised through hill top surface of sarsens and Tertiary deposits

B Downcutting and beginning of sarsen/solifluction movement

C Main period of sarsen movement downslope and downvalley

D Period of fluvial reworking

Figure 13.10. The probable sequence of periglacial slope evolution at Clatford Bottom, Wiltshire, central southern England. From Small et al. (1970). Reproduced by permission of The Geologists' Association.

sediment loads. At the same time, discharge fluctuated in accordance with both diurnal and seasonal melt of the ice sheets. Such conditions favor sediment deposition and braided-channel aggradation.

An example of the complex interaction between slope and fluvial processes is provided by a study of slope evolution at Walewice, in central Poland (Dylik, 1969a, b). It demonstrates how mass-wasting processes operated in conjunction with fluvial sediment aggradation and fluvio-thermal erosion to degrade, modify, and obliterate an earlier slope form. Aspects of thermokarst, slopewash, and fluvio-thermal erosion in current cold climates have been described, respectively, in Chapters 8, 9, and 10. Over a period of several summers during the mid-1960s, a series of trenches were systematically excavated along a gentle slope extending away from the northern edge of a low terrace towards the center of the Warsaw–Berlin pradolina, one of the major ice-marginal meltwater channels of the central North European Plain (Figure 13.11). A synthesis of the results of the excavations is shown in Figure 13.12.

The terrace is mantled with till (boulder clay) derived from the Middle Polish (Riss) glaciation. Therefore, modification of the terrace edge took place during the last (Wurm) cold phase when the pradolina was functioning normally as a meltwater channel. The slope that extends below the terrace consists of rhythmically-bedded silt and sand that has buried three older, and steeper, slopes. Ice-wedge pseudomorphs, present in the stratified sediments, indicate that deposition occurred under permafrost conditions. Lower down the slope, the stratified sediments are truncated by the second, completely buried, slope. At

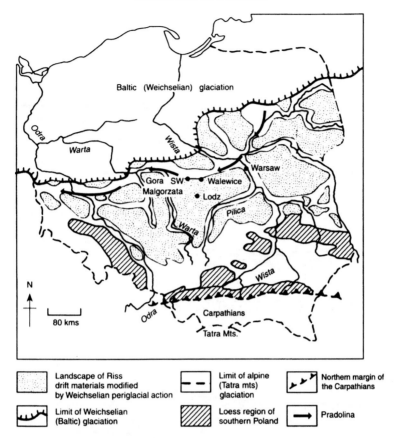

Baltic (Weichselian) glaciation

Odra

Warta

Wisla

Gora SW Malgorzata • Walewice

Warsaw

• Lodz

Pilica

Warta

Wisla

N

80 kms

Odra

Carpathians

Tatra Mts.

Landscape of Riss drift materials modified by Weichselian periglacial action

Limit of Weichselian (Baltic) glaciation

Limit of alpine (Tatra mts) glaciation

Loess region of southern Poland

Northern margin of the Carpathians

Pradolina

Figure 13.11. Map showing the limits of the last glaciation and the Pleistocene periglacial zone of central Poland. Also shown are the locations of Gora Sw Malgozata and Walewice (see Figure 13.12, 13.13).

the base of this buried slope, a large upturned block of stratified-slope sediment is, itself, buried by non-stratified slumped material.

The paleo-cryostratigraphy of the slope sediments at Walewice indicates that permafrost conditions were present throughout the period of time during which the terrace edge was degraded. The block appears to have fallen in a frozen state and to have been quickly buried because, if it had thawed, stratification would not have been preserved. In all probability, the block fell when a stream laterally undercut to form a thermo-erosional niche (see Chapters 8 and 10). Two generations of frost fissures are identified; an older set associated with collapse of the block and a younger set developed in material which buried the block. They prove that permafrost was present throughout this period of slope modification.

In terms of landscape modification, the slope at Walewice saw the elimination of the terrace edge and the development of a low-angled slope with a progressive reduction in relative relief through movement of material into the pradolina.

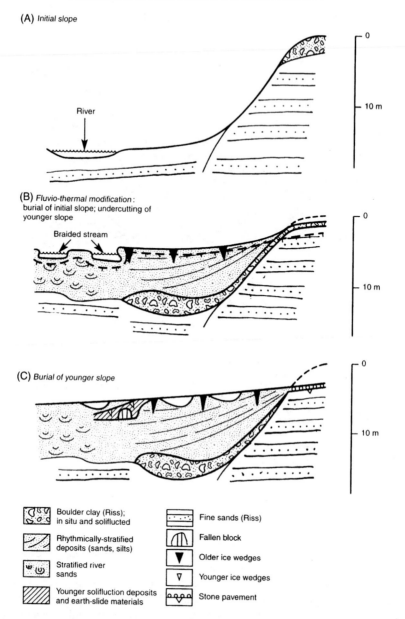

(A) *Initial slope*

River

0

10 m

(B) *Fluvio-thermal modification:*
burial of initial slope; undercutting of
younger slope

Braided stream

0

10 m

(C) *Burial of younger slope*

0

10 m

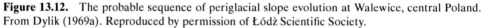

Boulder clay (Riss); in situ and soliflucted	Fine sands (Riss)
Rhythmically-stratified deposits (sands, silts)	Fallen block
Stratified river sands	Older ice wedges
Younger solifluction deposits and earth-slide materials	Younger ice wedges
	Stone pavement

Figure 13.12. The probable sequence of periglacial slope evolution at Walewice, central Poland. From Dylik (1969a). Reproduced by permission of Łódź Scientific Society.

13.5.3. Smoothing of Slopes

The overall flattening of relief that is thought to have occurred under periglacial conditions during the Pleistocene in the mid-latitudes is further illustrated by a second detailed slope study from central Poland (Dylik, 1963).

The isolated hill of Gora Sw Malgorzata is situated in the axial part of the Warsaw–Berlin pradolina (see Figure 13.11), and rises 20 m above the flat bottom of the broad depression. The hill possesses smooth concave slopes, grading from 25–35° in angle in their upper parts to 10° in their lower parts. The hill, composed of glaciofluvial sand, silt, and gravel which overlie till, is a glacial feature (kame) that formed during deglaciation of the penultimate ice sheet (Warta stage, Riss glaciation) (Dylik, 1963).

The slope deposits that mantle the foot of the hill enable reconstruction of the slope modifications that took place during the last (Vistulian-Weichslian) cold stage (Figure 13.13). The lowermost deposits are cross-bedded fluvial sand, indicating the previous existence of a stream at the foot of the slope. Overlying and obliterating the relief of these channel deposits are 5–10 m of solifluction debris and rhythmically-stratified deposits, all clearly derived from the glaciofluvial sediments which comprise the hill. Lobe-like solifluction structures constitute the lower section of the deposits. The change to rhythmically-

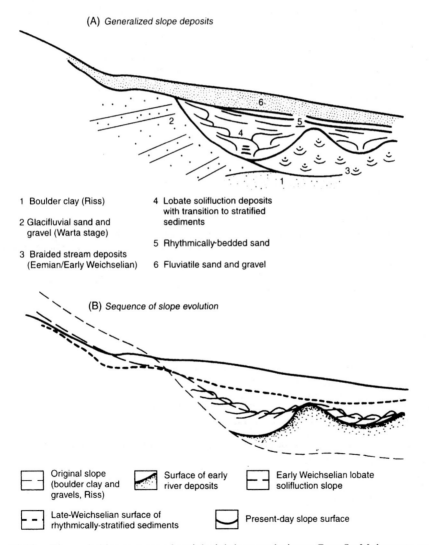

(A) Generalized slope deposits

1 Boulder clay (Riss)

2 Glacifluvial sand and
gravel (Warta stage)

3 Braided stream deposits
(Eemian/Early Weichselian)

4 Lobate solifluction deposits
with transition to stratified
sediments

5 Rhythmically-bedded sand

6 Fluviatile sand and gravel

(B) Sequence of slope evolution

Original slope
(boulder clay and
gravels, Riss)

Late-Weichselian surface of
rhythmically-stratified sediments

Surface of early
river deposits

Early Weichselian lobate
solifluction slope

Present-day slope surface

Figure 13.13. The probable sequence of periglacial slope evolution at Gora Sw Malgorzata, central Poland. From Dylik (1963). Reproduced by permission of Łódż Scientific Society.

bedded sediments indicates a change to sheet solifluction and slopewash on a progressively lower-angled slope. A second series of lobate deformations occur in an upper layer of stratified sand. Overlying all these sediments, and extending to the foot of the present slope, is a layer of cross-bedded sand and gravel with numerous pebble layers.

These slope deposits indicate a progressive change in the form of Gora Sw Malgorzata. In the earliest phase, probably at the beginning of the last cold period, solifluction and slopewash sediments obliterated the braided channel. As a result, the slope became longer and gentler. Then, solifluction filled the channel depressions and gave a step-like (lobate) profile. As slope gradient progressively decreased, stratified slopewash sediments were deposited and the slope assumed a smoother profile. Ultimately, the ancient channel relief was obliterated by a gentle depositional slope of 2–4° in angle. Further slope evolution was terminated by the deposition of bedded gravels which extend to the foot of the present-day hill. These suggest that the pradolina came into operation once again.

In terms of landscape modification, the sediments that flank the lower slopes of Gora Sw Malgorzata indicate continued slope modification throughout the last cold stage. Solifluction and sheetwash processes progressively reduced and extended the lower part of the slope while the upper part experienced erosion and retreat. Ultimately, a near-stable slope was probably attained. Then, during the climatic amelioration that accompanied the end of the cold stage, a second phase of solifluction was initiated before the lower slope was buried beneath fluvial sediments when discharge through the pradolina commenced.

ADVANCED READING

Ballantyne, C. K., Harris, C. (1994). *The Periglaciation of Great Britain*. Cambridge University Press, Cambridge, 330 pp.

Dylik, J. (1969). Slope development under periglacial conditions in the Łódż region. *Biuletyn Peryglacjalny*, **18**, 381–410.

Schokker, J., Koster, E. A. (2004). Sedimentology and facies distribution of Pleistocene cold-climate aeolian and fluvial deposits in the Roer Valley Graben (Southeastern Netherlands). *Permafrost and Periglacial Processes*, **15**, 1–20.

Vandenberghe, J. (1993). Changing fluvial processes under changing periglacial conditions. *Zeitschrift für Geomorphologie*, suppl. **88**, 17–28.

DISCUSSION TOPICS

1. Describe the typical landscape modifications that took place under periglacial conditions in the mid-latitudes during the Pleistocene.

2. Discuss the roles of snow, rain, wind, and aspect in the context of landscape modification during the cold periods of the Pleistocene.

3. What are the causes of asymmetrical valleys and what do they tell one about Pleistocene periglacial conditions?

4. What was the role played by fluvial processes in fashioning the Pleistocene periglacial landscapes of mid-latitudes?

PART IV
Applied Periglacial Geomorphology

14 Geotechnical and Engineering Aspects

Permafrost and intense frost action affect virtually all aspects of economic activity in periglacial environments. Geotechnical and engineering problems are caused primarily by frost heave and thaw settlement. The provision of housing, municipal services, water supply, roads, bridges, railways, and airstrips all require that the peculiarities of cold-climate terrain be taken into account. Similarly, natural resources exploration, development, and management present special problems, and the cost of both preventative and remedial measures is often significant.

14.1. INTRODUCTION

There is no denying the importance of permafrost in influencing human activities. Referring specifically to his experience in the construction of the Alaska Highway in 1942–1943, S. W. Muller wrote:

The destructive action of permafrost phenomena has materially impeded the colonization and development of extensive and potentially rich areas in the north. Roads, railways, bridges, houses and factories have suffered deformation, at times beyond repair, because the condition of permafrost ground was not examined beforehand, and because the behavior of frozen ground was little, if at all, understood. (Muller, 1943, pp. 1–2)

Cold-regions engineering is now a recognized specialty of engineering practice in North America and Scandinavia (Johnston, 1981; Senneset, 2000; Smith and Sego, 1994). In Russia and China, permafrost engineering is central to geocryology (Kamensky, 1998; Yershov, 1990). It is also realized that periglacial environments contains large quantities of hitherto untapped natural resources. In historic times, placer gold and mineral exploitation (zinc, copper, iron) activities were traditional activities. The last 30 years has seen discovery, in both Arctic North America and northern Russia, of large hydrocarbon (oil and gas) reserves and of precious minerals such as diamond, tungsten, and uranium.

This chapter outlines some of the geotechnical problems associated with the settlement, human activity, and natural resource development of periglacial environments. Special attention is given to oil and gas exploitation, the mining of precious metals, and the various preventative and remedial environmental measures that are commonly adopted.

The Periglacial Environment, Third Edition   Hugh M French
Copyright © 2007 John Wiley & Sons Ltd

14.2. COLD-REGIONS ENGINEERING

A number of geotechnical and engineering problems result from the occurrence of permafrost. For the most part they relate to the water and/or ice content of permafrost. These can be summarized as being either frost heave, thaw-subsidence, or hydrologic in nature.

14.2.1. General Principles

As explained in Chapter 4, pure water freezes at $0\,^{\circ}C$ and in doing so expands by approximately 9% of its volume. The most obvious result of soil freezing is the volume increase (heave) that results and the associated deformation of host sediment and rock. The heave has considerable practical significance since it causes displacement of buildings, foundations, and road surfaces. For soil to heave, the ice must first overcome the resistance to its expansion caused by the strength of overlying frozen soil. This usually occurs only when segregated ice lenses form. Frost heave causes significance damage to structures and foundations. In the case of pile foundations, repeated cycles of heave can progressively lift piles (Figure 14.1), in the same manner that frost-jacking of bedrock occurs (see Chapter 6). The annual cost of rectifying seasonal-frost damage in roads, utility foundations, and buildings in areas of permafrost and deep seasonal-frost, as present in areas of Canada, Alaska, Sweden, and northern Japan, is considerable. The costs of maintaining and upgrading some of the ageing and more primitive infrastructures in areas of northern Russia and northern China are even greater.

As explained in Chapter 7, ground ice is a major component of permafrost, particularly in fine-grained and unconsolidated sediments. Numerous case studies now document the ground subsidence that follows the thaw of ice-rich permafrost (see above). Thaw consolidation may occur as thawed sediments compact and settle under their own weight; the

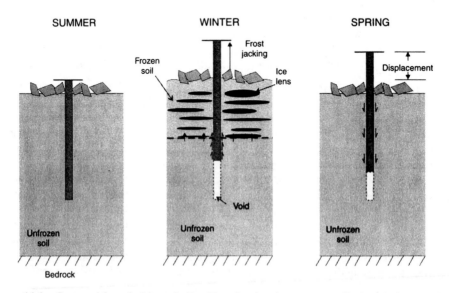

Figure 14.1. Seasonal frost jacking of piles. Freezing in winter causes pile foundations to lift. In spring and summer, piles do not return to their original position. From Wolfe (1998).

high pore-water pressures that are generated favor slope instability and slumping. These processes are described in Chapter 8. A related problem is that the physical properties of icy sediments, in which soil particles are cemented together by pore ice, may be considerably different to those of the same material in an unfrozen state. For example, in unconsolidated and/or soft sediments there is often a significant loss of bearing strength upon thawing. Beneath heated buildings, therefore, it is often essential to maintain the frozen state of the underlying material in order to support the structure.

As explained in Chapter 5, the hydrological and groundwater characteristics of permafrost terrain are different from those of non-permafrost terrain. For example, the presence of both perennially- and seasonally-frozen ground prevents the infiltration of water into the ground or, at best, confines it to the active layer. At the same time, subsurface flow is restricted to taliks. A high degree of mineralization in subsurface permafrost is often typical, caused by the restricted circulation imposed by permafrost and the concentration of dissolved solids in taliks. Thus, frozen ground eliminates many shallow depth aquifers, reduces the volume of unconsolidated deposits or bedrock in which water is stored, influences the quality of groundwater supply, and necessitates that wells be drilled deeper than in non-permafrost regions.

Engineers adopt a number of approaches that deal with these problems. If the site is underlain by hard igneous and metamorphic rock, as is the case for some regions of the Canadian Shield, ground ice is usually non-existent and permafrost problems can be largely ignored. In most areas, however, this simple approach is not feasible since an overburden of unconsolidated silty or organic sediment is usually present.

14.2.2. General Solutions

Modern construction techniques aim to maintain the thermal equilibrium of the permafrost and avoid the onset of thermokarst. The most common technique is the use of a pad or some sort of fill which is placed on the ground surface (Figure 14.2). This compensates for the increase in thaw which results from the warmth of the structure. By utilizing a pad of appropriate thickness, the thermal regime of the underlying permafrost is unaltered. It is possible, given the thermal conductivity of the materials involved and the mean air and ground temperatures at the site, to calculate the thickness of fill required. Too little fill, plus the increased conductivity of the compacted active layer beneath the fill, will result in thawing of permafrost and subsidence (Figure 14.2A). On the other hand, too much fill will provide too much insulation and the permafrost surface will aggrade on account of the reduced amplitude of the seasonal temperature fluctuation (Figure 14.2B, C). In northern Canada and Alaska, gravel is the most common aggregate used since it is reasonably widely available and is not as frost-susceptible as more fine-grained sediment.

Where large quantities of high-quality non-frost-susceptible aggregate are scarce and the structure justifies the cost, more sophisticated technologies are sometimes used. For example, the Yukon Government constructed two $350\,m^2$ multi-purpose municipal buildings in the communities of Ross River and Old Crow, using heat-pump chilled foundations (Goodrich and Plunkett, 1990). The aim was to prevent the thaw of permafrost beneath the buildings. At Ross River, located in the discontinuous permafrost zone, the permafrost is marginal and the mean annual ground temperature is between 0 °C and −0.5 °C. To prevent thaw, heat exchangers were placed in a sand layer within the granular fill used to level the site (Figure 14.3A). Heat flowing down through the floor is then captured by the heat exchangers and pumped back into the building. Thus, while the building is being

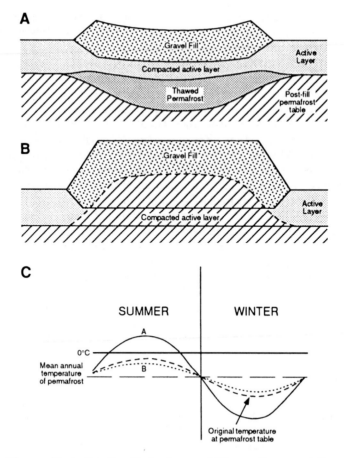

Figure 14.2. Diagram illustrating the effects of a gravel fill upon the ground-thermal regime and thickness of the active layer. (A) Too little fill; (B) too much fill; and (C) the effects of (A) and (B) upon the ground-thermal regime. From Ferrians et al. (1969), courtesy of the United States Geological Survey.

heated, the ground is being chilled. In the first two years following installation, comparisons of predicted and measured temperatures at the top and bottom of the gravel pad, and at the top of the ice-rich silt layer (Figure 14.3B, C), suggest the system works well and permafrost continues to be maintained beneath the structure.

In instances where the structure concerned is capable of supplying significant quantities of heat to the underlying permafrost, as in the case of a heated building or a warm oil pipeline, additional measures are frequently adopted. Usually the structure is mounted on wooden or concrete piles which are inserted into the permafrost. Because an air space is left between the ground surface and the structure, this allows the free circulation of air, which dissipates the heat emanating from the structure. In Russia, where cities with populations greater than 100 000 were built in Soviet times and large buildings were erected, construction usually took place on concrete pilings (Figure 14.4). Recent engineering studies now predict increasing failure of these piles in the years to come. This is partly because of warming permafrost (see Chapter 15, p. 388).

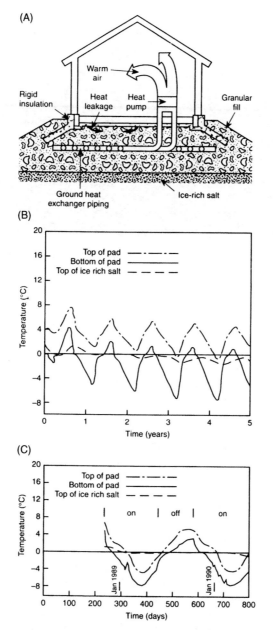

Figure 14.3. Diagram illustrating the geotechnical engineering incorporated into the structure of the Ross River School, Yukon Territory, Canada. (A) The insulated building uses heat-pump chilled foundations; (B) calculated temperatures; and (C) measured temperatures. From Baker and Goodrich (1990).

The disadvantage of pile foundations is their cost, which is especially significant for small buildings, such as private homes. Because the bearing capacity of piles largely depends on the adfreeze bond that develops either between the frozen slurry used as backfill and the surface of the pile or between the interface of the backfill and the native

Figure 14.4. Construction of apartment block complexes in Yakutsk, Siberia, use concrete pile foundations. Photograph was taken in July 1973. Many of these buildings are now predicted to fail by 2030 (see Chapter 15).

permafrost, the end-bearing capacity of piles in ice-rich permafrost is sometimes enhanced by the use of enlarged base ("belled") piles (Sego and Biggar, 2000). In coastal communities and elsewhere where permafrost can be saline, additional problems are encountered in pile usage because the adfreezing strength is accordingly reduced (Brouchkov, 2003).

Other techniques used to preserve the thermal regime of the permafrost include the insertion of open-ended culverts into the gravel pad, the placing of insulating matting immediately beneath both the building and the gravel pad, and, if the structure justifies it, the insertion of costly refrigeration units or "Cryo-Anchors" (Hayley, 1982).

Passive cooling is increasingly being incorporated into the foundation design of large structures and into the stability of waste rock-debris piles. This technique maintains permafrost by taking advantage of seasonal convection in high-permeability embankments and pads (Goering, 1998, 2000, 2003). For example, Figure 14.5A shows a typical roadway application where the embankment is built of a highly porous rock layer with high pore-air permeability. The arrows indicate the pattern of winter pore-air circulation that might occur. Because the surface would be capped with an impermeable asphalt layer, internal convection occurs beneath the center of the road. The net result is a decrease in mean soil temperatures and an upward movement of the permafrost table. The technique is especially applicable to road, rail, or airport embankments located in regions of warm permafrost. The same technique can also be applied to hazardous containment dykes such as around oil-storage facilities and to maintaining the stability of rock waste material discarded from opencast mining operations (Figure 14.5B). Another application is its use in maintaining the stability of earthen dams constructed to contain mine tailings. For example, at the BHP Ekati Diamond Mine at Lac de Gras, NWT, Canada, the installation of "thermosyphons" (Figure 14.6A, B) permits the transfer of heat between the atmosphere and the central core of the tailing dam, thereby maintaining it in a frozen state. As a result, the earthen dam is both impermeable to seepage through the dam and resistant to thaw effects from unfrozen lake water on the upstream side.

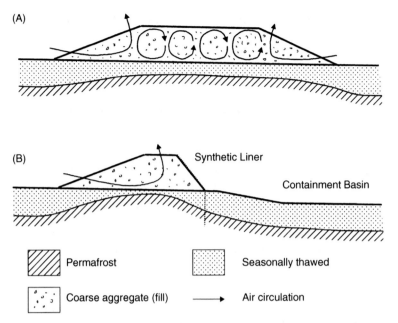

Figure 14.5. Schematic diagrams to illustrate passive cooling techniques that utilize free convection of air. (A) Convective embankment used for road and railway applications. (B) Convectively-cooled containment dyke used for dams and tailings ponds.

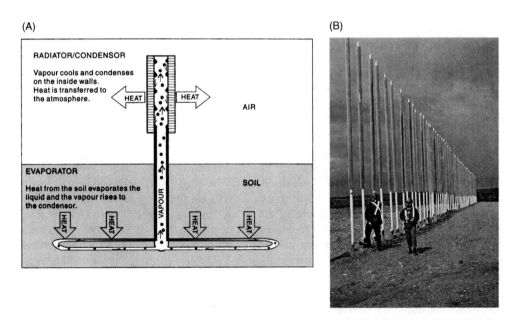

Figure 14.6. Application of thermosyphons in northern Canada. (A) Thermosyphon design used in Yellowknife. The device operates when the air temperature is colder than the ground temperature (usually in winter). (B) Thermosyphons installed atop an earthen dam, BHP Ekati Diamond Mine, Lac de Gras, NWT, approximately 300 km northeast of Yellowknife. The transfer of heat away from the central core of the dam allows it to remain in a frozen state. As a result, the rock-rubble dam is both impermeable to seepage through the dam and resistant to thaw effects from unfrozen water in the containment pond.

14.3. PROVISION OF MUNICIPAL SERVICES AND URBAN INFRASTRUCTURE

The provision of municipal services and urban infrastructure such as water supply and sewage disposal are particularly difficult in permafrost regions. This is because systems will freeze if left unprotected above ground yet it is costly and difficult to excavate trenches either to depths below the seasonal-frost level or within permafrost. Therefore, pipes to carry municipal services cannot be laid below ground, as is normally the case in non-permafrost regions, because heat from the pipes will promote thaw of enclosing permafrost and subsequent subsidence and fracture of the pipe.

The construction of the town of Inuvik in the Mackenzie Delta, NWT, Canada, in the early 1960s was an example of the careful manner in which the urban infrastructure must be planned and constructed in permafrost regions. A major factor that governed the location of the town was the presence of a large body of fluvio-glacial gravel a few kilometers to the south. This was used to place a gravel pad beneath the entire area of the proposed townsite. Today, this aggregate source has been exhausted and future growth of the community depends upon exploitation of more distant aggregate sources with associated higher costs of haulage.

At Inuvik, the provision of municipal services has been achieved through the use of continuously-insulated aluminum boxes that run above ground on supports and link each building to a central system (Figure 14.7A). These are termed utilidors. The cost of utilidor systems is high, involving a fair degree of town planning and constant maintenance, and can only be justified in large settlements.

Another example is provided by the town of Dawson City, Yukon Territory, Canada. There, the provision of municipal services was upgraded at considerable cost in the late 1970s and early 1980s by an alliance of Federal and Provincial government agencies. At issue was the restoration and maintenance of historic buildings, and the adequate provision of utilities to the town. Prior to 1980, the city water distribution and sewage systems were those that had been constructed in 1904. The town uses water from infiltration wells situated near the bank of the Klondike River. The distribution system consisted of wood-stave pipes within gravel in the active layer; all were emplaced at depths shallower than 1.2 m. In winter, the water was heated by electricity to +5.5 °C and enough flow was maintained to prevent freezing by bleeding into each house. At the end of the circulation system, the water temperature was about 1.1 °C. Needless to say, these antiquated water and sewage systems required frequent repair due to seasonal-frost heave, settlement of the pipes through thaw, and frost deterioration of the pipes. The electrical heating costs were also high. Accordingly, a new system of underground services was installed. Trenches were excavated to a minimum depth of 2.0 m and backfilled with coarse (frost-stable) gravel fill (Figure 14.7B) that was hauled to the site. Similar costly preventive measures are now regarded as normal practice in many communities elsewhere in Canada, and in Alaska, Svalbard, and northern Norway.

Sewage treatment is also problematic in many communities located in cold environments. Waste treatment plants are costly and most northern communities, in both North America and Russia, cannot afford them. Haulage and direct dilution in adjacent rivers or the sea are the solutions for many smaller settlements. Nearby lakes are sometimes used as settling ponds but these are not always available and there are environmental concerns since they may be used for fishing or by wildlife. There is also the inevitable risk of contamination of freshwater drinking supplies. Settling ponds, or sewage lagoons, are used by larger communities but are costly to excavate and do not function well because the low temperature inhibits bacterial decomposition. Furthermore, the lagoon may freeze

(A)

(B)

Figure 14.7. The provision of municipal services such as water and sewage. (A) At Inuvik, NWT, Canada, it is achieved by a utilidor system that links each building to a central plant. The photo shows typical utilidor topography. (B) At Dawson City, Yukon Territory, Canada, the installation of modern municipal services in 1980 was by means of trenches excavated to a minimum of 2.0 m and backfilled with coarse non-frost-susceptible aggregate.

completely during the winter months. In the city of Yakutsk, central Siberia, many high-rise buildings are connected to a large service tunnel which runs beneath the main street. This serves as a central sewage line that eventually empties raw sewage into the Lena River.

14.4. CONSTRUCTION OF BUILDINGS AND HOUSES

Problems associated with buildings usually relate to frost heaving and thaw subsidence. The problem of pile construction has been discussed earlier. The City of Yellowknife, NWT, Canada, illustrates the range of problems that are encountered (Wolfe, 1998). Similar problems have been described from Fairbanks, Alaska (Péwé, 1983c, pp. 11–62). Here, two specific examples from Yellowknife and the further example of Dawson City, Yukon Territory, illustrate the hazards presented by thawing permafrost and the costly remedial measures that are necessary.

Yellowknife is the largest community in the NWT and expanding rapidly, with a current population of approximately 20000. It owes its original existence to mining of gold and other minerals but is now also the administrative center of the NWT and a logistics base for diamond mining and other operations.

The Rockcliffe apartment building is a 2–3 story residential building, initially constructed in 1974 on piles drilled through surficial materials to underlying bedrock (Figure 14.8). A heated crawl space that was present beneath the building resulted in thaw

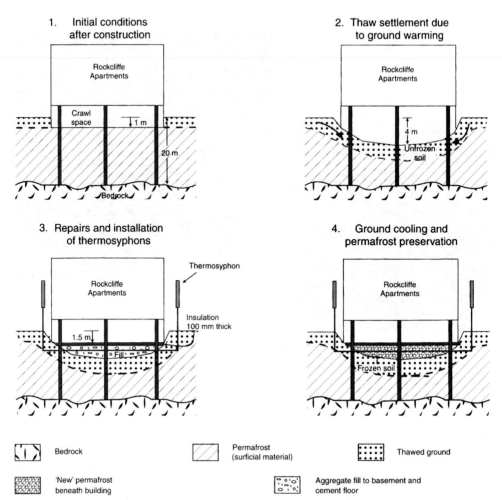

1. Initial conditions after construction

2. Thaw settlement due to ground warming

3. Repairs and installation of thermosyphons

4. Ground cooling and permafrost preservation

Bedrock

Permafrost (surficial material)

Thawed ground

'New' permafrost beneath building

Aggregate fill to basement and cement floor

Figure 14.8. At Yellowknife, NWT, Canada, the Rockcliffe residential apartment building experienced thaw subsidence and weakening of the pile foundations. Remedial action involved the installation of thermosyphons beneath the building. From Wolfe (1998).

settlement of underlying sediment. The crawl space enlarged from 1 m to 4 m in height over a 20-year period. The piles became stressed and the building began to shift. In 1994, the expanded crawl space was filled and leveled with aggregate and thermosyphon tubes were installed to remove heat from within the fill. By 1997, the infill had cooled to 0 °C and quasi-stability of the piles had been re-established.

One of the main streets of downtown Yellowknife is Franklin Avenue. This important street joins the old and new parts of the city. The deteriorating water and sewage systems that extended along the street were replaced in 1990. This followed upon many years of constant maintenance by city workers and several major episodes of leveling and road resurfacing. Since 1967, thaw subsidence had been generated by the service mains that were buried beneath the road. This was accentuated by the dark asphalt surface. In 1989–1990, geotechnical investigations were undertaken prior to reconstruction of the new sewer lines with a shallow drilling program (Figure 14.9A). It was found that ice-rich clay was present at several localities. During the subsequent trenching and excavations, it was established that the original service mains had sunk at least 2 m from their initial 1967 installation. At one locality, the thaw bulb was approximately 6 m deep. Accordingly, the new service mains were deliberately installed in thawed ground, and layers of 50 mm thick extruded polystyrene insulation were placed between the service mains and the underlying permafrost (Figure 14.9B). This has slowed further subsidence on this side of the road but subsidence continues on the other side and the ongoing saga of Franklin Avenue has yet to be completely eliminated.

Equally costly and site-specific measures have been undertaken in Dawson City, Yukon Territory, to maintain some of the historic buildings in that town. The underlying problem is that Dawson City was located, in 1898, on a restricted area of the flood plain of the Yukon River. The site is underlain by silt and alluvial gravel. Despite nearly a century of occupation, warm (between $-3\,°C$ and $-1\,°C$) permafrost is present in much of the town site to a depth of about 20 m. The presence of segregated and inactive (relict) ice wedges means that the soils are thaw-sensitive and subject to settlement if disturbed.

The earliest buildings were log structures or frame buildings placed on squared timbers laid at or near the surface. Virtually all the old buildings that still remain today have settled differentially, necessitating periodic jacking and leveling with additional cribbing, and/or eventual abandonment (Figure 14.10A). Since the early 1960s all new buildings in Dawson have been constructed on wooden piles or gravel pads. In restoring some of the historic buildings, Parks Canada (the Canadian Federal Government agency responsible for heritage affairs) has tried to maintain the original levels of the buildings with respect to the streets, ruling out the emplacement of thick gravel pads or the use of piles. Instead, the silty ice-rich material has been excavated and replaced by thaw-stable granular material to a depth of 5–7 m, and the historic buildings have been replaced in their original positions supported by adjustable jacks (Figure 14.10B). These examples illustrate the expense required to maintain the urban infrastructure.

In the Russian north, the urban infrastructure of many major cities is beginning to cause major engineering and geotechnical problems. Many of these cities have populations in excess of 100 000 persons and residential housing is in massive 5–6 story apartment blocks constructed upon piles in permafrost. Frost-jacking, thaw subsidence around buildings, deterioration of water and sewage facilities, and a general lack of maintenance due to fiscal concerns and poor management are now widespread. Unverified statistics provided by L. Khrustalev (2000) predict the number of major buildings that, on sound engineering practice criteria, will fail by 2030 (Table 14.1). In Vorkuta it is predicted that 50–60% of all major building constructed between 1950 and 1999 will have failed. Already, 9 out of 30 major buildings constructed in the 1990s are failing. In Yakutsk and Tiksi, similar situations exist, where it is predicted that nearly all building constructed perior to 1990 will have failed by 2030 (see pp. 388).

14.5. WATER-SUPPLY PROBLEMS

Because permafrost acts as an impermeable layer, the movement of groundwater is restricted to thawed zones or taliks (see Chapter 5). Given these hydrologic characteristics, a difficult problem for many northern settlements is the provision of drinking water.

Supra-permafrost water is subject to near-surface contamination and intra-permafrost water is often highly mineralized and difficult to locate. Therefore, the tapping of sub-permafrost water is often essential.

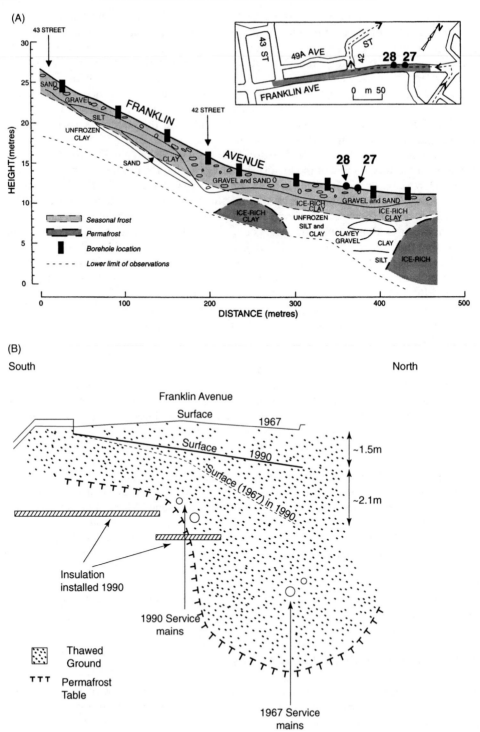

(A)

(B)

South

North

Franklin Avenue

The problem is least acute in regions of discontinuous permafrost, where extensive alluvial deposits provide an abundant source of groundwater. For example, in Fairbanks, Alaska, most houses rely on small diameter private wells (see Péwé, 1983c, pp. 67–85). However, in many areas of continuous permafrost, the permafrost is several hundred meters deep. In certain parts of Siberia, perennial springs that are fed by sub-permafrost water constitute the sole water source. In historic times, these sites assumed a religious significance from aboriginal (Yakut) peoples. In most areas of thick permafrost, drilling is either not possible, since the hole would freeze, or too costly. As a result, surface water bodies, particularly those which do not freeze to their bottoms in winter, must be utilized. Great care must be taken to prevent contamination. It follows that the supply of water is a severe limitation to any large-scale permanent settlement in many periglacial environments. For example, the water-supply problems at Sachs Harbour, a small Inuit community of approximately 250 people on southwest Banks Island, Canada, are reasonably typical. There, the water supply is derived from a lake approximately 3 km from the town. It is trucked every 3–4 days, by water tanker, to individual homes, which have indoor storage containers. Contamination is a problem and the size of the lake, one of the few deep enough not to freeze to its bottom during winter, limits growth of the community.

(A) (B)

Figure 14.10. Historic buildings at Dawson City, Yukon Territory, are built upon ice-rich and warm permafrost. (A) An abandoned building, approximately 85 years old, experienced severe thaw settlement in the central (previously heated?) part of the building. (B) A heritage building restored by Parks Canada and placed upon non-frost-susceptible granular fill following removal of ice-rich permafrost. Note the adjacent modern fire hydrant that required deep burial (see Figure 14.7B) of municipal services.

◄——

Figure 14.9. Thaw subsidence, road sagging, and sewer-pipe breakages are persistent problems along Franklin Avenue in downtown Yellowknife. (A) Cross-section along Franklin Avenue showing geology and permafrost conditions prior to reconstruction of sewer lines in 1990. From Wolfe (1998). (B) Diagram indicating the previous road surface levels, the location of the 1967 service mains, the locations of the new (1990) service mains, the limit of the thaw bulb beneath the road, and the locations of the insulated polystyrene layers installed. Source: unpublished data, NorthTech Consulting Ltd.

Table 14.1. The percentage of buildings that, on engineering grounds, are regarded as having already failed, or will fail, between 1900 and 2030, classified according to the year of initial commission, in Central Siberia.

City, year of construction	Year of predicted failure				
	1990	2000	2010	2020	2030
Vorkuta:					
1950	10	18	39	44	60
1960	9	17	31	43	60
1970	7	15	29	42	59
1980	3	12	26	40	57
1990	0	9	24	38	56
Tiksi:					
1950	9	23	63	90	98
1960	7	20	62	90	98
1970	4	18	61	90	98
1980	2	16	60	90	98
1990	0	14	60	90	97
Yakutsk:					
1950	8	28	72	94	100
1960	6	27	72	94	100
1970	4	25	71	94	100
1980	2	24	71	94	100
1990	0	22	70	94	100

Source: Khrustalev (2000).

14.6. ROADS, BRIDGES, RAILWAYS, AND AIRSTRIPS

Road construction in periglacial environments is no more difficult than elsewhere if there is an abundant supply of coarse, non-frost-susceptible aggregate for the roadbed. Unfortunately, this is not always the case because many areas lack either glacial outwash deposits or are distant from alluvial gravels. The result is that crushed bedrock is often hauled from distant sources to prevent the inevitable frost heaving that would occur if less suitable, but more local, materials were used. Gravel surfaces are the norm for many parts of Alaska, northern Canada, northern Russia, and northern Scandinavia. Black-topping (asphalt) is deliberately avoided except in the towns in order to reduce solar heat absorption. The most common technique is for the gravel pad to be placed directly upon the undisturbed ground surface by front-end loading.

A specific hazard as regards highway and railway construction relates to the formation of groundwater icings (see Chapter 5). They were first encountered during the building of the Alaskan Highway (Péwé, 1983c, pp. 74–80; Thomson, 1966). These are most common in the discontinuous permafrost zone where a road cutting, or any other man-made excavation, intersects with the supra-permafrost groundwater table. Groundwater seepage forms a sheet of ice, often several tens of square meters in extent. Unless precautions are taken, icings can occur on most northern highways wherever they traverse sloping terrain. Counter measures include the avoidance of road-cuts wherever possible, the installation of high-arch culverts to divert water from the source of the icing, and the provision of large drainage ditches adjacent to the road. Icings may also block culverts placed beneath road embankments and, by diverting snowmelt waters, initiate washouts in the thaw period. The costs of icing control and/or remedial measures can be considerable; for example, van Everdingen (1982) provides a conservative estimate of $20 000.00 for icing control at one locality studied on the Alaska Highway, Yukon, during the 1979–80 winter.

Traditional bridge construction methods are also a problem in permafrost regions. Traditionally, these are usually constructed using piles inserted across the river or stream channel. While in warmer climates the chief problem of piles is to obtain sufficient bearing strength, in permafrost regions the problem is to keep the piles in the ground because frost action tends to heave them upwards. Since heaving becomes progressively greater as the active layer freezes, it follows that the thicker the active layer the greater is the upward heaving force. In discontinuous permafrost, where the active layer may exceed 2 m in thickness, frost heaving of piles assumes critical importance. In parts of Alaska, for example, old bridge structures illustrate dramatically the effects of differential frost heave (see Péwé, 1983c) because it is not uncommon for a thawed zone to exist beneath the river channel. Thus, piles inserted in the stream bed experience little or no heave and piles inserted within permafrost on either side of the river are also unaffected. However, the piles adjacent to the river bank experience repeated heave since they are located in the zone of seasonal freezing. As a result, uparching of both ends of the bridge may occur.

In order to prevent these problems, alternative structures involving minimal pile support are now used. One example, that of the Eagle River Bridge on the Dempster Highway, northern Yukon (Figure 14.11A), is described here. The bridge was constructed in the late 1970s by Canadian Army engineers. It consists of a single 100 m long steel span with footings on the north side placed in permafrost. Drilling prior to construction had indicated that permafrost, present on the north bank to a depth of ~90 m, had a temperature of −3 °C. A deep near-isothermal talik existed beneath the Eagle River channel. On the opposing slope, the permafrost was only 8–9 m thick and marginal in temperature (−0.4 °C).

In order to maintain the delicate permafrost conditions and to provide structural integrity, 15 steel piles were inserted at each abutment (Figure 14.11B). Conventional adfreeze analysis indicated the optimum depth of emplacement of each pile was ~5 m. However, because of the warm permafrost at the south side, piles were driven to a depth of 30 m. On the north side, where the permafrost was colder, piles were driven to a depth of 12 m. The piles were backfilled with sand slurry to promote adfreeze. A further complexity was that construction had to be carried out during the winter (1976–77), in order to minimize surface terrain damage. Subsequent monitoring indicates that the piles experienced minimal heave, the thermal regime of the permafrost was maintained, and the bridge structure has performing satisfactorily (Johnston, 1980).

Several railways, constructed across permafrost terrain in Alaska and northern Canada, have also experienced costly maintenance problems on account of frost heave adjacent to bridge structures and thaw settlement along the railbed itself (Ferrians et al., 1969; Hayley, 1988; Hayley et al., 1983).

One illustration is provided by the Hudson Bay railway, which extends through northern Manitoba to the port of Churchill. During nearly 60 years of operation, thaw settlement of the railway embankment and destruction of bridge decks by frost heave have been perpetual problems. Test sections installed with heat pipes proved effective, but costly, measures that minimized thaw subsidence (Hayley, 1988). However, the numerous transitions from frozen to unfrozen terrain in discontinuous permafrost terrain made installation of such techniques along the entire 820 km route impractical. In Alaska, the

(A)

(B)

South   North

Deck elevation

cribs

River level
June

Stratified silt sand
and gravel

Sandy clay silt (stiff)

Sand, silt

Silt, sand
gravel (dense)

Sandy silty clay (stiff)

Elevation (m): 338, 320, 300, 285

Permafrost   Talik   Ice-rich silt clay   Pilings

0   15   30   Horizontal scale
Metres

Figure 14.11. The Eagle River Bridge, Dempster Highway, Northern Yukon Territory, Canada, was constructed in 1977–1978 as a single-span structure with minimal pile support in the river channel. (A) Oblique air view of the bridge, July 1979. (B) Engineering plan and permafrost conditions at the site, as determined by drilling prior to construction during the 1976–1977 winter. From Johnston (1980).

Copper River railway has experienced similar maintenance problems while other rail lines have been abandoned (Ferrians et al., 1969). The latest example comes from Tibet, where a number of experimental studies were undertaken in the mid-1970s to evaluate the feasibility of a railway across the plateau to Lhasa (Research Group on Experimental Roadbed Research, 1979). For a substantial portion of its length the proposed route crosses relatively warm and sometimes discontinuous permafrost (see Chapter 5). This major engineering project was completed in July, 2006. In order to maintain a frozen roadbed, the primary method adopted has been to construct an embankment that remains frozen (Cheng and Li, 2003). Numerical simulation shows that the annual thawing rate of permafrost, under natural conditions and assuming a ground surface temperature increase of 0.025 °C/year, requires a minimum height of the embankment of ~0.6 m for a 50-year period of serviceability of the railway (Ling et al., 2003). In one area where ground-ice bodies occur, duct-ventilation is being employed to keep ground temperatures below zero beneath the railbed (Niu et al., 2003).

The construction and maintenance of airstrips is also costly in permafrost regions since any slight heave or thaw settlement affects the runway grade. This is especially important with the use of large jet aircraft in northern regions, for both freight and passengers. In northern Alaska and Canada, gravel and rock-fill are used for most all-season airfields (Crory, 1988) but, in response to the safety issue of grade, black-top (asphalt) surfaces are now in place at large airports such as Barrow (Alaska), Inuvik, Yellowknife, Iqualuit (Canada), and Longyearbyen (Svalbard). This enhances the possibility of thaw settlement beneath the runway and increases the desirability of the runway subgrade to be of appropriate quality (i.e. composed of non-frost-susceptible aggregate) and sufficient quantity. At Inuvik, a program to monitor the progression of the frost-line was put in place following back-coating of the runway in 1981–1982 (Johnston, 1982). In northern Russia, runway maintenance is a little-discussed problem in the available literature. Because air transportation is vital to many northern settlements, to military usage, and to the support of resource exploitation activities, the importance of sound permafrost engineering cannot be underestimated. Recently, the US Federal Aviation Administration has requested that this issue of pavement roughness due to frost heave be addressed as a safety concern (Lenngren, 2000).

14.7. OIL AND GAS DEVELOPMENT

The periglacial environment contains appreciable quantities of natural resources. In the last 30 years, the major stimulus for much economic development in Arctic North America and in northern Russia has been the exploitation and development of energy reserves, notably hydrocarbons.

14.7.1. Exploration Problems

Some of the earliest exploration activity took place in northern Alaska in the late 1940s in the US Navy Petroleum Reserve (NPR-4). At that time there was little understanding of the sensitivity of tundra terrain to disturbance (Lawson et al., 1978). Drilling was undertaken in both summer and winter months, vehicles were moved randomly across the tundra, and waste-drilling mud was discarded at the site. Not surprisingly, long-lasting terrain damage was incurred (see Figure 8.16A). Subsequently, in the 1970s, the Prudhoe Bay oilfield was developed.

In Yukon Territory, Canada, similar exploration activity began in the mid-1960s. The boreal forest was removed for seismic surveys, haul roads, and for exploratory drilling sites. Similar disturbance resulted. A number of case studies of human-induced thermokarst from Northern Canada are described in Chapter 8.

By the mid-1970s, environmental and regulatory measures had been put in place in both Alaska and Northern Canada to minimize this sort of environmental damage. For example, in Canada, the Federal Government passed the Territorial Arctic Land Use Act and Regulations in 1971. These regulations largely restricted vehicle movement and drilling activity to the winter months, and imposed numerous other drilling procedures and management practices related to safety and environmental concerns. In both Alaska and Canada, the use of gravel pads and other environmentally-friendly engineering practices has now become standard for all operations, building, roads, and airstrips (Allmand, 1976; Crory, 1991). Today, procedures such as these are accepted standard international practice in the high latitudes (IUCN, 1993).

A similar history of exploration characterized northern Russia, where substantial oil and gas reserves occur in the Yamal and Gydan regions of western Siberia. Exploratory seismic and drilling activity commenced in the 1970s. There was little environmental concern until the early 1990s, when awareness of North American operating techniques and procedures became more widely available (Seligman, 2000).

In spite of modern environmental practice, terrain and environmental damage continues to occur in northern Siberia, where regulatory procedures are frequently avoided or minimized. In Arctic North America, the absence of easily-accessible gravel aggregate for pad construction in the Arctic islands remains a problem. A different set of problems relates to the increasing number of exploratory wells that are drilled to greater depths as deeper geological structures are tested. This increases the disposal problem of the much larger volume of waste-drilling fluids that is generated (see below).

14.7.2. Waste-Drilling-Fluid Disposal Problems

Some of the most difficult environmental problems associated with land-based drilling now relate to the disposal of waste-drilling fluids (French, 1980; Hardy BBT Ltd, 1988). These usually consist of kaolinite slurry. Their function is to provide lubrication to the drilling and, by varying the density of the mud, to counter any formation pressures encountered at depth, thereby preventing "blow-outs." To achieve these objectives, the composition of the mud system requires continual modification during drilling. Standard regulatory procedure requires that waste mud is contained within below-ground sumps that are subsequently buried upon abandonment of the hole. Problems arise because drilling muds contain chlorides and other salts, do not necessarily freeze within the sump following burial, and are often highly toxic because they contain weighting agents such as barites. In some instances, the sump is too small to contain the volume of waste mud that is generated and the muds spill onto adjacent terrain, causing damage to vegetation and pollution to adjacent water bodies. Efforts are continually being made to reduce the toxic components of the mud systems used, reduce the volumes of waste mud produced by recycling, and to experiment with other methods of containment or disposal of waste-drilling mud. For example, the author supervised a drilling operation in the Canadian High Arctic, sponsored jointly by the Federal Government of Canada and the Arctic Petroleum Operators Association, in which drilling mud, claimed to be non-toxic by the company concerned, was placed directly upon the tundra. The short-term effects were monitored (French, 1984, 1985) and the longer-term effects of this experimental procedure have yet to be determined. Other methods that are being tested include bioremediation (Peck, 1998; Reynolds et al., 1998).

14.7.3. Pipelines and Permafrost

The construction of pipelines through permafrost terrain further illustrates the complexity of frost heave and related permafrost problems. Inevitably, oil is "warm" and the ease of transmission through a pipeline varies with viscosity. The classic example was the construction of the Trans-Alaska Pipeline System (TAPS) from Prudhoe Bay on the North Slope to Valdez on the Pacific Coast between 1974 and 1977. This pipeline utilized many procedures designed to minimize permafrost problems (Heuer et al., 1982; Metz et al., 1982). Approximately half the route was elevated on vertical support members (VSMs), many with cooling devices ("heat tubes") to prevent heat transfer from the warm pipe to ice-rich (i.e. thaw-sensitive) permafrost (Figure 14.12). In Canada, a smaller oil pipeline from Norman Wells southwards has been in operation since 1985. Although this pipe is buried, monitoring of thaw penetration and surface settlement indicates that it has not experienced thaw-subsidence problems to the same extent as the Alaska line (Burgess and Smith, 2003). This is probably because it was designed to operate at, or close to, prevailing ground temperature. Instead, the main terrain problems associated with this pipeline have centered upon the stability of wood-chip-covered embankments along the right-of-way, and at stream crossings (Burgess and Harry 1990).

The construction of buried chilled gas pipelines presents other problems that are still not adequately resolved. For security reasons, the pipe must be buried wherever possible and the colder (denser) the gas, the easier is its transmission. The problem, therefore, is one of prolonged frost heave adjacent to the pipe (see Williams, 1979, pp. 75–82; Smith and Williams, 1990) with the possibility of eventual rupture (Figure 14.13). This might occur in the discontinuous permafrost zone wherever the pipe crosses unfrozen ground and where there would be relatively unlimited moisture migration towards the cold pipe. Equally, when the pipe passes from unfrozen (stable) to ice-rich (unstable) terrain, or vice-versa, thaw settlement may result.

The problems of operation of chilled gas pipelines are well illustrated by the 30-year Russian experience in northwest Siberia. The large gas fields of Yamburg, Urengoy, and Medvezh'ye now constitute an important energy source for western Russia and,

Figure 14.12. The Trans-Alaska Pipeline is elevated above ground for much of its route. This photograph, from near Fairbanks, shows the elevation of the pipe above the ground (~2.5 m), the cooling devices on the VSMs, and the gravel access and maintenance pad. The photograph was taken in May 1980.

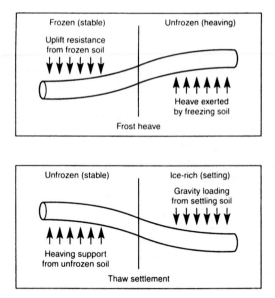

Figure 14.13. Conceptual illustration of the freezing and thawing effects of a buried, chilled gas pipeline crossing from unfrozen to frozen terrain, and vice versa, in an area where permafrost was discontinuous. Modified from Nixon (1990).

increasingly, Eastern Europe. A summary of the operating practices is given by Seligman (2000), who concludes that most problems can be mitigated, but not eliminated, by strict regulation (cooling) of gas temperatures from the moment a pipeline begins to operate. For example, in the early 1970s, there were no compressor stations or gas processing plants along the pipeline routes. As a result, large sections of trunk gas pipeline transmitted cool gas (below 0 °C) that promoted frost jacking. Subsequently, in the early 1980s, the introduction of multiple compressor stations together with inadequate gas cooling resulted in thawing of ground and creation of freeze–thaw thermal regimes around the pipe. Thaw-related displacement of the pipes followed together with upward flotation of buoyant, poorly-ballasted pipes in soils that had lost their load-bearing capacity. In some localities, the pipelines that had floated upwards became exposed at the surface to extreme fluctuations in seasonal temperatures, chilling the gas in winter and warming it in summer. The final result was that many pipelines now experience severe frost jacking and an unacceptable build-up of stresses in pipe steel. It is clear that the transmission of chilled gas in a safe and environmentally sound manner has still to be achieved.

One can be reasonably optimistic that solutions to these problems will be found. For example, in North America, several companies have undertaken natural-scale experiments that aim to understand the thermal regime around buried refrigerated pipelines. Much information is proprietary but it does appear that the magnitudes of heave, frost-penetration depths, and thaw settlement can be predicted at engineering levels of accuracy (Nixon, 1990). Therefore, regulatory issues, market demand, and world energy supply will largely determine the construction of new pipelines for oil and gas in both North America and Siberia. What is clear is that a thorough and in-depth understanding of the permafrost–terrain–vegetation relationship is required for modern resource development in northern regions. Today, most enlightened companies and regulatory agencies maintain "environmental" personnel and departments.

14.8. MINING ACTIVITIES

The environmental and geotechnical problems associated with other mineral exploitation activities center around (i) water supply, (ii) waste disposal, and (iii) contaminants migration through or over permafrost terrain. These are briefly discussed.

Many mining operations involve the disposal of waste material either in rock piles or in settling ponds ("tailings" ponds). Containment is required because of possible toxic components to the waste, therefore any interconnected surface water bodies such as lakes and streams are clearly inappropriate. In many cases, impoundments or dams are created for the specific purpose. In permafrost terrain this creates a problem because the water in the tailing pond is a heat sink and leads to the thaw of underlying permafrost, raising the possibility of eventual groundwater contamination. For example, this is a potential problem at the Red Dog lead–zinc mine in the Brooks Range, central Alaska (Weaver, 2003; Weaver and Kulas, 2003).

There are several possible procedures that minimize these sorts of problems. One solution is to divide the containment area into a number of sectors, which are used at different times of the year, allowing sediment to freeze under natural conditions before use once again the following year (Perlshtein and Pavlenkov, 2003). In this way, no water body, or heat sink, is created. In the case of the Ekati diamond mine, NWT, Canada, an earthen dam was constructed across a small drainage channel to create the containment pond. Here, the problem was not one of thaw of underlying bedrock, because this was impermeable Shield rock, but of possible leakage around and through the dam. To solve this problem, a series of thermosyphons were installed along the dam to transmit heat into the atmosphere and to keep the dam frozen (see Figure 14.6B).

Opencast mining presents stability problems for the large rock piles that are produced, together with issues as to the eventual closure of the pit following exhaustion of the resource. As regards rock-pile stability, passive methods (see Figure 14.5) are being used at Ekati to stabilize (i.e. freeze) the edge of the rock waste. The latter is placed in successive layers or terraces, each 15–20 m high. Convective air circulation through the rock piles is enhanced by the emplacement of less porous and less permeable layers at the base of each terrace level. The closure of an open pit associated with the Raglan Mine, Ungava Peninsula, northern Québec, was safely accomplished by a permanent freezing of the pit, which was backfilled with frozen rock waste from the adjacent rock-waste piles. The potential problem that was avoided was the creation of "acid-rock drainage" induced by the relatively high sulfide content of the waste rock (Szymanski et al., 2003).

A different set of problems related to opencast mining in permafrost occurs when salt-induced cryopegs, present in the bedrock strata that are being removed, are exposed in the pit wall. Seepage into the pit has to be pumped out but the problem remains to dispose of the water. Surface impoundment is not an option because of contamination of surface water systems, and the possible creation of a heat sink if impounded. One solution used at the Udachnaya kimberlite mine in Yakutia, Russia, has been to pump more than 10 million cubic meters of brine solution into a frozen aquifer at depth in the surrounding terrain (Alexeev and Alexeeva, 2003).

Containment problems are especially acute in the near-surface. Although perennially-frozen ground is an effective barrier against downward movement of solutes, thawing of the active layer and the ice contained at its base in the so-called "transient layer" greatly increases the lateral hydraulic conductivity in this zone during the thaw season. Leaching is also enhanced by snowmelt runoff. Therefore, a general problem is the leaching and movement of ions and contaminants away from containment areas. For example, the movement of potassium chloride from drilling mud contained in an unfrozen state in below-ground sumps in the Mackenzie Delta region is a concern of Canadian government regulatory agencies (Dyke, 2001). In Western Siberia, desalinization of the active layer following active-layer slumping in the Yamal Peninsula has been recorded (Streletskii et al., 2003).

There are many other problems associated with resource development in permafrost regions that are not dealt with in this chapter. Specific problems are associated with placer mining operations in both Alaska and Yukon, where elevated dissolved arsenic concentrations in streams is a concern for fish habitats. Groundwater seepage into the coal mine at Longyearbyen (Gruve 7), Svalbard, is a persistent problem (Holm, 1999), while mine ventilation and modification of the thermal regime within the mine may promote rock instability of the tunnel walls (Christiansen et al., 2005). A bigger and more general problem concerns the possibility of crude oil spills, either from pipelines on land or from oil spills offshore. Emergency measures are the subject of numerous impact assessments, and various biogeochemical methods of oil-spill remediation are being investigated. One must conclude that contaminant containment and remedial measures as regards groundwater and the terrestrial and marine ecosystems will become increasingly important issues in the future.

ADVANCED READING

Kamensky, R. M., ed. (1998). *Geocryological Problems of Construction in Eastern Russia and Northern China. Proceedings, International Symposium*, 23–25 September, Chita, Russia. SB, RAS Publishers, Yakutsk, vol. 1, 255 pp., vol. 2, 197 pp.

Péwé, T. L. (1983). Geologic hazards of the Fairbanks area, Alaska. Special Report 15, Division of Geological Surveys, College, Alaska, 109 pp.

Senneset, K., ed. (2000). *Proceedings. International Workshop on Permafrost Engineering*, Longyearbyen, Svalbard, Norway, 18–21 June. Norwegian University of Science and Technology (NTNU)/The University Courses on Svalbard (UNIS), 327 pp.

Smith, D. W., Sego, D. C. (1994). *Cold Regions Engineering. Proceedings, 7th International Cold Regions Engineering Specialty Conference*, March 7–9, Edmonton, Alberta, Canadian Society for Civil Engineering, Montreal, 869 pp.

Wolfe, S. A., ed. (1998). Living with frozen ground. A field guide to permafrost in Yellowknife, Northwest Territories. *Geological Survey of Canada*, Miscellaneous report 64, 71 pp.

DISCUSSION TOPICS

1. Why do permafrost and intense seasonal frost pose problems to geotechnical engineering?

2. How does permafrost affect man's economic activities?

3. Explain why the transport of oil and natural gas from permafrost regions gives rise to environmental and engineering concerns.

4. What are some of the environmental problems associated with mining activities in permafrost terrain?

15 Climate Change and Periglacial Environments

> This chapter examines the impact of predicted climate change upon periglacial environments. Emphasis is placed upon permafrost, seasonal snow cover, and vegetation changes. All will be significantly affected by enhanced atmospheric concentrations of greenhouse gases.
>
> Projected climate changes will most likely lead to decreases in the global extent of permafrost, seasonal snow cover, and sea ice. This will affect both local and global hydrology (meltwater flow, peak discharge, seasonal distribution of runoff, and sea-level rise), nature and distribution of plants and animals, local and regional terrain instability (thermokarst, slope failures, and other mass movements), and urban infrastructures.

15.1. GLOBAL CHANGE AND COLD REGIONS

Over 15 years ago, in 1990, The Intergovernmental Panel on Climate Change (IPCC) Scientific Assessment predicted an increase of global mean temperature during the following century of about 0.3 °C per decade (Houghton et al., 1990). This was based upon assumptions concerning present and future emissions of greenhouse gases, especially carbon dioxide (CO_2) and methane (CH^4). Several general circulation models (GCMs), based upon a doubling of CO_2, suggested that changes for the months of December–January–February would be of the order of 8–12 °C (Houghton et al., 1992) and that mean surface temperature may rise by 1.5 °C to 4.5 °C by the middle of the twenty-first century. More recently, in 2001, The Intergovernmental Panel on Climate Change (IPCC) Third Assessment Review (Anisimov et al., 2001) has confirmed these trends. It is generally accepted that human activity is the major cause of the increase in greenhouse gases. However, it is also being suggested that pre-industrial anthropogenic changes rival those of the industrial era (Figure 15.1) (Ruddiman, 2004). Long-term climatic variability must also be considered. Whatever the exact cause, it is clear that global temperatures rose in the twentieth century at a rate unprecedented in the last millennium (Bradley, 2000). In 2006, a National Oceanographic and Atmospheric Administration (NOAA) report indicates that the average concentration of CO_2 in the atmosphere has increased over the previous 40 years from 315 parts per million to 381 parts per million.

It is clear, therefore, that the periglacial environments of the world will be affected by these changes. Unknown is the speed at which these changes will occur. However, it is predicted that land surfaces will warm more rapidly then oceans and that high latitudes will warm more than the global mean, particularly in winter. All aspects of the landscape will be affected, including permafrost, cryogenic processes, snow, sea-ice cover,

The Periglacial Environment, Third Edition   Hugh M French
Copyright © 2007 John Wiley & Sons Ltd

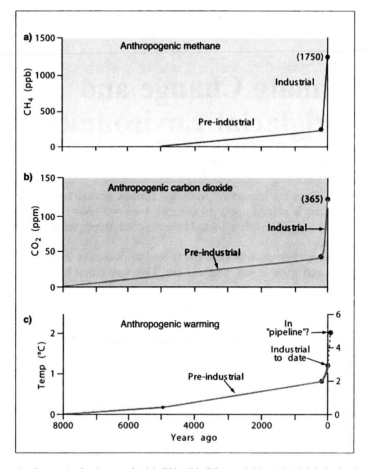

Figure 15.1.   Anthropogenic changes in (a) CH_4, (b) CO_2, and (c) global high-latitude temperature based on the 2001 Intergovernmental Panel on Climate Change, Third International Assessment Review estimate of the sensitivity of the climate system. According to Ruddiman (2004), pre-industrial anthropogenic changes are similar to those of the industrial era.

groundwater resources, and the water balance (see Slaymaker and French, 1993; Woo et al., 1992). The northern limits of the boreal forest, shrub-tundra, and tundra ecozones may shift northwards by several hundreds of kilometers. A number of feedback mechanisms will also come into play. These include: (i) increased methane flux due to the decomposition of organic matter frozen in permafrost and decomposition of methane hydrates, (ii) increased biomass production and decay in tundra and taiga zones, and (iii) decreased surface albedo as snow-cover extent and duration decrease.

In the last 15 years, a number of international science programs have been initiated that are of direct relevance to these predicted changes. They include the European Union 4th Framework Project entitled "Permafrost and Climate in Europe" (PACE), the Global Terrestrial Observing System (GTOS) of the Global Climate Observing System (GCOS), and several impact studies sponsored by the International Arctic Science Committee and The International Permafrost Association (The Bering Sea Impact Study – BESIS; The Barents Sea Impact Study – BASIS; Arctic Coastal Dynamics – ACD). A large number of reports and special journal issues are now available (e.g. Boer and Koster, 1992; Brown et al., 2000; Harris and Haeberli, 2003; Harris et al., 2001; Koster, 1993; Nelson, 2004; Rachold, 1999; Rachold et al., 2003; Weller and Anderson, 1998).

Not surprisingly, recent research has concentrated upon measurement of the cryospheric parameters likely to be sensitive to climate change. These include the depth of the active layer, the thermal condition of permafrost, the extent and duration of seasonal snow cover, and the freeze-up and break-up of lakes and rivers. Emphasis is also placed upon quantitative long-term studies that document the frequency and magnitude of cryogenic processes that are controlled, to varying degrees, by climatic factors. These processes include thermal-contraction cracking, thermokarst activity, active-layer slope failures, and solifluction movement.

15.2. CLIMATE CHANGE AND PERMAFROST

A recent simulation from the US National Center for Atmospheric Research (NCAR) examined the state of permafrost within a global climate model that includes interactions among the atmosphere, ocean, land, sea ice, and soil freezing (Lawrence and Slater, 2005). The NCAR-based Community Climate System Model (CCSM) utilized the projected high and low greenhouse-gas emissions for the twenty-first century as constructed by the IPCC. The simulations suggest that half of the area covered by the upper 3–4 m of permafrost could thaw by 2050 and as much as 90% by 2100. This would dramatically change the polar ecosystems. In northern Canada, Alaska, Greenland, and northern Scandinavia, there would be damage to buildings, roads and other structures, and mining and other activities would be severely affected. For example, certain winter (ice) roads would no longer be possible. The impact upon buildings and infrastructure in northern Russia would be even more catastrophic (see pp. 387–388). A general and qualitative assessment of the predicted impacts due to a 3 °C warming of permafrost is provided in Table 15.1.

Table 15.1. Qualitative assessment of projected impacts that might result from a 3 °C warming of permafrost.

FEATURE/PARAMETER	CONTINUOUS PERMAFROST TERRAIN				DISCONTINUOUS PERMAFROST TERRAIN		
	None	Low	Moderate	Severe	Low	Moderate	Severe
Thaw Lakes		X					X
Coastal Processes		X					X
Eolian Activity	X				?	<- X	
Vegetation			X	-> X	X	-> X	
Active-Layer Thickness		X					X
Permafrost Thawing (table and base)	X						X
Thaw Settlement	X						X
Slope Instability	X						X
Erosion		X					X
Solifluction		X					X
Engineering Impacts		X					X

Source: Weller and Lange (1999).

15.2.1. Ground-Thermal Regimes

Measurements in northern Canada, Alaska, Siberia, and Tibet all suggest that global climate change is already affecting permafrost temperatures and the thickness of the active layer (Osterkamp and Jorgenson, 2006; Pavlov, 1998; Pavlov and Moskalenko, 2002). A number of permafrost monitoring stations in the sub-arctic regions of Europe and Russia indicate that, during the period 1970–1990, there was a rise of 0.6–0.7°C in the temperature of permafrost at a depth of 3.0m (Pavlov, 1994). In northeast Siberia, permafrost at a depth of 10m increased by 0.03°C/year between 1980 and 1991. Data from the Qinghai-Xizang (Tibet) Plateau, China, indicate that temperatures at the 20m depth have risen by 0.2–0.3°C during the last two decades (Table 15.2). Because the Tibet Plateau is sparsely populated and human disturbances are minimal, it is clear this permafrost degradation is not caused by agriculture or human-induced terrain disturbance but by climatic amelioration. Air temperatures recorded at Wudaoliang indicate that the ten-year running means for 1971 and 1980 are 0.5°C and 0.7°C warmer than those for 1961 and 1970 (Wang, 1993). The lower altitudinal limit of permafrost in the Kunlun Shan and adjacent mountains has risen.

Table 15.2.   Recent trends in ground thermal regimes, Qinghai-Xizang (Tibet) Plateau.

(a) Rise in mean annual ground temperature at Fenguo Shan, 1962–1989

Year:	1962	1963	1967	1976	1979	1980	1984	1989
T(°C)	–3.5	–3.5	–3.5	–3.4	–3.3	–3.3	–3.3	–3.3

(b) Decrease in permafrost thickness along northern boundary of permafrost zone adjacent to the Qinghai-Xizang Highway

Year:	1974	1979	1985	1989
Average thickness (m):	15	14	12	10

Source: S. Wang (1993); quoted in Wang and French (1994).

Similar trends can be observed in many parts of northern Canada. For example, at Mayo, central Yukon Territory, the mean annual ground temperature at the depth of zero annual amplitude (10m) is –1.3°C, yet upward projection of the temperature gradient suggests a former mean temperature at the surface of permafrost of approximately –2.0°C (Burn, 1992). Geothermal modeling indicates that a period of 20 years is required for such a temperature change. It is suggested that increased snowfall and warmer winter temperatures are probably the cause of this ground warming. At the same time, some permafrost bodies along the southern boundaries of the permafrost zone in central Canada have either thinned or disappeared.

These trends, while widespread, are not universal in all areas of high latitude. For example, M. Allard et al. (1995) document a cooling trend in permafrost at the 20m depth of approximately 0.05°C/year for the period 1988–1993 at several coastal sites in northern Québec, Canada. Because the active-layer thickness remained unchanged during the period of analysis, it is suggested that the change is related to changes in the oceanic thermo-haline circulation system in the Arctic and North Atlantic oceans. They conclude that climate cooling takes place principally through longer and colder winters.

Collectively, these data mask considerable regional and site variability.

15.2.2. Thickness of the Active Layer

In general, the active layer responds to an increase in air temperature by an increase in thickness. Obviously, the exact amount will depend upon the magnitude of the thermal and nival offsets and the *n* factor (see Chapter 3). The Circumpolar-Active-Layer-Monitoring (CALM) program is designed to observe the response of the active layer and near-surface permafrost to climate change. Over 100 standardized sites have been established in the northern polar region and a similar observing system is being initiated in the southern hemisphere. The program is spearheaded by the International Permafrost Association.

As a generalization, increased thaw penetration and thaw subsidence characterize the various data sets. At some sites, thermokarst indicates degradation of warmer permafrost. The relationship with current climate is reasonably clear; for example, during the 1990s sites in both Alaska and northern Canada experienced maximum and minimum thaw depths in the years of warmest (1998) and coolest (2000) summers (Brown et al., 2000). Typical reports of active-layer monitoring programs are those for Alaska (Hinkel and Nelson, 2003), northern Canada (Smith et al., 2001; Tarnocai et al., 2004), West Greenland (Humlum, 1998c), Svalbard (Repelewska-Pekalowa and Pekala, 2004), northeast Russia (Zamolodchikov et al., 2004), Western Siberia (Melnikov et al., 2004), and Eastern Siberia (Fyodorov-Davydov et al., 2004).

15.2.3. Extent of Permafrost

If warming trends continue, dramatic changes will occur with respect to permafrost distribution. Figure 8.1 (see p. 187) illustrates the long-term effect of a rise in air temperature upon the ground thermal regime at two sites representative of continuous and discontinuous permafrost. In continuous permafrost, the active layer thickens and permafrost decreases in thickness from both top and bottom. In discontinuous permafrost, the permafrost may actually disappear. The time for this to occur is problematic. The increased thickness of the active layer would occur at approximately the same time as the climate warming; the change in permafrost thickness, however, may take several hundred years. For example, Kane et al. (1991) show that a surface warming of 4 °C over a 50-year period in northern Alaska would result in an increase in the active layer from 0.50 m to 0.93 m but that the permafrost temperature at a depth of 30 m would increase by only 1 °C. Likewise, Lachenbruch et al. (1988) demonstrated that the 2 °C increase in the temperature at the permafrost table over the last 100 years that has been experienced in Alaska has only penetrated approximately 100 m. This is one of the reasons why relict permafrost exists (see Chapter 11).

Current degradation (warming) of permafrost is convincingly demonstrated by temperature data obtained from a borehole in the Arctic Wildlife Refuge, Northern Alaska (Figure 15.2) (Osterkamp and Jorgenson, 2006). The curve in the initial (1985) borehole temperature plot indicates earlier warming, probably in the 1925–1950 period, while measurements made in 1998 and 2004 indicate ongoing warming.

Spatial changes in the extent of permafrost will also be very significant. Figures 15.3 and 15.4 indicate the extent to which permafrost may ultimately degrade in Canada and Siberia, based upon warming scenarios of +4 °C and +2 °C, respectively. In Siberia, it is estimated that there will be a 10% reduction in permafrost over a 50-year period.

A similar scale of change is predicted for northern Canada. For example, the influence of lakes upon permafrost has been studied in detail by Burn (2002, 2005) in connection with possible climate change. Today, lakes occupy approximately 25–30% of the surface area of the Pleistocene Mackenzie Delta region. Many probably originated through the melt-out of buried ice bodies. Today, only the larger water bodies possess taliks that perforate (i.e. penetrate) the permafrost. Geothermal modeling indicates that the critical dimensions for the development of open taliks are circular lakes of radius 180 m, and oval

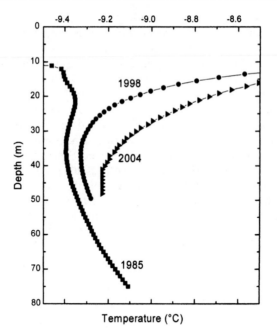

Figure 15.2.   Temperature profiles measured in 1985, 1998, and 2004 in the hole at Barter Island near the village of Kaktovik, Alaskan North Slope. From Osterkamp and Jorgenson (2006). Reproduced by permission of John Wiley & Sons Ltd.

lakes of half-width 130 m. By using a taiga lake near Inuvik (i.e. south of treeline and in a warmer climatic environment) as an analog for the potentially-warmer summer temperatures and higher winter snowfall, it was inferred that the critical size and half-width of round and oval lakes that would be required to penetrate permafrost would decrease to 165 m and 80 m, respectively. This would mean that about 20% of the Pleistocene delta would eventually be underlain by unfrozen ground.

In all these scenarios, it must be emphasized that degradation would be gradual and restoration of a new ground thermal equilibrium might take several millennia. Associated with these changes would be enhanced regional thermokarst activity. As described in Chapter 8, a range of processes would be involved, including subsidence, erosion, increased run-off, and slope instability.

15.2.4. Changes in Cryogenic Processes

Climate warming will also affect the magnitude and frequency of many of the geomorphic processes currently operating in periglacial environments. Current monitoring of thermal-contraction cracking, thaw-slumping, and active-layer detachments suggest that change is already occurring. For example, information on the frequency of current thermal-contraction cracking is now available from the Western Arctic Canada, northern Quebec, and Svalbard (see pp. 123–127). Continued monitoring should detect any ongoing climatic change. It is also possible to infer past climate change in the Western

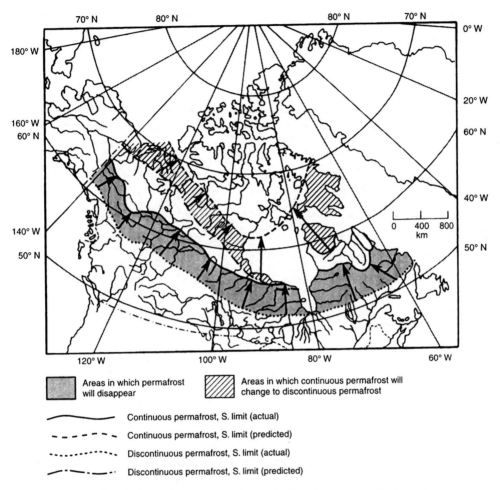

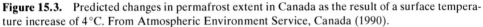

Figure 15.3. Predicted changes in permafrost extent in Canada as the result of a surface temperature increase of 4 °C. From Atmospheric Environment Service, Canada (1990).

Canadian Arctic because, while cracking has been measured at both Garry Island and Illisarvik, no cracking is recorded today from ice wedge sites at Inuvik, some 100 km inland and where temperatures are several degrees warmer than on the Arctic coast. Anecdotal evidence is also provided by the sound of cracking (Mackay, 1993a). Older residents along the Western Arctic coast clearly remember the audible sound of thermal contraction, as first reported by Leffingwell (1915) from northern Alaska. However, the sound of cracking is rarely reported from the mainland today and appears restricted to more northerly locations on Banks and Victoria islands, where winter temperatures are colder. Collectively, these observations support the conclusion that the climate of the Western Canadian Arctic has warmed over the last 100 years and that this is reflected in changes in the frequency of occurrence of thermal-contraction cracking.

Other processes likely to be immediately affected by climate warming include thaw-slumping and active-layer detachments. In the case of thaw-slumps, the rate of ablation of the ice and the rate of headwall retreat is directly linked to energy inputs from the

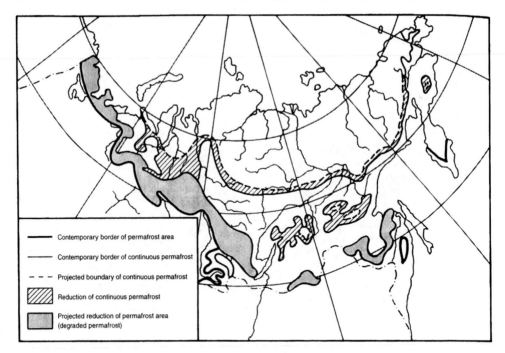

Legend:
— Contemporary border of permafrost area
— Contemporary border of continuous permafrost
– – – Projected boundary of continuous permafrost
▨ Reduction of continuous permafrost
▦ Projected reduction of permafrost area (degraded permafrost)

Figure 15.4. Predicted changes in permafrost distribution in Siberia as the result of a surface temperature increase of 2 °C. From Anisimov (1989, quoted in Street and Melnikov, 1990).

atmosphere (Lewkowicz, 1987, 1988b). In the absence of long-term data, however, it is unclear whether present rates of ablation are significantly higher than those that occurred in the late 19th century. The same is true for active-layer detachments, which are triggered as the active layer deepens. Studies on the Fosheim Peninsula of Ellesmere Island (Lewkowicz, 1990, 1992b) demonstrate their frequent occurrence today, and in the past. Whether the rate of current activity is greater or less than in the past, however, is open to debate, especially given the exceptionally hot summer of 1988 in the Canadian High Arctic (Edlund et al., 1989).

Although climate change will affect soils in all environments, cryosols (see Chapter 4) will be one of the most affected soil groups. This is because of increased thaw or localized disappearance of permafrost. Moreover, thawing of soil organic matter will trigger decomposition and release of carbon in the forms of CO_2 and CH_4. According to C. Tarnocai (2004b), approximately 48% of the total organic carbon in the cryosols of Canada will be severely affected by climate warming. Figure 15.5 illustrates the pathways of expected changes in organic and mineral cryosols as a result of climate warming. Carbon released in "wet" cryosols (i.e. organic terrain) will probably be released in the form of CH_4. This is 21 times more effective as a greenhouse gas than CO_2. Thus, the degradation of peaty permafrost terrain contains an extremely strong positive feedback mechanism for enhanced warming although this might be offset slightly by increased peat development (carbon sequestration) because of higher temperatures and elevated CO_2 levels.

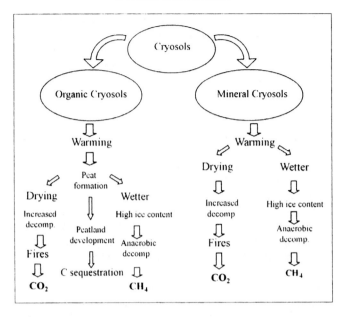

Figure 15.5. Diagram to illustrate the probable change in carbon balance in cryosols due to climate warming. From Tarnocai (2004a). Reproduced by permission of John Wiley & Sons Ltd.

15.3. OTHER RESPONSES

We can summarize the other potential changes which might occur in the periglacial environment under a number of headings.

15.3.1. Seasonal-Snow Cover

The variability of periglacial climates makes it difficult to generalize about seasonal-snow amounts and temporal distribution. For example, in environments of deep seasonal frost or marginal (discontinuous) permafrost, an increase in spring temperatures and rain events will decrease snow-cover duration and enhance summer thaw-penetration. But if warming occurs in colder regions of continuous permafrost where temperatures are projected to remain below or at freezing, winters would become warmer and the snow pack would increase in depth and longevity. For example, the latter scenario is probably the explanation for the warming of permafrost reported for the Mayo area, central Yukon Territory (Burn, 2002). In general, this is probably the most likely case for the majority of periglacial environments that currently experience continuous permafrost. By contrast, environments of deep seasonal frost, relict permafrost, and/or scattered or discontinuous permafrost may experience the former scenario, in which a decrease in the duration of snow cover will increase the period of time in which summer thaw can occur.

There is also a feedback mechanism associated with seasonal-snow cover. Any decrease in snow cover will lower the surface albedo and increase net radiation, thereby enhancing further warming. On the other hand, in areas where climate warming results in increased snow cover, albedo will also increase, leading to a reduction in atmospheric warming and probably increased cloudiness. Finally, climate warming may advance the melt season sufficiently that seasonal-snow accumulations, which would have provided late-season water sources for wetlands, may no longer exist. All these considerations make the impact of climate warming upon the seasonal-snow cover difficult to predict.

15.3.2. Sea Ice and Sea Level

As the world's atmosphere and oceans become warmer, sea levels are expected to rise. Projections by the Canadian Global Climate Model 1 indicate a rise in average global sea level due to thermal expansion of approximately 40 cm by 2090 (Hengeveld, 2000). Changes in the extent of the Greenland and Antarctic ice caps may also lead to a global sea-level increase of between 0.5 m and 1.0 m over the next century. Of special concern to periglacial environments will be the coastal lowlands that surround the Arctic basin in the western North American Arctic and in Siberia. A climate-induced rise in sea level will increase coastal erosion, flooding, and inshore marine sedimentation. There are significant applied implications. Coastal communities will experience repeated flooding and industrial structures built either at shoreline or offshore (drilling platforms and artificial islands for offshore oil or gas production) will have to deal with changes in water depth, shoreline configuration, and sea-ice conditions.

Sea ice is perhaps the single most powerful positive feedback that will determine the magnitude of global climate change in the coming century, particularly for the high northern latitudes and much of northwestern Europe (Miller et al., 2001). There is a strong possibility that the ice cover of the Arctic Ocean will disappear within the next 60–100 years. The US National Snow and Ice Data Center (NSIDC) reports that average surface air temperatures in the Arctic are as much as 2–3 °C higher than 50 years ago. Satellite records that extend back to 1978 indicate that the mean annual size of the Arctic ice cap is shrinking at approximately 8% a year and measurements indicate the thickness of the sea-ice cover has decreased by approximately 30% over the last 15 years. As sea ice is already floating, when it melts it does not raise global sea level to any significant extent. However, the change in net albedo, the increased source of atmospheric moisture from an ice-free ocean, and the consequent increase in cloudiness and precipitation will significantly alter the climates and terrestrial environments of the high northern latitudes.

These changes will have mixed economic implications. Any reduction in thickness, extent, and duration of Arctic sea ice would enhance coastal processes and shoreline modification, extend the shipping seasons along the Siberian, Alaskan, and Western Canadian Arctic coasts, and make feasible a year-round Arctic shipping route from Europe to Japan. Increased iceberg generation from Greenland tidewater glaciers would possibly interfere with marine transportation in the northern North Atlantic and offshore of eastern North America.

15.3.3. Gas Hydrates and Methane

Another important positive feedback related to the degradation of permafrost concerns to gas hydrates and the release of methane (CH_4) to the atmosphere.

Gas hydrates form where gas and water exist under high pressure and low temperature. These conditions exist within and beneath permafrost where solid crystals of gas hydrates can be found. Gas hydrates are a significant and important form of hydrocarbon. Large reserves occur in western Siberia (Makogon et al., 1972), the Mackenzie Delta and Arctic islands (Judge, 1982), and the Alaska North Slope (Collett, 1983). Since significant quantities of methane are trapped within gas hydrates, methane will be released to the atmosphere as permafrost degrades. This may lead to an additional 0.4 °C increase in global temperatures by 2020 and a 0.6–0.7 °C increase by 2050 (Street and Melnikov, 1990).

15.3.4. Seasonally-Frozen Ground

Climate warming will affect the extent of seasonally-frozen ground and the depth of frost penetration. Areas experiencing seasonal frost will be reduced in extent. For example, calculations based on degree days below 0 °C for Calgary and Toronto, two cities in Canada which experience seasonal frost, indicate that frost penetration will be reduced by 50–60% and 75–85%, respectively, as a result of an increase of 6 °C in the mean annual air temperature. Similar changes will occur in the northern United States, Sweden, northern Japan, Russia, and China, and all areas where seasonal frost is widespread. It can be anticipated that frost damage to roads and structures will be reduced significantly.

15.3.5. Boreal Forest, Tundra, and Polar Desert Ecosystems

Higher temperatures in winter and changed precipitation patterns will significantly affect vegetation zonation and agricultural practices in the sub-arctic. In particular, the treeline and the northern limit of the boreal forest will move northwards. Figure 15.6A shows the possible northward shift in position of the boreal forest in Western Canada. This is based upon the 600- and 1300-growing-degree-day isolines as approximations of the northern and southern boundaries. While the shift in the northern boundary ranges from about 100 km to 700 km, the shift in the southern boundary is much greater, about 250–900 km (Table 15.3). Such predictions are based on climate-change models assuming two-time CO_2 increases. One may question whether the rate of forest movement will be a direct response to a change in the thermal conditions and whether the growing-degree-day method is suitable for examining forest zone shifts. However, such predictions highlight the major changes in vegetation, land use, and agricultural practice that will follow upon warming in northern Canada. In general, in sub-arctic regions, the potential for agriculture would increase significantly if the growing season lengthened by 30–40%. Growing conditions in Whitehorse and Yellowknife, for example, would become similar to those today in Edmonton and Calgary, approximately 1000 km to the south.

Similar scenarios are predicted for Scandinavia (Boer et al., 1990) assuming a 3 °C mean annual temperature increase. One prediction indicates the northward movement of oak trees to areas currently experiencing only 116 growing days (Figure 15.6B). It is also predicted that the lower altitudinal limit of coniferous forest would rise by about 600 m and that boreal forest would disappear from all of Sweden except at elevations in excess of 1350 m.

In the polar deserts of the high latitudes, the limiting factors for viable plant communities relate to the availability of nutrients and water rather than to temperature. It follows that climate warming will influence arctic vegetation according to its effect on soil moisture and nutrients, especially nitrates. These changes are difficult to quantify at present, but we might expect a subtle but significant change in the distribution of tundra species, and their assemblages.

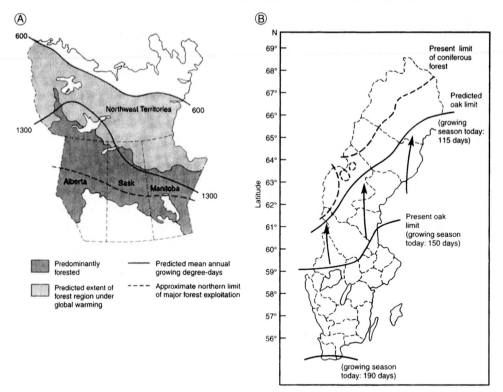

Figure 15.6. Predicted changes to the boreal forest following global climate warming. (A) Western Canada, based on a mean annual growing degree days (above 5 °C). From Wheaton and Singh (1988). (B) Northern Sweden, based on a 3 °C mean annual temperature increase, showing northward movement of northern limit of oak and the present limit of coniferous (boreal) forest. From Boer et al. (1990).

Closely linked to changes in vegetation zonation will be shifts in wildlife migratory patterns, and aquatic and marine habitats. Warmer and wetter winters, for example, may decimate populations of caribou and muskoxen, since heavier snowfall will bury the tundra mosses and lichens on which these animals depend. In contrast, most marine life and migratory birds are expected to flourish, and warmer temperatures would increase fish populations in rivers, lakes, and the ocean. Native settlements initially chosen for proximity to hunting or fishing grounds may no longer be ideally suited for such activities as wildlife and mammal migratory patterns evolve.

15.4. THE URBAN INFRASTRUCTURE

A conservative estimate is that between five and eight million people live in the periglacial domain. In addition to indigenous populations in Alaska, northern Canada, Greenland, northern Scandinavia, and northern Russia, there are a number of cities with populations in excess of 20000–50000 persons. In North America, these include administrative centers such as Fairbanks, Iqualuit, and Yellowknife. In Russia, there are nearly a dozen mining

Table 15.3. Predicted changes in vegetation zonation for western and northwestern Canada, as implied by climate change models.

(a) Northward shift (km) in northern and southern boundaries of boreal forest (delimited by 600- and 1300-growing-degree-day isolines)

Forest boundary	Climatic-change model	
	GISS	GFDL
Northern	80–720	100–730
Southern	470–920	250–900

(b) Changes in vegetation zonation according to GFDL-based model for three locations

Location	Present Zonation	GFDL Zonation
64°N, 120°W (southern NWT between Great Bear and Great Slave Lakes)	Sub-arctic	Aspen parkland
59°09′N, 120°W (northeastern BC\northwestern Alberta)	Boreal	Aspen parkland
55°5′N, 105°W (northern Saskatchewan)	Boreal	Boreal temperate

Source: Wheaton and Singh (1988).

cities, such as Norilsk and Vorkuta, with populations in excess of 50000–100000 persons, and there are the administrative cities of Yakutsk and Tyumen in addition to numerous smaller settlements and mining towns. No precise figures are readily available but the vast majority of the inhabitants of the periglacial environment live in Russia. Nearly all live in areas of warm or discontinuous permafrost. Yet it is these marginal permafrost bodies that will be first affected by climate warming. This will impact upon the urban infrastructures in many ways. Just two examples are given here.

First, during the 1959–1962 construction of Thompson, a mining town of ~10000 persons in Northern Manitoba, Canada, the city authorities denied building permits to certain residential lots because of the presence of randomly-distributed and shallow bodies of frozen ground (Johnston et al., 1963). Most were between 3.0 m and 4.0 m thick and, in all cases, the permafrost was marginal, possessing mean annual ground temperatures ranging between −0.5 °C and −1.5 °C. These plots remained vacant until the 1990s when, because of increased land values in the city, drilling was undertaken to ascertain whether permafrost still existed or not. It was found that frozen ground no longer existed on many plots (French, 1998). It was also determined that mean annual air temperatures at Thompson had probably risen by ~0.5 °C during the 1910–1993 period (French and Egorov, 1998) and, elsewhere in northern Manitoba, it was known that many peaty organic permafrost bodies had experienced decay during the early 1970s (Thie, 1974; Zoltai and Tarnocai, 1975). As a result, housing permits were issued, land values rose accordingly, and new houses were constructed in the 1990s.

A far different experience is predicted for many of the large cities that were constructed in northern Siberia during the Soviet era. Many have populations in excess of

Table 15.4. Decrease of bearing capacity (%) of frozen soil around pile foundations set in permafrost because of an increase in the mean annual air temperature at Yakutsk, Siberia.

Increase of soil temperature °C	Decrease of bearing capacity	
	(i) Buildings	(ii) Frozen soil under pipelines and power lines
0.0	0.0	0.0
0.5	7	9
1.0	15	19
1.5	23	31
2.0	50	63

Source: L. Khrustalov, 1999 (personal communication); quoted in Weller and Lange (1999).

100000 persons, living in drab five-story apartment buildings. Built between 1950 and 1990, many of these buildings are now predicted to fail because of a decrease in bearing capacity of the frozen soil into which the concrete piles are set (Table 15.4). This will result from permafrost warming in the upper few meters. Unverified statistics (see Table 14.1) indicate the number of major buildings that are predicted to fail by 2030. For example, in Vorkuta, the prediction is that 50–60% of all major building constructed between 1950 and 1999 will have failed by 2030. In the large cities of Tiksi and Yakutsk, the entire infrastructure may be lost by 2030 unless remedial measures are undertaken shortly. In many other cities in the Russian Arctic, building deformations are numerous. At Norilsk, approximately 250 of the 5–12-story apartment buildings are damaged and 100 need repair (Makarov et al., 2000; Weller and Lange, 1999). Additional problems in all these cities are frost-jacking, thaw subsidence around buildings, deterioration of water and sewage facilities, and a general lack of maintenance due to fiscal concerns and poor management.

15.5. CONCLUSIONS

Global climate warming will become one of the major environmental issues of the 21st century. The periglacial environments, which comprise approximately 20–25% of the Earth's land surface (see Chapter 1), will certainly be affected. There will be both positive and detrimental effects. A widespread warming of permafrost is already underway. Because of the positive feedbacks associated with the release of greenhouse gases as permafrost degrades, and because of the albedo effects associated with reduced snow and sea-ice covers, the periglacial environment will be more affected than most and will be a major contributor to enhanced climate warming.

ADVANCED READING

Anisimov, O., Fitzharris, B., Hagen, J. O., Jeffries, R., Marchant, H., Nelson, F. E., Prowse, T., Vaughan, D. G. (2001). Polar regions (Arctic and Antarctic). In: *Climate Change: Impacts, Adaptation, and Vulnerability, the Contribution of Working Group II of the Intergovernmental Panel on Climate Change*, Third Assessment Review. Cambridge University Press, Cambridge, pp. 801–840.

Boer, M., Koster, E. A., eds. (1992). Greenhouse impacts on cold-climate ecosystems and land-scapes. *Catena*, suppl. **22**, 151 pp.

DISCUSSION TOPICS

1. Why are periglacial environments particularly sensitive to predicted global changes?

2. What is the relationship between climate change and permafrost?

References

Academia Sinica (1975). *Permafrost*. Research Institute of Glaciology, Cryopedology and Desert Research, Lanzhou, China. (National Research Council of Canada, Ottawa, technical translation no. 2006, 1981, 224 pp.)

ACGR (Associate Committee on Geotechnical Research) (1988). Glossary of permafrost and related ground ice terms. Permafrost Subcommittee, National Research Council of Canada, Ottawa, technical memorandum 142, 156 pp.

Akerman, J. (1993). Solifluction and creep rates 1972–1991, Kapp Liné, West Spitsbergen. In: Frenzel, B., Matthews, J. A., Glaser, B., eds., *Solifluction and Climatic Variation in the Holocene*. Gustav Fischer Verlag, Stuttgart, pp. 225–250.

Aleshinskaya, Z. V., Bondarev, L. G., Gorbonov, A. P. (1972). Periglacial phenomena and some palaeo-geographical problems of Central Tien-Shan. *Biuletyn Peryglacjalny*, **21**, 5–14.

Alexeev, S. V., Alexeeva, L. P. (2003). The burial of drainage waters from the Udachnaya kimberlite pipe into permafrost. In: Phillips, M., Springman, S. A., Arenson, L. U., eds., *Permafrost, Proceedings of the Eighth International Conference on Permafrost*, 21–25 July, Zurich, Switzerland. Balkema, Lisse, vol. 1, pp. 1–4.

Allard, M., Kasper, J. N. (1998). Temperature conditions for ice-wedge cracking: field measurements from Salluit, northern Québec. In: Lewkowicz, A. G., Allard, M., eds., *Permafrost, Proceedings of the Seventh International Conference on Permafrost*, 23–27 June, Yellowknife, Canada. Centre d'études nordiques, Université Laval, Québec, Collection Nordicana no. 57, pp. 5–11.

Allard, M., Tremblay, G. (1983). Les processus d'érosion littorale périglaciaire de la région de Poste-de-la-Baleine et des îles Manitounuk sur la côte est de la mer d'Hudson, Canada. *Zeitschrift für Geomorphologie*, **47**, 27–60.

Allard, M., Fortier, R., Seguin, M. K. (1992). The thermal regime of intertidal permafrost, George River estuary, Ungava Bay, Québec. *Canadian Journal of Earth Sciences*, **29**, 249–259.

Allard, M., Wang, B., Pilon, J. A. (1995). Recent cooling along the southern shore of Hudson Strait, Québec, Canada, documented from permafrost temperature measurements. *Arctic and Alpine Research*, **27**, 157–166.

Allard, M., Caron, S., Bégin, Y. (1996). Climatic and ecological controls on ice segregation and thermokarst: the case history of a permafrost plateau in Northern Québec. *Permafrost and Periglacial Processes*, **7**, 207–227.

Allen, C. E., Darmody, R. G., Thorn, C. E., Dixon, J. C., Schlyter, P. (2001). Clay mineralogy, chemical weathering and landscape evolution in Arctic-Alpine Sweden. *Geoderma*, **99**, 277–294.

Allmand, W. (1976). Guidelines for scientific activities in northern Canada. Advisory Committee for Northern Development, Department of Indian Affairs and Northern Development, Ottawa, Ministry of Supply and Services, catalogue no. L 2-47/1976.

Andersson, J. G. (1906). Solifluction; a component of subaerial denudation. *Journal of Geology*, **14**, 91–112.

André, M.-F. (1993). *Les versants du Spitsberg*. Presses Universitaires de Nancy, Nancy, 361 pp.

André, M.-F. (1994). Rock glaciers in Svalbard. *Geografiska Annaler*, **76A**, 235–245.

André, M.-F. (1995a). Postglacial microweathering of granite roches moutonnées in northern Scandinavia (Riksgransen area, 68° N). In: Slaymaker, O., ed., *Steeplands Geomorphology*. John Wiley & Sons, Chichester, pp. 103–127.

André, M.-F. (1995b). Holocene climate fluctuations and geomorphic impact of extreme events in Svalbard. *Geografiska Annaler*, **77A**, 241–250.

André, M.-F. (1996). Rock weathering rates in arctic and subarctic environments (Abisko Mts, Swedish Lapland). *Zeitschrift für Geomorphologie*, **40**, 499–517.

André, M.-F. (1999). La livrée périglaciaire de paysages polaires: l'arbre qui cache la forêt? *Géomorphologie: relief, processus, environnement*, **1999**(3), 231–252.

André, M.-F. (2002). Rates of postglacial rock weathering on glacially-scoured outcrops (Abisko-Riksgransen area, 68° N). *Geografiska Annaler*, **84A**, 139–150.

André, M.-F. (2003). Do periglacial landscapes evolve under periglacial conditions? *Geomorphology*, **52**, 149–164.

André, M.-F., Hall K. (2005). Honeycomb development on Alexander Island, glacial history of George VI Sound, and palaeoclimatic implications (Two Step Cliffs/Mars Oasis, W. Antarctica). *Geomorphology*, **65**, 117–138.

Anisimov, O., Fitzharris, B., Hagen, J. O., Jeffries, R., Marchant, H., Nelson, F. E., Prowse, T., Vaughan, D. G. (2001). Polar regions (Arctic and Antarctic). In: *Climate Change: Impacts, Adaptation, and Vulnerability, the Contribution of Working Group II of the Intergovernmental Panel on Climate Change, Third Assessment Review*. Cambridge University Press, Cambridge, pp. 801–841.

Anisimova, N. P., Nikitina, N. M., Piguzova, V. M., Shepelyev, V. V. (1973). *Water Sources of Central Yakutia*. Guidebook, Second International Conference on Permafrost, Yakutsk, USSR, 47 pp.

Antoine, P., Lautridou, J.-P., Somme, J., Auguste, P., Aufret, J.-P., Baize, S., Clet-Pellerin, M., Coutard, J.-P., Dewolf, Y., Dugue, O., Joly, F., Laignel, B., Laurent, M., Lavoille, M., Lebret, P., Lecolle, F., Lefebvre, D., Limondin-Lozouet, N., Munaut, A.-V., Ozouf, J.-C., Quesnel, F., Rousseau, D.-D. (1998). Les formations Quaternaires de la France du nord-ouest: limites et correlations. *Quaternaire*, **9**, 227–241.

Antoine, P., Catt, J., Lautridou, J.-P., Somme, J. (2003). The loess and coversands of northern France and southern England. *Journal of Quaternary Science*, **18**, 309–318.

Archambeau, M., Archambeau, C. (1989). *Les Combarelles*. Editions Pierre Fanlac, Artevrafica Silva-Parma, 31 pp.

Are, F. (1972). The reworking of shores in the permafrost zone. In: Adams, W. P., Helleiner, F., eds., *International Geography*. University of Toronto Press, Toronto, vol. 1, pp. 78–79.

Are, F. E. (1973). *Development of Thermokarst Lakes in Central Yakutia*. Guidebook, Second International Conference on Permafrost, USSR Academy of Sciences, Section of Earth Sciences, Siberian Division, Yakutsk, 29 pp.

Are, F. (1983). Thermal abrasion of coasts. In: *Permafrost, Proceedings of the Fourth International Conference on Permafrost*, 17–22 July, Fairbanks, Alaska. National Academy Press, Washington, DC, pp. 24–28.

Are, F. E. (1988). Thermal abrasion of sea coasts (part 1). *Polar Geography and Geology*, **12**, 1–86.

Are, F. E. (1998). The thermoabrasion of Laptev Sea shores and its imput into sediment balance of the sea. *Earth Cryosphere*, **II**, 55–61 (in Russian).

Arnborg, L., Walker, H. J., Peippo, J. (1967). Suspended load in the Colville River, Alaska. *Geografiska Annaler*, **49A**, 131–144.

Astakhov, V. I. (1992). The last glaciation in West Siberia. *Sveriges Geologiska Undersökning*, **81**, 21–30.

Astakhov, V. I., Isayeva, L. L. (1988). The "Ice Hill": an example of "retarded deglaciation" in Siberia. *Quaternary Science Reviews*, **7**, 29–40.

Astakhov, V. I., Kaplyanskaya, F. A., Tarnogradsky, V. D. (1996). Pleistocene permafrost of West Siberia as a deformable glacier bed. *Permafrost and Periglacial Processes*, **7**, 165–191.

Aubekerov, B., Gorbunov, A. (1999). Quaternary permafrost and mountain glaciation in Kazakhstan. *Permafrost and Periglacial Processes*, **10**, 65–80.

Augustinus, P. C., Selby, M. J. (1990). Rock slope development in McMurdo Oasis, Antarctica, and implications for interpretations of glacial history. *Geografiska Annaler*, **72A**, 55–62.

Auer, V. (1927). Untersuchungen uber die waldgrenzen und torfboden in Lappland. Metsatieteellisen Koelaitoksen Julkaisuista, 12, Aus den Communicationes ex instituto quaestioneum forestalium Finlandiae, Helsinki, Valtioneuvoston Kirjapaino, 52 pp.

Avery, B. W. (1964). The soils and land use of the district around Aylesbury and Hemel Hempstead (Sheet 238), Memoirs, Soil Survey of Great Britain. HMSO, London.

Baer, K. E. von (1838). The ground ice or frozen soil of Siberia. *Journal of the Royal Geographical Society, London*, vol. 8. John Murray, London, pp. 210–213.

Babb, T. A., Bliss, L. C. (1974). Effects of physical disturbance on Arctic vegetation in the Queen Elizabeth Islands. *Journal of Applied Ecology*, **11**, 549–562.

Baker, T. H. W., Goodrich, L. E. (1990). Heat-pump chilled foundations for buildings on permafrost. *Geotechnical News*, **8**(3), 26–28.

Balescu, S., Dupuis, Ch., Quinif, Y. (1988). TL stratigraphy of pre-Weichselian loess from NW Europe using feldspar coarse grains. *Quaternary Science Reviews*, **7**, 309–313.

Balke, J., Haendel, D., Kruger, W. (1991). Contribution to the weathering-controlled removal of chemical elements from the active debris layer of the Schirmacher Oasis, East Antarctica. *Zeitschrift für Geologische Wissenschaften*, **19**, 153–158.

Balkwill, H. R., Roy, K. J., Hopkins, W. S., Sliter, W. V. (1974). Glacial features and pingos, Amund Ringnes Island, Arctic Archipelago. *Canadian Journal of Earth Sciences*, **11**, 1319–1325.

Ballantyne, C. K. (1978). The hydrologic significance of nivation features in permafrost areas. *Geografiska Annaler*, **60A**, 51–54.

Ballantyne, C. K., Harris, C. (1994). *The Periglaciation of Great Britain*. Cambridge University Press, Cambridge, 330 pp.

Ballantyne, C. K., Whittington, G. (1987). Niveo-aeolian sand deposits on An Teallach,Wester Ross, Scotland. *Transactions of the Royal Society of Edinburgh*, **78**, 51–63.

Baranov, I. Y. (1959). Geographical distribution of seasonally-frozen ground and permafrost. In: *General Geocryology*. V.A. Obruchev Institute of Permafrost Studies, Academy of Science, Moscow, part 1, chapter 7, pp. 193–219. National Research Council of Canada, Ottawa, technical translation no. 1121 (1964).

Barnes, P. W. (1982). Marine ice-pushed boulder ridges. *Arctic*, **35**, 312–316.

Baroni, C. (1996). Antarctic geomorphological and glaciological 1:250 000 map series, Mount Melbourne quadrangle (Victoria Land). PNRA, Museo Nazionale dell'Antartide, University of Siena, Italy.

Baroni, C., Carton, A., Seppi, R. (2004). Distribution and behaviour of rock glaciers in the Adamello-Presanella Massif (Italian Alps). *Permafrost and Periglacial Processes*, **15**, 243–260.

Barr, W. (1976). Retreating coasts and disappearing Islands in the Arctic. *Muskox*, **18**, 103–111.

Barsch, D. (1977). Eine Abschätzung von Schuttproduktion und Schutttransport in Bereich aktiver blockgletscher der Schweizer Alpen. Zeitschrift for Geomorphologie, **28**, 148–160.

Barsch, D. (1978). Active rock glaciers as indicators of discontinuous permafrost. An example from the Swiss Alps. In: *Permafrost, Proceedings of the Third International Conference on Permafrost*, 10–13 July, Edmonton, Alberta, Canada. National Research Council of Canada, Ottawa, vol. 1, pp. 349–352.

Barsch, D. (1988). Rockglaciers. In: Clark, M. J., ed., *Advances in Periglacial Geomorphology*. John Wiley & Sons, Chichester, pp. 69–90.

Barsch, D. (1993). Periglacial geomorphology in the 21st century. *Geomorphology*, **7**, 141–163.

Bateman, M. D. (1998). The origin and age of the coversand in North Lincolnshire, UK. *Permafrost and Periglacial Processes*, **9**, 313–325.

Bates, M. R., Keen, D. H., Lautridou, J.-P. (2003) Pleistocene marine and periglacial deposits of the English Channel. *Journal of Quaternary Science*, **18**, 319–337.

Bell, M., Walker, M. J. C. (1992). *Late Quaternary Environmental Change*. Addison Wesley Longman, Harlow, 273 pp.

Benedict, J. B. (1970). Downslope soil movement in a Colorado alpine region; rates, processes and climatic significance. *Arctic and Alpine Research*, **2**, 165–226.

Bennett, L. P., French, H. M. (1988). Observations on near-surface creep in permafrost, Eastern Melville Island, Arctic Canada. In: Senneset, K., ed., *Permafrost, Proceedings of the Fifth International Conference on Permafrost*, 2–5 August. Tapir, Trondheim, vol. 2, pp. 683–688.

Bennett, L. P., French, H. M. (1990). In situ permafrost creep, Melville Island, and implications for global change. In: *Permafrost-Canada, Proceedings of the Fifth Canadian Permafrost Confer-

ence, National Research Council of Canada – Centre d'études nordiques, Université Laval, Collection Nordicana no. 54, pp. 119–123.

Bennett, M. R., Glasser, N. F. (1996). *Glacial Geology: Ice Sheets and Landforms*. John Wiley & Sons, Chichester, 364 pp.

Berthling, I., Etzelmuller, B., Eiken, T., Sollid, J. L. (1998). Rock glaciers on Prins Karls Forland, Svalbard. 1: Internal structure, flow velocity and morphology. *Permafrost and Periglacial Processes*, **9**, 135–145.

Berthling, I., Etzelmuller, B., Isaksen, K., Sollid, J. L. (2000). Rock glaciers on Prins Karls Forland. II: GPR soundings and the development of internal structures. *Permafrost and Periglacial Processes*, **11**, 357–369.

Beschel, R. L. (1963). Suphur springs at Gypsum Hill. In: *Preliminary Report 1961–1962, Axel Heiberg Research Report*, McGill University, Montreal, pp. 183–187.

Beschel, R. L. (1966). Hummocks and their vegetation in the High Arctic. In: *Permafrost, International Conference Proceedings*, National Research Council of Canada publication 1287. National Academy of Sciences, Washington, DC, pp. 13–20.

Beyrich, A. A., Kolstrup, E., Linde, N., Pedersen, L. B., Thrysted, T., Gintz, D., Dynesius, L. (2003). Assessment of chemical denudation rates using hydrological measurements, water chemistry analysis and electromagnetic geophysical data. *Permafrost and Periglacial Processes*, **14**, 387–398.

Beyrich, A. A., Kolstrup, E., Thrysted, T., Gintz, D. (2004a). Water chemistry and its diversity in relation to local factors in the Latnjavagge drainage basin, arctic-oceanic Swedish Lapland. *Geomorphology*, **58**, 125–143.

Beyrich, A. A., Kolstrup, E., Thrysted, T., Linde, N., Pedersen, L. B., Dynesius, L. (2004b). Chemical denudation in arctic-alpine Latnjavagge (Swedish Lapland) in relation to regolith as assessed by radio magnetotelluric-geophysical profiles. *Geomorphology*, **57**, 303–319.

Bik, M. J. J. (1969). The origin and age of the prairie mounds of southern Alberta. *Biuletyn Peryglacjalny*, **19**, 85–130.

Billings, W. D., Mooney, H. A. (1968). The ecology of arctic and alpine plants. *Biological Review*, **43**, 481–529

Billings, W. D., Peterson, K. M. (1980). Vegetational change and ice-wedge polygons through the thaw-lake cycle in Arctic Alaska. *Arctic and Alpine Research*, **12**, 413–432.

Bird, E. (2000). *Coastal Geomorphology*. John Wiley & Sons, Chichester, 322 pp.

Bird, J. B. (1967). *The Physiography of Arctic Canada*. Johns Hopkins Press, Baltimore, MD, 336 pp.

Black, R. F. (1952). Growth of ice-wedge polygons in permafrost near Barrow, Alaska. *Bulletin, Geological Society of America*, **63**, 1235–1236.

Black, R. F. (1964). Periglacial phenomena of Wisconsin, north-central United States. In: *Report of INQUA Congress, Warsaw, Poland, 1961*, vol. IV. Łódź, pp. 21–28.

Black, R. F. (1969). Thaw depressions and thaw lakes; a review. *Biuletyn Peryglacjalny*, **19**, 131–150.

Black, R. F. (1973). Growth of patterned ground in Victoria Land, Antarctica. In: *Permafrost: The North American Contribution to the Second International Conference, Yakutsk, USSR*, publication 2115. National Academy of Sciences, Washington, DC, pp. 193–203.

Black, R. F. (1974). Ice-wedge polygons of northern Alaska. In: Coates, D. R., ed., *Glacial Geomorphology*. Publications in Geomorphology, State University of New York, Binghampton, pp. 247–275.

Black, R. F., Barksdale, W. L. (1949). Oriented lakes of northern Alaska. *Journal of Geology*, **57**, 105–118.

Blackwelder, E. (1925). Exfoliation as a phase of rock weathering. *Journal of Geology*, **33**, 793–806.

Boardman, J. (1978). Grèzes litées near Keswick, Cumbria. *Biuletyn Peryglacjalny*, **27**, 23–34.

Boch, S. G., Krasnov, I. I. (1943). O nagornykh terraskh i drevnikh poverkhnostyakh vyravnivaniya na Urale i svyazannykh s nimi problemakh. *Vsesoyuznogo Geograficheskogo obshchestva, Izvestiya*, **75**, 14–25 (English translation by A. Gladunova, 1994, On altiplanation terraces and ancient surfaces of levelling in the Urals and associated problems. In: Evans, D. J. A., ed., *Cold Climate Landforms*, John Wiley & Sons, Chichester, pp. 177–186).

Boer, M., Koster, E., eds. (1992). Greenhouse impact on cold-climate ecosystems and landscapes. *Catena*, suppl. **22**, 151 pp.

Boer, M. M., Koster, E. A., Lundberg, H. (1990). Greenhouse impact in Fennoscandia – Preliminary findings of a European Workshop on the effects of climatic change. *Ambio*, **19**(1), 2–10.

Bonfiglio, A., Cresson, J. A. (1982). Geomorphology and pinelands prehistory: a model into early aboriginal land use. In: Sinton, J. W., ed., *History, Culture and Archaeology of the New Jersey Pine Barrens*. Center for Environmental Research, Stockton State College, Ponoma, New Jersey, pp. 18–67.

Borchgrevink, C. E. (1901). *First on the Antarctic Continent. Being an Account of the British Antarctic Expedition, 1898–1900*. George Newnes Ltd, London, 333 pp.

Boulton, G. S. (1970). On the origin and transport of englacial debris in Svalbard glaciers. *Journal of Glaciology*, **9**, 213–229.

Bozhinskiy, A. N., Konishchev, V. N. (1982). On the possible formation mechanism of inclined ice wedges in the Yedoma sequence of North Yakutia. USSR Academy of Sciences, Section of Glaciology and Soviet Geophysical Committee and Institute of Geography, Data of Glaciological Studies, publication 43, pp. 139–142 (in Russian).

Bradley, R. S. (1999). *Palaeoclimatology. Reconstructing Climates of the Quaternary*, 2nd edn. Harcourt/Academic Press, Boston, MA, 613 pp.

Bradley, R. (2000). 1000 years of climate change. *Science*, **288**, 1353, 1355.

Bradshaw, R., Smith, I. (1963). Permafrost structures on Sully Island, Glamorgan. *Geological Magazine*, **100**, 556–564.

Braun, D. D. (1989). Glacial and periglacial erosion of the Appalachians. *Geomorphology*, **2**, 233–256.

Bray, M. T., French, H. M., Shur, Y. (2006). Further cryogenic observations in the CRREL permafrost tunnel, Fox, Alaska. *Permafrost and Periglacial Processes*, **17**, 233–234.

Brenning, A. (2005). Geomorphological, hydrological and climatic significance of rock glaciers in the Andes of Central Chile (33–35°S). *Permafrost and Periglacial Processes*, **16**, 231–240.

Brewer, M. C. (1958). The thermal regime of an Arctic lake. *Transactions, American Geophysical Union*, **39**, 278–284.

Brewer, R. (1976). *Fabric and Mineral Analysis of Soils* (revised). Kreiger, Huntington, NY, 482 pp.

Brink, V. C., Mackay, J. R., Freyman, S., Pearce, D. G. (1967). Needle ice and seedling establishment in southwestern British Columbia. *Canadian Journal of Plant Science*, **47**, 135–139.

Britton, M. E. (1967). Vegetation of the Arctic Tundra. In: Hansen, H. P., ed., *Arctic Biology*. Oregon State University Press, Corvallis, OR, pp. 67–130.

Brook, G. A., Ford, D. C. (1978). The nature of labyrinth karst and its implications for clima-specific models of tower karst. *Nature*, **280**, 383–385.

Brook, G. A., Ford, D. C. (1982). Hydrologic and geologic controls of carbonate water chemistry in the sub-Arctic Nahanni karst, Canada. *Earth Surface Processes and Landforms*, **7**, 1–16.

Brouchkov, A. (2003). Frozen saline soils of the Arctic coast: their distribution and engineering properties. In: Phillips, M., Springman, S. M., Arenson, L. U., eds., *Permafrost, Proceedings of the Eighth International Conference on Permafrost*, 21–25 July, Zurich, Switzerland. Balkema, Lisse, vol. 1, pp. 95–100.

Brown, E. H. (1969). Jointing, aspect, and the orientation of scarp-face dry valleys near Ivinghoe, Buckinghamshire. *Transactions, Institute of British Geographers*, **48**, 61–73.

Brown, J. (1967). An estimate of the volume of ground ice, Coastal Plain, northern Alaska. United States Army, Corps of Engineers, CRREL, Hanover, New Hampshire, memorandum, 22 pp.

Brown, J. (2001). International symposium on mountain and arid land permafrost and field excursion in Mongolia. *Frozen Ground*, **25**, 7–11.

Brown, J., Grave, N. A. (1979). Physical and thermal disturbance and protection of permafrost. United States Army, CRREL, Hanover, New Hampshire, special report 79–5, 42 pp.

Brown, J., Rickard, W., Vietor, D. (1969). The effect of disturbance on permafrost terrain, United States Army, CRREL, Hanover, New Hampshire, special report 138, 13 pp.

Brown, J., Ferrians, O. J., Heginbottom, J. A., Melnikov, E. S. (1997). *Circum-Arctic Map of Permafrost and Ground Ice Conditions*. United States Geological Survey, Circum-Pacific Map Series, CP-45, scale 1:10,000,000.

Brown, J., Hinkel, K. M., Nelson, F. E. (2000). The circumpolar active-layer monitoring (CALM) program: research designs and initial results. *Polar Geography*, **24**(3), 165–258.

Brown, R. J. E. (1960). The distribution of permafrost and its relation to air temperature in Canada and the USSR. *Arctic*, **13**, 163–177.

Brown, R. J. E. (1966). The relation between mean annual air and ground temperatures in the permafrost regions of Canada. In: *Permafrost, International Conference Proceedings*, National Research Council of Canada publication 1287. National Academy of Sciences, Washington, DC, pp. 241–246.

Brown, R. J. E. (1967a). *Permafrost in Canada*. Map 1246a, Geological Survey of Canada, National Research Council of Canada, Ottawa.

Brown, R. J. E. (1967b). Comparison of permafrost conditions in Canada and the USSR. *Polar Record*, **13**, 741–751.

Brown, R. J. E. (1970). *Permafrost in Canada: Its Influence on Northern Development*. University of Toronto Press, Toronto, 234 pp.

Brown, R. J. E. (1973a). Influence of climate and terrain factors on ground temperatures at three locations in the permafrost region of Canada. In: *Permafrost: The North American Contribution to the Second International Conference, Yakutsk, USSR*, publication 2115. National Academy of Sciences, Washington, DC, pp. 27–34.

Brown, R. J. E. (1973b). Permafrost distribution and relation to environmental factors in the Hudson Bay lowlands. In: *Proceedings; Symposium on the Physical Environment of the Hudson Bay Lowland*, University of Guelph, pp. 35–68 (research paper no. 576, Division of Building Research, National Research Council of Canada, Ottawa).

Brown, R. J. E. (1974). Ground ice as an initiator of landforms in permafrost regions. In: Fahey, B. D., Thompson, R. D., eds., *Research in Polar and Alpine Geomorphology, Proceedings, 3rd Guelph Symposium on Geomorphology*, Guelph, Ontario. GeoBooks, Norwich, pp. 25–42.

Brown, R. J. E. (1978). Influence of climate and terrain on ground temperature in the continuous permafrost zone of northern Manitoba and Keewatin District, Canada. In: *Permafrost, Proceedings of the Third International Conference on Permafrost*, 10–13 July, Edmonton, Alberta, Canada. National Research Council of Canada, Ottawa, vol. 1, pp. 15–21.

Brown, R. J. E., Péwé, T. L. (1973). Distribution of permafrost in North America and its relationship to the environment; a review 1963–1973. In: *Permafrost: The North American Contribution to the Second International Conference, Yakutsk, USSR*, publication 2115. National Academy of Sciences, Washington, DC, pp. 71–100.

Brown, W. G., Johnston, G. H., Brown, R. J. E. (1964). Comparison of observed and calculated ground temperatures with permafrost distribution under a northern lake. *Canadian Geotechnical Journal*, **1**, 147–154.

Brunschweiler, D. (1962). The periglacial realm in North America during the Wisconsin glaciation. *Biuletyn Peryglacjalny*, **11**, 15–27.

Bryan, K. (1946). Cryopedology – the study of frozen ground and intensive frost action with suggestions on nomenclature. *American Journal of Sciences*, **244**, 622–642.

Bryan, K. (1949). The geologic implications of cryopedology. *Journal of Geology*, **57**, 101–104.

Büdel, J. (1944). Die morphologischen Wirkungen des Eiszeitklimas im geltscherfreien Gebiet. *Geologische Rundschau*, **34**, 482–519.

Büdel, J. (1951). Die klimazonen des Eiszeitalters. *Eiszeitalter und Gegenwart*, **1**, 16–26 (English translation: *International Geology Review*, **1**(9), 72–79, 1959).

Büdel, J. (1953). Die "periglazial" morphologischen Wirkungen des Eiszeitklimas auf der Ganzen Erde. *Erdkunde*, **7**, 249–266.

Büdel, J. (1960). Die Frostschott-zone Südorst Spitzbergen. *Colloquium Geographica, Bonn*, **6**, 105 pp.

Büdel, J. (1963). Klimatische Geomorphologie. *Geographische Rundschau*, **15**, 269–285.

Büdel, J. (1970). Pedimente, rumpfflächen und rückland Steihänge; deren aktive und passive ruckverlegung in verschiedenen klimaten. *Zeitschrift für Geomorphologie*, **14**, 1–57.

Büdel, J. (1977). *Klima-Geomorphologie.* Gebruder Borntraeger, Berlin (English translation by Lenore Fischer and Detlef Busche, 1982, *Climatic Geomorphology.* Princeton University Press, Princeton, NJ, 443 pp.)

Bull, A. J. (1940). Cold conditions and landforms in the South Downs. *Proceedings, Geologist's Association,* **51**, 63–71.

Burbidge, G. H., French, H. M., Rust, B. R. (1988). Water-escape structures resembling ice-wedge casts in Late-Quaternary subaqueous outwash near St. Lazare, Québec, Canada. *Boreas,* **17**, 33–40.

Burgess, M., Harry, D. G. (1990). Norman Wells pipeline permafrost and terrain monitoring; geothermal and geomorphic observations. *Canadian Geotechnical Journal,* **27**, 233–244.

Burgess, M., Smith, S. L. (2003). 17 years of thaw penetration and surface settlement observations in permafrost terrain along the Norman Wells pipeline, Northwest Territories, Canada. In: Phillips, M., Springman, S. M., Arenson, L. U., eds., *Permafrost, Proceedings of the Eighth International Conference on Permafrost,* 21–25 July, Zurich, Switzerland. Balkema, Lisse, vol. 1, pp. 107–112.

Burn, C. R. (1988). The development of near-surface ground ice during the Holocene at sites near Mayo, Yukon Territory, Canada. *Journal of Quaternary Science,* **3**, 31–38.

Burn, C. R. (1989). Frost heave of subaqueous lake-bottom sediments, Mackenzie Delta, Northwest Territories. *Geological Survey of Canada,* paper 89-1D, 85–93.

Burn, C. R. (1990a). Implications for palaeoenvironmental reconstruction of recent ice-wedge development at Mayo, Yukon Territory. *Permafrost and Periglacial Processes,* **1**, 3–14.

Burn, C. R. (1990b). Frost-heave in lake-bottom sediments, Mackenzie Delta, Northwest Territories. In: *Permafrost-Canada, Proceedings of the Fifth Canadian Permafrost Conference,* National Research Council of Canada – Centre d'études nordiques, Université Laval, Collection Nordicana no. 54, pp. 103–109.

Burn, C. R. (1992). Recent ground warming inferred from the temperature in permafrost near Mayo, Yukon Territory. In: Dixon, J. C., Abrahams, A. D., eds., *Periglacial Geomorphology.* John Wiley & Sons, Chichester, pp. 327–350.

Burn, C. R. (1997). Cryostratigraphy, paleogeography, and climate change during the early Holocene warm interval, western Arctic coast, Canada. *Canadian Journal of Earth Sciences,* **34**, 912–925.

Burn, C. R. (1998a). The response (1958–1997) of permafrost and near-surface ground temperatures to forest fire, Takhini River valley, southern Yukon Territory. *Canadian Journal of Earth Sciences,* **35**, 184–199

Burn, C. R. (1998b). The active layer: two contrasting definitions. *Permafrost and Periglacial Processes,* **9**, 411–416.

Burn, C. R. (2002). Tundra lakes and permafrost, Richards Island, western Arctic coast, Canada. *Canadian Journal of Earth Sciences,* **39**, 1281–1298.

Burn, C. R. (2005). Lake-bottom thermal regimes, western Arctic coast, Canada. *Permafrost and Periglacial Processes,* **16**, 355–368.

Burn, C. R., Michel, F. A. (1988). Evidence for recent temperature-induced water migration into permafrost from the tritium content of ground ice near Mayo, Yukon Territory, Canada. *Canadian Journal of Earth Sciences,* **25**, 909–915.

Burn, C. R., Smith, C. A. S. (1988a). Observations of the "thermal offset" in near-surface mean annual ground temperatures at several sites near Mayo, Yukon Territory. *Arctic,* **41**, 99–104.

Burn, C. R., Smith, M. W. (1988b). Thermokarst lakes at Mayo, Yukon Territory, Canada. In: Senneset, K., ed., *Permafrost, Proceedings of the Fifth International Conference on Permafrost,* 2–5 August. Tapir, Trondheim, vol. 1, pp. 700–705.

Burn, C. R., Michel, F. A., Smith, M. W. (1986). Stratigraphic, isotopic, and mineralogical evidence for an early Holocene thaw unconformity at Mayo, Yukon Territory. *Canadian Journal of Earth Sciences,* **23**, 794–803.

Burt, T. P., Williams, P. J. (1976). Hydraulic conductivity in frozen soils. *Earth Surface Processes,* **1**, 349–360.

Butrym, J., Cegla, J., Dzulynski, S., Nakonieczny, S. (1964). New interpretation of "periglacial structures". *Folia Quaternaria*, **17**, 34 pp.

Cailleux, A. (1942). *Les actions éoliennes périglaciaires en Europe*. Memoire 46, Société Géologique de France, 176 pp.

Cailleux, A. (1956). Mares, mardelles et pingos. *Comptes Rendus, Académie des Sciences, Paris*, **242**, 1912–1914.

Cailleux, A. (1957). Les mares du sud-est de Sjaelland (Danemark). *Comptes Rendus, Académie des Sciences, Paris*, **245**, 1074–1076.

Cailleux, A. (1974). Formes précoces et albédos du nivéo-éolien. *Zeitschrift für Geomorphologie*, **18**, 437–459.

Cailleux, A. (1978). Niveo-eolian deposits. In: Fairbridge, R. W., Bourgeois, J., eds., *Encyclopedia of Earth Sciences*. Reinhold, New York; *The Encyclopedia of Sedimentology*, vol. 6, pp. 501–503.

Cailleux, A., Calkin, P. (1963). Orientation of hollows in cavernously-weathered boulders in Antarctica. *Biuletyn Peryglacjalny*, **12**, 147–150.

Cailleux, A., Taylor, G. (1954). *Cryopédologie, études des sols gelés. Expéditions Polaires Françaises*. Hermann & Cie, Paris, 218 pp.

Caine, T. N. (1967). The tors of Ben Lomond, Tasmania. *Zeitschrift für Geomorphologie*, **11**, 418–429.

Caine, T. N., Thurman, E. M. (1990). Temporal and spatial variations in the solute content of an alpine stream. *Geomorphology*, **4**, 55–72.

Cairnes, D. D. (1912). Differential erosion and equiplanation in portions of Yukon and Alaska. *Geological Society of America, Bulletin*, **23**, 333–348.

Calkin, P., Cailleux, A. (1962). A quantitative study of cavernous weathering (taffoni) and its application to glacial chronology in Victoria Valley, Antarctica. *Zeitschrift für Geomorphologie*, **6**, 317–324.

Campbell, I. B., Claridge, G. G. C. (1987). *Antarctic: Soils, Weathering, Processes and Environment*. Elsevier, Amsterdam, 368 pp.

Canada Soil Survey Committee (1978). *The Canadian System of Soil Classification*. Canada Department of Agriculture, Ottawa, publication 1646, 164 pp.

Capello, C. F. (1959). *Periglaciale o crionivale?* Societa Geografica Italiana, Rome, 24 pp. (English translation by M. Tosella and D. Verdiani-Noero, 1960, Periglacial or cryonival? General terminology of morphological phenomena due to discontinuous frost. Instituto di Geografia, Universita di Torino, Torino, 23 pp.)

Capps, S. R. (1919). The Kantishna region, Alaska. *United States Geological Survey, Bulletin*, **687**, 7–112.

Carey, K. L. (1970). Icing occurrence, control and prevention, an annotated bibliography. US Army Corps of Engineers, CRREL, Hanover, New Hampshire, special report, 151 pp.

Carey, S. K., Woo, M.-K. (2000). The role of soil pipes as a slope runoff mechanism, Subarctic Yukon, Canada. *Journal of Hydrology*, **233**, 206–222.

Carson, C. E., Hussey, K. M. (1962). The oriented lakes of Arctic Alaska. *Journal of Geology*, **70**, 417–439.

Carson, C. E., Hussey, K. M. (1963). The oriented lakes of Arctic Alaska; a reply. *Journal of Geology*, **71**, 532–533.

Carter, L. D. (1981). A Pleistocene sand sea on the Alaskan Arctic Coastal Plain. *Science*, **211**, 381–383.

Carton, A., Dramis, F., Smiraglia, C. (1988). A first approach to the systematic study of the rock glaciers in the Italian Alps. In: Senneset, K., ed., *Permafrost, Proceedings of the Fifth International Conference on Permafrost*, 2–5 August. Tapir, Trondheim, vol. 1, pp. 712–717.

Catt, J. A. (1977). Loess and coversands. In: Shotton, F. W., ed., *British Quaternary Studies: Recent Advances*. Clarendon Press, Oxford, pp. 221–229.

CCGM/CGMW-ANDRA (Commission de la Carte Géologique du monde/Commission for the Geological Map of the World–Agence nationale pour la Gestion des Déchets radioactifs)) (1999). *Maps of the World Environments During the Last Two Climatic Extremes: (1) The Last Glacial maximum (ca 18000 +/–2000 yrs B.P.); (2) The Holocene Optimum (ca 8000 +/–1000 yrs B.P.).*

CGMW, Paris and ABDRA, Chatenay-Malabry cedex, France. Scale: 1:25,000,000, plus Explanatory Notes, 26 pp.

Cederstrom, D. J., Johnston, P. M., Subitzky, S. (1953). Occurrence and development of ground water in permafrost regions. *United States Geological Survey, circular 275*, 30 pp.

Chambers, M. J. G. (1966). Investigations of patterned ground at Signy Island, South Orkney Islands. II: Temperature regimes in the active layer. *Bulletin, British Antarctic Survey*, **10**, 71–83.

Chambers, M. J. G. (1967). Investigations of patterned ground at Signy Island, South Orkney islands. III: Miniature patterns, frost heaving and general conclusions. *Bulletin, British Antarctic Survey*, **12**, 1–22.

Chandler, R. J. (1970a). The degradation of Lias Clay slopes in an area of the East Midlands. *Quarterly Journal Engineering Geology*, **2**, 161–181.

Chandler, R. J. (1970b). A shallow slab slide in the Lias Clay near Uppingham, Rutland. *Géotechnique*, **20**, 253–260.

Chandler, R. J. (1972). Periglacial mudslides in Vestspitsbergen and their bearing on the origin of fossil "solifluction" shears in low-angled clay slopes. *Quarterly Journal of Engineering Geology*, **5**, 223–241.

Chandler, R. J., Kellaway, G. A., Skempton, A. W., Wyatt, R. J. (1976). Valley slope sections in Jurassic strata near Bath, Somerset. *Philosophical Transactions of the Royal Society, London*, **A283**, 527–555.

Chatwin, S. C., Rutter, N. W. (1978). *Upper Mackenzie River Valley. Field Trip No. 2*, Third International Conference on Permafrost, Edmonton, Alberta, 53 pp.

Cheng, G. (1983). The mechanism of repeated-segregation for the formation of thick layered ground ice. *Cold Regions Science and Technology*, **8**, 57–66.

Cheng, G., Dramis, F. (1992). Distribution of mountain permafrost and climate. *Permafrost and Periglacial Processes*, **3**, 83–91.

Cheng, G., Li, X. (2003). Constructing the Qinghai–Tibet Railroad: new challenges to Chinese permafrost scientists. In: Phillips, M., Springman, S. A., Arenson, L. U., eds., *Permafrost, Proceedings of the Eighth International Conference on Permafrost*, 21–25 July, Zurich, Switzerland. Balkema, Lisse, vol. 1, pp. 131–137.

Christiansen, H. H. (2005). Thermal regime of ice-wedge cracking in Adventdalen, Svalbard. *Permafrost and Periglacial Processes*, **16**, 87–98.

Christiansen, H. H., Svensson, H. (1998). Windpolished boulders as indicators of a Late Weichselian wind regime in Denmark in relation to neighbouring areas. *Permafrost and Periglacial Processes*, **9**, 1–21.

Christiansen, H. H., Svensson, H. (1999). Windpolish evidence: an important direct indicator of geomorphologically active palaeo-winds. A reply to the discussion by Vandenberghe, Isarin and Renssen. *Permafrost and Periglacial Processes*, **10**, 203–204.

Christiansen, H. H., French, H. M., Humlum, O. (2005). Permafrost in the Gruve-7 mine, Adventdalen, Svalbard. *Norsk Geografisk Tidsskrift*, **59**, 109–115.

Chueca, J. (1992). A statistical analysis of the spatial distribution of rock glaciers, Spanish Central Pyrenees. *Permafrost and Periglacial Processes*, **3**, 261–265.

Church, M. (1972). Baffin Island sandurs; a study of Arctic fluvial processes. *Geological Survey Canada, Bulletin*, **216**, 208 pp.

Church, M. (1974). Hydrology and permafrost with reference to northern North America. In: *Permafrost Hydrology; Proceedings of Workshop Seminar*, Canadian National Committee, International Hydrological Decade. Environment Canada, Ottawa, pp. 7–20.

Church, M. (2005). Continental drift. *Earth Surface Processes and Landforms*, **30**, 129–130.

Church, M., Ryder, J. M. (1972). Paraglacial sedimentation: a consideration of fluvial processes conditioned by glaciation. *Bulletin, Geological Society of America*, **83**, 3059–3072.

Clark, G. M., Ciolkosz, E. J. (1988). Periglacial geomorphology of the Appalachian Highlands and Interior Highlands south of the glacial border – a review. *Geomophology*, **10**, 475–477.

Clark, G. M., Behling, R. E., Braun, D. D., Ciolkosz, E. J., Kite, J. S., Marsh, B. (1992). Central Appalachian periglacial geomorphology. Pennsylvanian State University College of Agriculture, agronomy series no. 120, 248 pp.

Clark, I. D., Lauriol, B. (1997). Aufeis of the Firth River basin, Northern Yukon, Canada: insights into permafrost hydrology and karst. *Arctic and Alpine Research*, **29**, 240–252.

Clark, M. J. (1988). Periglacial hydrology. In: Clark, M. J., ed., *Advances in Periglacial Geomorphology*. John Wiley & Sons, Chichester, pp. 415–462.

Clark, M. J., French, H. M., Harry, D. G. (1984). Reconnaisance techniques for the estimation of Arctic coastal sediment budget and process, In: Clark, M. W., ed., *Coastal Research: UK Perspectives*. GeoBooks, Norwich, pp. 1–14.

Collett, T. S. (1983). Detection and evaluation of natural gas hydrates from well logs, Prudhoe Bay, Alaska. In: *Permafrost, Proceedings of the Fourth International Conference on Permafrost*, 17–22 July, Fairbanks, Alaska. National Academy Press, Washington, DC, pp. 169–174.

Conant, L. C., Black, R. F., Hosterman, J. W. (1976). Sediment-filled pots in upland gravels of Maryland and Virginia. *Journal of Research, United States Geological Survey*, **4**, 353–358.

Conca, J. L., Astor, A. M. (1987). Capillary moisture flow and the origin of cavernous weathering in dolerites of Bull Pass, Antarctica. *Geology*, **15**, 151–154.

Cook, F. A. (1967). Fluvial processes in the high Arctic. *Geographical Bulletin*, **9**, 262–268.

Cook, F. A., Raiche, V. G. (1962). Freeze–thaw cycles at Resolute, N.W.T. *Geographical Bulletin*, **18**, 64–78.

Cooke, R. U., Warren, A., Goudie, A. S. (1993). *Desert Geomorphology*. University College London Press, London, 526 pp.

Corte, A. E. (1966). Particle sorting by repeated freezing and thawing. *Biuletyn Peryglacjalny*, **15**, 175–240.

Corte, A. E. (1971). Laboratory formation of extrusion features by multicyclic freeze-thaw in soils. In: *Étude des phénomènes périglaciaires en laboratoire*, Colloque International de Géomorphologie, Liège-Caen, Centre de Géomorphologie à Caen, bulletin no. 13-14-15, pp. 117–131.

Corte, A. E. (1978). Rock glaciers as permafrost bodies with a debris cover on an active layer. A hydrological approach in the Andes of Mendoza, Argentina. In: *Permafrost, Proceedings of the Third International Conference on Permafrost*, 10–13 July, Edmonton, Alberta, Canada. National Research Council of Canada, Ottawa, vol. 1, pp. 263–269.

Corte, A. E. (1988). Geocryology of the Central Andes and rock glaciers. In: Senneset, K., ed., *Permafrost, Proceedings of the Fifth International Conference on Permafrost*, 2–5 August. Tapir, Trondheim, vol. 1, pp. 718–723.

Côté, M. M., Burn, C. R. (2002). The oriented lakes of Tuktoyaktuk Peninsula, western Arctic coast, Canada: a GIS-based approach. *Permafrost and Periglacial Processes*, **13**, 61–70.

Coultish, T. L., Lewkowicz, A. G. (2003). Palsa dynamics in a subarctic mountainous environment, Wolf Creek, Yukon Territory, Canada. In: *Permafrost, Proceedings of the Eighth International Conference on Permafrost*, 21–25 July, Zurich, Switzerland. Balkema, Lisse, vol. 1, 163–168.

Couture, N. J., Pollard, W. H. (1998). An assessment of ground ice volumes near Eureka, Northwest Territories. In: Lewkowicz, A. G., Allard, M., eds., *Permafrost, Proceedings of the Seventh International Conference on Permafrost*, 23–27 June, Yellowknife, Canada. Centre d'études nordiques, Université Laval, Québec, Collection Nordicana no. 57, pp. 195–200.

Craig, B. G. (1959). Pingo in the Thelon Valley, Northwest Territories; radiocarbon age and historical significance of the contained organic material. *Bulletin, Geological Society of America*, **70**, 509–510.

Crory, F. E. (1988). Airfields in arctic Alaska. In: Senneset, K., ed., *Permafrost, Proceedings of the Fifth International Conference on Permafrost*, 2–5 August. Tapir, Trondheim, vol. 3, pp. pp. 49–55.

Crory. F. E. (1991). Construction guidelines for oil and gas exploration in Northern Alaska. United States Army Corps of Engineers, CRREL, Hanover, New Hampshire, report 91-21, 83 pp.

Cui, Zhijiu (1983). An investigaton of rock glaciers in the Kunlun Shan, China. In: *Permafrost, Proceedings of the Fourth International Conference on Permafrost*, 17–22 July, Fairbanks, Alaska. National Academy Press, Washington, DC, pp. 208–211.

Currey, D. R. (1964). A preliminary study of valley asymmetry in the Ogotoruk Creek area, Northwest Alaska. *Arctic*, **17**, 85–98.

Czeppe, Z. (1964). Exfoliation in a periglacial climate. *Geographia Polonica*, 2, 5–10.

Czeppe, Z. (1965). Activity of running water in south-western Spitsbergen. *Geographia Polonia*, **6**, 141–150.

Czudek, T. (1964). Periglacial slope development in the area of the Bohemian Massif in Northern Moravia. *Biuletyn Peryglacjalny*, **14**, 169–194.

Czudek, T. (1990). Zum problem der kryoplanationsterrassen. *Petermanns Geographische Mitteilungen*, **134**, 226–238.

Czudek, T., Demek, J. (1970). Thermokarst in Siberia and its influence on the development of lowland relief. *Quaternary Research*, **1**, 103–120.

Czudek, T., Demek, J. (1973). The valley cryopediments in Eastern Siberia. *Biuletyn Peryglacjalny*, **22**, 117–130.

Dahl, R. (1966). Blockfields and other weathering forms in the Narvik Mountains. *Geografiska Annaler*, **48A**, 224–227.

Dallimore, S. R., Wolfe, S. A. (1988). Massive ground ice associated with glaciofluvial sediments, Richards Island, NWT, Canada. In: Senneset, K., ed., *Permafrost, Proceedings of the Fifth International Conference on Permafrost*, 2–5 August. Tapir, Trondheim, vol. 1, pp. pp. 132–137.

Dallimore, S. R., Wolfe, S. A., Soloman, S. M. (1996a). Influence of ground ice and permafrost on coastal evolution, Richards Island, Beaufort Sea coast, NWT. *Canadian Journal of Earth Sciences*, **33**, 664–675.

Dallimore, S. R., Nixon, F. M., Egginton, P. A., Bisson, J. G. (1996b). Deep-seated creep of massive ground ice, Tuktoyaktuk, NWT, Canada. *Permafrost and Periglacial Processes*, **7**, 337–348.

Dallmann, W. K., Kjaernet, T., Nottvedt, A. (2001). Geological Map of Svalbard 1:100,000, Sheet C9G Adventdalen. Temakarst No 31/32, Norwegian Polar Institute, Tromso, 55 pp.

Danilov, I. D., Komarov, I. A., Vlasenko, A. Yu. (1998). Pleistocene-Holocene permafrost of the East Siberian Eurasian Arctic shelf. In: Lewkowicz, A. G., Allard, M., eds., *Permafrost, Proceedings of the Seventh International Conference on Permafrost*, 23–27 June, Yellowknife, Canada. Centre d'études nordiques, Université Laval, Collection Nordicana no. 57, pp. 207–211.

Danilova, N. S. (1956). Soil wedges and their origin. In: *Data on the Principles of the Study of the Frozen Zones in the Earth's Crust*, issue 111. V. A. Obruchev Insitute of Permafrost Studies, Academy Science, Moscow (National Research Council of Canada, technical translation no. 1088, Ottawa, 1964, pp. 90–99.).

Darmody, R. G., Thorn, C. E., Harder, R. L., Schlyter, P., Dixon, J. C. (2000). Weathering implications of water chemistry in an arctic-alpine environment, northern Sweden. *Geomorphology*, **34**, 89–100.

Darmody, R. G., Allen, C. E., Thorn, C. E., Dixon, J. C. (2001). The poisonous rocks of Karkevagge. *Geomorphology*, **41**, 53–62.

Datsko, P. S., Rogov, V. V. (1988). Transformation of dispersed deposits under cycles of freezing–thawing. In: *Microstructure of Frozen Ground*. Moscow University Press, Moscow, pp. 132–149 (in Russian).

Davies, K. H., Keen, D. H. (1985). The age of Pleistocene marine deposits at Portland, Dorset. *Proceedings, Geologist's Association*, **96**, 217–225.

Dawson, A. G. (1992). *Ice Age Earth: Late Quaternary Geology and Climate*. Routledge, London, 293 pp.

De Gans, W. (1988). Pingo scars and their identification. In: Clark, M. J., ed., *Advances in Periglacial Geomorphology*. John Wiley & Sons, Chichester, pp. 299–322.

De Gans, W., Sohl, H. (1981). Weichselian pingo remnants and permafrost on the Drente plateau (The Netherlands). *Geologie en Mijnbouw*, **60**, 447–452.

de Groot, Th., Cleveringa, P., Klijnstra, B. (1987). Frost-mound scars and the evolution of a Late Dryas environment (northern Netherlands). *Geologie en Mijnbouw*, **66**, 239–250.

De La Beche, H. T. (1839). *Report on the Geology of Cornwall, Devon, and West Somerset*. Memoirs, Geological Survey, United Kingdom.

Delisle, G., Allard, M., Fortier, R., Calmels, F., Larrivée, E. (2003). Umiujaq, Northern Québec: innovative techniques to monitor the decay of a lithalsa in response to climate change. *Permafrost and Periglacial Processes*, **14**, 375–385.

Delluc, B., Delluc, G., Roussot, A., Roussot-Larroque, J. (2001). *Discovering Perigord Prehistory*. Editions Sud Ouest, Bordeaux, 95 pp.

Demek, J. (1964). Castle koppies and tors in the Bohemian Highland (Czechoslovakia). *Biuletyn Peryglacjalny*, **14**, 195–216.

Demek, J. (1969a). Cryoplanation terraces, their geographical distribution, genesis and development. *Ceskoslovenski Akademie Ved Rozpravy, Rad Mathematickych A Prirodnich Ved, Rocnik*, **79**(4), 80 pp.

Demek, J. (1969b). Cryogene processes and the development of cryoplanation terraces. *Biuletyn Peryglacjalny*, **18**, 115–125.

Demek, J. (1972a). Die pedimentation im subnivalen Bereich. *Göttingen Geographische Abhandlungen*, **60**, 145–154.

Demek, J. (1972b). *Manual of Detailed Geomorphological Mapping*. Publishing House of the Czechoslovak Academy of Sciences, Prague, 344 pp.

Demek, J. (1978). Periglacial geomorphology. In: Embleton, C., Brunsden, D., Jones, D. K. C., eds., *Geomorphology, Present Problems and Future Prospects*. Oxford University Press, Oxford, pp. 139–155.

Denny, C. S. (1936). Periglacial phenomena in southern Connecticut. *American Journal of Science*, **32**, 322–342.

Derbyshire, E. (1972). Tors, rock weathering, and climate in southern Victoria Land, Antarctica. In: *Polar Geomorphology*, Institute of British Geographers Special Publication no. 4, pp. 93–105.

Derbyshire, E. (1973). Periglacial phenomena in Tasmania. *Biuletyn Peryglacjalny*, **22**, 131–148.

Derbyshire, E., Love, M. A., Martin, J. E., (1985). Fabrics of probable segregated ground-ice origin in some sediment cores from the North Sea Basin. In: Boardman, J., ed., *Soils and Quaternary Landscape Evolution*. John Wiley & Sons., Chichester, pp. 261–280.

DeWolf, Y. (1988). Stratified slope deposits. In: Clark, M. J., ed., *Advances in Periglacial Geomorphology*. John Wiley & Sons, Chichester, pp. 91–110.

Dijkmans, J. W. A. (1989). Frost wedges in an eolian sand sheet near Sondre Stromfjord, West Greenland, and the palaeoenvironmental implications. *Zeitschrift für Geomorphologie*, **33**, 339–353.

Dijkmans, J. W. A., Koster, E. A. (1990). Morphological development of dunes in a subarctic environment, Central Kobuk Valley, Northwestern Alaska. *Geografiska Annaler*, **72A**, 93–109.

Dijkmans, J. W. A., Mucher, H. J. (1989). Niveo-aeolian sedimentation of loess and sand: an experimental and micromorphological approach. *Earth Surface Processes and Landforms*, **14**, 303–315.

Dijkmans, J. W. A., Koster, E. A., Galloway, J. P., Mook, W. G. (1986). Characteristics and origin of calcretes in a subarctic environment, Great Kobuk Sand Dunes, Northwestern Alaska, USA. *Arctic and Alpine Research*, **18**, 377–387.

Dines, H. G., Hollingworth, S. E., Edwards, W., Buchan, S., Welch, F. B. A. (1940). The mapping of head deposits. *Geological Magazine*, **77**, 198–226.

Dingman, S. L., Koutz, F. R. (1974). Relations among vegetation, permafrost and potential insolation in central Alaska. *Arctic and Alpine Research*, **6**, 37–42.

Dionne, J.-C. (1975). Blocs soulevés par le froid dans les schorres de la baie de James. *Revue de Géographie de Montréal*, **29**, 161–166.

Dionne, J.-C. (1989). The role of ice and frost in tidal marsh development – a review with particular reference to Québec, Canada. *Essener Geographisk Arbeiten*, **18**, 171–210.

Dionne, J.-C. (1994). Les cordons de blocs frangeants (Boulder barricades). Mise au point avec exemples du Québec. *Revue de Géomorphologie dynamique*, **43**, 49–70.

Dionne, J.-C. (2002). The boulder barricade at Cap à la Baleine, North Shore of Gaspé Peninsula (Québec): nature of boulders, origin and significance. *Journal of Coastal Research*, **18**, 652–661.

Doerr, S. H. (1999). Karst-like landforms and hydrology in quartzites of the Venezuelan Guyana shield: pseudokarst or "real" karst?. *Zeitschrift für Geomorphologie*, **43**, 1–17.

Dokuchaev, W. W. (1900). *Zones naturelles de sols*. St Petersburg.

Dorn, R., Oberlander, T. M. (1982). Rock varnish. *Progress in Physical Geography*, **6**, 317–367.

Dorn, R., Krinsley, D. H., Lin, T., Anderson, S., Clark, J., Cahill, T. A., Gill, T. E. (1992). Manganese-rich rock varnish does occur in Antarctica. *Chemical Geology*, **99**, 289–298.

Dostovalov, B. N., Kudryavtsev, V. A. (1967). *Obshcheye mierzlotovedeniya* [General Permafrost Science]. Moscow State University, Moscow, 463 pp. (in Russian).

Dostovalov, B. N., Popov, A. I. (1966). Polygonal systems of ice wedges and conditions of their development. In: *Permafrost, International Conference Proceedings*, National Research Council of Canada publication 1287. National Academy of Sciences, Washington, DC, pp. 102–105.

Douglas, G. R., McGreevy, J. P., Whalley, W. B. (1983). Rock weathering by frost shattering processes. In: *Permafrost, Proceedings of the Fourth International Conference on Permafrost*, 17–22 July, Fairbanks, Alaska. National Academy Press, Washington, DC, pp. 244–248.

Douglas, G. R., Whalley, W. B., McGreevy, J. P. (1991). Rock properties as controls on free-face debris fall activity. *Permafrost and Periglacial Processes*, **2**, 311–319.

Dramis, F., Kotarba, A. (1992). Southern limit of relict rock glaciers, Central Apennines, Italy. *Permafrost and Periglacial Processes*, **3**, 253–257.

Dredge, L. A. (1992). Breakup of limestone bedrock by frost shattering and chemical weathering, Eastern Canadian Arctic. *Arctic and Alpine Research*, **24**, 314–323.

Dredge, L. A., Kerr, D. E., Wolfe, S. A. (1999). Surficial materials and related ground ice conditions, Slave Province, NWT, Canada. *Canadian Journal of Earth Sciences*, **36**, 1227–1238.

Dresch, J. (1982). *Géographie des régions arides*. Presses Universitaires du France, Paris, 277 pp.

Dubikov, G. I. (1982). Origin of tabular ice in permafrost in Western Siberia. In: Popov, A. I., ed., *Tabular Ice in the Cryolithozone*. Permafrost Insitute, Yakutsk, USSR, pp. 24–42 (in Russian).

Dubikov, G. I. (2002). Composition and cryogenic structure of permafrost in West Siberia. GEOS, Moscow, 246 pp. (in Russian, with English summary, pp. 231–233).

Duk-Rodkin, A., Barendregt, R. W., Froese, D. G., Weber, F., Enkin, R., Smith, I. R., Zazula, G. D., Waters, P., Klassen, R. (2004). Timing and extent of Plio-Pleistocene glaciations in north-western Canada and east-central Alaska. In: Ehlers, J., Gibbard, P. L., eds., *Quaternary Glaciations – Extent and Chronology*, part II. Elsevier, Amsterdam, pp. 313–345.

Dunbar, M., Greenaway, K. R. (1956). *Arctic Canada from the Air*. Queen's Printer, Ottawa, 541 pp.

Dury, G. H. (1959). *The Face of the Earth*. Penguin, Harmondsworth, 226 pp.

Dyke, A. S. (1976). Tors and associated weathering phenomena, Somerset Island, District of Franklin. *Geological Survey of Canada*, paper 76-1B, 209–216.

Dyke, A. S. (1978). Qualitative rates of frost heaving in gneissic bedrock on Southeastern Baffin Island, District of Franklin. *Geological Survey of Canada*, paper 78-1A, 501–502.

Dyke, A. S., Prest, V. K. (1987). The Late Wisconsinan and Holocene history of the Laurentide ice sheet. *Géographie physique et Quaternaire*, **41**, 247–264.

Dyke, A. S., Zoltai, S. C. (1980). Radiocarbon-dated mudboils, Central Canadian Arctic. *Geological Survey of Canada*, paper 80-1B, 271–275.

Dyke, L. S. (1984). Frost heaving of bedrock in permafrost regions. *Bulletin, Association of Engineering Geologists*, **XXI**(4), 389–405.

Dyke, L. S. (1991). Temperature changes and thaw of permafrost adjacent to Richards Island, Mackenzie Delta, NWT. *Canadian Journal of Earth Sciences*, **28**, 1834–1842.

Dyke, L. S. (2001). Contaminant migration through the permafrost active layer, Mackenzie Delta area, Northwest Territories, Canada. *Polar Record*, **37**, 215–228.

Dyke, L., Wolfe, S. (1993). Ground temperatures and recent coastal change at the north end of Richards Island, Mackenzie Delta, Northwest Territories. In: *Geological Survey of Canada*, paper 93-1E, 83–91.

Dylik, J. (1953). Periglacial investigations in Poland. *Bulletin, Société des Sciences et des Lettres de Łódź*, **4**, 1–16.

Dylik, J. (1956). Coup d'oeil sur la Pologne périglaciaire. *Biuletyn Peryglacjalny*, **4**, 195–238. (English translation by A. Guilcher, 1992, A general view of periglacial Poland. In: Evans, D. J. A., ed., 1994, *Cold Climate Landforms*, John Wiley & Sons, Chichester, pp. 45–81).

Dylik, J. (1957). Tentative comparison of planation surfaces occurring under warm and under cold semi-arid conditions. *Biuletyn Peryglacjalny*, **5**, 175–186.

Dylik, J. (1960). Rhythmically stratified slope waste deposits. *Biuletyn Peryglacjalny*, **8**, 31–41.

Dylik, J. (1963). Periglacial sediments of the Sw. Malgorzata hill in the Warsaw-Berlin pradolina. *Bulletin, Société des Sciences et des Lettres de Łódź*, **14**, 1–16.

Dylik, J. (1964a). Eléments essentiels de la notion de "périglaciaire". *Biuletyn Peryglacjalny*, 14, 111–132.

Dylik, J. (1964b). Le thermokarst, phénomène négligé dans les études du Pleistocene. *Annales de Géographie*, **73**, 513–523.

Dylik, J. (1966). Problems of ice wedge structure and frost fissure polygons. *Biuletyn Peryglacjalny*, **15**, 241–291.

Dylik, J. (1968). Thermokarst. In: Fairbridge, R. W., ed., *Encyclopedia of Geomorphology*. Reinhold, New York, pp. 1149–1151.

Dylik, J. (1969a). Slope development under periglacial conditions in the Łódź region. *Biuletyn Peryglacjalny*, **18**, 381–410.

Dylik, J. (1969b). Slope development affected by frost fissures and thermal erosion. In: Péwé, T. L., ed., *The Periglacial Environment*. McGill-Queen's University Press, Montreal, pp. 365–386.

Dylik, J. (1971). L'érosion thermique actuelle at ses traces figées dans le paysage de la Pologne Centrale. *Bulletin de l'Académie Polonaise des Science, Série des Sciences de la Terre*, **xix**, 55–61.

Dylik, J. (1972). Rôle du ruisellement dans le modèle périglaciaire. In: Hoverman, J., Oberback, G., eds., *Sonderdrück aus Heft 60 der Hans-Poser-festschrift, Göttinger Geographische Abhandlungen*, **60**, 169–180.

Dylikowa, A. (1962). Notion et terme "périglaciaire". *Biuletyn Peryglacjalny*, **11**, 149–163.

Dylikowa, A., Gozdzik, J. S., Jahn, A. (1978). Methodology of field studies: fossil frost- and ice wedges. *Biuletyn Peryglacjalny*, **27**, 171–179.

Eakin, W. M. (1916). The Yukon-Koyukuk region, Alaska. *United States Geological Survey, Bulletin*, **631**, 67–88.

Edelman, C. H., Maarleveld, G. C. (1949). De asymmetrische dalen va de Veluwe. *Tijdschrift Koninklijk Nederland Aardrijkskundig Genootschap*, **66**, 143–146.

Edelman, C. H., Tavernier, R. (1940). Periglacial verschijnselen, meer in het bijzonder in der Antwerpsche kempen. *Natuurwetenschappelijk Tijdschrift*, **XXII**, 139–153.

Edelman, C. H., Florschütz, F., Jeswiet, J. (1936). Über spätpleistozäne und frühholozäne kryoturbate Ablagerungen in den östlichen Niederlanden. Geologisch-Mijnbouwkundig Genootschap voor Nederland en Kolonien., *Verhandelingen, Geologisch Series*, **11**, 301–360.

Edlund, S. A., Alt, B. T., Young, K. (1989). Interaction of climate, vegetation, and soil hydrology at Hot Weather Creek, Fosheim Peninsula, Ellesmere Island, Northwest Territories. *Geological Survey of Canada*, paper 89-1D, 125–133.

Egginton, P. A., Dyke, L. S. (1982). Density gradients and injection structures in mudboils in central District of Keewatin. *Geological Survey of Canada*, paper 82-1B, 173–176.

Egginton, P. A., French, H. M. (1985). Solifluction and related processes, Eastern Banks Island, NWT. *Canadian Journal of Earth Sciences*, **22**, 1671–1678.

Egginton, P. A., Shilts, W. W. (1978). Rates of movement associated with mudboils, central District of Keewatin. *Geological Survey of Canada*, paper 78-1B, 203–206.

Ehlers, J. (1996). *Quaternary and Glacial Geology* (English version by P. L. Gibbard). John Wiley & Sons, Chichester, 578 pp.

Eissmann, L. (1978). Mollisoldiapirismus. *Zeitschrift für Angewandte Geologie*, **24**(3), 130–138.

Eissmann, L. (1994). Grundzüge der Quartärgeologie Mitteldeutschlands (Sachsen, Sachsen-Anhalt, Südbrandengurg, Thüringen). *Altenburger Naturwissenschaftliche Forschungen*, **7**, 55–135.

Eissmann, L. (2002). Quaternary geology of eastern Germany (Saxony, Saxon-Anhalt, South Brandenburg, Thuringia), type area of the Elsterian and Saalian stages in Europe. *Quaternary Science Reviews*, **21**, 1275–1346.

Elder, K., Kattelman, R. (1993). A low-angle slushflow in the Kirghiz Range, Kirgizstan. *Permafrost and Periglacial Processes*, **4**, 301–310.

Ermolaev, M. M. (1932a). Geological and geomorphological description of Bol'shoi Lyakhovskii Island. *Trudy SOPS AN SSSR*, Yakut ser., issue 7 (in Russian).

Ermolaev, M. M. (1932b). *Directions for Expedition Research of Ancient Ice as a Geographical Factor (Primarily in Arctic Regions)*. All-Union Arctic Institute, USSR Academy of Sciences, Leningard, 42 pp. (in Russian).

Ershov, E. D. (1984). Transformation of dispersed deposits under repeated freezing–thawing. *Engineering Geology*, **3**, 59–66 (in Russian).

Ershov, E. D., Cheverev, V. G., Lebedenko, Yu P., Shevchenko, L. V. (1980). Water migration, formation of texture and ice segregation in freezing and thawing clayey soils. In: *Permafrost, Proceedings of the Third International Conference on Permafrost*, 10–13 July, Edmonton, Alberta, part 1: English translations of twenty-six of the Soviet papers. National Research Council of Canada, publication 18119, Ottawa, Canada, pp. 159–175.

Etienne, S. (2002). The role of biological weathering in periglacial areas: a study of weathering rinds in south Iceland. *Geomorphology*, **47**, 75–86.

Evans, D. J. A. (1993). High-latitude rock glaciers: a case study of forms and processes in the Canadian Arctic. *Permafrost and Periglacial Processes*, **4**, 17–36.

Everett, K. R. (1965). Slope movement and related phenomena. In: Wilimovsky, N. J., ed., *The Environment of the Cape Thompson Region, Alaska*. United States Atomic Energy Commission, PNE-481, pp. 175–220.

Everett, K. R. (1967). Mass-wasting in the Taseriaq area, West Greenland. *Meddelelser om Gronland*, **165**, 1–32.

Fahey, B. D. (1973). An analysis of diurnal freeze–thaw and frost heave cycles in the Indian Peaks region of the Colorado Front Range. *Arctic and Alpine Research*, **5**, 269–281.

Fahey, B. D. (1981). Origin and age of upland schist tors in Central Otago, New Zealand. *New Zealand Journal of Geology and Geophysics*, **24**, 399–413.

Fahey, B. D. (1985). Salt weathering as a mechanism of rock breakup in cold climates: an experimental approach. *Zeitschrift für Geomorphologie*, **29**, 99–111.

Farrell, S. C., Gagnon, K., Malinousky, T., Colombo, R., Mujica, K., Mitrocsak, J., Cozzi, A., Van Woudenberg, E., Weisbecker, T. (1985). Pleistocene? braided stream deposits in the Atsion Quadrangle, northwestern Atlantic County, New Jersey. In: Talkington, R. W., ed., *Geological Investigations on the Coastal Plain of Southern New Jersey*, part 1, *Field Guide*, Second Annual Meeting, Geological Association of New Jersey, Stockton State College, Ponoma, New Jersey, pp. A-1–A-11.

Faucher, D. (1931). Note sur la dissymetrie des vallons de l'Armagnal. *Bulletin Societé d'histoire naturelle de Toulouse*, **61**, 262–268.

Federoff, N. (1966). Les cryosols. *Sciences du Sol*, **2**, 77–110.

Fernald, A. T. (1964). Surficial geology of the Central Kobuk River Valley, Northwestern Alaska. *United States Geological Survey, Bulletin*, **1181-K**, 1–31.

Ferrians, O. J. (1965). Permafrost map of Alaska. United States Geological Survey, Miscellaneous Map, 1–445.

Ferrians, O. J., Kachadoorian, R., Green, G. W. (1969). Permafrost and related engineering problems in Alaska. United States Geological Survey, professional paper 678, 37 pp.

Fisher, T. G. (1996). Sand-wedge and ventifact palaeoenvironmental indicators in North West Saskatchewan, Canada, 11 ka to 9.9 ka BP. *Permafrost and Periglacial Processes*, **7**, 391–408.

Fitzpatrick, E. A. (1956). An indurated soil horizon formed by permafrost. *Journal of Soil Science*, **7**, 248–254.

Flemal, R. C. (1976). Pingos and pingo scars: their characteristics, distribution, and utility in reconstructing former permafrost environments. *Quaternary Research*, **6**, 37–53.

Flemal, R. C., Hinkley, K. C., Hesler, J. L. (1973). The Dekalb Mounds; a possible Pleistocene (Woodfordian) pingo field in north central Illinois. *Geological Society of America, memoir* 136, 229–250.

Forbes, D. L. (1989). Maximum storm surge elevations in the Tuktoyaktuk region of the Canadian Beaufort Sea. *Arctic*, **42**, 182.

Ford, D. C. (1984). Karst groundwater activity and landform genesis in modern permafrost regions of Canada. In: LaFleur, R. G., ed., *Groundwater as a Geomorphic Agent*, Allen & Unwin, London, pp. 340–350.

Ford, D. C. (1987). Effects of glaciations and permafrost upon the development of karst in Canada. *Earth Surface Processes and Landforms*, **12**, 507–521.

Ford, D. C. (1996). Karst in a cold climate: effects of glaciation and permafrost conditions upon the karst landfrom systems of Canada. In: McCann, S. B., Ford, D. C., eds., *Geomorphology Sans Frontieres*. John Wiley & Sons, Chichester, pp. 153–179.

Ford, D. C., Williams, P. W. (1989). *Karst Geomorphology and Hydrology*. Unwin Hyman, London, 601 pp.

Forman, S. L., Pierson, J. (2002). Late-Pleistocene luminescence chronology of loess deposition in the Missouri and Mississippi river valleys, United States. *Palaeogeography, Palaeoclimatology, Palaeoecology*, **186**, 25–46.

Forman, S. L., Oglesby, R., Markgraf, V., Stafford, T. (1995). Palaeoclimatic significance of Late-Quaternary eolian deposition on the Piedmont and High Plains, central United States. *Global and Planetary Change*, **11**, 35–55.

Fortier, D., Allard, M. (2005). Frost-cracking conditions, Bylot Island, Eastern Canadian Arctic Archipelago. *Permafrost and Periglacial Processes*, **16**, 145–161.

Fox, C. A. (1994). Micromorphology of permafrost-affected soils. In: Kimble, J. M., Ahrens, R. J., eds., *Proceedings, Meeting on Classification, Correlation and Management of Permafrost-Affected Soils*, July. USDA, SCS, NSSC, Lincoln, Nebraska, pp. 51–62.

Fox, C. A., Protz, R. (1981). Definition of fabric distributions to characterize the rearrangement of soil particles in the Turbic Cryosols. *Canadian Journal of Soil Science*, **61**, 29–34.

Francou, B. (1988). Eboulis stratifiés dans les Hautes Andes Centrales du Pérou. *Zeitschrift für Geomorphologie*, **32**, 47–76.

Francou, B. (1990). Stratification mechanisms in slope deposits in high subequatorial mountains. *Permafrost and Periglacial Processes*, **1**, 249–263.

Fraser, T. A., Burn, C. R. (1997). On the nature and origin of "muck" deposits in the Klondike area, Yukon Territory. *Canadian Journal of Earth Sciences*, **34**, 1333–1344.

French, H. M. (1970). Soil temperatures in the active layer, Beaufort Plain. *Arctic*, **23**, 229–239.

French, H. M. (1971a). Ice-cored mounds and patterned ground, southern Banks Island, Western Canadian Arctic. *Geografiska Annaler*, **53A**, 32–38.

French, H. M. (1971b). Slope asymmetry of the Beaufort Plain, northwest Banks Island, NWT, Canada. *Canadian Journal of Earth Sciences*, **8**, 717–731.

French, H. M. (1972a). Asymmetrical slope development in the Chiltern Hills. *Biuletyn Peryglacjalny*, **21**, 51–73.

French, H. M. (1972b). Proglacial drainage of northwest Banks Island, District of Franklin, NWT. *Musk-Ox*, **10**, 26–31.

French, H. M. (1973). Cryopediments on the Chalk of Southern England. *Biuletyn Peryglacjalny*, **22**, 149–156.

French, H. M. (1974a). Mass-wasting at Sachs Harbour, Banks Island, NWT, Canada. *Arctic and Alpine Research*, **6**, 71–78.

French, H. M. (1974b). Active thermokarst processes, eastern Banks Island, Western Canadian Arctic. *Canadian Journal of Earth Sciences*, **11**, 785–794.

French, H. M. (1975a). Man-induced thermokarst, Sachs Harbour airstrip, Banks Island, NWT. *Canadian Journal of Earth Sciences*, **12**, 132–144.

French, H. M. (1975b). Pingo investigations and terrain disturbance studies, Banks Island, District of Franklin. *Geological Survey of Canada*, paper 75-1, 459–464.

French, H. M. (1976a). *The Periglacial Environment*. Longman, London, 308 pp.

French, H. M. (1976b). Geomorphological processes and terrain disturbance studies, Banks Island, District of Franklin. *Geological Survey of Canada*, paper 76-1A, 289–292.

French, H. M. (1976c). Pingo investigations, Banks Island, District of Franklin. *Geological Survey of Canada*, paper 76-1A, 235–238.

French, H. M. (1978). Terrain and environmental problems of Canadian Arctic oil and gas exploration. *Musk-Ox*, **21**, 11–17.

French, H. M. (1980). Terrain, land use and waste-drilling fluid disposal problems, Arctic Canada. *Arctic*, **33**, 794–806.

French, H. M. (1981). Sump studies IV: permafrost terrain disturbances adjacent to exploratory wellsites, Northern Yukon Territory. *Environmental Studies 19*, Indian and Northern Affairs Canada, Ottawa, 41 pp.

French, H. M. (1984). Terrain and environmental problems associated with exploratory drilling, Northern Canada. In: *Permafrost, Fourth International Conference – Final Proceedings*. National Academy Press, Washington, DC, 20418, pp. 129–132.

French, H. M. (1985). Surface disposal of waste drilling fluids, Ellef Ringnes Island, NWT: short-term observations. *Arctic*, **38**, 292–302.

French, H. M. (1986). Periglacial involutions and mass displacement structures, Banks Island, Canada. *Geografiska Annaler*, **68A**(3), 167–174.

French. H. M. (1987). Periglacial processes and landforms in the Western Canadian Arctic. In: Boardman, J., ed., *Periglacial Processes and Landforms in Britain and Ireland*. Cambridge University Press, Cambridge, pp. 27–43.

French, H. M. (1988). The active layer. In: Clark, M. J., ed., *Advances in Periglacial Geomorphology*. John Wiley & Sons, Chichester, pp. 151–177.

French, H. M. (1996). *The Periglacial Environment*, 2nd edn. Addison Wesley Longman, Harlow, 341 pp.

French, H. M. (1998). Permafrost, real estate and climate change: the case of Thompson, Northern Manitoba, Canada. *Biuletyn Peryglacjalny*, **37**, 35–44.

French, H. M. (1999). Past and present permafrost as an indicator of climate change. *Polar Research*, **18**, 269–274.

French, H. M. (2000). Does Lozinski's periglacial realm exist today? A discussion relevant to modern usage of the term "periglacial". *Permafrost and Periglacial Processes*, **11**, 35–42.

French, H. M. (2003). The development of periglacial geomorphology: 1-up to 1965. *Permafrost and Periglacial Processes*, **14**, 29–60.

French, H. M., Demitroff, M. (2001). Cold-climate origin of the enclosed depressions and wetlands ("spungs") of the Pine Barrens, Southern New Jersey, USA. *Permafrost and Periglacial Processes*, **12**, 337–350.

French, H. M., Dutkiewicz, L. (1976). Pingos and pingo-like forms, Banks Island, Western Canadian Arctic. *Biuletyn Peryglacjalny*, **26**, 211–222.

French, H. M., Egginton, P. (1973). Thermokarst development, Banks Island, Western Canadian Arctic. In: *Permafrost: The North American Contribution to the Second International Conference, Yakutsk, USSR*, publication 2115. National Academy of Sciences, Washington, DC, pp. 203–212.

French, H. M., Egorov, I. E. (1998). 20th century variations in the southern limit of permafrost near Thompson, Northern Manitoba, Canada. In: Lewkowicz, A. G., Allard, M., eds., *Permafrost, Proceedings of the Seventh International Conference on Permafrost*, 23–27 June, Yellowknife, Canada. Université Laval, Québec, Collection Nordicana no. 57, pp. 297–304.

French, H. M., Gozdzik, J. (1988). Pleistocene epigenetic and syngenetic frost fissures, Belchatow, Poland. *Canadian Journal of Earth Sciences*, **25**, 2017–2027.

French, H. M., Guglielmin, M. (1999). Observations on the ice-marginal periglacial geomorphology of Terra Nova Bay, Northern Victoria Land, Antarctica. *Permafrost and Periglacial Processes*, **10**, 331–347.

French, H. M., Guglielmin, M. (2000). Cryogenic weathering of granite, Northern Victoria Land, Antarctica. *Permafrost and Periglacial Processes*, **11**, 305–314.

French, H. M., Guglielmin, M. (2002a). Observations on granite weathering phenomena, Mount Keinath, Northern Victoria Land, Antarctica. *Permafrost and Periglacial Processes*, **13**, 231–236.

French, H. M., Guglielmin, M. (2002b). Cryogenic grooves on a granite nunatak, Northern Victoria Land, Antarctica. *Norsk Geografisk Tidsskrift*, **56**, 112–116.

French, H. M., Harry, D. G. (1983). Ground ice conditions and thaw lakes, Sachs River Lowlands, Banks Island, Canada. In: Poser, H., Schunke, E., eds., *Mesoformen des Reliefs im heutigen Periglazialraum*, Abhandlungen der Akademie der Wissenschaften in Gottingen, Math.-Phys. Klasse, no. 35, 70–81.

French, H. M., Harry, D. G. (1988). Nature and origin of ground ice, Sandhills Moraine, southwest Banks Island, Western Canadian Arctic. *Journal of Quaternary Science*, **3**, 19–30.

French, H. M., Harry, D. G. (1990). Observations on buried glacier ice and massive segregated ice, western Arctic coast, Canada. *Permafrost and Periglacial Processes*, **1**, 31–43.

French, H. M., Harry, D. G. (1992). Pediments and cold-climate conditions, Barn Mountains, unglaciated northern Yukon, Canada. *Geografiska Annaler*, **74A**, 145–157.

French, H. M., Pollard, W. H. (1986). Ground-ice investigations, Klondike District, Yukon Territory. *Canadian Journal of Earth Sciences*, **23**, 550–560.

French, H. M., Slaymaker, O. (1993). Canada's cold landmass. In: French, H. M., Slaymaker, O., eds., *Canada's Cold Environments*. McGill-Queen's Press, Montreal, pp. 3–27.

French, H. M., Thorn, C. E. (2006). The changing nature of periglacial geomorphology. *Géomorphologie: relief, processus, environnement*, **2006**(3), 165–174.

French, H. M., Harry, D. G., Clark, M. J. (1982). Ground ice stratigraphy and late Quaternary events, southwest Banks Island, Western Canadian Arctic. In: French, H. M., ed., *Proceedings Fourth Canadian Permafrost Conference*, Calgary, Alberta. National Research Council of Canada, Ottawa, pp. 81–90.

French, H. M., Harris, S. A., van Everdingen, R. O. (1983). The Klondike and Dawson. In: French, H. M., Heginbottom, J. A., eds., *Northern Yukon Territory and Mackenzie Delta, Canada. Guidebook to Permafrost and Related Features*. Fourth International Conference on Permafrost, Guidebook 3, Division of Geological and Geophysical Surveys, Fairbanks, Alaska, pp. 35–63.

French, H. M., Bennett, L., Hayley, D. W. (1986). Ground ice conditions near Rea Point and on Sabine Peninsula, eastern Melville Island. *Canadian Journal of Earth Sciences*, **23**, 1389–1400.

French, H. M., Demitroff, M., Forman, S. L. (2003). Evidence for Late-Pleistocenene permafrost in the New Jersey Pine Barrens (latitude 39°N), Eastern USA. *Permafrost and Periglacial Processes*, **14**, 259–274.

French, H. M., Demitroff, M., Forman, S. L. (2005). Evidence for Late-Pleistocene thermokarst in the New Jersey Pine Barrens (latitude 39°N), Eastern USA. *Permafrost and Periglacial Processes*, **16**, 173–186.

Friedman, J. D., Johansson, C. E., Oskarsson, N., Svensson, H., Thorarinsson, S., Williams, R. S. (1971). Observations on Ielandic polygon surfaces and palsa areas. Photo interpretation and field studies. *Geografiska Annaler*, **53A**, 115–145.

Fristrup, B. (1953). Wind erosion within the Arctic deserts. *Geografisk Tidsskrift*, **52**, 51–56.

Froese, D. G., Barendregt, R. W., Enkin, R. J., Baker, J. (2000). Paleomagnetic evidence for multiple Late Pliocene – Early Pleistocene glaciations in the Klondike area, Yukon Territory. *Canadian Journal of Earth Sciences*, **37**, 863–877.

Froese, D. G., Westgate, J. A., Preece, S. J., Reyes, A. V. (2006). The ancient cryosphere? Pliocene appearance and relict 700,000 year old permafrost in Yukon Territory. Poster, J. R. Mackay 90th Birthday Symposium, February 17, 2006, University of British Columbia, Vancouver.

Fry, E. J. (1927). The mechanical action of crustaceous lichens on substrata of shale, schist, limestone and obsidian. *Annals of Botany*, **41**, 437–460.

Fujino, K., Horiguchi, K., Shinbori, M., Kato, K. (1983). Analysis and characteristics of cores from a massive ice body in Mackenzie Delta, NWT, Canada. In: *Permafrost, Proceedings of the Fourth International Conference on Permafrost*, 17–22 July, Fairbanks, Alaska. National Academy Press, Washington, DC, pp. 316–321.

Fujino, K., Sato, S., Matsuda, K., et al. (1988). Characteristics of the massive ground ice body in the western Canadian Arctic. In: Senneset, K., ed., *Permafrost, Proceedings of the Fifth International Conference on Permafrost*, 2–5 August. Tapir, Trondheim, vol. 1, pp. 143–147.

Fukuda, M. (1983). The pore-water pressure in porous rocks during freezing. In: *Permafrost, Proceedings of the Fourth International Conference on Permafrost*, 17–22 July, Fairbanks, Alaska. National Academy Press, Washington, DC, vol. 1, pp. 322–327.

Fulton, R. J., ed. (1989). *Quaternary Geology of Canada and Greenland*. Geology of Canada no. 1. Geological Survey of Canada, Ottawa, 839 pp.

Furrer, G., Bachmann, F., Fitze, P. (1971). Erdströme als Formelemente von Soliflukhousdecken im Raum Munt Chevagi/Munt Buffalora. *Ergebnisse der Wissenschaftern Untersuchungen im Schwizerischen Nationalpark*, **11**, 188–269.

Fyles, J. G. (1962). Physiography. In: Thorsteinsson, R., Tozer, E. T., *Banks, Victoria and Stefansson Islands, Arctic Archipelago*. Geological Survey of Canada, memoir 330, pp. 8–17.

Fyodorov-Davydov, D. G., Sorokovikov, V. A., Ostoumov, V. E., Kholodov, A. L., Mitroshin, I. A., Mergelov, N. S., Davydov, S. P., Zimov, S. A., Davydova, A. I. (2004). Spatial and temporal observations of seasonal thaw in the Northern Kolyma Lowlands. *Polar Geography*, **28**, 308–325.

Galloway, R. (1985). Fossil ice wedges in Patagonia and their palaeoclimatic significance. *Acta Geocriogenica (Mendoza, Argentina)*, **3**, 106–113.

Gamper, M. (1983). Controls and rates of movement of solifluction lobes in the Eastern Swiss Alps. In: *Permafrost, Proceedings of the Fourth International Permafrost Conference*, 17–22 July, Fairbanks, Alaska. National Academy Press, Washington, DC, vol. 1, pp. 328–333.

Gasanov, S. S. (1969). *Structure and History of Formation of Permafrost in Eastern Chukotka.* Nauka, Moscow, 168 pp. (in Russian).

Gasanov, S. S. (1978). Cryolithogenesis as an independent hydrothermal type of sedimentary process [in Russian]. In: *Permafrost, Proceedings of the Third International Conference on Permafrost*, 10–13 July, Edmonton, Alberta, Canada. National Research Council of Canada, Ottawa, vol. 1, pp. 271–276. Translated (1980) by the National Research Council of Canada, publication 18119, Ottawa, pp. 225–244.

Geikie, J. (1894). *The Great Ice Age*, 3rd edn. Stanford, London, 850 pp.

Gerard, R. (1990). Hydrology of floating ice. National Hydrology Research Institute, science report no. 1, Saskatoon, pp. 103–134.

Gerasimov, I. P., Markov, K. K. (1968). *Permafrost and Ancient Glaciation.* Defence Research Board, Ottawa, translation T499R, pp. 11–19.

Geukens, F. (1947). De asymmetrie der droge dalen van Haspengouw. *Natuurwetenschappelijk. Tijdschrift*, **29**, 13–18.

Gibbard, P. L. (1988). The history of the great northwest European rivers during the past three million years. *Philosophical Transactions of the Royal Society of London*, **B318**, 559–602.

Gilichinsky, D., ed. (1994). *Viable Microorganisms in Permafrost.* Puschino Scientific Center, Russian Academy of Sciences, Institute of Soil Science and Photosynthesis, Puschino, 115 pp.

Gilichinsky, D. (2002a). Permafrost. In: Bitton, G., ed., *Encyclopedia of Environmental Microbiology.* John Wiley & Sons, New York, pp. 2367–2385.

Gilichinsky, D. (2002b). Planetary permafrost and astrobiology. *Frozen Ground*, Newsletter of the International Association of Permafrost (UNIS, Norway), **26**, 52–54.

Gilichinsky, D., Wagener, S. (1995). Microbial life in permafrost: a historical review. *Permafrost and Periglacial Processes*, **6**, 243–250.

Gilichinsky, D., Wagener, S., Vishnevetskaya, T. A. (1995). Permafrost microbiology. *Permafrost and Periglacial Processes*, **6**, 281–291.

Glasby, G. P., McPherson, J. G., Kohn, B. P., Johnson, J. H., Kemp, J. R., Freeman, A. G., Tricker, M. J. (1981). Desert varnish in southern Victoria Land, Antarctica. *New Zealand Journal of Geology and Geophysics*, **24**, 389–397.

Gleason, K. J., Krantz, W. B., Caine, N., George, J. H., Gunn, R. D. (1986). Geometrical aspects of sorted patterned ground in recurrently frozen soil. *Science*, **232**, 216–220.

Gloriad, A., Tricart, J. (1952). Etude statistique des vallées asymmétriques de la feuille St. Pol, au 1:50,000. *Revue de Géomorphologie Dynamique*, **3**, 88–98.

Goering, D. (1998). Experimental investigation of air convection embankments for permafrost-resistant roadway design. In: Lewkowicz, A. G., Allard, M., eds., *Permafrost, Proceedings of the Seventh International Conference on Permafrost*, 23–27 June, Yellowknife, Canada. Centre d'études nordiques, Université Laval, Québec, Collection Nordicana no. 57, pp. 319–326.

Goering, D. (2000). Passive cooling for foundation stabilization and containment. In: Senneset, K., ed., *Proceedings, International Workshop on Permafrost Engineering*, Longyearbyen, Svalbard, Norway, 18–21 June. Norwegian University of Science and Technology (NTNU)/University Courses on Svalbard (UNIS), pp. 37–56.

Goering, D. (2003). Thermal response of air convection embankments to ambient temperature fluctuations. In: Phillips, M., Springman, S. M., Arenson, L. U., eds., *Permafrost: Proceedings of the Eighth International Conference on Permafrost*, 21–25 July, Zurich, Switzerland. Balkema, Lisse, vol. 1, pp. 291–296.

Gold, L. W., Lachenbruch, A. (1973). Thermal conditions in permafrost – a review of North American literature. In: *Permafrost: The North American Contribution to the Second International*

Conference, Yakutsk, USSR, publication 2115. National Academy of Sciences, Washington, DC, pp. 3–26.

Gold, L. W., Johnston, G. H., Slusarchuk, W. A., Goodrich, L. E. (1972). Thermal effects in permafrost. In: *Proceedings, Canadian Northern Pipeline Conference*, Ottawa, Associate Committee on Geotechnical Research, National Research Council of Canada, technical memorandum 104, pp. 25–45.

Good, T. R., Bryant, I. D. (1985). Fluvio-aeolian sedimentation-an example from Banks Island, NWT, Canada. *Geografiska Annaler*, **67**, 33–46.

Goodrich, L. E. (1982). The influence of snow cover on the ground thermal regime. *Canadian Geotechnical Journal*, **19**, 421–432.

Goodrich, L. E., Plunkett, J. C. (1990). Performance of heat pump chilled foundations. In: French, H. M., ed., *Proceedings, Fourth Canadian Permafrost Conference*. Calgary, National Research Council Canada, Ottawa, pp. 409–418.

Gorbunov, A. P. (1988a). The alpine permafrost zone of the USSR. In: Senneset, K., ed., *Permafrost, Proceedings of the Fifth International Conference on Permafrost*, 2–5 August. Tapir, Trondheim, vol. 1, pp. 154–158.

Gorbunov, A. P. (1988b). *Rock Glaciers*. Siberian Division, Soviet Academy of Sciences, Novosibirsk, Nauka, 108 pp. (in Russian).

Gorbunov, A. P., Tytkov, C. H. (1989). *Rock Glaciers in the Mid-Asia Mountains*. Siberian Division, Soviet Academy of Sciences, Yakutsk, Nauka, 163 pp. (in Russian).

Goudie, A. (1974). Further experimental investigation of rock weathering by salt and other mechanical processes. *Zeitschrift für Geomorphologie*, suppl. **21**, 1–12.

Goudie, A. (1995). *The Changing Earth, Rates of Geomorphological Processes*. Blackwell, Oxford, 302 pp.

Gowan, R. J., Dallimore, S. R. (1990). Ground ice associated with granular deposits in the Tuktoyaktuk Coastlands area, NWT. In: *Permafrost-Canada, Proceedings of the Fifth Canadian Permafrost Conference*, National Research Council of Canada – Centre d'études nordiques, Université Laval, Collection Nordicana no. 54, pp. 283–290.

Gozdzik, J. (1973). Origin and stratigraphic position of periglacial structures in Middle Poland. *Acta Geographica Łódziensia*, **31**, 104–117.

Granberg, H. B. (1973). Indirect mapping of the snowcover for permafrost prediction at Schefferville, Québec. In: *Permafrost: The North American Contribution to the Second International Conference, Yakutsk, USSR*, publication 2115. National Academy of Sciences, Washington, DC, pp. 113–120.

Grantz, A., White, D. E., Whithead, H. C., Tagg, A. R. (1962). Saline springs, Copper River lowland, Alaska. *Bulletin, American Association of Petroleum Geologists*, **46**, 1990–2002.

Grave, N. A., Nekrasov, I. A. (1961). Some observations of thermokarst in the vicinity of the settlement of Anadyr. *Problemy Severa*, **94**, 157–164 (in Russian; English translation, 1962, *Problems of the North*, **4**, 165–172).

Gravis, G. F. (1969). Fossil slope deposits in the northern Arctic asymmetrical valleys. *Biuletyn Peryglacjalny*, **20**, 239–257.

Gray, J. T. (1973). Geomorphic effects of avalanches and rock falls on steep mountain slopes in the central Yukon Territory. In: Fahey, B. D., Thompson, R. D., eds., *Research in Polar and Alpine Geomorphology, Proceedings, Third Guelph Symposium on Geomorphology*. Geo Abstracts, Norwich, pp. 107–117.

Grechishchev, S. E. (1970). *Basis of Method for Predicting Thermal Stresses and Deformations in Frozen Soils* (in Russian). Ministerstvo Geologii SSSR, Vsesoyuznyi Nauchno-Issledovatel'skii Institut Gidrogeologii I Inzhenernoi Geologii (VSEGINGEO), Moscow. (English translation, 1976, National Research Council Canada, technical translation 1986).

Gregory, K. J., Walling, D. E. (1973). *Drainage Basin Form and Process: A Geomorphological Approach*. Edward Arnold, London, 458 pp.

Griggs, D. T. (1936). The factor of fatigue in rock exfoliation. *Journal of Geology*, **44**, 783–796.

Grimberbieux, J. (1955). Origine et asymmétrie des vallees sèches de Hesbaye. *Annales, Société Géologique de Belgique*, **78**, 267–286.

Gripp, K. (1926). Über frost und strukturboden auf Spitzbergen. *Zeitschrift Gesellschaft für Erd-kunde (Berlin)*, 351–354.

Grosse, G., Schirrmeister, L., Kunitsky, V. V., Hubberten, H.-W. (2005). The use of CORONA images in remote sensing of periglacial geomorphology: an illustration from the NE Siberian coast. *Permafrost and Periglacial Processes*, **16**, 163–172.

Grosso, S. A., Corte, A. E. (1989). Pleistocene ice-wedge casts at 34°S, eastern Andes piedmont, southwest of South America. *Geografiska Annaler*, **71A**, 125–136.

Grosso, S. A., Corte, A. E. (1991). Cryoplanation surfaces in the Central Andes at latitude 35°S. *Permafrost and Periglacial Processes*, **2**, 49–58.

Guglielmin, M., Smiraglio, C., eds. (1997). Rock glacier inventory of the Italian Alps. Italian Glaciological Committee Archive, Italian Glaciological Committee, Torino, 3, 103 pp.

Guillien, Y. (1951). Les grèzes litées de Charente. *Revue Géographique de Pyrénées et de Sud-Ouest*, **22**, 154–162.

Guo, D., Huang, Y., Xu, S., Zhang, L. (1982). Taliks in the Buqu river valley area, northern slope of Tangual Mountain. In: *Proceedings, First Chinese Conference on Glaciology and Geocryology*. China Science Press, Beijing, pp. 11–18 (in Chinese).

Guthrie, R. D. (1990). *Frozen Fauna of the Mammoth Steppe: A Story of Blue Babe*. University of Chicago Press, Chicago, IL, 323 pp.

Guthrie, R. D. (2001). Origin and causes of the mammoth steppe: a story of cloud cover, woolly mammal tooth pits, buckles, and inside-out Beringia. *Quaternary Science Reviews*, **20**, 549–574.

Hack, J. T., Goodlett, J. C. (1960). Geomorphology and forest ecology of a mountain region in the Central Appalachians. United States Geological Survey, professional paper 347, 65 pp.

Haeberli, W. (1973). Die basis temperatur der winterlichen schneedecke als möglicher indikator für die verbreitung von permafrost in den Alpen. *Zeitschrift für Gletschologie und Glazialgeologie*, **9**, 221–227.

Haeberli, W. (1978). Special aspects of high mountain permafrost methology and zonation in the Alps. In: *Permafrost, Proceedings of the Third International Conference on Permafrost*, 10–13 July, Edmonton, Alberta, Canada. National Research Council of Canada, Ottawa, vol. 1, pp. 379–384.

Haeberli, W. (1985). Creep of mountain permafrost; internal structure and flow of alpine rock glaciers. *Mittgeilugen der Versuchsanstalt fur Wassenbau Hydrologie and Glaziologie*, no. 77, 142 pp.

Haeberli, W., Burn, C. R. (2002). Natural hazards in forests: glacier and permafrost effects as related to climate change. In: Sidle, R. C., ed., *Environmental Change and Geomorphic Hazards in Forests*. IUFRO Research Series 9, CABI Publishing, Wallingford, pp. 167–202.

Haesaerts, P. (1983). Stratigraphic distribution of periglacial features indicative of permafrost in the Upper Pleistocene loesses of Belgium. In: *Permafrost, Proceedings of the Fourth International Conference on Permafrost*, 17–22 July, Fairbanks, Alaska. National Academy Press, Washington, DC, vol. 1, pp. 421–426.

Haesaerts, P., Van Vliet-Lanoë, B. (1981). Phénomènes périglaciaires observés à Maisières-Canal, à Harmignies et à Rocourt (Belgique). *Biuletyn Peryglacjalny*, **28**, 291–324

Hall, K. (1995). Freeze–thaw weathering: the cold region "panacea". *Polar Geography and Geology*, **19**, 79–87.

Hall, K. (1997a). Rock temperatures and implications for cold region weathering: 1. New data from Viking Valley, Alexander Island (Antarctica). *Permafrost and Periglacial Processes*, **8**, 69–90.

Hall, K. (1997b). Observations on "cryoplanation" benches in Antarctica. *Antarctic Science*, **9**, 181–187.

Hall, K. (1998). Rock temperatures and implications for cold region weathering: II. New data from Rothera, Adelaide Island (Antarctica). *Permafrost and Periglacial Processes*, **9**, 47–55.

Hall, K. (1999). The role of thermal stress fatigue in the breakdown of rock in cold regions. *Geomorphology*, **31**, 47–63.

Hall, K. (2004). Evidence for freeze–thaw events and their implications for rock weathering in Northern Canada. *Earth Surface Processes and Landforms*, **29**, 43–58.

Hall, K., André, M.-F. (2001). New insights into rock weathering as deduced from high frequency rock temperature data: an Antarctic study. *Geomorphology*, **41**, 23–35.

Hall, K., André, M.-F. (2003). Rock thermal data at the grain scale: applicability to granular disintegration in cold environments. *Earth Surface Processes and Landforms*, **28**, 823–836.

Hall, K., Otte, W. (1990). A note on biological weathering on nunataks of the Juneau Icefield, Alaska. *Permafrost and Periglacial Processes*, **1**, 189–196.

Hall, K., Thorn, C. E., Matsuoka, N., Prick, A. (2002). Weathering in cold regions: some thoughts and perspectives. *Progress in Physical Geography*, **26**, 577–604.

Hallet, B. (1990). Self-organisation in freezing soils: from microscopic ice lenses to patterned ground. *Canadian Journal of Physics*, **68**, 842–852.

Hallet, B., Prestrud, S. (1986). Dynamics of periglacial sorted circles in Western Spitzbergen. *Quaternary Research*, **26**, 81–99.

Hallet, B., Anderson, S. P., Stubbs, C. W., Gregory, E. C. (1988). Surface soil displacements in sorted circles, Western Spitzbergen. In: Senneset, K., ed., *Permafrost, Proceedings of the Fifth International Conference on Permafrost*, 2–5 August. Tapir, Trondheim, vol. 1, pp. 770–775.

Hallet, B., Walder, J. S., Stubbs, C. W. (1991). Weathering by segregation ice growth in microcracks at sustained subzero temperatures: verification from an experimental study using acoustic emissions. *Permafrost and Periglacial Processes*, **2**, 283–300.

Hamelin, L.-E. (1964). La famille du mot "périglaciaire". *Biuletyn Peryglacjalny*, **14**, 133–152.

Hamilton, T. D., Galloway, J. P., Koster, E. A. (1988). Late Wisconsin eolian activity and related alluviation, central Kobuk River Valley. In: Galloway, J. P., Hamilton, T. D., eds., *Geologic Studies in Alaska by the U.S. Geological Survey During 1987*, United States Geological Survey, circular 1016, pp. 39–43.

Hansell, R. I. C., Scott, P. A., Staniforth, R., Svoboda, J. (1983). Permafrost development in the intertidal zone at Churchill, Manitoba. *Arctic*, **36**, 198–203.

Hanson, S., Hoelzle, M. (2004). The thermal regime of the active layer at the Murtel rock glacier based on data from 2002. *Permafrost and Periglacial Processes*, **15**, 273–282.

Hardy BBT Ltd (1988). Handling and disposal of waste drilling fluids from on-land sumps in the Northwest Territories and Yukon. Environmental Studies Research Funds, Indian Affairs and Northern Development, Ottawa, report no. 093, 58 pp.

Harington, C. R. (2003). *Annotated Bibliography of Quaternary Vertebrates of Northern North America – With Radiocarbon Dates*. University of Toronto Press, Toronto, 539 pp.

Harper, J. R., Henry, R. F., Stewart, G. G. (1988). Maximum storm surge elevations in the Tuktoyaktuk region of the Canadian Beaufort Sea. *Arctic*, **41**, 48–52.

Harris, C. (1972). Processes of soil movement in turf-banked solifluction lobes, Okstindan, northern Norway. In: *Polar Geomorphology*, Insitute of British Geographers Special Publication, no. 4, pp. 155–174.

Harris, C. (1977). Engineering properties, groundwater conditions, and the nature of soil movement on a solifluction slope in North Norway. *Quaternary Journal of Engineering Geology*, **10**, 27–43.

Harris, C. (1981). *Periglacial Mass-Wasting: A Review of Research*. British Geomorphological Research Group, Research Monograph 4. Geo Abstracts, Norwich, 204 pp.

Harris, C. (1985). Geomorphological applications of soil micromorphology with particular reference to periglacial sediments and processes. In: Richards, K. S., Arnett, R. R., Ellis, S., eds., *Geomorphology and Soils*. Allen & Unwin, London, pp. 219–232.

Harris, C. (1987). Mechanisms of mass movement in periglacial environments. In: Anderson, M. G., Richards, K. S., eds., *Slope Stability*. John Wiley & Sons, Chichester, pp. 531–559.

Harris, C., Lewkowicz, A. G. (1993a). Form and internal structure of active-layer detachment slides, Fosheim Peninsula, Ellesmere Island, Northwest Territories, Canada. *Canadian Journal of Earth Sciences*, **30**, 1708–1714.

Harris, C., Lewkowicz, A. G. (1993b). Micromorphological investigations of active-layer detachment slides, Ellesmere Island, Canadian Arctic. In: *Permafrost, Proceedings of the Sixth International Conference on Permafrost*, 5–9 July, Beijing, China. South China University of Technology Press, pp. 232–237.

Harris, C., Lewkowicz, A. G. (2000). An analysis of the stability of thawing slopes, Ellesmere Island, Nunavut, Canada. *Canadian Geotechnical Journal*, **37**, 449–462.

Harris, C., Davies, M. C. R. (2000). Gelifluction: observations from large-scale laboratory simulations. *Arctic, Antarctic and Alpine Research*, **32**, 202–207.

Harris, C., Haeberli, W. (2003). Warming permafrost in the mountains of Europe. *World Meteorological Organization, Bulletin*, **52**, 1–5.

Harris, C., Wright, M. D. (1980). Some last glaciation drift deposits near Pontypridd, South Wales. *Geological Journal*, **15**, 7–20.

Harris, C., Gallop, M., Coutard, J.-P. (1993). Physical modelling of gelifluction and frost creep: some results of a large scale laboratory experiment. *Earth Surface Processes and Landforms*, **18**, 383–398.

Harris, C., Davies, M. R. C., Coutard, J.-P. (1995). Laboratory simulation of periglacial solifluction: significance of porewater pressures, moisture contents and undrained shear strengths during soil thawing. *Permafrost and Periglacial Processes*, **6**, 293–312.

Harris, C., Davies, M. C. R., Coutard, J.-P. (1997). Rates and processes of periglacial solifluction: an experimental approach. *Earth Surface Processes and Landforms*, **22**, 849–868.

Harris, C., Rea, B. R., Davies, M. C. R. (2000). Geotechnical centrifuge modelling of gelifluction processes: validation of a new approach to periglacial slope studies. *Annals of Glaciology*, **31**, 263–268.

Harris, C., Haeberli, W., Vonder Muhll, D., King, L. (2001). Permafrost monitoring in the high mountains of Europe: the PACE Project in its global context. *Permafrost and Periglacial Processes*, **12**, 3–12.

Harris, S. A. (1979). Ice caves and permafrost zones in southwest Alberta. *Erdkunde*, **33**, 61–70.

Harris, S. A. (1983). Cold air drainage west of Fort Nelson, British Columbia. *Arctic*, **35**, 539–541.

Harris, S. A., Brown, R. J. E. (1978). Plateau Mountain: a case study of alpine permafrost in the Canadian Rocky Mountains. In: *Permafrost, Proceedings of the Third International Conference on Permafrost*, 10–13 July, Edmonton, Alberta, Canada. National Research Council of Canada, Ottawa, vol. 1, pp. 385–391.

Harris, S. A., Brown, R. J. E. (1982). Permafrost distribution along the Rocky Mountains in Alberta. In: French, H. M., ed., *Proceedings, Fourth Canadian Permafrost Conference*, Calgary, Alberta. National Research Council Canada, Ottawa, pp. 59–67.

Harris, S. A., van Everdingen, R. O., Pollard, W. H. (1983). The Dempster Highway – Dawson to Eagle Plain. In: French, H. M., Heginbottom, J. A., eds., *Northern Yukon Territory and Mackenzie Delta, Canada. Guidebook to Permafrost and Related Features*, Fourth International Conference on Permafrost, guidebook no. 3. Alaska Division of Geological and Geophysical Surveys, Fairbanks, Alaska, pp. 65–86.

Harry, D. G., French, H. M. (1983). The orientation and evolution of thaw lakes, southwest Banks Island, Canadian Arctic. In: *Permafrost, Proceedings of the Fourth International Conference on Permafrost,* 17–22 July, Fairbanks, Alaska. National Academy Press, Washington, DC, vol. 1, pp. 456–461.

Harry, D. G., Gozdzik, J. S. (1988). Ice wedges: growth, thaw transformation, and paleoenvironmental significance. *Journal of Quaternary Science*, **3**, 39–55.

Harry, D. G., McInnes, K. (1988). The effect of forest fires on permafrost terrain stability, Little Chicago–Travaillant Lake area, Mackenzie Valley, NWT. In: *Geological Survey of Canada*, paper 88-1D, 91–94.

Harry, D. G., French, H. M., Clark, M. J. (1983). Coastal conditions and processes, Sachs Harbour, Banks Island, Western Canadian Arctic. *Zeitschrift für Geomorphologie*, suppl. **47**, 1–26.

Harry, D. G., French, H. M., Pollard, W. H. (1985). Ice wedges and permafrost conditions near King Point, Beaufort Sea coast, Yukon Territory. *Geological Survey of Canada,* paper 85-1A, 111–116.

Harry, D. G., French, H. M., Pollard, W. H. (1988). Massive ground ice and ice-cored terrain near Sabine Point, Yukon Coastal Plain. *Canadian Journal of Earth Sciences*, **25**, 1846–1856.

Hauck, C., Vonder Muhll, D. (2003). Inversion and interpretation of two-dimensional geoelectrical measurements for detecting permafrost in mountainous regions. *Permafrost and Periglacial Processes*, **14**, 305–318.

Hauck, C., Isaksen, K., Vonder Muhll, D., Sollid, J. L. (2004). Geophysical surveys designed to delineate the altitudinal limit of mountain permafrost: an example from Jotunheimen, Norway. *Permafrost and Periglacial Processes*, **15**, 191–206.

Haugen, R. K., Brown, J. (1970). Natural and man-induced disturbances of permafrost terrain. In: Coates, D. R., ed., *Environmental Geomorphology*. State University of New York, Binghampton, NY, pp. 139–149.

Hawkins, A. B., Privett, K. D. (1981). A building site on cambered ground at Radstock, Avon. *Quarterly Journal of Engineering Geology*, **14**, 151–167.

Hayley, D. W. (1982). Application of heat pipes to design of shallow foundations on permafrost. In: French, H. M., ed., *Proceedings, Fourth Canadian Permafrost Conference*, Calgary, Alberta. National Research Council Canada, Ottawa, pp. 535–544.

Hayley, D. W. (1988). Maintenance of a railway grade over permafrost in Canada. In: Senneset, K., ed., *Permafrost, Proceedings of the Fifth International Conference on Permafrost*, 2–5 August. Tapir, Trondheim, vol. 1, pp. 43–48.

Hayley, D. W., Roggensack, W. D., Jubien, W. E., Johnson, P. V. (1983). Stabilization of sinkholes on the Hudson Bay Railway. In: *Permafrost, Proceedings of the Fourth International Conference on Permafrost*, 17–22 July, Fairbanks, Alaska. National Academy Press, Washington, DC, vol. 1, pp. 468–473.

Heginbottom, J. A. (1973). Effects of surface disturbance upon permafrost. Environmental-Social Committee Northern Pipelines, Task Force on Northern Oil Development, Information Canada, report 73-16, 9 pp.

Heginbottom, J. A., Radburn, L. K. (1993). Permafrost and ground ice conditions of northwestern Canada. Map 1691, Geological Survey of Canada, scale 1:1,000,000.

Helbig, K. (1965). Asymmetrische eiszeittaler in Süddeutschland und Osterreich. *Wurzburger Geographisches Arbeiten*, **14**, 103 pp.

Hengeveld, H. G. (2000). *Projections for Canada's Future Climate*. Meteorological Service of Canada, Environment Canada, Downsview, Ontario, 27 pp.

Henriksen, M. A., Mangerud, J., Matiouchkov, A., Paus, A., Svendsen, J. I. (2003). Lake stratigraphy implies an 80,000 yr delayed melting of buried dead ice in northern Russia. *Journal of Quaternary Science*, **18**, 663–679.

Hequette, A., Barnes, P. W. (1990). Coastal retreat and shoreface profile variations in the Canadian Beaufort Sea. *Marine Geology*, **91**, 113–132.

Hequette, A., Hill, P. R. (1993). Storm-generated currents and offshore sediment transport on a sandy shoreface, Tibjak Beach, Canadian Beaufort Sea. *Marine Geology*, **113**, 283–304.

Hequette, A., Ruz, M.-H. (1991). Spit and barrier island migration in the southeastern Canadian Beaufort Sea. *Journal of Coastal Research*, **7**, 677–698.

Hétu, B. (1995). Le tilage des éboulis stratifiés cryonivaux en Gaspésie (Québec, Canada); Rôle de la sédimentation nivéo-éolienne et des transits supranivaux. *Permafrost and Periglacial Processes*, **6**, 147–171.

Hétu, B., Van Steijn, H., Vandelac, P. (1994). Les coulées de pierres glacées: un nouveau type de coulées de pierraille sur les talus d'éboulis. *Géographie physique et Quaternaire*, **48**, 3–22.

Hétu, B., van Steijn, H., Bertran, P. (1995). Le rôle des coulées de pierres sèches dans la genèse d'un certain type d'éboulis stratifiés. *Permafrost and Periglacial Processes*, **6**, 173–194.

Hétu, B., Gray, J. T., Gangloff, P., Archambault, B. (2003). Postglacial talus-derived rock glaciers in the Gaspé Peninsula, Québec (Canada). In: Phillips, M., Springman, S. M., Arenson, L. U., eds., *Permafrost, Proceedings of the Eighth International Conference on Permafrost*, 21–25 July, Zurich, Switzerland. Balkema, Lisse, vol. 1, pp. 389–394.

Heuer, C. E., Krzewinski, T. G., Metz, M. C. (1982). Special thermal design to prevent thaw settlement and liquefaction. In: French, H. M., ed., *Proceedings, Fourth Canadian Permafrost Conference*, Calgary, Alberta. National Research Council Canada, Ottawa, pp. 507–522.

Hill, D. E., Tedrow, J. C. F. (1961). Weathering and soil formation in the Arctic environment. *American Journal of Science*, **259**, 84–101.

Hinkel, K. M. (2005). Discussion of paper by J. D. Pelletier entitled "Formation of oriented thaw lakes by thaw slumping". *Journal of Geophysical Research*, **110**, FO2018, doi: 10.1029/2004FJ000158.

Hinkel, K. M., Nelson, F. E. (2003). Spatial and temporal patterns of active-layer thickness at Circumpolar-Active-Layer-Monitoring (CALM) sites in northern Alaska, 1995–2000. *Journal of Geophysical Research-Atmospheres*, **108**, D2; doi: 10.1029/2001JD000927.

Hinkel, K. M., Outcalt, S. I., Taylor, A. E. (1997). Seasonal patterns of coupled flow in the active layer at three sites in northwest North America. *Canadian Journal of Earth Sciences*, **34**, 667–678.

Hinkel, K. M., Frohn, R. C., Nelson, F. E., Eisner, W. R., Beck, R. A. (2005). Morphometric and spatial analysis of thaw lakes and drained thaw-lake basins in the Western Arctic Coastal Plain, Alaska. *Permafrost and Periglacial Processes*, **16**, 327–342.

Hodgson, D. A. (1982). Surficial materials and geomorphological processes, Western Sverdrup and adjacent islands, District of Franklin. *Geological Survey of Canada*, paper 81–9, 44 pp.

Hodgson, D. A., Nixon, F. M. (1998). Ground ice volumes determined from shallow cores from western Fosheim Peninsula, Ellesmere Island, Northwest Territories. *Geological Survey of Canada, Bulletin*, no. 507, 178 pp.

Hoekstra, P. (1969). Water movement and freezing pressures. *Soil Science Society of America Proceedings*, **33**, 512–518.

Hoelzle, M. (1992). Permafrost occurrence from BTS measurements and climatic parameters in the Eastern Swiss Alps. *Permafrost and Periglacial Processes*, **3**, 143–147.

Högbom, B. (1914). Über die geologische bedeutung des frostes, Uppsala Universitet. *Geological Institute Bulletin*, **12**, 257–389.

Hollingworth, S. E., Taylor, J. H., Kellaway, G. A. (1944). Large-scale superficial structures in the Northampton Ironstone field. *Quaternary Journal Geological Society, London*, **100**, 1–44.

Holm, K. (1999). *Longyearbyen – Svalbard historisk veiviser*. Nor-Trykk Narvik AS, Narvik.

Holmes, G. W., Hopkins, D. M., Foster, H. L. (1968). Pingos in central Alaska. *United States Geological Survey Bulletin*, 1241-H, 40 pp.

Hopkins, D. M. (1949). Thaw lakes and thaw sinks in the Imuruk Lake area, Seward Peninsula, Alaska. *Journal of Geology*, **57**, 119–131.

Hopkins, D. M., Karlstrom, T. D., and others (1955). Permafrost and ground water in Alaska. United States Geological Survey, professional paper 264-F, pp. 113–146.

Hopkins, D. M., Matthews, C. E., Schweger, C. E., Young, S. B., eds. (1982). *Palaeoecology of Beringia*. Academic Press, New York, 489 pp.

Horswill, P., Horton, A. (1976). Cambering and valley bulging in the Gwash valley at Empingham, Rutland. *Philosophical Transactions of the Royal Society of London*, **A283**, 427–451.

Houghton, J. T., Jenkins, G. J., Ephraums, J. J., eds. (1990). *Climate Change. The IPCC Scientific Assessment*. Cambridge University Press, Cambridge, 364 pp.

Houghton, J. T., Callandar, B. A., Varney, S. K., eds. (1992). *Climate Change 1992. The Supplementary Report to the IPCC Scientific Assessment*. Cambridge University Press, Cambridge, 199 pp.

Hugenholtz, C. H., Lewkowicz, A. G. (2002). Morphometry and environmental characteristics of turf-banked solifluction lobes, Kluane Range, Yukon Territory, Canada. *Permafrost and Periglacial Processes*, **13**, 301–313.

Hughes, O. L. (1969). Distribution of open-system pingos in central Yukon Territory with respect to glacial limits. *Geological Survey of Canada*, paper 69-34, 8 pp.

Hughes, O. L. (1972). Surficial geology of northern Yukon Territory and northwestern District of Mackenzie, Northwest Territories. *Geological Survey of Canada*, paper 69-36, Map 1319 A.

Hughes, O. L., Veillette, J. J., Pilon, J., Hanley, P. T., van Everdingen, R. O. (1973). Terrain evaluation with respect to pipeline construction, Mackenzie transportation corridor, central part, Lat. 64°–68°N. Environmental-Social Committee Northern Pipelines, Task Force on Northern Oil Development, report no. 73-37, 74 pp.

Hume, J. D., Schalk, M. (1964). The effects of beach borrow in the Arctic. *Shore and Beach*, April, 5 pp.

Hume, J. D., Schalk, M. (1967). Shoreline processes near Barrow, Alaska; a comparison of the normal and the catastrophic. *Arctic*, **20**, 86–103.

Hume, J. D., Schalk, M., Hume, P. W. (1972). Short-term climatic changes and coast erosion, Barrow, Alaska. *Arctic*, **25**, 272–279.

Humlum, O. (1996). Origin of rock glaciers: observations from Mellemfjord, Disko Island, Central West Greenland. *Permafrost and Periglacial Processes*, **7**, 361–380.

Humlum, O. (1997). Active-layer thermal regime at three rock glaciers in Greenland. *Permafrost and Periglacial Processes*, **8**, 383–408.

Humlum, O. (1998a). The climatic significance of rock glaciers. *Permafrost and Periglacial Processes*, **9**, 375–395.

Humlum, O. (1998b). Rock glaciers on the Faeroe Islands, the North Atlantic. *Journal of Quaternary Science*, **13**, 293–307.

Humlum, O. (1998c). Active-layer thermal regime 1991–1996 at Qeqertarsuuq, Disko Island, Central West Geenland. *Arctic and Alpine Research*, **30**, 295–305.

Humlum, O. (1999). Late-Holocene climate in central West Greenland: meteorological data and rock-glacier isotope evidence. *The Holocene*, **9**, 581–594.

Humlum, O. (2000). The geomorphic significance of rock glaciers: estimates of rock glacier debris volumes and headwall recession rates in West Greenland. *Geomorphology*, **35**, 41–67.

Humlum, O., Elberling, B., Hormes, A., Fjord-heim, K., Hansen, O. H., Heinemeier, J. (2005). Late-Holocene glacier growth in Svalbard, documented by subglacial relict vegetation and living soil microbes. *The Holocene*, **15**, 396–407.

Hunter, J. A., Judge, A. S., Macaulay, H. A., Good, R. L., Gagne, R. M., Burns, R. A. (1976). The occurrence of permafrost and frozen sub-seabottom materials in the southern Beaufort Sea. Beaufort Sea Technical Report no. 22, Environment Canada, Ottawa, 174 pp.

Hussey, K. M., Michelson, R. W. (1966). Tundra relief features near Point Barrow, Alaska. *Arctic*, **19**, 162–184.

Hustich, I. (1966). On the forest-tundra and the northern tree-lines. Annals, Turku University, A, II, 36 (Kevo Subarctic Station, report 3), 7–47.

Hutchinson, J. N. (1974). Periglacial solifluxion; an approximate mechanism for clayey soils. *Géotechnique*, **24**, 438–443.

Hutchinson, J. N., Gostelow, T. P. (1976). The development of an abandoned cliff in London Clay at Hadleigh, Essex. *Philosophical Transactions of the Royal Society of London*, **A283**, 557–604.

Hyatt, J. A. (1998). The origin of lake-bed ground ice at Water Supply Lake, Pond Inlet, Nunavut, Canada. In: Lewkowicz, A. G., Allard, M., eds., *Permafrost, Proceedings of the Seventh International Conference on Permafrost*, 23–27 June, Yellowknife, Canada. Centre d'études nordiques, Université Laval, Québec, Collection Nordicana no. 57, pp. 487–493.

Ikeda, A., Matsuoka, N. (1999). Measurement of bottom temperature of the winter snow cover (BTS) in relation to rock-glacier activity, Corviglia, Swiss Alps: a preliminary report. Annual report, Institute of Geosciences, University of Tsukuba, 25, 13–19.

IPCC (Intergovernmental Panel on Climate Change) (2001). *Third Assessment Report. WGI. Climate Change 2001: The Scientific Basis* (Houghton, J. T., et al., eds.), 881 pp., and *WGII. Climate Change 2001: Impacts, Adaptations and Vulnerability* (McCarthy, J. J., et al., eds.), 1032 pp. Cambridge: Cambridge University Press.

Isaksen, K., Holmlund, P., Sollid, J. L., Harris, C. (2001). Three deep alpine-permafrost boreholes in Svalbard and Scandinavia. *Permafrost and Periglacial Processes*, **12**, 13–26.

Ishikawa, M., Hirakawa, K. (2000). Mountain permafrost distribution based on BTS measurements and DC resistivity soundings in the Daisetsu Mountains, Hokkaido, Japan. *Permafrost and Periglacial Processes*, **11**, 109–123.

IUCN (The World Conservation Union) (1993). *Oil and Gas Exploration and Production in Arctic and Subarctic Onshore Regions*. IUCN Gland, Switzerland and Cambridge, UK, with E&P Forum, London, 56 pp.

Ives, J. D. (1966). Blockfields, associated weathering forms on mountain tops, and the nunatak hypothesis. *Geografiska Annaler*, **48A**, 220–223.

Iwata, S. (1987). Debris-mantled rectilinear slopes in the western Sor Rondane mountains, East Antarctica. In: *Proceedings, National Institute of Polar Research (NIPR), Symposium Antarctic Geoscience*, **1**, 178–192.

Jahn, A. (1960). Some remarks on evolution of slopes on Spitsbergen. *Zeitschrift für Geomorphologie*, suppl. **1**, 49–58.

Jahn, A. (1961). Quantitative analysis of some periglacial processes in Spitsbergen. *Nauka O Ziemi II*, seria B, **5**, 3–34.

Jahn, A. (1962). The origin of granite tors. *Czasopismo Geograficzne*, **33**, 41–44.

Jahn, A. (1972). Niveo-eolian processes in the Sudetes Mountains. *Geographia Polonica*, **23**, 93–110.

Jahn, A. (1975). *Problems of the Periglacial Zone*. PWN Polish Scientific Publishers, Warsaw, 219 pp.

Jahn, A. (1976). Contemporaneous geomorphological processes in Longyeardalen, Vestspitsbergen (Svalbard). *Biuletyn Peryglacjalny*, **26**, 253–268.

Jahn, A. (1977). Periglacial forms produced by shore ice at Hornsund (Spitsbergen). *Acta Universitatis Wratislaviensis*, **387**, 19–29.

Jahn, A. (1983). Soil wedges on Spitsbergen. In: *Permafrost, Proceedings of the Fourth International Conference on Permafrost*, 17–22 July, Fairbanks, Alaska. National Academy Press, Washington, DC, vol. 1, pp. 525–530.

Jennings, J. N. (1971). *Karst*. MIT Press, Boston, MA, 252 pp.

Jersak, J. (1973). Eemian and early Würmian soils in loess of Poland. *Biuletyn Peryglacjalny*, **22**, 169–184.

Jersak, J. (1977). Cyclic development of the loess cover in Poland. *Biuletyn Instytut Geologicznego*, **305**, 83–97.

Jerwood, L. A., Robinson, D. A., Williams, R. B. G. (1990a). Experimental frost and salt weathering of chalk – I. *Earth Surface Processes and Landforms*, **15**, 611–624.

Jerwood, L. A., Robinson, D. A., Williams, R. B. G. (1990b). Experimental frost and salt weathering of chalk – II. *Earth Surface Processes and Landforms*, **15**, 699–708.

John, B. S., Sugden, D. E. (1975). Coastal geomorphology of high latitudes. *Progress in Geography*, **7**, 53–132.

Johnsen, S. J., Dansgaard, W. S., Clausen, H. B., Langway, C. C., Jr. (1972). Oxygen isotope profiles through the Antarctic and Greenland ice sheets. *Nature*, **235**, 429–434.

Johnson, D. W. (1932). Rock planes in arid regions. *Geographical Review*, **22**, 656–665.

Johnson, P. G. (1974). Mass movement of ablation complexes and their relationship to rock glaciers. *Geografiska Annaler*, **56A**, 93–101.

Johnson, W. H. (1990). Ice wedges and relict patterned ground in Central Illinois and their environmental significance. *Quaternary Research*, **33**, 51–72.

Johnsson, G. (1959). True and false ice wedges in southern Sweden. *Geografiska Annaler*, **44**, 378–404.

Johnston, G. H. (1980). Permafrost and the Eagle River bridge, Yukon Territory, Canada. In: *Proceeding, Permafrost Engineering Workshop*. National Research Council of Canada, Ottawa, technical memorandum 130, pp. 12–28.

Johnston, G. H., ed. (1981). *Permafrost: Engineering Design and Construction*. John Wiley & Sons, New York, 340 pp.

Johnston, G. H. (1982). Design and performance of the Inuvik, NWT airstrip. In: French, H. M., ed., *Proceedings, Fourth Canadian Permafrost Conference*, Calgary, Alberta. National Research Council of Canada, Ottawa, pp. 577–585.

Johnston, G. H., Brown, R. J. E. (1964). Some observations on permafrost distribution at a lake in the Mackenzie Delta. *Arctic*, **17**, 162–175.

Johnston, G. H., Brown, R. J. E., Pickersgill, D. N. (1963). Permafrost investigations at Thompson, Manitoba: terrain studies. Division of Building Research, National Research Council of Canada, Ottawa, technical paper 158, 51 pp.

Jorgenson, M. T., Kreig, R. A. (1988). A model for mapping permafrost distribution based on landscape component maps and climatic variables. In: Senneset, K., ed., *Permafrost, Proceedings of the Fifth International Conference on Permafrost*, 2–5 August. Trondheim, Tapir, vol. 1, pp. 176–182.

Jorré, G. (1933). Probleme des "terrasses goletz" sibériennes. *Revue de Géographie Alpine*, **21**, 347–371.

Judge, A. S. (1982). Natural gas hydrates in Canada. In: French, H. M., ed., *Proceedings Fourth Canadian Permafrost Conference*, Calgary, Alberta. National Research Council of Canada, Ottawa, pp. 320–328.

Kaab, A., Kneisel, C. (2006). Permafrost creep within a recently deglaciated glacier forefield: Muragl, Swiss Alps. *Permafrost and Periglacial Processes*, **17**, 79–85.

Kaab, A., Reichmuth, T. (2005). Advance mechanisms of rock glaciers. *Permafrost and Periglacial Processes*, **16**, 187–193.

Kaab, A., Haeberli, W., Gudmundsson, G. H. (1997). Analysing the creep of mountain permafrost using high precision aerial photogrammetry: 25 years of monitoring Gruben rock glacier, Swiss Alps. *Permafrost and Periglacial Processes*, **8**, 409–426.

Kachurin, S. P. (1938). Retreat of permafrost. *Reports, USSR Academy of Sciences*, **XIX**(8), 593–597 (in Russian).

Kachurin, S. P. (1955). Is the development of thermokarst always indicative of a recession of the permafrost table? In: *Data on the Principles of the Study of Frozen Zones in the Earth's Crust*, issue II, Academy of Sciences of the USSR, V. A. Obruchev Institute of Permafrost Studies, Moscow, 74 pp. (in Russian; English translation: National Research Council of Canada, technical translation 1006, Ottawa, 1962, pp. 25–32).

Kachurin, S. P. (1962). Thermokarst within the territory of the USSR. *Biuletyn Peryglacjalny*, **11**, 49–55.

Kaiser, K. (1960). Klimazeugen des periglazialen dauerfrostbodens in Mittel- und West-Europa. *Eiszeitalter und Gegenwart*, **11**, 121–141.

Kallio, A., Reiger, S. (1969). Recession of permafrost in a cultivated soil of interior Alaska. *Proceedings, Soil Science Society of America*, **33**, 430–432.

Kamensky, R. M., ed. (1998). *Geocryological Problems of Construction in Eastern Russia and Northern China. Proceedings, International Symposium*, 23–25 September, Chita, Russia. SB RAS, Yakutsk, vol. 1, 255 pp., vol. 2, 197 pp.

Kane, D. L., Slaughter, C. W. (1972). Seasonal regime and hydrological significance of stream icing in central Alaska. In: *The Role of Snow and Ice in Hydrology. Proceedings of the Banff Symposium*. IAHS Publication 107, pp. 528–540.

Kane, D. L., Hinzman, L. D., Zarling, J. P. (1991). Thermal response of the active layer to climatic warming in a permafrost environment. *Cold Regions Science and Technology*, **19**, 111–122.

Kaplanskaya, F. A., Tarnogradskiy, V. D. (1986). Remnants of the Pleistocene ice sheets in the permafrost zone as an object for paleoglaciological research. *Polar Geography and Geology*, **10**, 257–266.

Kariya, Y. (2002). Geomorphic processes at a snowpatch hollow on Gassan Volcano, Northern Japan. *Permafrost and Periglacial Processes*, **13**, 107–116.

Karrasch, H. (1970). Das phanomen der klimabedingten reliefasymmetrie in Mitteleuropa, *Gottingen Geographische Abhandlungen*, **56**, 299 pp.

Karte, J. (1979). Raumliche abgrenzung und regionale differenzierung des periglaziärs. *Bochumer Geographische Arbeiten*, **35**, 211 pp.

Karte, J. (1983). Grèzes litées as a special type of periglacial slope sediments in the German Highlands. *Polarforschung*, **53**(2), 67–74.

Karte, J. (1987). Pleistocene periglacial conditions in north and central Europe. In: Boardman, J., ed., *Periglacial Processes and Landforms in Britain and Ireland*. Cambridge University Press, Cambridge, pp. 67–75.

Karte, J., Liedtke, H. (1981). The theoretical and practical definition of the term 'periglacial' in its geographical and geological meaning. *Biuletyn Peryglacjalny*, **28**, 123–135.

Karunaratne, K. C., Burn, C. R. (2003). Freezing n-factors in discontinuous permafrost terrain, Takhini River valley, Yukon Territory, Canada. In: Phillips, M., Springman, S. M., Arenson, L. U., eds., *Proceedings, Eighth International Conference on Permafrost*, 21–25 July 2003, Zurich, Switzerland. Balkema, Lisse, vol. 1, pp. 519–524.

Karunaratne, K. C., Burn, C. R. (2004). Relations between air and surface temperature in discontinuous permafrost terrain near Mayo, Yukon Territory. *Canadian Journal of Earth Sciences*, **41**, 1437–1451.

Kasper, J., Allard, M. (2001). Late Holocene climatic change as detected by the growth and decay of ice wedges on the southern shore of Hudson Strait, northern Québec, Canada. *The Holocene*, **11**, 563–577.

Kasse, C. (1997). Cold-climate aeolian sand-sheet formation in North-western Europe (c 14-12.4 ka); a response to permafrost degradation and increased aridity. *Permafrost and Periglacial Processes*, **8**, 295–311.

Katasonov, E. M. (1969). Composition and cryogenic structure of permafrost. National Research Council of Canada, Ottawa, technical translation 1358, 25–36.

Katasonov, E. M. (1973). Present-day ground and ice veins in the region of the Middle Lena. *Biuletyn Peryglacjalny*, **23**, 81–89.

Katasonov, E. M. (1975). Frozen-ground and facial analysis of Pleistocene deposits and paleogeography of central Yakutia. *Biuletyn Peryglacjalny*, **24**, 33–40.

Katasonov, E. M., Ivanov, M. S. (1973). *Cryolithology of Central Yakutia*, Guidebook, Second International Permafrost Conference, Yakutsk, USSR, 38 pp.

Keen, D. H. (1985). Late-Pleistocene deposits and mollusca from Portland, Dorset. *Geological Magazine*, **122**, 181–186.

Kellaway, G. A. (1972). Development of non-diastrophic Pleistocene structures in relation to climate and physical relief in Britain. In: *Proceedings, 24th International Geological Congress*, Montreal, section 12, pp. 136–146.

Kerfoot, D. E. (1974). Thermokarst features produced by man-made disturbances to the tundra terrain. In: Fahey, B. D., Thompson, R. O., eds., *Research in Polar and Alpine Geomorphology, Proceedings, Third Guelph Symposium on Geomorphology*, Guelph, Ontario. GeoBooks, Norwich, pp. 60–72.

Kerney, M. P. (1963). Late-glacial deposits on the Chalk of south-east England. *Philosophical Transactions of the Royal Society, London*, **B246**, 203–254.

Kerney, M. P., Brown, E. H., Chandler, T. J. (1964). The late-glacial and post-glacial history of the Chalk escarpment near Brook, Kent. *Philosophical Transactions of the Royal Society, London*, **B248**, 135–204.

Kershaw, G. P., Gill, D. (1979). Growth and decay of palsas and peat plateaux in the Macmillan Pass – Tsichu River area, Northwest Territories, Canada. *Canadian Journal of Earth Sciences*, **16**, 1362–1367.

Khrustalev, L. (2000). Allowance for climate change in designing foundation on permafrost grounds. In: Senneset, K., ed., *Proceedings, International Workshop on Permafrost Engineering*, Longyearbyen, Svalbard, Norway, 18–21 June. Norwegian University of Science and Technology (NTNU)/University Courses on Svalbard (UNIS), pp. 25–36.

Kimble, J. M., ed. (2004). *Cryosols. Permafrost-Affected Soils*. Springer-Verlag, Berlin, 726 pp.

King, L. (1983). High mountain permafrost in Scandinavia. In: *Permafrost, Proceedings of the Fourth International Conference on Permafrost*, 17–22 July, Fairbanks, Alaska. National Academy Press, Washington, DC, pp. 612–617.

King, L. (1986). Zonation and ecology of high mountain permafrost in Scandinavia. *Geografiska Annaler*, **68A**, 131–139.

King. L. (2000). Mountain permafrost in Europe: occurrence, characteristics, prospecting, mapping and monitoring. In: Pena, J. J., Sanchez-Fabre, Lozano, M. V., eds., *Procesos y formas periglaciares en la montana mediterranea*. Instituto de Etudios Turolenses, Teruel, pp. 3–24.

King, L. C. (1953). Canons of landscape evolution. *Geological Society of America, Bulletin*, **64**, 721–752.

Kinnard, C., Lewkowicz, A. G. (2006). Frontal advance of turf-banked solifluction lobes, Kluane Range, Yukon Territory, Canada. *Geomorphology*, **73**, 261–276.

Klassen, R., Shilts, W. (1987). *Bylot Island, Eastern Canadian Arctic*. XII INQUA Congress Field Excursion A-I, Guidebook, National Research Council of Canada, Ottawa, 54 pp.

Klatkowa, H. (1965). Vallons en berceau et vallées sèches aux environs de Łódź. *Acta Geographica Łódziensia*, **19**, 124–142.

Kleman, J., Borgström, I. (1990). The boulder fields of Mt. Fulufjället, west-central Sweden – Late-Weichselian boulder blankets and interstadial periglacial phenomena. *Geografiska Annaler*, **72A**, 63–78.

Klene, A. E., Nelson, F. E., Shiklomanov, N. I. (2001). The N-factor in natural landscapes: variability of air and soil-surface temperatures, Kuparuk River basin, Alaska, USA. *Arctic, Antarctic and Alpine Research*, **33**, 140–148.

Kneisel, C. (2004). New insights into mountain permafrost occurrence and characteristics in glacier forefields at high altitude through the application of 2D resistivity imaging. *Permafrost and Periglacial Processes*, **15**, 221–228.

Kokelj, S. V., Burn, C. R. (2004). Tilt of spruce trees near ice wedges, Mackenzie Delta, Northwest Territories, Canada. *Arctic, Antarctic and Alpine Research*, **36**, 615–623.

Kokelj, S. V., Burn, C. R. (2005). Near-surface ground ice in sediments of the Mackenzie Delta, Northwest Territories, Canada. *Permafrost and Periglacial Processes*, **16**, 291–304.

Kolstrup, E. (1980). Climate and stratigraphy in northwestern Europe between 30000 BP and 13000 BP with special reference to the Netherlands. *Mededelingen Rijks Geologische Dienst*, **32**(15), 181–253.

Kolstrup, E. (1987). Frost-wedge casts in western Jutland and their possible implications for European periglacial research. *Zeitschrift für Geomorphologie*, **31**, 449–461.

Komar, P. D. (1976). *Beach Processes and Sedimentation*. Prentice-Hall, Englewood Cliffs, NJ, 429 pp.

Kondratjeva, K. A., Khrutzky, S. F., Romanovskii, N. N. (1993). Changes in the extent of permafrost during the Late-Quaternary period in the territory of the Former Soviet Union. *Permafrost and Periglacial Processes*, **4**, 113–119.

Konishchev, V. N. (1982). Characteristics of cryogenic weathering in the permafrost zone of the European USSR. *Arctic and Alpine Research*, **14**, 261–265.

Konishchev, V. N., Rogov, V. V. (1993). Investigations of cryogenic weathering in Europe and Northern Asia. *Permafrost and Periglacial Processes*, **4**, 49–64.

Konishchev, V. N., Rogov, V. V., Schurina, G. N. (1976). Cryogenic factor influence on primary minerals (results of experimental investigation). *Problems of Cryolithology*, **5**, 50–61 (in Russian).

Konrad, J.-M., Morgenstern, N. R. (1983). Frost susceptibility of soils in terms of their segregation potential. In: *Permafrost, Proceedings of the Fourth International Conference on Permafrost*, 17–22 July, Fairbanks, Alaska. National Academy Press, Washington, DC, pp. 660–665.

Konrad, J.-M., Morgenstern, N. R. (1984). Frost-heave prediction of chilled pipelines buried in unfrozen soils. *Canadian Geotechnical Journal*, **21**, 100–115.

Köppen, W. V. (1923). *Die klimate der Erde*. De Gruyter, Berlin, 369 pp.

Korotayev, V. N. (1986). Geomorphology of river deltas on the Arctic coast of Siberia. *Polar Geography and Geology*, **10**, 139–147.

Koster, E. A. (1988). Ancient and modern cold-climate aeolian sand deposition: a review. *Journal of Quaternary Science*, **3**, 69–83.

Koster, E. A. (1993). Global warming and periglacial landscapes. In Roberts, N., ed., *The Changing Global Environment*. Blackwell, Oxford, pp. 127–149.

Koster, E. A., Dijkmans, J. W. A. (1988). Niveo-aeolian deposits and denivation forms, with special reference to the Great Kobuk Sand Dunes, Northwestern Alaska. *Earth Surface Processes and Landforms*, **13**, 153–170.

Kotler, E., Burn, C. R. (2000). Cryostratigraphy of the Klondike 'muck' deposits, west-central Yukon Territory. *Canadian Journal of Earth Sciences*, **37**, 849–861.

Kotzebue, O. von, (1821). *A Voyage of Discovery into the South Sea and Bering's Strait for the Purpose of Exploring a Northeast Passage*. London (English translation, 3 vols).

Krantz, W. B., Gleason, K. J., Caine, N. (1988). Patterned ground. *Scientific American*, **259**(6), 68–75.

Krigstrom, A. (1962). Geomorphological studies of sandur plains and their braided rivers in Iceland. *Geografiska Annaler*, **44**, 328–346.

Kriuchkov, V. V. (1968). Necessity of soil conservation in the Far North. *Priroda*, **12**, 72–74 (in Russian; English translation by E. R. Hope, 1969, Defence Research Board, Ottawa, Canada, T 523 R, 5 pp.).

Kudryavtsev, V. A. (1965). Temperature, thickness and discontinuity of permafrost. In: *Principles of Geocryology (Permafrost Studies)*, Part 1, *General Geocryology*, USSR Academy of Sciences, Moscow, 1959, chapter VIII, pp. 219–273 (National Research Council of Canada, Ottawa, technical translation 1187, 75 pp.).

Kudryavtsev, V. A. (1978). *Obshcheye merzlotovedenya (gyeokreologeya)* [General Permafrost Science – Geocryology]. Moscow State University, Moscow, 464 pp. (in Russian).

Kudryavtsev, V. A., Kondrat'yeva, K. A., Romanovskii, N. N. (1978). Zonal and regional patterns of formation of the permafrost region in the USSR. In: *Permafrost, Proceedings of the Third International Conference on Permafrost*, 10–13 July, Edmonton, Alberta, Canada. National Research Council of Canada, Ottawa, vol. 1, pp. 419–426.

Kwon, S.-S. (1978). Block field in the vicinity of Bomosa Temple. *Journal of Geography (Korea)*, **5**, 49–54 (in Korean).

Kwon, S.-S. (1979). Fossil periglacial phenomena on the southern parts of Geoje Island, Korea. *Journal of Geography (Korea)*, **6**, 151–153 (in Korean).

Laban, C., van der Meer, J. J. M. (2004). Pleistocene glaciation in The Netherlands. In: Ehlers, J., Gibbard, P. L., eds., *Quaternary Glaciations – Extent and Chronology*, part II. Elsevier, Amsterdam, pp. 251–260.

Lachenbruch, A. (1957). Thermal effects of the ocean on permafrost. *Bulletin, Geological Society of America*, **68**, 1515–1529.

Lachenbruch, A. (1962). Mechanics of thermal-contraction cracks and ice-wedge polygons in permafrost. *Geological Society of America*, special paper 70, 69 pp.

Lachenbruch, A. (1966). Contraction theory of ice-wedge polygons; a qualitative discussion. In: *Permafrost, International Conference Proceedings*, National Research Council of Canada publication 1287. National Academy of Sciences, Washington, DC, pp. 63–71.

Lachenbruch, A. H. (1968). Permafrost. In: Fairbridge, R. W., ed., *Encyclopedia of Geomorphology*. Reinhold, New York, pp. 833–838.

Lachenbruch, A. H., Cladouhos, T. T., Saltus, R. W. (1988). Permafrost temperature and the changing climate. In: Senneset, K., ed., *Permafrost, Proceedings of the Fifth International Conference on Permafrost*, 2–5 August. Trondheim, Tapir, vol. 3, pp. 9–17.

Ladanyi, B., Johnston, G. H. (1973). Evaluation of in-situ creep properties of frozen soils with the pressuremeter. In: *Permafrost, Proceedings of the Second International Conference on Permafrost*, Yakutsk, USSR. National Academy of Sciences, Washington, DC, publication 2115, pp. 310–317.

Lambiel, C., Delaloye, R. (2004). Contribution of real-time kinematic GPS in the study of creeping mountain permafrost: examples from the western Swiss Alps. *Permafrost and Periglacial Processes*, **15**, 229–242.

Lamirande, I., Lauriol, B., Lalonde, A. E., Clark, I. D. (1999). La production de limon sur des terrasses de cryoplanation dans les monts Richardson, Canada. *Canadian Journal of Earth Sciences*, **36**, 1645–1654.

Laroque, S. J., Hétu, B., Filion, L. (2001). Geomorphic and dendrochronological impacts of slush-flows in Central Gaspé Peninsula (Québec, Canada). *Geografiska Annaler*, **83A**, 191–201.

Lauriol, B. M., Godbout, L. (1988). Les terrasses de cryoplanation dans le nord du Yukon: distribution, genèse et age. *Géographie physique et Quaternaire*, **42**, 303–313.

Lauriol, B. M., Gray, J. T. (1980). Processes responsible for the concentration of boulders in the intertidal zone in Leaf Basin, Ungava. *Geological Survey of Canada*, paper 80-10, 281–292.

Lauriol, B. M., Gray, J. T. (1990). Drainage karstique en milieu de pergélisol: le cas de l'île d'Akpatok, T.N.O., Canada. *Permafrost and Periglacial Processes*, **1**, 129–144.

Lauriol, B. M., Carrier, L., Thibaudeau, P. (1988). Topoclimate zones and ice dynamics in the caves of the Northern Yukon, Canada. *Arctic*, **41**, 215–220.

Lauriol, B. M., Duchesne, C., Clark, I. D. (1995). Sytématique du remplissage en eau des fentes de gel: les résultats d'une étude oxygène-18 et deutérium. *Permafrost and Periglacial Processes*, **6**, 47–55.

Lauriol, B. M., Lalonde, A. E., Dewez, V. (1997). Weathering of quartzite on a cryoplanation terrace in northern Yukon, Canada. *Permafrost and Periglacial Processes*, **8**, 147–153.

Lautridou, J.-P., Giresse, P. (1981). Genèse et signification paléoclimatique des limons à doublets de Normandie. *Biuletyn Peryglacjalny*, **28**, 149–161.

Lawrence D. M., Slater, A. G. (2005). A projection of severe near-surface permafrost degradation during the 21st century. *Geophysical Research Letters*, **32**, L24401, doi: 10.1029/20005GL025080.

Lawson, D. E., Brown, J., Everett, K. R., Johnson, A. W., Komarkova, V., Murray, B. M., Murray, D. F., Webber, P. J. (1978). Tundra disturbance and recovery following the 1949 exploratory drilling, Fish Creek, northern Alaska. US Army, CRREL, Hanover, New Hampshire, report 78-28, 81 pp.

LeDrew, E., Barber, D., Agnew, T., Dunlop, D. (1992). Canadian sea-ice atlas from microwave remotely-sensed imagery: July 1987 to June 1990. Atmospheric Environment Service, Environment Canada, Climatological Studies no. 44, 80 pp.

Leffingwell, E. de K. (1915). Ground-ice wedges, the dominant form of ground-ice on the north coast of Alaska. *Journal of Geology*, **23**, 635–654.

Leffingwell, E. de K. (1919). The Canning River region, Northern Alaska. *United States Geological Survey*, professional paper 109, 251 pp.

Lefroy, J. H. (1887). Second Report of a Committee for enquiring into the depth of permanently frozen soil in the polar region. *Proceedings, Royal Geographical Society*, **IX**, new series, 769–774.

Lefroy, J. H. (1889a). Report upon the depth of permanently-frozen soil in the polar regions, its geographical limits, and relations to the present poles of greatest cold. *Proceedings of the Geographical Section of the British Association, London*, 740–746.

Lefroy, J. H. (1889b). On the depth of the permanently-frozen soil in British North America. *Proceedings of the Geographical Section of the British Association, London*, 761–763.

Lemmens, M., Lorrain, R., Haren, J. (1982). Isotopic composition of ice and subglacially precipitated cacite in an alpine area. *Zeitschrift für Gletscherkunde und Glazialgeologie*, **18**, 151–159.

Lenngren, C. A. (2000). Guidelines for airport runway roughness induced by frost heave. In: Senneset, K., ed., *Proceedings, International Workshop on Permafrost Engineering*, Longyearbyen, Svalbard, 18–21 June. Norwegian University of Science and Technology (NTNU)/University Courses on Svalbard (UNIS), pp. 139–154.

Leopold, L. B., Wolman, M. G., Miller, J. P. (1964). *Fluvial Processes in Geomorphology*. Freeman, San Francisco, CA, 522 pp.

Leshikov, F. N. (1999). Specific development of cryogenic phenomena in seismic districts of southern East Siberia. *Earth Cryosphere*, **III**, 31–37 (in Russian).

Lewis, C. A. (1988). Periglacial landforms. In: Moon, B. P., Dardis, G. F., eds., *The Geomorphology of Southern Africa*. Southern Book Publishers, Johannesburg, pp. 103–119.

Lewis, C. R. (1962). Icing mound on the Sadlerochit River, Alaska. *Arctic*, **15**, 145–150.

Lewkowicz, A. G. (1983). Erosion by overland flow, central Banks Island, Western Canadian Arctic. In: *Permafrost, Proceedings of the Fourth International Conference on Permafrost*, 17–22 July, Fairbanks, Alaska. National Academy Press, Washington, DC, vol. 1, pp. 701–706.

Lewkowicz, A. G. (1987). Nature and importance of thermokarst processes, Sandhills moraine, Banks Island, Canada. *Geografiska Annaler*, **69A**, 1077–1085.

Lewkowicz, A. G. (1988a). Ablation of massive ground ice, Mackenzie Delta. In: Senneset, K. ed., *Permafrost, Proceedings of the Fifth International Conference on Permafrost*, 2–5 August, Trondheim, Norway. Tapir, Trondheim, vol. 1, pp. 605–610.

Lewkowicz, A. G. (1988b). Slope processes. In: Clark, M. J., ed., *Advances in Periglacial Geomorphology*. John Wiley & Sons, Chichester, pp. 325–368.

Lewkowicz, A. G. (1990). Morphology, frequency and magnitude of active-layer detachment slides, Fosheim Peninsula, Ellesmere Island, N.W.T. In: *Permafrost-Canada, Proceedings of the Fifth Canadian Permafrost Conference*, National Research Council of Canada – Centre d'études nordiques, Université Laval, Collection Nordicana no. 54, pp. 111–118.

Lewkowicz, A. G. (1992). Factors influencing the distribution and initiation of active-layer detachment slides on Ellesmere Island, Arctic Canada. In: Dixon, J. C., Abrahams, A. D., eds., *Periglacial Geomorphology*. John Wiley & Sons, Chichester, pp. 223–250.

Lewkowicz, A. G. (1994). Ice-wedge rejuvenation, Fosheim Peninsula, Ellesmere Island, Canada. *Permafrost and Periglacial Processes*, **5**, 251–268.

Lewkowicz, A. G. (1998). Aeolian sediment transport during winter, Black Top Creek, Fosheim Peninsula, Ellesmere Island, Canadian Arctic. *Permafrost and Periglacial Processes*, **9**, 35–46.

Lewkowicz, A. G., Coultish, T. L. (2004). Beaver damming and palsa dynamics in a subarctic mountainous environment, Wolf Creek, Yukon Territory, Canada. *Arctic, Antarctic, and Alpine Research*, **36**, 208–218.

Lewkowicz, A. G., Ednie, M. (2004). Probability mapping of mountain permafrost using the BTS method, Wolf Creek, Yukon Territory, Canada. *Permafrost and Periglacial Processes*, **15**, 67–80.

Lewkowicz, A. G., French, H. M. (1982a). The hydrology of small runoff plots in an area of continuous permafrost, Banks Island, NWT. In: French, H. M., ed., *Proceedings, Fourth Canadian Permafrost Conference*, Calgary, Alberta. National Research Council of Canada, Ottawa, pp. 151–162.

Lewkowicz, A. G., French, H. M. (1982b). Downslope water movement and solute concentrations within the active layer, Banks Island, NWT. In: French, H. M., ed., *Proceedings, Fourth Canadian Permafrost Conference*, Calgary, Alberta. National Research Council of Canada, Ottawa, pp. 163–172.

Lewkowicz, A. G., Gudjonsson, K. A. (1992). Slope hummocks on Fosheim Peninsula, Northwest Territories. *Geological Survey of Canada*, paper 92-1B, 97–102.

Lewkowicz, A. G., Harris, C. (2005). Frequency and magnitude of active-layer detachment failures in discontinuous and continuous permafrost, Northern Canada. *Permafrost and Periglacial Processes*, **16**, 115–130.

Lewkowicz, A. G., Kokelj, S. V. (2002). Slope-sediment yield in arid lowland continuous permafrost environments, Canadian Arctic Archipelago. *Catena*, **46**, 261–283.

Lewkowicz, A. G., Wolfe, P. M. (1994). Sediment transport in Hot Weather Creek, Ellesmere Island, NWT, Canada, 1990–1991. *Arctic and Alpine Research*, **26**, 213–226.

Lewkowicz, A. G., Young, K. L. (1991). Observations of aeolian transport and niveo-aeolian deposition at three lowland sites, Canadian Arctic Archipelago. *Permafrost and Periglacial Processes*, **2**, 197–210.

Liebman, M. O. (1996). Cryogenic landslides on the Yamal Peninsula, Russia: preliminary observations. *Permafrost and Periglacial Processes*, **6**, 259–264.

Liebman, M. O., Kizyakov, A. I., Sulerzhitsky, L. D., Zaretskaia, N. E. (2003). Dynamics of landslide slopes and their development on Yamal Peninsula. In: Phillips, M., Springman, S. A., Arenson, L. U., eds., *Permafrost, Proceedings of the Eighth International Conference on Permafrost*, 21–25 July, 2003, Zurich, Switzerland. Balkema, Lisse, vol. 1, pp. 651–656.

Liestol, O. (1976). Pingos, springs and permafrost in Spitsbergen. In: *Norsk Polarinstitutt Arbok, 1975*. Oslo, pp. 7–29.

Lindsay, J. F. (1973). Ventifact evolution in Wright Valley, Antarctica. *Geological Society of America Bulletin*, **84**, 1791–1798.

Ling, F., Zhang, T., Zhang, L. X. (2003). Critical embankment height: Fenghuoshan mountains, Qinghai-Xizang (Tibet) Railway, China. In: Phillips, M., Springman, S. A., Arenson, L. U., eds., *Permafrost, Proceedings of the Eighth International Conference on Permafrost*, 21–25 July, Zurich, Switzerland. Balkema, Lisse, vol. 1, pp. 685–689.

Linton, D. L. (1955). The problem of tors. *Geographical Journal*, **121**, 470–487.

Linton, D. L. (1964). The origin of the Pennine tors; an essay in analysis. *Zeitschrift für Geomorphologie*, **8**, 5–24.

Linton, D. L. (1969). The abandonment of the term "periglacial". *Palaeoecology of Africa and of the surrounding islands and Antarctica*, **5**, 65–70.

Lokrantz, H., Ingolsson, O., Forman, S. L. (2003). Glacio-tectonised Quaternary sediments at Cape Shpindler, Yugorski Peninsula, Arctic Russia: implications for glacial history, ice movements and Kara Sea Ice-Sheet configuration. *Journal of Quaternary Science*, **18**, 527–543.

Lorrain, R. D., Demeur, P. (1985). Isotopic evidence for relic Pleistocene glacier ice on Victoria Island, Canadian Arctic Archipelago. *Arctic and Alpine Research*, **17**, 89–98.

Loveday, J. (1962). Plateau deposits of the southern Chiltern Hills. *Proceedings, Geologists' Association*, **73**, 83–102.

Lozinski, W. von (1909). Über die mechanische Verwitterung der Sandsteine im gemässigten klima. *Bulletin International de l'Academie des Sciences de Cracovie class des Sciences Mathematique et Naturalles*, **1**, 1–25. (English translation by T. Mrozek (1992). On the mechanical weathering of sandstones in temperate climates. In: Evans, D. J. E. ed., 1994, Cold climate landform, John Wiley and Sons Ltd, pp. 119–134).

Lozinski, W. von (1912). Die periglaziale fazies der mechanischen Verwitterung. *Comptes Rendus, XI Congrès Internationale Géologie, Stockholm 1910*, 1039–1053.

Luckman, B. H. (1977). The geomorphic activity of snow avalanches. *Geografiska Annaler*, **59A**, 31–48.

Luckman, B. H. (1988). Debris accumulation patterns on talus slopes in Surprise valley, Alberta. *Géographie physique et Quaternaire*, **42**(3), 247–278.

Lunardini, V. J. (1978). Theory of n-factors and correlation of data. In: *Permafrost, Proceedings of the Third International Conference on Permafrost*, 10–13 July, Edmonton, Alberta, Canada. National Research Council of Canada, Ottawa, vol. 1, pp. 40–46.

Luoto, M., Seppälä, M. (2002). Characteristics of earth hummocks (pounus) with and without permafrost in Finnish Lapland. *Geografiska Annaler*, **84A**, 127–136.

Maarleveld, G. C. (1960). Wind directions and cover sands in the Netherlands. *Biuletyn Peryglacjalny*, **8**, 49–58.

Maarleveld, G. C. (1976). Periglacial phenomena and the mean annual temperature during the last glacial time in The Netherlands. *Biuletyn Peryglacjalny*, **26**, 57–78.

Maarleveld, G. C. (1981). Summer thaw depth in cold regions and fossil cryoturbation. *Geologie en Mijnbouw*, **60**, 347–352.

Maarleveld, G. C., van den Toorn, J. C. (1955). Pseudo-solle in Noord-Nederland. *Tijdschrift Koninklijk Nederlands Aardrijkskundig Genootschap*, **72**, 334–360.

Mackay, J. R. (1962). Pingos of the Pleistocene Mackenzie Delta area. *Geographical Bulletin*, **18**, 21–63.

Mackay, J. R. (1963). *The Mackenzie Delta area*. Geographical Branch Memoir, no. 8, 202 pp.

Mackay, J. R. (1966). Segregated epigenetic ice and slumps in permafrost, Mackenzie Delta area, NWT. *Geographical Bulletin*, **8**, 59–80.

Mackay, J. R. (1970). Disturbances to the tundra and forest tundra environment of the Western Arctic. *Canadian Geotechnical Journal*, **7**, 420–432.

Mackay, J. R. (1971). The origin of massive icy beds in permafrost, Western Arctic coast, Canada. *Canadian Journal of Earth Sciences*, **8**, 397–422.

Mackay, J. R. (1972a). The world of underground ice. *Annals, Association of American Geographers*, **62**, 1–22.

Mackay, J. R. (1972b). Offshore permafrost and ground ice, southern Beaufort Sea, Canada. *Canadian Journal of Earth Sciences*, **9**, 1550–1561.

Mackay, J. R. (1973a). The growth of pingos, Western Arctic Coast, Canada. *Canadian Journal of Earth Sciences*, **10**, 979–1004.

Mackay, J. R. (1973b). Problems in the origin of massive icy beds, Western Arctic, Canada. In: *Permafrost, Proceedings of the Second International Conference on Permafrost*, Yakutsk, USSR. National Academy of Sciences, Washington, DC, vol. 1, pp. 223–228.

Mackay, J. R. (1974a). Ice-wedge cracks, Garry Island, Northwest Territories. *Canadian Journal of Earth Sciences*, **11**, 1366–1383.

Mackay, J. R. (1974b). Reticulate ice veins in permafrost, Northern Canada. *Canadian Geotechnical Journal*, **11**, 230–237.

Mackay, J. R. (1974c). The rapidity of tundra-polygon growth and destruction, Tuktoyaktuk Peninsula-Richards Island area, NWT. *Geological Survey of Canada*, paper 74-1A, 391–392.

Mackay, J. R. (1975a). The closing of ice-wedge cracks in permafrost, Garry Island, Northwest Territories. *Canadian Journal of Earth Sciences*, **12**, 1668–1674.

Mackay, J. R. (1975b). Relict ice wedges, Pelly Island, NWT (107C/12). *Geological Survey of Canada*, paper 75-1, part A, 469–470.

Mackay, J. R. (1977a). Pulsating pingos, Tuktoyaktuk Peninsula, N.W.T. *Canadian Journal of Earth Sciences*, **14**, 209–222.

Mackay, J. R. (1977b). Changes in the active layer from 1968 to 1976 as a result of the Inuvik fire. *Geological Survey of Canada*, paper 77-1B, 273–275.

Mackay, J. R. (1978a). The use of snow fences to reduce ice-wedge cracking, Garry Island, Northwest Territories. *Geological Survey of Canada*, paper 78-1, 523–524.

Mackay, J. R. (1978b). Sub-pingo water lenses, Tuktoyaktuk Peninsula, Northwest Territories. *Canadian Journal of Earth Sciences*, **15**, 461–462.

Mackay, J. R. (1979a). Pingos of the Tuktoyaktuk Peninsula area, Northwest Territories. *Géographie physique et quaternaire*, **33**, 3–61

Mackay, J. R. (1979b). An equilibrium model for hummocks (non-sorted circles), Garry Island, Northwest Territories. *Geological Survey of Canada*, paper 79-1A, 165–167.

Mackay, J. R. (1980a). Deformation of ice-wedge polygons, Garry Island, Northwest Territories. *Geological Survey of Canada*, paper 80-1A, 287–291.

Mackay, J. R. (1980b). The origin of hummocks, western Arctic coast. *Canadian Journal of Earth Sciences*, **17**, 996–1006.

Mackay, J. R. (1981a). Active-layer slope movement in a continuous permafrost environment, Garry Island, Northwest Territories, Canada. *Canadian Journal of Earth Sciences*, **18**, 1666–1680.

Mackay, J. R. (1981b). An experiment in lake drainage, Richards Island, Northwest territories: a progress report. *Geological Survey of Canada*, paper 81-1A, 63–68.

Mackay, J. R. (1981c). Aklisuktuk (growing fast) pingo, Tuktoyaktuk Peninsula, Northwest Territories. *Arctic*, **34**, 270–273.

Mackay, J. R. (1982). Active-layer growth, Illisarvik experimental drained lake site, Richards Island, Northwest Territories. *Geological Survey of Canada*, paper 82-1A, 123–126.

Mackay, J. R. (1983a). Downward water movement into frozen ground, western arctic coast, Canada. *Canadian Journal of Earth Sciences*, **20**, 120–134.

Mackay, J. R. (1983b). Oxygen-isotope variations in permafrost, Tuktoyaktuk Peninsula area, Northwest Territories. *Geological Survey of Canada*, paper 83-1B, 67–74.

Mackay, J. R. (1984a). The frost heave of stones in the active layer above permafrost with downward and upward freezing. *Arctic and Alpine Research*, **16**, 439–446.

Mackay, J. R. (1984b). Lake-bottom heave in permafrost; Illisarvik drained lake site, Richards Island, Northwest Territories. *Geological Survey of Canada*, paper 84-1B, 173–177.

Mackay, J. R. (1984c). The direction of ice-wedge cracking in permafrost: upwards or downwards? *Canadian Journal of Earth Sciences*, **21**, 516–524.

Mackay, J. R. (1985a). Pingo ice of the western Arctic coast, Canada. *Canadian Journal of Earth Sciences*, **22**, 1452–1464.

Mackay, J. R. (1985b). Permafrost growth in recently-drained lakes, Western Arctic Coast. *Geological Survey of Canada*, paper 85-1B, 177–189.

Mackay, J. R. (1986a). Fifty years (1935–1985) of coastal retreat west of Tuktoyaktuk, District of Mackenzie. *Geological Survey of Canada*, paper 86-1A, 727–735.

Mackay, J. R. (1986b). Growth of Ibyuk Pingo, Western Arctic Coast, Canada and some implications for environmental reconstructions. *Quaternary Research*, **26**, 68–80.

Mackay, J. R. (1986c). Frost mounds. In: French, H. M., ed., *Focus: Permafrost Geomorphology, The Canadian Geographer*, **30**, 363–364.

Mackay, J. R. (1986d). The first 7 years (1978–1985) of ice-wedge growth, Illisarvik experimental drained lake site, western Arctic coast. *Canadian Journal of Earth Sciences*, **23**, 1782–1795.

Mackay, J. R. (1988a). The birth and growth of Porsild Pingo, Tuktoyaktuk Peninsula, District of Mackenzie. *Arctic*, **41**, 267–674.

Mackay, J. R. (1988b). Catastrophic lake drainage, Tuktoyaktuk peninsula area, District of Mackenzie. *Geological Survey of Canada*, paper 88-1D, 83–90.

Mackay, J. R. (1989a). Massive ice: some field criteria for the identification of ice types. *Geological Survey of Canada*, paper 89-1G, 5–11.

Mackay, J. R. (1989b). Ice-wedge cracks, western Arctic coast. *Canadian Geographer*, **33**, 365–368.

Mackay, J. R. (1990a). Some observations on the growth and deformation of epigenetic, syngenetic and anti-syngenetic ice wedges. *Permafrost and Periglacial Processes*, **1**, 15–29.

Mackay, J. R. (1990b). Seasonal growth bands in pingo ice. *Canadian Journal of Earth Sciences*, **27**, 1115–1125.

Mackay, J. R. (1992a). The frequency of ice-wedge cracking (1967–1987) at Garry Island, western Arctic coast, Canada. *Canadian Journal of Earth Sciences*, **29**, 236–248.

Mackay, J. R. (1992b). Lake stability in an ice-rich permafrost environment: examples from the Western Arctic Coast. In: Robarts, R. D., Bothwell, M. L., eds., *Arctic Eco-Systems in Semi-Arid*

Regions: Implications for Resource Management, National Hydrology Research Institute Symposium Series, Environment Canada, Saskatoon, pp. 1–26.

Mackay, J. R. (1993a). The sound and speed of ice-wedge cracking, Arctic Canada. *Canadian Journal of Earth Sciences*, **30**, 509–518.

Mackay, J. R. (1993b). Air temperature, snow cover, creep of frozen ground, and the time of ice-wedge cracking, western Arctic Coast. *Canadian Journal of Earth Sciences*, **30**, 1720–1729.

Mackay, J. R. (1995a). Active-layer changes (1968 to 1993) following the forest-tundra fire near Inuvik, NWT, Canada. *Arctic and Alpine Research*, **27**, 323–336.

Mackay, J. R. (1995b). Ice wedges on hillslopes and landform evolution in the late Quaternary, western Arctic coast. *Canadian Journal of Earth Sciences*, **32**, 1093–1105.

Mackay, J. R. (1997). A full-scale field experiment (1978–1995) on the growth of permafrost by means of lake drainage, western Arctic coast: a discussion of the method and some results. *Canadian Journal of Earth Sciences*, **34**, 17–33.

Mackay, J. R. (1998). Pingo growth and collapse, Tuktoyaktuk Peninsula area, western Arctic coast, Canada: a long-term field study. *Géographie physique et Quaternaire*, **52**, 271–323.

Mackay, J. R. (1999). Cold-climate shattering (1974 to 1993) of 200 glacial erratics on the exposed bottom of a recently drained Arctic lake, western Arctic coast, Canada. *Permafrost and Periglacial Processes*, **10**, 125–136.

Mackay, J. R. (2000). Thermally-induced movements in ice-wedge polygons, western Arctic coast. *Géographie physique et Quaternaire*, **54**, 41–68.

Mackay, J. R., Burn, C. R. (2002). The first 20 years (1978–1979 to 1998–1999) of active layer development, Illisarvik experimental drained lake site, western Arctic coast, Canada. *Canadian Journal of Earth Sciences*, **39**, 1657–1674.

Mackay, J. R., Burn, C. R. (2005). A long-term field study (1951–2003) of ventifacts formed by katabatic winds at Paulatuk, western Arctic coast, Canada. *Canadian Journal of Earth Sciences*, **42**, 1615–1635.

Mackay, J. R., Dallimore, S. R. (1992). Massive ice of the Tuktoyaktuk area, western Arctic coast, Canada. *Canadian Journal of Earth Sciences*, **29**, 1235–1249.

Mackay, J. R., Lavkulich, L. M. (1974). Ionic and oxygen isotopic fractionation in permafrost growth. *Geological Survey of Canada*, paper 74-1, 255–256.

Mackay, J. R., MacKay, D. K. (1974). Snow cover and ground temperatures, Garry Island, NWT. *Arctic*, **27**, 287–296.

Mackay, J. R., Mathews, W. H. (1974). Needle ice striped ground. *Arctic and Alpine Research*, **6**, 79–84.

Mackay, J. R., Matthews, J. V. (1983). Pleistocene ice and sand wedges, Hooper Island, Northwest Territories. *Canadian Journal of Earth Sciences*, **20**, 1087–1097.

Mackay, J. R., Slaymaker, O. (1989). The Horton River breakthrough and resulting geomorphic changes in a permafrost environment, Western Arctic coast, Canada. *Geografiska Annaler*, **71A**, 171–184.

Mackay, J. R., Rampton, V. N., Fyles, J. G. (1972). Relic Pleistocene permafrost, Western Arctic, Canada. *Science*, **176**, 1321–1323.

Makarov, V. I., Kadkina, E. L., Pikulev, V. P., Kolesnikova, O. V. (2000). The evolution of town Norilsk natural-technical system. In: Senneset, K., ed., *Proceedings, International Workshop on Permafrost Engineering*, Longyearbyen, Svalbard, Norway, 18–21 June. Norwegian University of Science and Technology (NTNU)/University Courses on Svalbard (UNIS), pp. 225–243.

Makogon, Y. F., Trebin, F. A., Trofimuk, A. A., Tsarev, V. P., Cherskiy, N. V. (1972). Detection of a pool of natural gas in a solid (hydrated gas) state. *Doklady Academy of Sciences of the USSR, Earth Science Section*, **196**, 199–200 (English translation).

Malaurie, J. N., Guillien, Y. (1953). Le modèle cryo-nival des versants meubles de Skansen (Disko, Groenland). Interprétation général des grèzes litées. *Bulletin, Société Géologique de France*, **3**, 703–721.

Malin, M. C. (1987). Abrasion in ice-free areas of Southern Victoria Land. *Antarctic Journal of the United States*, **22**, 38–40.

Marovelli, R. L., Chen, T. S., Veith, K. F. (1966). Thermal fragmentation of rock. *American Institute of Mining, Metallurgical and Petroleum Engineers*, **235**, 1–15.

Marsh, B. (1987). Pleistocene pingo scars in Pennsylvania. *Geology*, **15**, 945–947.

Marsh, B. (1998). Wind-transverse corrugations in Pleistocene periglacial landscapes in central Pennsylvania. *Quaternary Research*, **49**, 149–156.

Marsh, B. (1999). Paleoperiglacial landscapes of Central Pennsylvania. Sixty-Second Annual Reunion, Northeast Friends of the Pleistocene, 1999 Trip, Bucknell University, Lewisburg, Pennsylvania, 69 pp.

Marshall, P., Brown, M. C. (1974). Ice in Coulthard Cave, Alberta. *Canadian Journal of Earth Sciences*, **11**, 510–518.

Matsuoka, N. (1990). The rate of bedrock weathering by frost action: field measurements and a predictive model. *Earth Surface Processes and Landforms*, **15**, 73–90.

Matsuoka, N. (1991). A model of the rate of frost shattering: application to field data from Japan, Svalbard and Antarctica. *Permafrost and Periglacial Processes*, **2**, 271–281.

Matsuoka, N. (1995). Rock weathering and landform development in the Sør Rondane Mountains, Antarctica. *Geomorphology*, **12**, 323–339.

Matsuoka, N. (2001a). Direct observation of frost wedging in alpine bedrock. *Earth Surface Processes and Landforms*, **26**, 601–614.

Matsuoka, N. (2001b). Microgelivation versus macrogelivation: towards bridging the gap between laboratory and field frost weathering. *Permafrost and Periglacial Processes*, **12**, 299–313.

Matsuoka, N. (2001c). Solifluction rates, processes and landforms: a global review. *Earth Science Reviews*, **55**, 107–134.

Matsuoka, N., Hirakawa, K. (2000). Solifluction resulting from one-sided and two-sided freezing: field data from Svalbard. *Polar Geoscience*, **13**, 187–201.

Matsuoka, N, Ikeda, A. (1998). Some observations regarding mountain permafrost in the Japanese Alps. Annual Report, Institute of Geosciences, University of Tsukuba, 24, 19–25.

Matsuoka, N., Moriwaki, K., Hirakawa, K. (1988). Diurnal frost-heave activity in the Sør-Rondane Mountains, Antarctica. *Arctic and Alpine Research*, **20**, 422–428.

Matsuoka, N., Morikawa, K., Hirakawa, K. (1996). Field experiments on physical weathering and wind erosion in an Antarctic cold desert. *Earth Surface Processes and Landforms*, **21**, 687–699.

Matsuoka, N., Hirakawa, K., Watanabe, T., Morikawa, K. (1997). Monitoring of periglacial slope processes in the Swiss Alps: the first two years of frost shattering, heave and creep. *Permafrost and Periglacial Proceses*, **8**, 155–177.

Matsuoka, N., Ikeda, A., Hirakawa, K., Watanabe, T. (2003). Contemporary periglacial processes in the Swiss Alps: seasonal, inter-annual and long-term variations. In: Phillips, M., Springman, S. A., Arenson, L. U., eds., *Permafrost, Proceedings of the Eighth International Conference on Permafrost*, 21–25 July, Zurich, Switzerland. Balkema, Lisse, vol. 1, pp. 735–740.

Matsuoka, N., Ikeda, A., Date, T. (2005). Morphometric analysis of solifluction lobes and rock glaciers in the Swiss Alps. *Permafrost and Periglacial Processes*, **16**, 99–114.

Matthes, F. E. (1900). Glacial sculpture of the Bighorn Mountains, Wyoming. United States Geological Survey, 21st Annual report 1899–1900, pp. 167–190.

McCann, S. B. (1972). Magnitude and frequency of processes operating on Arctic beaches, Queen Elizabeth Islands, NWT, Canada. In: Adams, P. W., Helleiner. F., eds., *International Geography*. University of Toronto Press, Toronto, vol. 1, pp. 41–43.

McCann, S. B., Cogley, J. G. (1973). The geomorphic significance of fluvial activity in high latitudes. In: Fahey, B. D., Thompson, R. D., eds., *Research in Polar and Alpine Geomorphology: Proceedings, Third Guelph Symposium on Geomorphology*. GeoAbstracts, Norwich, pp. 118–135.

McCann, S. B., Owens, E. H. (1969). The size and shape of sediments in three Arctic beaches, southwest Devon Island, NWT, Canada. *Arctic and Alpine Research*, **1**, 267–278.

McCarroll, D. (1990). Differential weathering of feldspar and pyroxene in an Arctic-Alpine environment. *Earth Surface Processes and Landforms*, **15**, 641–651.

McCarroll, D., Vines, H. (1995). Rock weathering by the lichen *Lecidea auricculata* in an arctic alpine environment. *Earth Surface Processes and Landforms*, **20**, 199–206.

McConnell, R. G., Tyrrell, J. B. (1898). Preliminary note on the gold deposits and gold mining in the Klondike region. *Geological Survey of Canada*, memoir 284 (1957), 17–23.

McDonald, B. C., Lewis, C. P. (1973). Geomorphic and sedimentologic processes of rivers and coasts, Yukon Coastal Plain. Report 73-39, Environmental-Social Committee Northern

Pipelines, Task Force on Northern Oil Development, Information Canada, Ottawa, 245 pp.

McKenna-Neuman, C., Gilbert, R. (1986). Aeolian processes and landforms in glaciofluvial environments of southeastern Baffin Island, NWT, Canada. In: Nickling, W. G., ed., *Aeolian Geomorphology.* Allen & Unwin, Boston, MA, pp. 213–235.

McRoberts, E. C., Morgenstern, N. R. (1974). The stability of thawing slopes. *Canadian Geotechnical Journal*, **11**, 447–469.

McRoberts, E. C., Nixon, J. F. (1975). Reticulate ice veins in permafrost, northern Canada: Discussion. *Canadian Geotechnical Journal*, **12**, 1509–1562.

Meinardus, W. (1912). *Beobachtungen über Detritussortierung und Strukturböden auf Spitzbergen.* Gesell, Erdkunde, Berlin, pp. 250–259.

Meinardus, W. (1930). *Arktische boden. Handbuch der Bodenlehre.* Herausgegeben von E. Blank, 3, 27–96.

Melnikov, E. S., Leibman, M. O., Moskalenko, N. G., Vasiliev, A. A. (2004). Active-layer monitoring in the cryolithozone of West Siberia. *Polar Geography*, **28**, 267–285.

Melnikov, V. P., Spesivtsev, V. I. (2000). *Cryogenic Formations in the Earth's Lithosphere.* Novosibirsk Scientific Publishing Center UIGGM, SB RAS Publishing House, 343 pp. (in Russian and English).

Metz, M. C., Krzewinski, T. G., Clarke, E. S. (1982). The Trans-Alaska Pipeline workpad – an evaluation of present conditons. In: French, H. M., ed., *Proceedings, Fourth Canadian Permafrost Conference*, Calgary, Alberta. National Research Council of Canada, Ottawa, pp. 523–534.

Michaud, Y., Dyke, L. D. (1990). Mechanism of bedrock frost heave in permafrost regions. In: *Proceedings Fifth Canadian Permafrost Conference*, Université Laval, Collection Nordicana 54, pp. 125–130.

Michaud, Y., Dionne, J.-C., Dyke, L. D. (1989). Frost bursting: a violent expressioin of frost action in rock. *Canadian Journal of Earth Sciences*, **26**, 2075–2080.

Michel, F. A. (1990). Isotopic composition of ice-wedge ice in Northwestern Canada. In: *Proceedings Fifth Canadian Permafrost Conference*, Université Laval, Collection Nordicana 54, pp. 5–9.

Michel, F. A., Fritz, P. (1982). Significance of isotope variations in permafrost waters at Illisarvik, NWT. In: French, H. M., ed., *Proceedings, Fourth Canadian Permafrost Conference*, Calgary, Alberta. National Research Council of Canada, Ottawa, pp. 173–181.

Middendorf, A. T. E. von (1862). *Sibirien Reise.* Part 1. *Uebersicht der Natur-Nord und Ost-Sibiriens.* Kaiserlichen Akademie der Wissenschaften, St Petersburg, 783 pp.

Miller, G. H., Geirsdottir, A., Koerner, R. M. (2001). Climate implications of changing Arctic sea ice. *Eos, Transactions, American Geophysical Union*, **82**, 97, 103.

Miller, R. D. (1972). Freezing and heaving of saturated and unsaturated soils. *Highway Research Record*, **393**, 1–11.

Minervin, A. V. (1982). The role of cryogenic processes in forming of loess deposits. *Problems of Cryolithology*, **10**, 41–61 (in Russian).

Moorman, B. J., Michel, F. A., Wilson, A. (1996). C-14 dating of trapped gases in massive ground ice, Western Canadian Arctic. *Permafrost and Periglacial Processes*, **7**, 257–266.

Moorman, B. J., Michel, F. A., Wilson, A. T. (1998). The development of tabular massive ground ice at Peninsula Point, NWT, Canada. In: Lewkowicz, A. G., Allard, M., eds., *Permafrost, Proceedings of the Seventh International Conference on Permafrost*, 23–27 June, Yellowknife, Canada. Centre d'études nordiques, Université Laval, Québec, Collection Nordicana no. 57, pp. 757–761.

Morgan, A. V. (1971). Polygonal patterned ground of Late-Weichselian age in the area north and west of Wolverhampton, England. *Geografiska Annaler*, **54A**, 146–156.

Morgan, A. V., Morgan, A. (1980). Beetle bits – the science of paleoentomology. *Geoscience Canada*, **7**(1), 22–29.

Morgan, A. V., Morgan, A., Ashworth, A. C., Matthews, Jr., J. V. (1983). Late-Wisconsin fossil beetles in North America. In: Porter, S. C., ed., *Late-Quaternary Environments of the United States*, vol. 1, *The Late Pleistocene.* Longman, London, pp. 354–363.

Morgenstern, N. R. (1981). Geotechnical engineering and frontier resource development. *Géotechnique*, **31**, 305–365.

Morgenstern, N. R., Nixon, J. F. (1971). One dimensional consolidation of thawing soils. *Canadian Geotechnical Journal*, **8**, 558–565.

Mottershead, D. N. (1971). Coastal head deposits between Start Point and Hope Cove, Devon. *Field Studies*, **5**, 433–453.

Mottershead, D. N. (1977). *South West England; Guidebook for excursions A6 and C6*, X INQUA Congress. GeoAbstracts, Norwich, 59 pp.

Mullenders, W., Gullentops, F. (1969). The age of the pingos of Belgium. In: Péwé, T. L., ed., *The Periglacial Environment*. McGill-Queen's University Press, pp. 321–336.

Müller, F. (1959). Beobachtung uber pingos. *Meddelelser om Gronland*, **153**(3), 127 pp. (English translation, National Research Council of Canada, technical translation TT-1073, 117 pp.)

Muller, S. W. (1943). Permafrost or permanently frozen ground and related engineering problems. Special Report, Strategic Engineering Study, Intelligence Branch, Office, Chief of Engineers, no. 62, 136 pp. Second printing, 1945, 230 pp. (Reprinted in 1947, J. W. Edwards, Ann Arbor, Michigan, 231 pp.)

Murton, J. B. (1996a). Near-surface brecciation of Chalk, Isle of Thanet, south-east England: a comparison with ice-rich brecciated bedrocks in Canada and Spitsbergen. *Permafrost and Periglacial Processes*, **7**, 153–164.

Murton, J. B. (1996b). Thermokarst-lake-basin sediments, Tuktoyaktuk Coastlands, western arctic Canada. *Sedimentology*, **43**, 737–760.

Murton, J. B. (2001). Thermokarst sediments and sedimentary structures, Tuktoyaktuk Coastlands, western Arctic Canada. *Global and Planetary Change*, **28**, 175–192.

Murton, J. B., French, H. M. (1993a). Sand wedges and permafrost history, Crumbling Point, Pleistocene Mackenzie Delta, Canada. In: *Permafrost, Proceedings of the Sixth International Conference on Permafrost*, Beijing, China. Science Press, Beijing, vol. 1, pp. 482–487.

Murton, J. B., French, H. M. (1993b). Thaw modification of frost-fissure wedges, Richards Island, Pleistocene Mackenzie Delta, western Arctic Canada. *Journal of Quaternary Science*, **8**, 185–196.

Murton, J. B., French, H. M. (1993c). Thermokarst involutions, Summer Island, Pleistocene Mackenzie Delta, Western Canadian Arctic. *Permafrost and Periglacial Processes*, **4**, 217–F229.

Murton, J. B., French, H. M. (1994). Cryostructures in permafrost, Tuktoyaktuk coastlands, western Arctic, Canada. *Canadian Journal of Earth Sciences*, **31**, 737–747.

Murton, J. B., Kolstrup, E. (2003). Ice-wedge casts as indicators of palaeotemperatures: precise proxy or wishful thinking? *Progress in Physical Geography*, **27**, 155–170.

Murton, J. B., Lautridou, J.-P. (2003). Recent advances in the understanding of Quaternary periglacial features of the English Channel coastlands. *Journal of Quaternary Science*, **18**, 301–307.

Murton, J. B., Whiteman, C. A., Allen, P. (1995). Involutions in the Middle Pleistocene (Anglian) Barham Soil, Eastern England: a comparison with thermokarst involutions from Arctic Canada. *Boreas*, **24**, 269–280.

Murton, J. B., Worsley, P., Gozdzik, J. (2000). Sand veins and wedges in cold aeolian environments. *Quaternary Science Reviews*, **19**, 899–922.

Murton, J. B., Coutard, J.-P., Lautridou, J.-P., Ozouf, J.-C., Robinson, D. A., Williams, R. B. G., Guillemet, G., Simmons, P. (2000). Experimental design for a pilot study on bedrock weathering near the permafrost table. *Earth Surface Processes and Landforms*, **25**, 1281–1294

Murton, J. B., Coutard, J.-P., Lautridou, J.-P., Ozouf, J.-C., Robinson, D. A., Williams, R. B. G. (2001). Physical modelling of bedrock brecciation by ice segregation in permafrost. *Permafrost and Periglacial Processes*, **12**, 255–266.

Murton, J. B., Bateman, M. D., Baker, C. A., Knox, R., Whiteman, C. A. (2003). The Devensian periglacial record on Thanet, Kent, U.K. *Permafrost and Periglacial Processes*, **14**, 217–246.

Murton, J. B., Waller, R. I., Hart, J. K., Whiteman, C. A., Pollard, W. H., Clark, I. D.(2004). Stratigraphy and glaciotectonic structures of permafrost deformed beneath the northwest margin of the Laurentide ice sheet, Tutktoyaktuk Coastlands, Canada. *Journal of Glaciology*, **50**, 399–412.

Murton, J. B., Whiteman, C. A., Waller, R. I., Pollard, W. H., Clark, I. D., Dallimore, S.R. (2005). Basal ice facies and supraglacial melt-out till of the Laurentide ice sheet, Tuktoyaktuk Coastlands, western Arctic Canada. *Quaternary Science Reviews*, **24**, 681–708.

Muscoe, G. E. (1982). The origin of honeycomb weathering. *Bulletin, Geological Society of America*, **93**, 108–115.

Natural Resources Canada (1995). *Canada Permafrost. The National Atlas of Canada*, 5th edn. Ottawa, MCR 4177, scale 1:75000000.

Nekrasov, I. A. (1963). K voprosu o klassifikatsii talikov [On the question of classification of talik]. *Izvestiya Vsesoyuzhogo Geografii Obshchestva*, **95**, 143–153 (English translation: Soviet hydrology: selected papers, *American Geophysical Union*, **2**, 192–200).

Nekrasov, I. A., Gordeyev, P. P. (1973). *The Northeast of Yakutia*. Guidebook, Second International Permafrost Conference, Yakutsk, 46 pp.

Nelson, F. E. (1989). Cryoplanation terraces: periglacial cirque analogs. *Geografiska Annaler*, **71A**, 31–41.

Nelson, F. E., ed. (2004). Eurasian contributions from the circumpolar active-layer monitoring (CALM) workshop. *Polar Geography*, **28**(4), 253–340.

Newell, W. L., Powers, D. S., Owens, J. P., Stanford, S. D., Stone, B. D. (2000). Surficial geologic map of central and southern New Jersey. United States Geological Survey, Miscellaneous Investigations Series, Map 1-2540-D.

Nichols, R. L. (1966). Geomorphology of Antarctica. In: Tedrow, J. C. F., ed., *Antarctic Soils and Soil Forming Processes*. American Geophysical Union, Antarctic Research Series, 8, pp. 1–59.

Nicholson, F. H., Granberg, H. B. (1973). Permafrost and snowcover relationships near Schefferville. In: *North American Contribution, Second International Conference on Permafrost*, Yakutsk, USSR. National Academy of Science, publication 2115, pp. 151–158.

Nicholson, F. H., Thom, B. G. (1973). Studies at the Timmins 4 permafrost experimental site. In: *North American Contribution, Second International Conference on Permafrost*, Yakutsk, USSR. National Academy of Science, publication 2115, pp. 159–166.

Nickling, W. G. (1978). Eolian sediment transport during dust storms: Slims River Valley, Yukon Territory. *Canadian Journal of Earth Sciences*, **15**, 1069–1084.

Nikiforoff, C. C. (1955). Hardpan soils of the coastal plain of southern Maryland. United States Geological Survey, professional paper 267-B, 45–62.

Niu, F. J., Cheng, G.-D., Lai, Y. M. (2003). Laboratory study on a duct-ventilation roadbed of the Qinghai-Tibet railway. In: Phillips, M., Springman, S. A., Arenson, L. U., eds., *Permafrost, Proceedings of the Eighth International Conference on Permafrost*, 21–25 July, Zurich, Switzerland. Balkema, Lisse, vol. 2, pp. 815–820.

Nixon, J. F. (1990). Northern pipelines in permafrost terrain. *Geotechnical News*, **8**, 25–26.

Nixon, J. F., McRoberts, E. C. (1973). A study of some factors affecting the thawing of frozen soils. *Canadian Geotechnical Journal*, **10**, 439–452.

Norris, K. (1977). Blow River and Davidson Mountains, Yukon Territory – District of Mackenzie. *Geological Survey of Canada*, Map 1516 A.

Nowaczyk, B. (1976). Eolian coversands in central-west Poland. *Quaestiones Geographicae*, **3**, 57–77.

Nyberg, R. (1985). Debris flows and slush avalanches in northern Swedish Lappland. University of Lund, Sweden, Department of Geography, *Avhandlingar*, **XCVII**, 222 pp.

Nyberg, R. (1991). Geomorphic processes at snowpatch sites in the Abisko mountains, northern Sweden. *Zeitschrift für Geomorphologie*, **35**, 321–345.

Oches, E. A., McCoy, W. D. (1995). Amino-acid geochronology applied to the correlation and dating of central European loess deposits. *Quaternary Science Reviews*, **14**, 767–782.

Odegard, R. S., Isaksen, K., Eiken, T., Sollid, J. L. (2003). Terrain analyses and surface velocity measurements of the Hiorthfjellet rock glacier, Svalbard. *Permafrost and Periglacial Processes*, **14**, 359–366.

Ollier, C. D. (1963). Insolation weathering: examples from central Australia. *American Journal of Science*, **261**, 376–381.

Ollier, C. D. (1984). *Weathering*. Longman, London, 270 pp.

Ollier, C. D., Thomasson, A. J. (1957). Asymmetrical valleys of the Chiltern Hills. *Geographical Journal*, **123**, 71–80.

Onesti, L. (1985). Meteorological conditions that initiate slushflows in the central Brooks Range, Alaska. *Annals of Glaciology*, **6**, 23–25.

Onesti, L., Hestnes, E. (1989). Slush-flow questionnaire. *Annals of Glaciology*, **13**, 226–230.

Osterkamp, T. E., Jorgenson, J. C. (2006). Warming of permafrost in the Arctic National Wildlife Refuge, Alaska. *Permafrost and Periglacial Processes*, **17**, 65–69.

Ostrem, G. (1963). Comparative crystallographic studies on ice from ice-cored moraines, snow-banks and glaciers. *Geografiska Annaler*, **45**, 210–240.

Outcalt, S. I., Nelson, F. E., Hinkel, K. M., Martin, G. D. (1986). Hydrostatic-system palsas at Toolik Lake, Alaska: field observations and simulation. *Earth Surface Processes and Landforms*, **11**, 79–94.

Owens, E. H., McCann, S. B. (1970). The role of ice in the Arctic beach environment with special reference to Cape Ricketts, Southwest Devon Island, NWT. *American Journal of Science*, **268**, 397–414.

Palmer, J., Nielson, R. A. (1962). The origin of granite tors on Dartmoor, Devonshire. *Proceedings, Yorkshire Geological Society*, **33**, 315–340.

Palmer, J., Radley, J. (1961). Gritstone tors of the English Pennines. *Zeitschrift für Geomorphologie*, **5**, 37–52.

Parameswaran, V. R., Mackay, J. R. (1983). Field measurements of electrical freezing potentials in permafrost areas. In: *Permafrost, Proceedings of the Fourth International Conference on Permafrost*, 17–22 July, Fairbanks, Alaska. National Academy Press, Washington, DC, pp. 962–967.

Pavlov, A. V. (1994). Current changes of climate and permafrost in the Arctic and Sub-Arctic of Russia. *Permafrost and Periglacial Processes*, **5**, 101–110.

Pavlov, A. V. (1998). Active-layer monitoring in Northern West Siberia. In: Lewkowicz, A. G., Allard, M., eds., *Permafrost, Proceedings of the Seventh International Conference on Permafrost*, 23–27 June, Yellowknife, Canada. Université Laval, Québec, Collection Nordicana no. 57, pp. 875–880.

Pavlov, A. V., Moskalenko, N. G. (2002). The thermal regime of soils in the North of Western Siberia. *Permafrost and Periglacial Processes*, **13**, 43–51.

Pawluk, S. (1988). Freeze–thaw effects on granular structure reorganization for soil materials of varying texture and moisture content. *Canadian Journal of Soil Science*, **7**, 1333–1342.

Payette, S., Morneau, C., Sirois, L., Desponts, M. (1989). Recent fire history of the northern Québec biomes. *Ecology*, **70**, 656–673.

Pearce, A. J., Elson, J. A. (1973). Postglacial rates of denudation by soil movement, free face retreat, and fluvial erosion, Mont St. Hilaire, Québec. *Canadian Journal of Earth Sciences*, **10**(1), 91–101.

Peck, L. (1998). Remediation of wastewater by land treatment. Consideration of soil temperature in winter. U.S. Army Corps of Engineers, CRREL, Hanover, New Hampshire, report 98-8, 18 pp.

Pecsi, M. (1964). Chronological problems of the patterned soils of Hungary. *Biuletyn Peryglacjalny*, **14**, 279–293.

Pelletier, J. D. (2005). Formation of oriented thaw lakes by thaw slumping. *Journal of Geophysical Research*, **110**, F02018, doi: 10.1029/4004JF000158.

Peltier, L. C. (1950). The geographic cycle in periglacial regions as it is related to climatic geomorphology. *Annals, Association of American Geographers*, **40**, 214–236.

Perlshtein, G. Z., Pavlenkov, D. A. (2003). The use of natural cold for impounding the toxic wastes of the mining industry. In: Phillips, M., Springman, S. A., Arenson, K. U., eds., *Permafrost, Proceedings of the Eighth International Conference on Permafrost*, 21–25 July, Zurich, Switzerland. Balkema, Lisse, vol. 2, pp. 879–883.

Perov, V. F. (1969). Block fields in the Khibiny Mts. *Biuletyn Peryglacjalny*, **19**, 381–387.

Perrett, L. (1912). Prospecting frozen ground. *Mineralogical Science Press*, **104**, 856–857.

Peteet, D. M., Daniels, R., Heusser, L. E., Vogel, J. S., Southon, J. R., Nelson, D. E. (1994). Wisconsinan Late-glacial environmental change in southern New England: a regional synthesis. *Journal of Quaternary Science*, **9**, 151–154.

Pethick, J. (1984). *An Introduction to Coastal Geomorphology*. E. Arnold, London, 260 pp.

Péwé, T. L. (1948). Origin of the Mima Mounds. *Scientific Monthly*, **46**, 293–296.

Péwé, T. L. (1954). Effect of permafrost upon cultivated fields. *United States Geological Survey Bulletin*, **989F**, 315–351.

Péwé, T. L. (1955). Origin of the upland silt near Fairbanks, Alaska. *Geological Society of America Bulletin*, **66**, 699–724.

Péwé, T. L. (1959). Sand-wedge polygons (tesselations) in the McMurdo Sound region, Antarctica. *American Journal of Science*, **257**, 545–552.

Péwé, T. L. (1966a). Permafrost and its effect on life in the north. In: Hansen, H. P., ed., *Arctic Biology*, 2nd edn. Oregon State University Press, Corvallis, OR, pp. 27–66.

Péwé, T. L. (1966b). Ice wedges in Alaska – classification, distribution and climatic significance. In: *Permafrost, International Conference Proceedings*, National Research Council of Canada publication 1287. National Academy of Sciences, Washington, DC, pp. 76–81.

Péwé, T. L. (1969). The periglacial environment. In: Péwé, T. L., ed., *The Periglacial Environment*. McGill-Queen's University Press, Montreal, pp. 1–9.

Péwé, T. L. (1970). Altiplanation terraces of early Quaternary age near Fairbanks, Alaska. *Acta Geographica Łódziensia*, **24**, 357–363.

Péwé, T. L. (1974). Geomorphic processes in polar deserts. In: Smiley, T. L., Zumberge, J. H., eds., *Polar Deserts and Modern Man*. University of Arizona Press, Tucson, AZ, pp. 33–52.

Péwé, T. L. (1975). Quaternary geology of Alaska. United States Geological Survey, professional paper 835, 1–145.

Péwé, T. L. (1983a). Alpine permafrost in the contiguous United States: a review. *Arctic and Alpine Research*, **15**, 145–156.

Péwé, T. L. (1983b). The periglacial environment in North America during Wisconsin time. In: Porter, S. C., ed., *Late-Quaternary Environments of the United States*, vol. 1, *The Late Pleistocene*. Longman, London, pp. 157–189.

Péwé, T. L. (1983c). Geologic hazards of the Fairbanks area, Alaska. Division of Geological and Geophysical Surveys, Fairbanks, Alaska, special report 15, 109 pp.

Péwé, T. L., Journaux, A. (1983). Origin and character of loesslike silt in unglaciated south-central Yakutia, Siberia, USSR. United States Geological Survey, professional paper 1262, 46 pp.

Péwé, T. L., Rowan, D. E., Péwé, R. H., Stuckenrath, R. (1982). Glacial and periglacial geology of northwest Blomesletta Peninsula, Spitsbergen, Svalbard. *Norsk Polarinstitutt, Skrifter*, **177**, 32 pp.

Péwé, T. L., Tungsheng, L., Slatt, R. M., Bingyuan, L. (1995). Origin and character of loesslike silt in the southern Qinghai-Xizang (Tibet) Plateau, China. United States Geological Survey, professional paper 1549, 55 pp.

Péwé, T. L., Berger, G. W., Westgate, J. A., Brown, P. A., Leavitt, S. W. (1997). Eva Interglaciation Forest Bed, unglaciated east-central Alaska: global warming 125000 years ago. Geological Society of America, special publication 319, 54 pp.

Pihlainen, J. A., Johnston, G. H. (1963). Guide to a field description of permafrost. National Research Council, Associate Committee on Soil and Snow Mechanics, technical memorandum 79, 23 pp.

Pissart, A. (1953). Les coulées pierreuses du Plateau des Hautes Fagnes. *Annales, Société Géologique de Belgique*, **76**, 203–219.

Pissart, A. (1956). L'origine périglaciaire des viviers des Hautes Fagnes. *Annales, Société Géologique de Belgique*, **79**, 119–131.

Pissart, A. (1958). Les dépressions fermées dans la région parisienne; Le problème de leur origine. *Revue de Géomorphologie Dynamique*, **9**, 73–83.

Pissart, A. (1960). Les dépressions fermées de la région parisienne; les difficultés d'admettre une origine humaine. *Revue de Géomorphologie Dynamique*, **11**, 12 pp.

Pissart, A. (1963). Les traces de "pingos" du Pays de Galles (Grande Bretagne) et du Plateau des Hautes Fagnes (Belgique). *Zeitschrift für Geomorphologie*, **7**, 147–165.

Pissart, A. (1966a). Etude de quelques pentes de l'île Prince Patrick. *Annales, Société Géologique de Belgique*, **89**, 377–402.

Pissart, A. (1966b). Le rôle géomorphologique du vent dans la région de Mould Bay (Ile Prince Patrick, N.W.T., Canada). *Zeitschrift für Gomorphologie*, **10**, 226–236.

Pissart, A. (1967a). Les pingos de l'île Prince Patrick (76°N–120°W). *Geographical Bulletin*, **9**, 189–217. (English translation, National Research Council of Canada, technical translation TT-1401, Ottawa, 46 pp.).

Pissart, A. (1967b). Les modalités de l'écoulement de l'eau sur l'île Prince Patrick. *Biuletyn Peryglacjalny*, **16**, 217–224.

Pissart, A. (1968). Les polygons de fente de gel de l'île Prince Patrick (Arctique Canadien, 76° Lat. N). *Biuletyn Peryglacjalny*, **17**, 171–180.

Pissart, A. (1970). Les phénomènes physiques éssentiels liés au gel; les structures périglaciaires qui en résultent et leur signification climatique. *Annales, Société Géologique de Belgique*, **93**, 7–49.

Pissart, A. (1975). Sols à buttes, cercles non triés et sols striés non triés de l'île de Banks (Canada, N.W.T.). *Biuletyn Peryglacjany*, **26**, 275–285.

Pissart, A. (1977). Apparition et évolution des sols structuraux périglaciaires de haute montagne. Expériences de terrain au Chambeyron (Alpes, France). *Abhandlungen der Akademie der Wissenschaften in Göttingen, Math.-Physik Klasse*, **31**, 142–156.

Pissart, A. (1990). Advances in periglacial geomorphology. *Zeitschrift für Geomorphologie*, **79**, 119–131.

Pissart, A. (2000). Remnants of lithalsas of the Hautes Fagnes, Belgium: a summary of present-day knowledge. *Permafrost and Periglacial Processes*, **11**, 327–356.

Pissart, A. (2002). Palsas, lithalsas and remnants of these periglacial mounds: a progress report. *Progress in Physical Geography*, **26**, 605–621.

Pissart, A., French, H. M. (1976). Pingo investigations, north-central Banks Island, Canadian Arctic. *Canadian Journal of Earth Sciences*, **13**, 937–946.

Pissart, A., Gangloff, P. (1984). Les palses minérales et organiques de la vallée de l'Aneveau, près de Kuujjuaq, Québec subarctique. *Géographie physique et quaternaire*, **38**, 217–228.

Pissart, A., Vincent, J. S., Edlund, S. A. (1977). Dépôts et phénomènes éoliens sur l'île de Banks, Territoires du Nord-Ouest, Canada. *Canadian Journal of Earth Sciences*, **14**, 2452–2480.

Plug, L. J., Werner, B. (2001). Fracture networks in frozen ground. *Journal of Geophysical Research*, **106**, 8599–8613.

Plug, L. J., Werner, B. (2002). Nonlinear dynamics of ice-wedge networks and resultant sensitivity to severe cooling events. *Nature*, **417**, 929–933.

Pollard, W. H. (1990). The nature and origin of ground ice in the Herschel Island area, Yukon Territory. In: *Proceedings, Fifth Canadian Permafrost Conference*, Université Laval, Centre d'etudes nordiques, Québec, Collection Nordicana **54**, 23–30.

Pollard, W. H. (2000a). Distribution and characterization of ground ice on Fosheim Peninsula, Ellesmere Island, Nunavut. In: Garneau, M., Alt, B., eds., *Environmental Response to Climate Change in the Canadian High Arctic*. Geological Survey of Canada, Bulletin **529**, 207–233.

Pollard, W. H. (2000b). Ground-ice aggradation on Fosheim peninsula, Ellesmere Island, Nunavut. In: Garneau, M., Alt, B., eds., *Environmental Response to Climate Change in the Canadian High Arctic*, Geological Survey of Canada, Bulletin, **529**, 325–333.

Pollard, W. H. (2005). Icing processes associated with High Arctic perennial springs, Axel Heiberg Island, Nunavut, Canada. *Permafrost and Periglacial Processes*, **16**, 51–68.

Pollard, W. H., French, H. M. (1980). A first approximation of the volume of ground ice, Richards Island, Pleistocene Mackenzie Delta, Northwest Territories, Canada. *Canadian Geotechnical Journal*, **17**, 509–516.

Pollard, W. H., French, H. M. (1983). Seasonal frost mound occurrence, North Fork Pass, Ogilvie Mountains, northern Yukon, Canada. In: *Permafrost, Proceedings of the Fourth International Conference on Permafrost*, 17–22 July, Fairbanks, Alaska. National Academy Press, Washington, DC, pp. 1000–1004.

Pollard, W. H., French, H. M. (1984). The groundwater hydraulics of seasonal frost mounds, Northern Yukon. *Canadian Journal of Earth Sciences*, **21**, 1073–1081.

Pollard, W. H., French, H. M. (1985). The internal structure and ice crystallography of seasonal frost mounds. *Journal of Glaciology*, **31**, 157–162.

Pollard, W. H., McKay, C. P. (1997). Mineralized spring occurrence in the Expedition Fiord area, Axel Heiberg Island, NWT. In: Lewkowicz, A. G., Kokelj, S. V., eds., *Program and Abstracts, 27 Arctic Workshop*, Department of Geography, University of Ottawa, Ottawa, pp. 185–186.

Pope, G. A., Dorn, R. I., Dixon, J. C. (1995). A new conceptual model of understanding geographical variations in weathering. *Annals, Association of American Geographers*, **85**, 38–64.

Popov, A. I. (1956). Le thermokarst. *Biuletyn Peryglacjalny*, **4**, 319–330.

Popov, A. I. (1961). Cartes des formations périglaciaires actuelles et pleistocenes en territoire de l'U.R.S.S. *Biuletyn Peryglacjalny*, **10**, 87–96.

Popov, A. I. (1962). The origin and development of massive fossil ice. Issue 11, Academy of Sciences of the USSR, V. A. Obruchev Institute of Permafrost Studies, Moscow. (National Research Council of Canada, Ottawa, technical translation 1006, 5–24).

Popov, A. I. (1973). *Album of Cryogenic Formations in the Earth's Crust and Relief.* Idz-vo MGU (Moscow State University), Moscow, 56 pp. (in Russian).

Popov, A. I., Kachurin, S. P., Grave, N. A. (1966). Features of the development of frozen geomorphogy in northern Eurasia. In: *Proceedings, 1st International Conference on Permafrost*. National Academy of Science–National Research Council of Canada, publication 1287, pp. 181–185.

Popov, A. I., Rozenbaum, G. E., Tumel, N. V. (1985). *Cryolithology*. Moscow State University Press, Moscow, 238 pp. (in Russian).

Popov, I. V., Gvozdetskiy, N. A., Chiksishev, A. G., Kudelin, B. I. (1972). Karst in U.S.S.R. In: Herak, M., Stingfield, V. T., eds., *Karst*. Elsevier, Amsterdam, pp. 355–416.

Porsild, A. E. (1938). Earth mounds in unglaciated arctic northwestern America. *Geographical Review*, **28**, 46–58.

Porsild, A. E. (1955). The vascular plants of the western Canadian Arctic archipelago. *National Museum of Canada, Bulletin*, **135**, 226 pp.

Porter, S. C., Singhvi, A., Zhisheng, A., Zhongping, L. (2001). Luminescence age and palaeoenvironmental implications of a Late-Pleistocene ground wedge on the northeastern Tibet Plateau. *Permafrost and Periglacial Processes*, **12**, 203–210.

Poser, H. (1948). Boden- und klimaverhältnisse in Mittel- und Westeuropa während der Würmeiszeit. *Erdkunde*, **2**, 53–68.

Potter, N. (1972). Ice-cored rock glacier, Galena Creek, Northern Absaroka Mountains, Wyoming. *Bulletin, Geological Society of America*, **83**, 3025–3068.

Preece, S. J., Westgate, J. A., Stemper, B. A., Péwé, T. L. (1999). Tephrochronology of late Cenozoic loess at Fairbanks, central Alaska. *Bulletin, Geological Society of America*, **111**, 71–90.

Prest, V. K. (1984). Late Wisconsinan glacier complex. In: Fulton, R. J., ed., *Quaternary Stratigraphy of Canada, a Canadian contribution to IGCP Project 24*. Geological Survey of Canada, paper 84-10, pp. 21–38.

Prestwich, J. (1892). The raised beaches and "head" or rubble drift of the south of England. *Quarterly Journal of the Geological Society*, **48**, 263–343.

Price, L. W. (1973). Rates of mass wasting in the Ruby Range, Yukon Territory. In: *Permafrost; North American Contribution, Second International Permafrost Conference*, Yakutsk, USSR. National Academy of Science, publication 2115, 235–245.

Price, W. A. (1968). Oriented lakes. In: Fairbridge, R. W., ed., *Encyclopedia of Geomorphology*. Reinhold, New York, pp. 784–796.

Prick, A. (1997). Critical degree of saturation as a threshold moisture level in frost weathering of limestone. *Permafrost and Periglacial Processes*, **8**, 91–99.

Prick, A. (2003). Frost weathering and rock fall in an arctic environment, Longyearbyen, Svalbard. In: Phillips, M., Springman, S. M., Arenson, L. U., eds., *Permafrost, Proceedings of the Eighth International Conference on Permafrost*, 21–25 July, Zurich, Switzerland. Balkema, Lisse, vol. 2, pp. 907–912.

Prick, A., Pissart, A., Ozouf, J. C. (1993). Variations dilatométriques de cylindres de roches calcaires subissant des cycles de gel-dégel. *Permafrost and Periglacial Processes*, **4**, 1–15.

Priesnitz, K. (1981). Fussflächen und taler in der Arktis NW-Kanadas und Alaskas. *Polarforschung*, **51**, 145–159.

Priesnitz, K. (1988). Cryoplanation. In: Clark, M. J., ed., *Advances in Periglacial Geomorphology*. John Wiley & Sons, Chichester, pp. 49–67.

Priesnitz, K., Schunke, E. (1983). Periglaziale pediplanation in der Kanadischen Kordillere. In: Poser, H., Schunke, E., eds., *Mesoformen des reliefs im heutigen Periglazialraum. Abhandlungen Akademie Wissenschaften in Göttingen, Math.-Phys. Klasse*, no. 35, 266–280.

Priesnitz, K., Schunke, E. (2002). The fluvial morphodynamics of two small permafrost drainage basins, Richardson Mountains, Northwestern Canada. *Permafrost and Periglacial Processes*, **13**, 207–218.

Priestley, R. E. (1914). *Antarctic Adventure: Scott's Northern Party*. Fisher Unwin, London. Reprint, with a new foreword by Sir Vivian Fuchs, McClelland and Stewart, 1974, 382 pp.

Prince, H. (1961). Some reflexions on the origin of hollows in Norfolk compared with those in the Paris region. *Revue de Géomorphologie Dynamique*, **12**, 110–117.

Qiu, G., Cheng, G. (1995). Permafrost in China: past and present. *Permafrost and Periglacial Processes*, **6**, 3–14.

Qiu, G., Gorbunov, A. P. (1993). Studies of alpine permafrost in central Asia. 1. Northern Tian Shan. Permafrost Institute, Siberian Branch, Russian Academy of Sciences (Yakustsk) and Institute of Glaciology and Geocryology, Academia Sinica (Lanzhou), 79 pp.

Rabassa, J., Gomez, P. C., Diaz, M. V., Coronato, A., Alberti, A. P., Bujalesky, G. G., Vega, J. M. R. (1994). Periglacial features in Northern Tierra del Fuego, the Argentinian Fuegian Andes and Northern James Ross Island, Antarctic Peninsula: a comparison. *Bollettino di Geofisica*, **45**, 205–208.

Rachold, V., ed. (1999). Expeditions in Siberia in 1998. Reports on Polar Research, Alfred-Wegener-Institute for Polar and Marine Research, Bremerhaven, no. 315, 268 pp.

Rachold, V., Brown, J., Solomon, S., Sollid, J.-L., eds. (2003). Reports on Polar and Marine Research, Arctic Coastal Dynamics. Report of the 3rd International Workshop, University of Oslo, 2–5 December 2002. Alfred-Wegener-Institute for Polar and Marine Research, Bremerhaven, no. 443, 127 pp.

Rampton, V. N. (1974). The influence of ground ice and thermokarst upon the geomorphology of the Mackenzie-Beaufort region. In: Fahey, B. D., Thompson, R. D., eds., *Research in Polar and Alpine Geomorphology, Proceedings, 3rd Guelph Symposium on Geomorphology*. GeoBooks, Norwich, pp. 43–59.

Rampton, V. N. (1982). *Quaternary Geology of the Yukon Coastal Plain*. Geological Survey of Canada, Bulletin 317, 49 pp.

Rampton, V. N. (1988). *Quaternary geology of the Tuktoyaktuk Coastlands, Northwest Territories*. Geological Survey of Canada, memoir 423, 98 pp.

Rampton, V. N., Walcott, R. I. (1974). Gravity profiles across ice-cored topography. *Canadian Journal of Earth Sciences*, **11**, 110–122.

Rapp, A. (1960a). Recent development of mountain slopes in Karkevagge and surroundings, northern Sweden. *Geografiska Annaler*, **42**, 71–200.

Rapp, A. (1960b). Talus slopes and mountain walls at Tempelfjorden, Spitsbergen. *Norsk Polarinstitutt Skrifter*, **119**, 96 pp.

Rapp, A. (1967). Pleistocene activity and Holocene stability of hillslopes, with examples from Scandinavia and Pennsylvania. *Les Congrès et Colloques de L'Université de Liège*, **40**, 229–244.

Rapp, A. (1985). Extreme rainfall and rapid snowmelt as causes of mass movements in high latitude mountains. In: Church, M., Slaymaker, O., eds., *Field and Theory: Lectures in Geocryology*. University of British Columbia Press, Vancouver, pp. 35–56.

Raup, H. (1966). Turf hummocks in the Mesters Vig District, northeast Greenland. In: *Permafrost, International Conference Proceedings*, National Research Council of Canada publication 1287. National Academy of Sciences, Washington, DC, pp. 43–50.

Ray, R. J., Krantz, W. B., Caine, T. N., Gunn, R. D. (1983). A model for sorted patterned ground regularity. *Journal of Glaciology*, **29**, 317–337.

Reanier, R. E., Ugolini, F. C. (1983). Gelifluction deposits as sources of palaeoenvironmental information. In: *Permafrost, Proceedings of the Fourth International Conference on Permafrost*, 17–22 July, Fairbanks, Alaska. National Academy Press, Washington, DC, vol. 1, pp. 1042–1047.

Reedyk, S., Woo, M.-K., Prowse, T. D. (1995). Contribution of icing ablation to streamflow in a discontinuous permafrost area. *Canadian Journal of Earth Sciences*, **32**, 13–20.

Reger, R. D., Péwé, T. L. (1976). Cryoplanation terraces; indicators of a permafrost environment. *Quaternary Research*, **6**, 99–109.

Reid, C. (1887). On the origin of dry valleys and of coombe rock. *Quarterly Journal Geological Society*, **43**, 364–373.

Reimnitz, E., Maurer, D. K. (1979). Effects of storm surges on the Beaufort Sea coast, Northern Alaska. *Arctic*, **32**, 329–344.

Reimnitz, E., Eicken, H., Martin, T. (1995). Multi-year fast ice along the Taymyr Peninsula, Siberia. *Arctic*, **48**, 359–367.

Repelewska-Pekalowa, J., Pekala, K. (2004). Active-layer dynamics at the Calypsostranda CALM site, Recherche Fiord region, Spitsbergen. *Polar Geography*, **28**, 326–340.

Research Group on Experimental Roadbed Research (1979). Experimental roadbed in an area with a thick layer of ground ice. In: *Permafrost, Proceedings of the Third International Conference on Permafrost*, 10–13 July, Edmonton, Alberta, Canada. National Research Council of Canada, Ottawa, vol. 2, pp. 187–197.

Reynolds, C. M., Braley, W. A., Travis, M. D., Perry, L. B., Isandar, I. K. (1998). Bioremediation of hydrocarbon-contaminated soils and groundwater in northern climates. U.S. Army Corps of Engineers, CRREL, Hanover, New Hampshire, report no. 98-5, 18 pp.

Rice, A. (1976). Insolation warmed over. *Geology*, **4**, 61–62.

Richard, P. J. H. (1994). Wisconsinan Late-glacial environmental change in Québec: a regional synthesis. *Journal of Quaternary Science*, **9**, 165–170.

Richardson, J. (1839). Notice of a few observations which it is desirable to make on the frozen soil of British North America; drawn up for distribution among the Officers of the Hudson's Bay Company. *Journal of the Royal Geographical Society, London*, **9**, 117–120.

Richardson, J. (1841). On the frozen soil of North America. *Edinburgh New Philosophical Journal*, **30**, 110–123.

Richardson, J. (1851). *Arctic Searching Expedition*, vol. 1. Longman, Brown, Green and Longman, London, 413 pp.

Richter, D., Simmons, G. (1974). Thermal expansion behaviour of igneous rocks. *International Journal of Rock Mechanics, Mining Science, and Geomechanical Abstracts*, **11**, 403–411.

Richter, H., Haase, G., Barthel, H. (1963). Die Goletzterrassen. *Petermanns Geographische Mitteilungen*, **107**, 183–192.

Richter, W. (1985). Remarkable morphological forms in the Schirmacher Oasis, Dronning Maud Land, East Antarctica. *Zeitschrift für geologische Wissenschaften*, **13**, 389–398.

Riseborough, D. W. (2006). Discussion of C. R. Burn "Lake-bottom thermal regimes, western Arctic coast, Canada". *Permafrost and Periglacial Processes*, **17**, 87–89.

Riseborough, D. W., Smith, M. W. (1998). Exploring the limits of permafrost. In: Lewkowicz, A. G., Allard, M. eds., *Permafrost, Proceedings of the Seventh International Conference on Permafrost*, 23–27 June, Yellowknife, Canada. Centre d'études nordiques, Université Laval, Québec, Collection Nordicana no. 57, pp. 935–942.

Roberts, M. C. (1985). The geomorphology and stratigraphy of the Lizard Loess in south Cornwall, England. *Boreas*, **14**, 75–82.

Robinson, S. D., Pollard, W. H. (1998). Massive ground ice within Eureka Sound bedrock, Ellesmere Island, Canada. In: Lewkowicz, A. G., Allard, M., eds., *Permafrost, Proceedings of the Seventh International Conference on Permafrost*, 23–27 June, Yellowknife, Canada. Centre d'études nordique, Université Laval, Québec, Collection Nordicana no. 57, pp. 949–953.

Rochette, J.-C., Cailleux, A. (1971). Dépôts nivéo-èoliens annuels à Poste-de-la-Baleine, Nouveau-Québec. *Revue Géographie de Montréal*, **25**, 35–41.

Rockie, W. A. (1942). Pitting on Alaskan farms; a new erosion problem. *Geographical Review*, **32**, 128–134.

Rogov, V. V. (1987). The role of gas–liquid inclusions in mechanism of cryogenic disintegration of quartz. Vestnik, Moscow University, *Geography*, **3**, 81–85 (in Russian).

Romanovskii, N. N. (1961). Erosion-thermokarst depressions in the north of coastal lowlands of Yakutia and Novosibirsk islands. In: *Cryotic Studies*, no. 1, Izd. MGU, Moscow, pp. 124–144 (in Russian).

Romanovskii, N. N. (1973). Regularities in formation of frost-fissures and development of frost-fissure polygons. *Biuletyn Peryglacjalny*, **23**, 237–277.

Romanovskii, N. N. (1974). Development principles of the polygonal-wedge microrelief. *Vestnik Moskovskogo Universiteta, Geologiya*, **29**(5), 86–101 (in Russian).

Romanovskii, N. N. (1977a). On the theory of thermokarst. *Vestnik Moskovskogo Universiteta, Geologiya*, **32**, 65–71 (in Russian; English translation: UDC 554.343.001.11, pp. 48–53).

Romanovskii, N. N. (1977b). *Formation of Polygonal-Wedge Structures*. Nauka SSSR, Novosibirsk, 212 pp. (in Russian).

Romanovskii, N. N. (1978). Principles of classification of polygonal-vein structures. In: *Permafrost, Proceedings of the Third International Conference on Permafrost*, 10–13 July, Edmonton, Alberta, Canada. National Research Council of Canada, Ottawa, vol. 1, 318–324 (in Russian).

Romanovskii, N. N. (1980). *The Frozen Earth*. Moscow University Press, Moscow, 188 pp. (in Russian).

Romanovskii, N. N. (1985). Distribution of recently active ice and soil wedges in the USSR. In: Church, M., Slaymaker, O., eds., *Field and Theory; Lectures in Geocryology*. University of British Columbia Press, Vancouver, pp. 154–165.

Romanovskii, N. N. (1993). *Principles of Cryogenesis of Lithozone*. Moscow University Press, Moscow, 344 pp. (in Russian).

Romanovskii, N. N., Hubberten, H.-W. (2001). Results of permafrost modelling of the lowlands and shelf of the Laptev Sea region, Russia. *Permafrost and Periglacial Processes*, **12**, 191–202.

Romanovskii, N. N., Tyurin, A. I. (1983). Rock stream deserption. In: *Permafrost, Proceedings of the Fourth International Conference on Permafrost*, 17–23 July, Fairbanks, Alaska. National Academy Press, Washington, DC, vol. 1, pp. 1078–1082.

Romanovskii, N. N., Tyurin, A. I. (1986). Kurums. *Biuletyn Peryglacjalny*, **31**, 249–259.

Romanovskii, N. N., Zaitsev, V., Lisitrine, O., Tyurin, A. (1989). *Kurums and Cryolithogical Facies*. Moscow University Press, Moscow, 152 pp. (in Russian).

Romanovskii, N. N., Hubberten, H.-W., Gavrilov, A. V., Tumskoy, V. E., Tipenko, G. S., Grigoriev, M. N. (2000). Thermokarst and land–ocean interactions, Laptev Sea region, Russia. *Permafrost and Periglacial Processes*, **11**, 137–152.

Root, J. D. (1975). Ice-wedge polygons, Tuktoyaktuk area, NWT. *Geological Survey of Canada*, paper 75-1, B, 181.

Rosen, P. S. (2005). Boulder barricades. In: Schwartz, M. L., ed., *The Encyclopaedia of Coastal Science*, Springer, 204–206.

Rösli, A., Harnik, A. B. (1980). Improving the durability of concrete to freezing and de-icing salts. In: Sereda, P. J., Livan, G. G., eds., *Durability of Building Materials and Components*. ASTM STP 691. American Society for Testing and Materials, Philadelphia, PA, pp. 464–473.

Roth, E. S. (1965). Temperature and water content as factors in desert weathering. *Journal of Geology*, **73**, 454–468.

Rott, H. (1983). SAR data analysis for an Alpine test site. The European SAR-580 Experiment investigation preliminary report. Publication JRC-ESA, Brussels, September, pp. 1–14.

Rozenbaum, G. E., Shpolyanskaya, N. A. (1998a). Late Cenozoic history of the Russian Arctic. *Permafrost and Periglacial Processes*, **9**, 247–273.

Rozenbaum, G. E., Shpolyanskaya, N. A. (1998b). A model of Quaternary permafrost evolution in the Arctic. In: Lewkowicz, A. G., Allard, M., eds., *Permafrost, Proceedings of the Seventh International Conference on Permafrost*, 23–27 June, Yellowknife, Canada. Centre d'études nordiques, Université Laval, Québec, Collection Nordicana no. 57, pp. 973–978.

Rozenbaum, G. E., Arkhangelov, A. A., Koniakhin, M. A. (1978). Thermokarst-cave ice in Yana-Kolyma Lowland. *Moscow State University, Problems of Geocryology*, **7**, 74–92 (in Russian).

Ruddiman, W. F. (2004). Early anthropogenic overprints on Holocene climate. PAGES News, PAGES International Project Office, Bern, Switzerland, no. 12, 18–19.

Rudberg, S. (1963). Morphological processes and slope development on Axel Heiberg Island, NWT, Canada. *Abhandlungen der Akademie der Wissenschaften, Gottingen*, KL.11, 14, 218–228.

Ruegg, G. H. J. (1983). Periglacial eolian evenly-laminated sandy deposits in the Late Pleistocene of N.W. Europe, a facies unrecorded in modern sedimentological handbooks. In: Brookfield. M. E., Ahlbrandt, T. S., eds., *Eolian Sediments and Processes*. Elsevier, Amsterdam, pp. 455–482.

Ruz, M.-H., Hequette, A., Hill, P. R. (1992). A model of coastal evolution in a transgressed thermokarst topography, Canadian Beaufort Sea. *Marine Geology*, **106**, 251–278.

St-Onge, D. A. (1959). Note sur l'érosion du gypse en climat périglaciaire. *Revue Canadienne de Géographie*, **XIII**, 155–162.

St-Onge, D. A. (1965). La géomorphologie de l'île Ellef Ringnes, Territoires du Nord-Ouest, Canada. *Étude Géographique, Direction de la Géographie*, no. 38, Ottawa, 46 pp.

St-Onge, D. A. (1969). Nivation landforms. *Geological Survey Canada*, paper 69-30, 12 pp.

St-Onge, D. A., McMartin, I. (1995). Quaternary geology of the Inman River area, Northwest Territories. *Geological Survey of Canada*, Bulletin 446, 59 pp.

St-Onge, D. A., Pissart, A. (1990). Un pingo en système fermé dans des dolomies Palaeozoiques de l'arctique Canadien. *Permafrost and Periglacial Processes*, **1**, 275–282.

Salvigsen, O., Elgersma, A. (1985). Large-scale karst features and open taliks at Vardeborgsletta, outer Isfjorden, Svalbard. *Polar Research*, **3**, 145–153.

Savigny, K. W., Morgenstern, N. R. (1986a). Geotechnical conditions of slopes at a proposed pipeline crossing, Great Bear River valley, Northwest Territories. *Canadian Geotechnical Journal*, **23**, 490–503.

Savigny, K. W., Morgenstern, N. R. (1986b). In-situ creep properties in ice-rich permafrost soil. *Canadian Geotechnical Journal*, **23**, 504–514.

Savigny, K. W., Morgenstern, N. R. (1986c). Creep behaviour of undisturbed clay permafrost. *Canadian Geotechnical Journal*, **23**, 515–527.

Schafer, J. P. (1949). Some periglacial features in central Montana. *Journal of Geology*, **57**, 154–174.

Schlyter, P. (1995). Ventifacts as palaeo-wind indicators in southern Scandinavia. *Permafrost and Periglacial Processes*, **6**, 207–219.

Schokker, J., Koster, E. A. (2004). Sedimentology and facies distribution of Pleistocene cold-climate aeolian and fluvial deposits in the Roer Valley Graben (Southeastern Netherlands). *Permafrost and Periglacial Processes*, **15**, 1–20.

Schunke, E. (1977). The ecology of thufurs in Iceland. *Berichte ans der forschurgsstelle Nedri As, Hveragerdi (Island)*, no. 26, 39–69.

Schwan, J. (1986). The origin of horizontal alternating bedding in Weichselian aeolian sands in Northwestern Europe. *Sedimentary Geology*, **49**, 73–108.

Schweger, C. E. (1997). Late-Quaternary palaeoecology of the Yukon: a review. In: Danks, H. V., Downes, J. A., eds., *Insects of the Yukon, Biological Survey of Canada (Terrestrial Arthropods)*. Ottawa, pp. 59–72.

Scott, P. A., Hansell, R. I. C., Erickson, W. R. (1993). Influences of wind and snow on northern tree-line environments at Churchill, Manitoba, Canada. *Arctic*, **46**, 316–323.

Sego, D., Biggar, K. (2000). Enlarged base (belled) piles for use in ice or ice-rich permafrost. In: Senneset, K, ed., *Proceedings, International Workshop on Permafrost Engineering*, Longyearbyen, Svalbard, Norway, 18–21 June. Norwegian University of Science and Technology (NTNU)/ University Courses on Svalbard (UNIS), pp. 155–182.

Séguin, M. K., Allard, M. (1984). Le pergélisol et les processus thermokarstiques de la région de la rivière Nastapoca, Nouveau-Québec. *Géographie physique et quaternaire*, **XXXVIII**, 11–25.

Sekyra, J. (1956). The development of cryopedology in Czechoslovakia. *Biuletyn Peryglacjalny*, **4**, 351–369.

Sekyra, J. (1969). Periglacial phenomena in the oases and the mountains of the Enderby Land and the Dronning Maud Land (East Antarctica). *Biuletyn Peryglacjalny*, **19**, 277–289.

Selby, M. J. (1971a). Salt weathering of landforms, and an Antarctic example. In: *Proceedings, Sixth Geography Conference*, New Zealand Geographical Society, Christchurch, pp. 30–35.

Selby, M. J. (1971b). Slopes and their development in an ice-free, arid area of Antarctica. *Geografiska Annaler*, **53**, 235–245.

Selby, M. J. (1974). Slope evolution in an Antarctic Oasis. *New Zealand Geographer*, **30**, 18–34.

Selby, M. J. (1977). Transverse erosional marks on ventifacts from Antarctica. *New Zealand Journal of Geology and Geophysics*, **20**, 949–969.

Selby, M. J., Rains, R. B., Palmer, R. W. P. (1974). Eolian deposits of the ice-free Victoria Valley, Southern Victoria Land, Antarctica. *New Zealand Journal of Geology and Geophysics*, **17**, 543–562.

Seligman, B. J. (2000). Long-term variability of pipeline-permafrost interactions in North-West Siberia. *Permafrost and Periglacial Processes*, **11**, 5–22.

Sellman, P. V., Brown, J., Lewellen, R. I., McKim, H., Merry, C. (1975). The classification and geomorphic implications of thaw lakes on the Arctic coastal plain, Alaska. United States Army, CRREL, Hanover, New Hampshire, research report 344, 21 pp.

Senneset. K., ed. (2000). *Proceedings, International Workshop on Permafrost Engineering*, Long-yearbyen, Svalbard, Norway, 18–21 June. Norwegian University of Science and Technology (NTNU)/The University Courses on Svalbard (UNIS), 327 pp.

Seppälä, M. (1971). Evolution of eolian relief of the Kaamasjoki-Kiell-ajoki River Basin in Finnish Lapland. *Fennia*, **104**, 543–562.

Seppälä, M. (1972a). Location, morphology and orientation of inland dunes in Northern Sweden. *Geografiska Annaler*, **54A**, 85–104.

Seppälä, M. (1972b). The term "palsa": *Zeitschrift für Geomorphologie*, **16**, 463.

Seppälä, M. (1974). Some quantitative measurements of the present-day deflation on Hietatievat, Finnish Lapland. *Abhandlungen der Akademie der Wissenschaften Göttingen, Math.-Phys. Klasse*, III, **29**, 208–220.

Seppälä, M. (1982). An experimental study of the formation of palsas. In: French, H. M., ed., *Proceedings, Fourth Canadian Permafrost Conference*, Calgary, Alberta. National Research Council of Canada, Ottawa, pp. 36–42.

Seppälä, M. (1995a). How to make a palsa: a field experiment on permafrost formation. *Zeitschrift für Geomorphologie*, suppl. **99**, 91–96.

Seppälä, M. (1995b). Deflation and redeposition of sand dunes in Finnish Lapland. *Quaternary Science Reviews*, **14**, 799–809.

Seppälä, M. (1997). Piping causing thermokarst in permafrost, Ungava Peninsula, Québec, Canada. *Geomorphology*, **20**, 313–319.

Seppälä, M. (1998). New permafrost formed in peat hummocks (pounus), Finnish Lapland. *Permafrost and Periglacial Processes*, **9**, 367–374.

Seppälä, M. (2004). *Wind as a Geomorphic Agent in Cold Climates*. Cambridge University Press, Cambridge, 368 pp.

Seppälä, M. (2005). Frost heave on earth hummocks (pounus) in Finnish Lapland. *Norsk Geografisk Tidsskrift*, **59**, 171–176.

Seppälä, M., Gray, J., Richard, J., (1988). The development of depressed-centre ice-wedge polygons in the northernmost Ungava Peninsula, Québec, Canada. In: Senneset, K., ed., *Permafrost, Proceedings of the Fifth International Conference on Permafrost*, 2–5 August, Trondheim, Norway. Tapir, Trondheim, vol. 1, pp. 862–866.

Serrano, E., Martinez de Pison, E., Lopez-Martinez, J. (1996). Periglacial and nival landforms and deposits. In: Lopez-Martinez, J., Thomson, M. R. A., Thomson, J. W., eds., *Geomorphological Map of Byers Peninsula, Livingston Island*, BAS GEOMAP series, sheet 5-A, map (scale 1:25000) and text. British Antarctic Survey, Natural Environmental Research Council, pp. 28–34.

Shackleton, N. J., Opdyke, N. D. (1973). Oxygen isotope and palaeomagnetic stratigraphy of equatorial Pacific core V28-239: oxygen isotope temperatures and ice volumes on a 10-5 and 10-6 year scale. *Quaternary Research*, **3**, 39–55.

Shackleton, N. J., Berger, A., Peltier, W. A. (1990). An alternative astronomical calibration of the lower Pleistocene timescale based on ODP Site 677. *Transactions of the Royal Society of Edinburgh: Earth Sciences*, **81**, 251–261.

Shamanova, I. I. (1971). Manifestation of erosion in permafrost on the Yamal Peninsula. *Izvestiya Akad. nauk SSSR*, Ser. Geograf. [News of the USSR Academy of Sciences, Geographic Series], no. 2, 92–98 (in Russian; English translation, 1971, *Soviet Hydrology: Selected Papers*, no. 2, 198–203).

Shang, J. (1982). Some characteristics of permafrost along the Qinghai-Xizang Plateau. In: *Proceedings, First Chinese Conference on Glaciology and Geocryology*. China Science Press, Beijing, pp. 38–43 (in Chinese).

Sharp, R. P. (1942a). Ground ice mounds in tundra. *Geographical Review*, **32**, 417–423.

Sharp, R. P. (1942b). Periglacial involutions in northeastern Illinois. *Journal of Geology*, **50**, 113–133.

Sharp, R. P. (1949). Pleistocene ventifacts east of the Big Horn mountains, Wyoming. *Journal of Geology*, **57**, 175–195.

Shepherd, A., Wingham, D. J., Mansley, J. A., Corr, H. F. (2001). Inland thinning of Pine Island Glacier, West Antarctica. *Science*, **291**, 862–864.

Sher, A. V., Kaplina, T. N., Giterman, R. E., Lozhkin, A. V., Arkhangelov, A. A., Kiselyov, S. V., Kouznetsov, Yu V., Virina, E. I., Zazhigin, V. S. (1979). *Late-Cenozoic of the Kolyma Lowlands*, Guidebook, XIV Pacific Science Congress. Academy of Sciences, USSR, Moscow, 115 pp.

Shi, Y., ed. (1988). *Map of Snow, Ice and Frozen Ground in China (1:4000000), with Explanatory Notes*. China Cartographic Publishing House, Beijing.

Shilts, W. W. (1974). Physical and chemical properties of unconsolidated sediments in permanently frozen terrain, District of Keewatin. *Geological Survey of Canada*, paper 74-A, 229–235.

Shilts, W. W. (1975). Tundra fires, southeast District of Keewatin. *Geological Survey of Canada*, paper 75-1, part B, pp. 187–195.

Shilts, W. W. (1978). Nature and genesis of mudboils, central Keewatin, Canada. *Canadian Journal of Earth Sciences*, **15**, 1053–1068.

Shostakovitch, W. B. (1927). Der ewig gefrorene boden Siberiens. *Gessel Erdkunde Berlin Zeitschrift*, 394–427.

Shotton, F. W. (1960). Large-scale patterned ground in the valley of the Worcestershire Avon. *Geological Magazine*, **97**, 404–408.

Shumskii, P. A. (1959). Ground (subsurface) ice. In: *Principles of Geocryology*, Part 1, *General Geocryology*. Academy of Sciences of the USSR, Moscow, chapter IX, pp. 274–327 (in Russian; English translation: C. de Leuchtenberg (1964), National Research Council of Canada, Ottawa, technical translation 1130, 118 pp.).

Shumskiy, P. A., Vtyurin, B. I. (1966). Underground ice. In: *Proceedings, 1st International Permafrost Conference*. National Academy of Science–National Research Council of Canada, publication 1287, pp. 108–113.

Shur, Y. L. (1977). *Thermo-Physical Principles Behind the Thermokarst Process*. Nedra, Moscow, 80 pp. (in Russian).

Shur, Y. L. (1988a). *The Upper Horizon of Permafrost and Thermokarst*. Akademia Moscow, Nauka, 210 pp. (in Russian).

Shur, Y. L. (1988b). The upper horizon of permafrost soils. In: Senneset, K., ed., *Permafrost, Proceedings of the Fifth International Conference on Permafrost*, 2–5 August. Tapir, Trondheim, vol. 1, pp. 867–871.

Shur, Y., Hinkel, K. M., Nelson, F. E. (2005). The transient layer: implications for geocryology and climate-change science. *Permafrost and Periglacial Processes*, **16**, 5–18.

Shur, Y., French, H. M., Bray, M. T., Anderson, D. A. (2004). Syngenetic permafrost growth: cryostratigraphic observations from the CRREL Tunnel near Fairbanks, Alaska. *Permafrost and Periglacial Processes*, **15**, 339–347.

Shur, Y., Vasiliev, A., Kanevsky, M., Maximov, V., Pokrovsky, S., Zaikanov, V. (2002). Shore erosion in Russian Arctic. In: Merrill, K. S., ed., *Proceedings, Eleventh International Conference, Cold Regions Impacts on Transportation and Infrastructure*, May 20–22, Anchorage, Alaska. ASCE, VA, pp. 736–747.

Sigafoos, R. S. (1951). Soil instability in tundra vegetation. *Ohio Journal of Science*, **51**, 281–298.

Sissons, J. B. (1981) The Loch Lomond stadial in the British Isles. *Nature*, **280**, 199–203.

Slaymaker, O., French, H. M. (1993). Cold environments and global change. In: French, H. M., Slaymaker, O., eds., *Canada's Cold Environments*. McGill-Queen's University Press, Montreal, pp. 313–334.

Sletten, R. S., Hallet, B., Fletcher, R. C. (2003). Resurfacing time of terrestrial surfaces by the formation and maturation of polygonal patterned ground. *Journal of Geophysical Research*, **108** (E4), 8044, doi: 10.1029/2002JE001914, 2003; GDS 25-1-10.

Sloan, C. E., van Everdingen, R. O. (1988). Region 28, Permafrost region. In: Back, W., Rosenshein, J. C., Seaber, P. R., eds., *Hydrology, The Geology of North America*. Geological Society of America, Boulder, CO, vol. 0-2, pp. 263–270.

Small, R. J., Clark, M. J., Lewin, J. (1970). The periglacial rock-stream at Clatford Bottom, Marlborough Downs, Wiltshire. *Proceedings, Geologist's Association*, **81**, 87–98.

Smith, D. I. (1972). The solution of limestone in an arctic environment. In: *Polar Geomorphology*, Insitute of British Geographers Special Publication no. 4, pp. 187–200.

Smith, D. J. (1987). Frost-heave activity in the Mount Rae Area, Canadian Rocky Mountains. *Arctic and Alpine Research*, **19**, 155–166.

Smith, D. W., Sego, D. C., eds. (1994). *Cold Regions Engineering: A Global Perspective. Proceedings, 7th International Cold Regions Engineering Specialty Conference*, March 7–9, 1994, Edmonton, Alberta. Canadian Society for Civil Engineering, Montreal, 869 pp.

Smith, H. T. U. (1949). Physical effects of Pleistocene climatic changes in non-glaciated areas; eolian phenomena, frost action and stream terracing. *Bulletin, Geological Society of America*, **60**, 1485–1516.

Smith, H. T. U. (1962). Periglacial frost features and related phenomena in the United States. *Biuletyn Peryglacjalny*, **11**, 325–342.

Smith, H. T. U. (1964). Periglacial eolian phenomena in the United States. In: *6th International Congress on the Quaternary (VI–INQUA)*, Warsaw, 1961. *Report*, **4**, 177–186.

Smith, M. W. (1977). Computer simulation of microclimate and ground thermal regimes: test results and programme description. ALUR 1975-76-72. Department of Indian Affairs and Northern Development, Ottawa, 74 pp.

Smith, M. W. (1985a). Models of soil freezing. In: Church, M., Slaymaker, O., eds., *Field and Theory: Lectures in Geocryology*. University of British Columbia Press, Vancouver, BC, pp. 96–120.

Smith, M. W. (1985b). Observations of soil freezing and frost heave at Inuvik, Northwest Territories, Canada. *Canadian Journal of Earth Sciences*, **22**, pp. 283–290.

Smith, M. W., Hwang, C. T. (1973). Thermal disturbance due to channel shifting, Mackenzie Delta area, NWT, Canada. In: *Permafrost: North American Contribution, Second International Conference*, Yakutsk, USSR. National Academy of Science–National Research Council of Canada Publication 2115, pp. 51–60.

Smith, M. W., Riseborough, D. W. (2002). Climate and the limits of permafrost: a zonal analysis. *Permafrost and Periglacial Processes*, **13**, 1–15.

Smith, S. L., Williams, P. J. (1990). Ice lens orientation around a chilled buried pipe. In: *Proceedings, 5th Canadian Permafrost Conference*, Université Laval, Nordicana 54, 83–87.

Smith, S. L., Burgess, M. M., Nixon, F. M. (2001). Response of active-layer and permafrost temperatures to warming during 1998 in the Mackenzie Delta, Northwest Territories, and at Canadian Forces Station Alert and Baker Lake, Nunavut. *Geological Survey of Canada, Current Research* 20001-E5, 8 pp.

Soil Classification Working Group (1998). *The Canadian System of Soil Classification*, 3rd edn. Research Branch, Agriculture and Agri-Food Canada, Publication 1646, 187 pp.

Sokolov, B. L. (1978). Regime of naleds. In: *Permafrost; The USSR Contribution to the Second International Conference*. National Academy of Sciences, Washington, DC, pp. 408–411.

Sollid, J. L., Sörbel, L. (1992). Rock glaciers in Svalbard and Norway. *Permafrost and Periglacial Processes*, **3**, 215–220.

Solomatin, V. I. (1986). *Petrology of Underground Ice*. Academy Nauka, Novosibirsk, 215 pp. (in Russian).

Soloviev, P. A. (1962). *Alasnyye relief Centralnoj Jakutii i ego proiskhozhdeniye. Mnoholetnemerzlyye Porody I Soputstvuyuhchiye yim Yavleniya na territorii JASSR*. Izdatelstvo AN SSSR, Moscow, pp. 38–53 (in Russian).

Soloviev, P. A. (1973a). *Alas Thermokarst Relief of Central Yakutia*. Guidebook, Second International Permafrost Conference, Yakutsk, USSR, 48 pp.

Soloviev, P. A. (1973b). Thermokarst phenomena and landforms due to frost heaving in Central Yakutia. *Biuletyn Peryglacjalny*, **23**, 135–155.

Soons, J. M., Price, L. W. (1990). Periglacial phenomena in New Zealand. *Permafrost and Periglacial Processes*, **1**, 145–160.

Souchez, R. (1966). Réflexions sur l'évolution des versants sous climat froid. *Revue de Géographie Physique et de Géologie Dynamique*, **VIII**, 317–334.

Souchez, R. (1967a). Gélivation et évolution des versants en bordure de l'Islandsis d'Antartides orientale. *Les Congrès et Colloques de L'Université de Liège*, **40**, 291–298.

Souchez, R. (1967b). Le recul des verrous-gradins et les rapports glaciaire-périglaciaire en Antarctique. *Revue de Géomorphologie dynamique*, no. 2, 6 pp.

Sparks, B. W., Williams, R. G. B., Bell, F. G. (1972). Presumed ground ice depressions in East Anglia. *Proceedings of the Royal Society of London*, **A327**, 329–343.

Spate, A. P., Burgess, J. S., Shevlin, J. (1995). Rates of rock surface lowering, Princess Elizabeth Land, Eastern Antarctica. *Earth Surface Processes and Landforms*, **20**, 567–573.

Stalker, A. (1960). Ice-pressed drift forms and associated deposits in Alberta. *Geological Survey of Canada, Bulletin*, **57**, 38 pp.

Stalker, A. (1984). Ice age bones – a clue. *Geos*, **2**, 11–14.

Stalker, A., Churcher, C. S. (1982). Ice-Age deposits and animals from the southwestern part of the Great Plains of Canada. *Geological Survey of Canada*, miscellaneous report, no. 31.

Stanford, S. D., Ashley, G. M., Russell, E. W. B., Brenner, G. J. (2002). Rates and patterns of late-Cenozoic denudation in the northernmost Atlantic coastal plain and piedmont. *Bulletin, Geological Society of America*, **114**, 1422–1437.

Stangl, K. O., Roggensack, W. D., Hayley, D. W. (1982). Engineering geology of surficial soils, eastern Melville Island. In: French, H. M., ed., *Proceedings, Fourth Canadian Permafrost Conference*, Calgary, Alberta. National Research Council of Canada, Ottawa, pp. 136–147.

Street, R. B., Melnikov, P. I. (1990). Seasonal snow cover, ice and permafrost. In: Tegart, W. J. McG., Sheldon, G. W., Griffiths, D. C., eds., *Climate Change, The IPCC Impacts Assessment*, Report prepared for IPCC by Working Group II, WMO-UNEP, Australian Govenment Publishing Service, Canberra, chapter 7, pp. 7-1–7-33.

Streletskii, D. A., Streletskaya, I. D., Rogov, V. V., Leibman, M. O. (2003). Redistribution of ions within the active layer and upper permafrost, Yamal, Russia. In: Phillips, M., Springman, S. A., Arenson, L. U., eds., *Permafrost, Proceedings of the Eighth International Conference on Permafrost*, 18–21 July, Zurich, Switzerland. Balkema, Lisse, vol. 2, pp. 1117–1122.

Sumgin, M. I. (1927). *Soil permafrost within the USSR*, 1st edn. Vladivostok (2nd edn, Acad. Nauka, Moscow, 372 pp.) (in Russian).

Sutcliffe, A. J. (1985). *On the Track of Ice Age Mammals*. British Museum (National History), London, 224 pp.

Svensson, H. (1977). Observations on polygonal fissuring in non-permafrost areas of the Norden countries. *Abhandlungen der Akademie der Wissenschaften in Gottingen, Mathematische-Physikalische Klasse*, **31**, 63–76.

Svensson, H. (1983). Ventifacts as paleowind indicators in a former periglacial area of southern Scandinavia. In: *Permafrost, Proceedings of the Fourth International Conference*, 17–22 July, Fairbanks, Alaska. National Academy Press, Washington, DC, vol. 1, pp. 1217–1220.

Svensson, H. (1988a). Ice-wedge casts and relict polygonal patterns in Scandinavia. *Journal of Quaternary Science*, **3**, 57–67.

Svensson, H. (1988b). Recent frost fissuring in a coastal area of southwestern Sweden. *Norsk Geografisk Tidsskrift*, **42**, 271–277.

Sverdrup, H. V. (1938). Notes on erosion by drifting snow and transport of solid material by sea ice. *American Journal of Science*, **35**, 370–373.

Sweeting, M. M. (1972). *Karst Landforms*. Macmillan, London, 362 pp.

Synge, F. M. (1981). Quaternary glaciation and changes of sea level in the south of Ireland, *Geologie en Mijnbouw*, **650**, 305–351.

Szymanski, M. B., Zivkovic, A., Tchekhovski, A., Swarbrick, B. (2003). Designing for closure of an open pit in the Canadian Arctic. In: Phillips, M., Springman, S. A., Arenson, K. U., eds., *Permafrost, Proceedings of the Eighth International Conference on Permafrost*, 21–25 July, Zurich, Switzerland. Balkema, Lisse, vol. 2, pp. 1123–1128.

Taber, S. (1929). Frost heaving. *Journal of Geology*, **37**, 428–461.

Taber, S. (1930). The mechanics of frost heaving. *Journal of Geology*, **38**, 303–317.

Taillefer, F. (1944). La dissymétrie des vallées Gasgonnes. *La revue de Géographie de Pyrénnees et du Sud-Ouest*, **xv**, 153–181.

Tarnocai, C. (2004a). Classification of cryosols in Canada. In: Kimble, J. M., ed., *Cryosols: Permafrost-Affected Soils*. Springer-Verlag, Berlin, pp. 599–610.

Tarnocai, C. (2004b). Organic carbon in Cryosolic soils in the northern circumpolar region. *Frozen Ground*, News Bulletin, International Permafrost Association, no. 28, 6–7.

Tarnocai, C., Netterville, J. A. (1976). Some characteristics of a pingo in the Simpson Peninsula, N. W. T. *Canadian Journal of Earth Sciences*, **13**, 490–492.

Tarnocai, C., Pettapiece, W. W., Zoltai, S. C. (1973). Cryoturbated soils in Northern Canada. Canada Soil Survey, Department of Agriculture, Winnipeg, 64 pp.

Tarnocai, C., Nixon, F. M., Kutny, L. (2004). Circumpolar-Active-Layer-Monitoring (CALM) sites in the Mackenzie Valley, northwestern Canada. *Permafrost and Periglacial Processes*, **15**, 141–153.

Taylor, A. E. (1995). Field measurements of n-factors for natural forest areas, Mackenzie Valley, Northwest Territories. *Geological Survey of Canada*, paper 1995-B, 89–98.

Taylor, A. E. (2000). Relationship of ground temperatures to air temperatures in forests. *Geological Survey of Canada, Bulletin*, **547**, 111–117.

Taylor, G. (1916). *With Scott: The Silver Lining*. Smith, Elder and Co., London, 464 pp.

Taylor, G. (1922). *The Physiography of The McMurdo Sound and Granite Harbour Region. British Antarctic (Terra Nova) Expedition, 1910–1913*. Harrison and Sons Ltd, London, 246 pp.

Taylor, R. B., McCann, S. B. (1976). The effect of sea and nearshore ice on coastal processes in Canadian Arctic Archipelago. *Revue Géographie de Montréal*, **30**, 123–132.

Tedrow, J. C. F. (1969). Thaw lakes, thaw sinks and soils in northern Alaska. *Biuletyn Peryglacjalny*, **20**, 337–345.

Tedrow, J. C. F. (1977). *Soils of the Polar Landscapes*. Rutgers Press, New Brunswick, NJ, 638 pp.

Te Punga, M. T. (1956). Altiplanation terraces in southern England. *Biuletyn Peryglacjalny*, **4**, 331–338.

Te Punga, M. T. (1957). Periglaciation in southern England. *Tijdschrift Koninklijk Nederlands Aardrijkskundig Genootschap*, **74**, 400–412.

Thidaudeau, P., Roberge, J., Lauriol, B. (1988). Agressivité chimique des eaux dans les massifs calcaires du nord du Yukon, Canada. *Revue de géomorphologie dynamique*, **XXXVII**, 61–70.

Thie, J. (1974). Distribution and thawing of permafrost in the southern part of the discontinuous zone in Manitoba. *Arctic*, **27**, 189–200.

Thomasson, A. J. (1961). Some aspects of the drift deposits and geomorphology of southeast Hertfordshire. *Proceedings, Geologist's Association*, **72**, 287–302.

Thompson, E. C., Sayles, F. H. (1972). In-situ creep analysis of a room in frozen soil. *ASCE Journal of Soil Mechanics and Foundation Division*, **98**, 899–916.

Thomson, S. (1966). Icings on the Alaskan Highway. In: *Proceedings, First International Permafrost Conference*. National Academy of Sciences-National Research Council of Canada, publication 1287, pp. 526–529.

Thorn, C. E. (1976). Quantitative evaluation of nivation in the Colorado Front Range. *Geological Society of America Bulletin*, **87**, 1169–1178.

Thorn, C. E. (1979a). Ground temperatures and surficial transport in colluvium during snow-patch meltout, Colorado Front Range. *Arctic and Alpine Research*, **11**, 41–52.

Thorn, C. E. (1979b). Bedrock freeze–thaw weathering regime in an alpine environment, Colorado Front Range. *Earth Surface Processes*, **4**, 211–228.

Thorn, C. E. (1988). Nivation: a geomorphic chimera. In: Clark, M. J., ed., *Advances in Periglacial Geomorphology*. John Wiley & Sons, Chichester, pp. 3–31.

Thorn, C. E. (1992). Periglacial geomorphology. What? Where? When? In: Dixon, J. C., Abrahams, A. D., eds, *Periglacial Geomorphology*. John Wiley & Sons, Chichester, pp. 1–30.

Thorn, C. E., Hall, K. (1980). Nivation: an arctic–alpine comparison and reappraisal. *Journal of Glaciology*, **25**, 109–124.

Thorn, C. E., Hall, K. (2002). Nivation and cryoplanation: the case for scrutiny and integration. *Progess in Physical Geography*, **26**, 533–550.

Thorn, C. E., Loewenherz, D. S. (1987). Spatial and temporal trends in alpine periglacial studies: implications for paleo-reconstruction. In: Boardman, J., ed., *Periglacial Processes and Landforms in Britain and Ireland*. Cambridge University Press, Cambridge, pp. 57–65.

Thorn, C. E., Darmody, R. G., Dixon, J. C., Schlyter, P. (2001). The chemical weathering regime of Karkevagge, arctic-alpine Sweden. *Geomorphology*, **41**, 37–52.

Thorn, C. E., Darmody, R. G., Dixon, J. C., Schlyter, P. (2002). Weathering rates of buried machine-polished rock disks, Karkevagge, Swedish Lapland. *Earth Surface Processes and Landforms*, **27**, 831–845.

Thorsteinsson, R., Tozer, E. T. (1962). *Banks, Victoria and Stefansson Islands, Arctic Archipelago*. Geological Survey of Canada, Memoir 330, 83 pp.

Tikhomirov, B. A. (1959). The origin of mounds associated with thermokarst and the dynamics of vegetation in northern Siberia. In: *Fundamentals of Permafrostology, Proceedings, 7th Inter-department Committee*. USSR Academy of Sciences, Moscow, pp. 162–167 (in Russian).

Tolstov, A. N. (1961). The region of intensive erosion and thermokarst. *Problemy Severa*, **4**, 151–156 (in Russian; English translation, 1962, *Problems of the North*, **4**, 159–163).

Tomirdiaro, S. V., Ryabchun, V. K. (1978). Lake thermokarst on the Lower Anadyr Lowland. In: *Permafrost: USSR Contribution to the Second International Conference*, Yakutsk, USSR. Washington, DC, National Academy of Sciences, pp. 94–100.

Tong, B., Li, S. (1983). Permafrost characteristics and factors affecting them, Qinghai-Xizang Plateau. In: *Professional Papers on Qinghai-Xizang Plateau Permafrost*. China Science Press, Beijing, pp. 1–11 (in Chinese).

Trenhaile, A. S., Rudakas, P. A. (1981). Freeze–thaw and shore platform development in Gaspé, Québec. *Géographie physique et quaternaire*, **35**, 171–181.

Tricart, J. (1950). Le modélé périglaciaire. *Cours de Géomorphologie*, 1, Fasc. 1, CDU, Paris, 270 pp.

Tricart, J. (1963). Géomorphologie des régions froides. Presses Universitaires de France, Paris, 289 pp.

Tricart, J. (1968). Periglacial landscapes. In: Fairbridge, R. W., ed., *Encyclopedia of Geomorphology*. Reinhold, New York, pp. 829–833.

Tricart, J. (1970). *Geomorphology of Cold Environments*. Translation of *Géomorphologie des regions froides*, revised by J. Tricart in 1968, by E. Watson. Macmillan, St Martin's Press, London, 320 pp.

Tricart, J., Cailleux, A. (1967). *Le modélé des régions périglaciaires*. Traité de Géomorphologie, Paris, SEDES, 512 pp.

Troll, C. (1944). Strukturböden, Solifluktion und Frostklimate de Erde. *Geologische Rundschau*, **34**, 545–694. (English translation, 1958, Structure soils, solifluction and frost climate of the Earth. Translation 43, United States Army Snow Ice and Permafrost Research Establishment, Corps of Engineers, Willmette, Illinois, 121 pp.).

Troll, C. (1962). "Solle" and "Mardelle". *Erdkunde*, **16**, 31–34.

Tsytovich, N. A. (1973). Mekhanika merzlykh gruntov [The mechanics of frozen ground]. Vysshaya Shkola Press, Moscow, 446 pp. (in Russian; English translation: Swinzow, G. K., Tschebotarioff, G. P., eds., 1975, Scripta/McGraw-Hill, New York, 426 pp.).

Tsytovich, N. A., Sumgin, M. I. (1937). *Principles of Frozen Ground Mechanics*. Moscow, Nauka, 423 pp. (in Russian).

Tyrtikov, A. P. (1964). The effect of vegetation on perennially frozen soil. National Research Council of Canada, technical translation 1088, Ottawa, pp. 69–90.

Tyurin, A. J., Romanovskii, N. N., Poltev, N. F. (1982). *Frost Facies Analysis of Rock Streams*. Nauka, Moscow, 148 pp. (in Russian).

Urdea, P. (1992). Rock glaciers and periglacial phenomena in the Southern Carpathians. *Permafrost and Periglacial Processes*, **3**, 267–273.

Ushakova, L. F. (1986). Changing of intensity of weathering as function of depth. *Izvestiya VNII Gidrotechniky*, **193**, 79–82 (in Russian).

Van Vliet-Lanoë, B. (1982). Structures et microstructures associées à la formation de glace de ségrégation: leur conséquences. In: French, H. M., ed., *Proceedings, Fourth Canadian Permafrost Conference*, Calgary, Alberta. National Research Council of Canada, Ottawa, pp. 116–122.

Van Vliet-Lanoë, B. (1985). Frost effects in soils. In: Boardman, J., ed., *Soils and Quaternary Landscape Evolution*. John Wiley & Sons, Chichester, pp. 117–158.

Van Vliet-Lanoë, B. (1988). The genesis of cryoturbations and their significance in environmental reconstruction. *Journal of Quaternary Science*, **3**, 85–96.

Van Vliet-Lanoë, B. (1991). Differential frost heave, load casting and convection: converging mechanisms; a discussion of the origin of cryoturbations. *Permafrost and Periglacial Processes*, **2**, 123–139.

Van Vliet-Lanoë, B. (1998). Frost and soils: implications for palaeosols, paleo-climates and stratigraphy. *Catena*, **34**, 157–183.

Van Vliet-Lanoë, B., Seppälä, M. (2002). Stratigraphy, age and formation of peaty earth hummocks (pounus), Finnish Lapland. *The Holocene*, **12**, 187–199.

Van Vliet-Lanoë, B., Coutard, J.-P., Pissart, A. (1984). Structures caused by repeated freezing and thawing in various loamy sediments. A comparison of active, fossil and experimental data. *Earth Surface Processes and Landforms*, **9**, 553–566.

Van Vliet-Lanoë, B., Seppälä, M., Kayhko, J. (1993). Dune dynamics and cryoturbation features controlled by Holocene water level changes, Hietatievat, Finnish Lapland. *Geologie en Mijnbouw*, **72**, 211–224.

Van Vliet-Lanoë, B., Fox, C. A., Gubn, S. V. (2004). Micromorphology of cryosols. In: Kimble, J. M., ed., *Cryosols: Permafrost-Affected Soils*. Springer-Verlag, New York, chapter 2, pp. 365–390.

Vandenberghe, J. (1983). Some periglacial phenomena and their stratigraphical position in the Weichselian deposits in the Netherlands. *Polarforschung*, **53**, 97–107.

Vandenberghe, J. (1993). Changing fluvial processes under changing periglacial conditions. *Zeitschrift für Geomorphologie*, suppl. **88**, 17–28.

Vandenberghe, J., Krook, L. (1982). Statigraphy and genesis of Pleistocene deposits at Alphen (southern Netherlands). *Gologie en Mijnbouw*, **60**, 417–426.

Vandenberghe, J., Nugteren, G. (2001). Rapid climatic changes recorded in loess successions. *Global and Planetary Change*, **28**, 1–9.

Vandenberghe, J., Pissart, A. (1993). Permafrost changes in Europe during the Last Glacial. *Permafrost and Periglacial Processes*, **4**, 121–135.

Vandenberghe, J., van den Broek, P. (1982). Weichselian convolution phenomena and processes in fine sediments. *Boreas*, **11**, 299–315.

Vandenberghe, J., Kasse, C., Bohncke, S., Kozarski, S. (1994). Climate-related river activity at the Weichselian-Holocene transition: a comparative study of the Warta and Maas rivers. *Terra Nova*, **6**, 476–485.

Vandenberghe, J., Isarin, R. F. B., Renssen, H. (1999). Comments on "Windpolished boulders as indicators of a Late Weichselian wind regime in Denmark in relation to neighbouring areas" by Christiansen and Svensson. (9, (1): 1–21, 1998). *Permafrost and Periglacial Processes*, **10**, 199–201.

Vandenberghe, J., Zhijiu, C., Liang, Z., Wei, Z. (2004). Thermal-contraction crack networks as evidence for Late-Pleistocene permafrost in Inner Mongolia, China. *Permafrost and Periglacial Processes*, **15**, 21–29.

van der Vinne, G., Prowse, T. D., Andres, D. (1991). Economic impact of river ice jams in Canada. In: Prowse, T. D., Ommanney, C. S. L., eds., *Northern Hydrology: Selected Perspectives*, National Hydrology Research Institute Symposium no. 6, Saskatoon, pp. 333–352.

van Everdingen, R. O. (1978). Frost mounds at Bear Rock near Fort Norman, NWT, 1975–1976. *Canadian Journal of Earth Sciences*, **15**, 263–276.

van Everdingen, R. O. (1981). Morphology, hydrology and hydrochemistry of karst in permafrost terrain near Great Bear Lake, Northwest Territories. National Hydrology Research Institute, paper no. 11, Inland Waters Directorate, 53 pp.

van Everdingen, R. O. (1982). Management of groundwater discharge for the solution of icing problems in the Yukon. In: French, H. M., ed., *Proceedings, Fourth Canadian Permafrost Conference*, Calgary, Alberta. National Research Council of Canada, Ottawa, pp. 212–228.

van Everdingen, R. O. (1987). The importance of permafrost in the hydrological regime. In: Healey, M. C., Wallace, R. R., eds., *Canadian Bulletin of Aquatic Resources*, **215**, pp. 243–276.

van Everdingen, R. O. (1990). Ground-water hydrology. In: Prowse, T. D., Ommaney, C. S. L., eds., *Northern Hydrology: Canadian Perspectives*, National Hydrology Research Institute Report no. 1, Saskatoon, pp. 77–101.

van Everdingen, R. O., compiler and editor (1998). Multi-language glossary of permafrost and related ground-ice terms, version 2. International Permafrost Association. Arctic Institute of North America, University of Calgary, Alberta, mimeographed.

van Huissteden, J. (1990). Tundra rivers of the last glacial: sedimentation and geomorphological processes during the middle pleniglacial in Twente, eastern Netherlands. *Mededelingen Rijks Geologische Dienst*, no. 44-3, 3–138.

van Huissteden, J., Vandenberghe, J. (1988). Changing fluvial style of periglacial lowland rivers during the Weichselian Pleniglacial in the eastern Netherlands. *Zeitschrift für Geomorphologie*, suppl. **71**, 131–146.

van Steijn, H., Bertran, P., Francou, B., Hétu, B., Texier, J.-P. (1995). Models for the genetic and environmental interpretation of stratified slope deposits: a review. *Permafrost and Periglacial Processes*, **6**, 125–146.

Vasil'chuk, Y. K., Vasil'chuk, A. C. (1997). Radiocarbon dating and oxygen-isotope variations in Late-Pleistocene syngenetic ice wedges, Northern Siberia. *Permafrost and Periglacial Processes*, **8**, 335–345.

Velichko, A. A. (1975). Paragenesis of a cryogenic (periglacial) zone. *Biuletyn Peryglacjalny*, **24**, 89–110.

Velichko, A. A. (1982). *Paleogeography of Europe During the Last One Hundred Thousand Years*, (gen. ed. Gerasimov, I. P.) Nauka, Moscow, 156 pp. (in Russian with abstract and legends in English), 156 pp.

Vernon, P., Hughes, O. L. (1966). Surficial geology of Dawson, Larsen Creek and North Creek map areas, Yukon Territory. *Geological Survey of Canada Bulletin*, **136**, 25 pp.

Viereck, L. A. (1965). Relationship of white spruce to lenses of perennially frozen ground, Mount McKinley National Park. *Arctic*, **18**, 262–267.

Viereck, L. A. (1973a). Ecological effects of river flooding and forest fires on permafrost in the taiga of Alaska. In: *Permafrost; North American Contribution, Second International Permafrost Conference*, Yakutsk, USSR. National Academy of Science, publication 2115, pp. 60–67.

Viereck, L. A. (1973b). Wildfire in the taiga of Alaska. *Quaternary Research*, **3**, 465–495.

Viles, H., Spencer, T., eds. (1995). *Coastal Problems*. Edward Arnold, London, 350 pp.

Vincent, J.-S., (1982). The Quaternary history of Banks Island, NWT, Canada. *Géographie physique et Quaternaire*, **36**, 209–232.

Vincent, J.-S. (1983). *La géologie du Quaternaire et la géomorphologie de l'île de Banks, Arctique Canadien*. Geological Survey of Canada, Memoir 405, 118 pp.

Vincent, J.-S. (1989). Quaternary geology of the northern Canadian Interior Plains. In: Fulton, R. J., ed., *Quaternary Geology of Canada and Greenland*, Geology of Canada, No. 1, Geological Survey of Canada, Ottawa, pp. 100–137.

Vogt, T., Corte, A. E. (1996). Secondary precipitates in Pleistocene and present cryogenic environments (Mendoza Precordillera, Argentina, Transbaikalia, Siberia, and Seymour Island, Antarctica). *Sedimentology*, **43**, 53–64.

Vogt, T., del Valle, H. F. (1994). Calcretes and cryogenic structures in the area of Puerto Madryn (Chubut, Patagonia, Argentina). *Geografiska Annaler*, **76A**, 57–75.

Vogt, T., Larqué, P. (1998). Transformations and neotransformations of clay in the cryogenic environment: examples from Transbaikalia (Siberia) and Patagonia (Argentina). *European Journal of Soil Science*, **49**, 367–376.

Vogt, T., Erbajeva, M. A., Vogt, H. (1995). Premières preuves de conditions périglaciaires au Pléistocene inférieur en Transbaikalie (Sibérie, Russe). *Comptes Rendus, Academie des Sciences Paris*, Série II, **320**, 861–866.

Vtyurin, B. I. (1975). *Underground ice in the USSR*. Nauka, Moscow, 212 pp. (in Russian).

Vtyurina, E. A. (1974). *The Cryogenic Structures of the Active Layer*. Nauka, Moscow, 126 pp. (in Russian).

Wagner, S. (1992). Creep of alpine permafrost, investigated on the Murtel rock glacier. *Permafrost and Periglacial Processes*, **3**, 157–162.

Wahl, H. E., Fraser, D. B, Harvey, R. C., Maxwell, J. B. (1987). *Climate of Yukon*. Environment Canada, Atmospheric Environment Service, Supply and Services, Canada, Ottawa, 323 pp.

Walder, J. S., Hallet, B. (1985). A theoretical model of the fracture of rock during freezing. *Geological Society of America, Bulletin*, **96**, 336–346.

Walder, J. S., Hallet, B. (1986). The physical basis of frost weathering: toward a more fundamental and unified perspective. *Arctic and Alpine Research*, **18**, 27–32.

Walker, D. A., Walker, M. D., Everett, K. R., Weber, P. J. (1985). Pingos of the Prudhoe Bay region, Alaska. *Arctic and Alpine Research*, **17**, 321–336.

Walker, H. J. (1974). The Colville River and the Beaufort Sea: some interactions. In: Reed, J. C., Sater, J. E., eds., *The Coast and Shelf of the Beaufort Sea*. The Arctic Institute of North America, Washington, DC, pp. 513–540.

Walker, H. J. (1978), Lake tapping in the Colville River delta. In: *Permafrost, Proceedings of the Third International Conference on Permafrost*, 10–13 July, Edmonton, Alberta, Canada. National Research Council of Canada, Ottawa, vol. 1, pp. 233–238.

Walker, H. J. (1983). Erosion in a permafrost-dominated delta. In: *Permafrost, Proceedings of the Fourth International Conference on Permafrost*, 17–22 July, Fairbanks, Alaska. National Academy Press, Washington, DC, pp. 1344–1349.

Walker, H. J. (1991). Bluff erosion at Barrow and Wainwright, Arctic Alaska. *Zeitschrift für Geomorphologie*, suppl. **81**, 53–61.

Walker, H. J. (1998). Arctic deltas. *Journal of Coastal Research*, **14**, 718–738.

Walker, H. J. (2005a). Arctic, coastal geomorphology. In: Schwartz, M. L., ed., *Encyclopaedia of Coastal Science*, Springer Verlag, New York, 49–55.

Walker, H. J. (2005b). Ice-bordered casts. In: Schwartz, M. L., ed., *Encyclopaedia of Coastal Science*, Springer Verlag, New York, 542–545.

Walker, H. J., Arnborg, L. (1966). Permafrost ice wedge effect on riverbank erosion, In: *Proceedings, 1st International Permafrost Conference*. National Academy of Science–National Research Council, Publication 1287, pp. 164–171.

Wallace, R. E. (1948). Cave-in lakes in the Nebesna, Chisana and Tanana river valleys, eastern Alaska. *Journal of Geology*, **56**, 171–181.

Walters, J. C. (1978). Polygonal patterned ground in central New Jersey. *Quaternary Research*, **10**, 42–54.

Walters, J. C. (1994) Ice-wedge casts and relict polygonal patterned ground in north-east Iowa, USA. *Permafrost and Periglacial Processes*, **5**, 269–282.

Wang, B., French, H. M. (1994). Climate controls and high-altitude permafrost, Qinghai-Xizang (Tibet) Plateau, China. *Permafrost and Periglacial Processes*, **5**, 87–100.

Wang, B., French, H. M. (1995a). Frost heave and its implications for patterned ground, Tibet Plateau, China. *Arctic and Alpine Research*, **27**, 337–344.

Wang, B., French, H. M. (1995b). In-situ creep of frozen soil, Fenghuo Shan, Tibet Plateau, China. *Canadian Geotechnical Journal*, **32**(3), 545–552.

Wang, B., French, H. M. (1995c). Permafrost on the Tibet Plateau, China. *Quaternary Science Reviews*, **14**, 255–274.

Wang, S. (1990). Thaw slides and subsidence in Fenghuo Shan area, Qinghai-Xizang Plateau. *Chinese Journal of Glaciology and Geocryology*, **12**, 63–70 (in Chinese).

Wang, S. (1993). Permafrost changes along the Qinghai-Xizang Highway during the last decades. *Arid Land Geography*, **16**, 1–8 (in Chinese).

Wankiewicz, A. (1984). Analysis of winter heat flow in an ice-covered Arctic stream. *Canadian Journal of Civil Engineering*, **11**, 430–443.

Washburn, A. L. (1956). Classification of patterned ground and review of suggested origins. *Bulletin, Geological Society of America*, **67**, 823–865.

Washburn, A. L. (1967). Instrumental observations on mass wasting in the Mesters Vig District, Northeast Greenland. *Meddelelser om Gronland*, **166**, 318 pp.

Washburn, A. L. (1969). Weathering, frost action and patterned ground in the Mesters Vig District, Northeast Greenland. *Meddelelser om Gronland*, **176**, 303 pp.

Washburn, A. L. (1979). *Geocryology: A Survey of Periglacial Processes and Environments*. Edward Arnold, London, 406 pp.

Washburn, A. L. (1983a). Palsas and continuous permafrost. In: *Permafrost, Proceedings of the Fourth International Conference on Permafrost*, 17–22 July, Fairbanks, Alaska. National Academy Press, Washington, DC, pp. 1372–1377.

Washburn, A. L. (1983b). What is a palsa? In: Poser, H., Schunke, E., eds., *Mesoformen des reliefs im heutigen Periglazialraum*. Abhandlungen der Akademie der Wissenschaften in Gottingen, Math.-Phys. Klasse, no. 35, pp. 34–47.

Washburn, A. L. (1989). Near-surface soil displacement in sorted circles, Resolute Area, Cornwallis Island, Canadian High Arctic. *Canadian Journal of Earth Sciences*, **25**, 941–955.

Washburn, A. L. (1997). Plugs and plug circles: a basic form of patterned ground, Cornwallis Island, Arctic Canada – origin and implications. Geological Society of America, memoir 190, 87 pp.

Washburn, A. L., Smith, D. D., Goddard, R. H. (1963). Frost cracking in a middle-latitude climate. *Biuletyn Peryglacjalny*, **12**, 175–189.

Waters, R. S. (1962). Altiplanation terraces and slope development in west-Spitsbergen and southwest England. *Biuletyn Peryglacjalny*, **11**, 89–101.

Watson, E. (1977). The periglacial environment of Great Britain during the Devensian. *Philosophical Transactions of the Royal Society of London*, **B280**, 183–198.

Watt, A. S., Perrin, R. M. S., West, R. G. (1966). Patterned ground in Breckland: structure and composition. *Journal of Ecology*, **54**, 239–258.

Watts, S. H. (1983). Weathering pit formation in bedrock near Cory Glacier, southeastern Ellesmere Island, Northwest Territories. *Geological Survey of Canada*, paper 83-1A, 487–491.

Wayne, W. J. (1967). Periglacial features and climatic gradient in Illinois, Indiana, and western Ohio. In: Cushing, E. J., Wright, H. E. Jr., eds., *Quaternary Palaeoecology*. Yale University Press, New Haven, CT, pp. 393–414.

Wayne, W. J. (1991). Ice-wedge casts of Wisconsinan age in eastern Nebraska. *Permafrost and Periglacial Processes*, **2**, 211–233.

Weaver, J. (2003). Assessment of sub-permafrost groundwater conditions at the Red Dog Mine, Alaska. In: Phillips, M., Springman, S. A., Arenson, L. U., eds., *Permafrost, Proceedings of the Eighth International Conference on Permafrost*, 21–25 July, Zurich, Switzerland. Balkema, Lisse, vol. 2, pp. 1223–1228.

Weaver, J., Kulas, J. (2003). Development of a numerical model to assess the potential long-term thermal impact on permafrost temperatures at the Red Dog Mine, Alaska. In: Phillips, M., Springman, S. A., Arenson, L. U., eds., *Permafrost, Proceedings of the Eighth International Conference on Permafrost*, 21–25 July, Zurich, Switzerland. Balkema, Lisse, vol. 2, pp. 1229–1233.

Wedel, J. H., Thorne, G. A., Baracos, P. C. (1977). Site-intensive hydrologic study of a small catchment on Bathurst Island (interim report 1976). Hydrologic Regimes Freshwater Project No 1 (FP-1-76-1). Ottawa, Inland Waters Directorate, Environment Canada.

Weeks, A. G. (1969). The stability of slopes in south-east England as affected by periglacial activity. *Quarterly Journal Engineering Geology*, **5**, 223–241.

Weeks, W. S. (1920). Thawing frozen gravel with cold water. *Mineralogical Science Press*, **120**, 367–370.

Wegmann, M., Keusen, H. R. (1998). Recent geophysical investigations at a high alpine permafrost construction site in Switzerland. In: Lewkowicz, A. G., Allard, M., eds., *Permafrost, Proceedings of the Seventh International Conference on Permafrost*, 23–27 June, Yellowknife, Canada. Centre d'études nordiques, Université Laval, Québec, Collection Nordicana no. 57, pp. 1119–1123.

Weigand, G. (1965). Fossile pingos in Mitteleuropa. *Wurzburger Geographische Arbeiten*, **16**, 152 pp.

Wein, R. W. (1976). Frequency and characteristics of Arctic tundra fires. *Arctic*, **29**, 213–222.

Wein, R. W., Bliss, L. C. (1973). Changes in Arctic *Eriophorum* tussock communities following fire. *Ecology*, **54**, 845–852.

Wein, R. W., Shilts, W. W. (1976). Tundra fires in the District of Keewatin. *Geological Survey of Canada*, paper 76-1A, 511–515.

Weller, G., Anderson, P. A., eds. (1998). *Implications of global change in Alaska and the Bering Sea region. Proceedings of a Workshop*, June 1997, Center for Global Change and Arctic System Research, University of Alaska, Fairbanks, Alaska, 157 pp.

Weller, G., Lange, M. (1999). Impacts of global climate change in the Arctic regions: an initial assessment. Discussion Paper, Workshop on the Impacts of Global Change, 25–26 April, 1999, Tromso, Norway, International Arctic Science Committee, Oslo, Norway, 30 pp.

Weller, M. W., Derksen, D. V. (1979). The geomorphology of Teshepuk Lake in relation to coastline configuration of Alaska's coastal plain. *Arctic*, **32**, 152–160.

Wernecke, L. (1932). Glaciation, depth of frost and ice veins at Keno Hill and vicinity. *Engineering and Mining Journal*, **133**, 33–43.

Werner, B. T., Hallet, B. (1993). Numerical simulation of self-organized stone stripes. *Nature*, **361**, 142–145.

West, R. G. (1977). *Pleistocene Geology and Biology*, 2nd edn. Longman, London, 440 pp.

Westgate, J., Preece, S. J., Froese, D. G., Walter, R. C., Sandu, A. S., Schweger, C. E. (2001). Dating Early and Middle (Reid) Pleistocene glaciations in central Yukon. *Quaternary Research*, **56**, 288–306.

Westin, B., Zuidhoff, F. S. (2001). Ground-thermal conditions in a frost-crack polygon, a palsa and a mineral palsa (lithalsa) in the discontinuous permafrost zone, Northern Sweden. *Permafrost and Periglacial Processes*, **12**, 325–335.

Whalley, W. B., Azizzi, F. (1994). Rheological models of active rock glaciers: evaluation, critique and a possible test. *Permafrost and Periglacial Processes*, **5**, 37–51.

Whalley, W. B., Martin, H. E. (1994). Rock glaciers in Trollaskagi: their origin and climatic significance. In: Stotter, J., Wilhelm, F., eds., *Environmental change in Iceland*, Munchener Geographische Abhandlungen, Band B12, 289–308.

Wheaton, E. E., Singh, T. (1988). Exploring the implications of climatic change for the boreal forest and forestry economics of western Canada. *Climate Change Digest*, 89-02, Environment Canada, 18 pp.

Whintle, A. G. (1990). A review of current research on TL dating of loess. *Quaternary Science Reviews*, **9**, 385–397.

White, S. E. (1971). Rock glacier studies in the Colorado Front Range, 1961 to 1968. *Arctic and Alpine Research*, **3**, 43–64.

White, S. E. (1976). Is frost action really only hydration shattering? *Arctic and Alpine Research*, **8**, 1–6.

Whiteman, C. A., Kemp, R. A. (1990). Pleistocene sediments, soils, and landscape evolution at Stebbing, Essex, England. *Journal of Quaternary Science*, **5**, 145–161.

Whitney, M. I., Dietrich, R. V. (1973). Ventifact sculture by windblown dust. *Geological Society of America Bulletin*, **84**, 1561–2582.

Williams, M. W., Knauf, M., Caine, N., Liu, F., Verplanck, P. L. (2006). Geochemistry and source waters of rock glacier outflow, Colorado Front Range. *Permafrost and Periglacial Processes*, **17**, 13–34.

Williams, P. J. (1961). Climatic factors controlling the distribution of certain frozen ground phenomena. *Geografiska Annaler*, **43**, 339–347.

Williams, P. J. (1968). Ice distribution in permafrost profiles. *Canadian Journal of Earth Sciences*, **5**, 1381–1386.

Williams, P. J. (1976). *Volume change in frozen soils. Laurits Bjerrum Memorial Volume.* Norwegian Geotechnical Insitute, Oslo, pp. 233–346.

Williams, P. J. (1977). General properties of freezing soils. In: Williams, P. J., Fremond, M., eds., *Soil Freezing and Highway Construction*. Carleton University, Ottawa, 702 pp.

Williams, P. J. (1979). *Pipelines and Permafrost: Physical Geography and Development in the Circumpolar North*. Longman, London, 98 pp.

Williams, P. J. (1986). *Pipelines and Permafrost. Science in a Cold Climate*. Carleton University Press, Ottawa, 137 pp.

Williams, P. J., Smith, M. W. (1989). *The Frozen Earth. Fundamentals of Geocryology*. Cambridge University Press, Cambridge, 306 pp.

Williams, R. B. G. (1964). Fossil patterned ground in eastern England. *Biuletyn Peryglacjalny*, **14**, 337–349.

Williams, R. B. G. (1987). Frost weathered mantles on the Chalk. In: Boardman, J., ed., *Periglacial Processes and Landforms in Britain and Ireland*. Cambridge University Press, Cambridge, pp. 127–133.

Williams, R. B. G., Robinson, D. A. (2001). Experimental frost weathering of sandstone by various combinations of salts. *Earth Surface Processes and Landforms*, **26**, 811–818.

Winkler, E. M. (1977). Insolation warmed over: comment and reply. *Geology*, **5**, 188–190.

Wolfe, P. E. (1953). Periglacial frost–thaw basins in New Jersey. *Journal of Geology*, **61**, 113–141.

Wolfe, S. A. (1998). Living with frozen ground. A field guide to permafrost in Yellowknife, Northwest Territories. *Geological Survey of Canada*, miscellaneous report 64, 71 pp.

Woo, M.-K. (1986). Permafrost hydrology in North America. *Atmosphere-Ocean*, **24**(3), 201–234.

Woo, M.-K. (1990). Permafrost hydrology. In: Prowse, T. D., Ommanney, C. S. L., eds., *Northern Hydrology: Canadian Perspectives*. National Hydrology Research Institute, science report 1, Environment Canada, pp. 63–76.

Woo, M.-K. (1991). Arctic streamflow. In: Woo, M.-K., Gregor, D. J., eds., Arctic environment: Past, Present & Future, Department of Geography, McMaster University, Hamilton, Ontario, 105–111.

Woo, M.-K., Sauriol, J. (1980). Channel development in snow-filled valleys, Resolute, NWT, Canada. *Geografiska Annaler*, **62A**, 37–56.

Woo, M.-K., Steer, P. (1982). Occurrence of surface flow on arctic slopes, southwestern Cornwallis Island. *Canadian Journal of Earth Sciences*, **19**, 2368–2377.

Woo, M.-K., Steer, P. (1983). Slope hydrology as influenced by thawing of the active layer, Resolute, NWT. *Canadian Journal of Earth Sciences*, **20**, 978–986.

Woo, M.-K., Xia, Z. (1995). Suprapermafrost groundwater seepage in gravelly terrain, Resolute, NWT, Canada. *Permafrost and Periglacial Processes*, **6**, 57–72.

Woo, M.-K., Edlund, S. A., Young, K. L. (1991). Occurrence of early snow-free zones on Fosheim Peninsula, Ellesmere Island, Northwest Territories. *Geological Survey of Canada*, paper 91-1B, 9–14.

Woo, M.-K., Lewkowicz, A. G., Rouse, W. R. (1992). Response of the Canadian permafrost environment to climatic change. *Physical Geography*, **13**, 287–317.

Wood, B. L. (1969). Periglacial tor topography in southern New Zealand. *New Zealand Journal of Geology and Geophysics*, **12**, 361–375.

Worsley, P., Gurney, S. D. (1996). Geomorphology and hydrogeological significance of the Holocene pingos in the Karup Valley area, Traill Island, northern East Greenland. *Journal of Quaternary Science*, **11**, 249–262.

Wright, H. E. (1961). Late-Pleistocene climate of Europe; a review. *Bulletin, Geological Society of America*, **72**, 933–984.

Xing, Z., Wu, X., Qu, R. (1980). Determination of the ancient permafrost table, based on the variation in the content of clay minerals. *Bingchuan Dongtu*, **2**, 39–44 (in Chinese; English translation: 1984, National Research Council of Canada, technical translation 253, pp. 231–240).

Yatsu, E. (1988). *The Nature of Weathering*. Sozosha, Tokyo, 624 pp.

Yershov, E. D. (1990). *Obshcheya Geokriologiya*. Nedra, Moscow. (English translation, Williams, P. J., ed., 1998, *General Geocryology*, Cambridge University Press, Cambridge, 580 pp.).

Yoshikawa, K. (1993). Notes on open-system pingo ice, Adventdalen, Spitsbergen. *Permafrost and Periglacial Processes*, **4**, 327–334.

Yoshikawa, K., Harada, K. (1995). Observations on nearshore pingo growth, Adventdalen, Spitsbergen. *Permafrost and Periglacial Processes*, **6**, 361–372.

Young, A. (1972). *Slopes*. Oliver and Boyd, Edinburgh, 288 pp.

Young, A. (1974). The rate of slope retreat. In: Brown, E. H., Waters, R. S., eds., *Progress in Geomorphology – Papers in Honour of David L. Linton*, Institute of British Geographers, special publication 7, 65–78.

Zamolodchikov, Kotov, A. N., Karelin, D. V., Razzhivin, V. Yu. (2004). Active-layer monitoring in Northeast Russia: spatial, seasonal and interannual variability. *Polar Geography*, **28**, 286–307.

Zazula, G. D., Froese, D. G., Schweger, C. E., Mathewes, R. W., Beaudoin, A. B., Telka, A. M., Harington, C. R., Westgate, J. A. (2003). Ice-age steppe vegetation in east Beringia. *Nature*, **423**, 603.

Zazula, G. D., Froese, D. G., Westgate, J. A., Farge, C. La, Mathewes, R. W. (2005). Palaeoecology of Beringian "packrat" middens from central Yukon Territory, Canada. *Quaternary Research*, **63**, 189–198.

Zhou, Y., Guo, D. (1982). Principal characteristics of permafrost in China. *Journal of Glaciology and Geocryology*, **4**, 1–19 (in Chinese).

Zhou Youwu, Guo Dongxin, Qiu Guoqing, Cheng Guodong, Li Shude (2000). *Geocryology in China*. Cold and Arid Regions Environmental and Engineering Research Institute (CAREERI), Chinese Academy of Sciences, 450 pp. (in Chinese).

Zoltai, S. C. (1971). Southern limit of permafrost features in peat landforms, Manitoba and Saskatchewan. Geological Association of Canada, special paper no. 9, pp. 305–310.

Zoltai, S. C. (1972). Palsas and peat plateaus in Central Manitoba and Saskatchewan. *Canadian Journal of Forest Research*, **2**, 291–302.

Zoltai, S. C. (1973). Vegetation, surficial deposits and permafrost relationships in the Hudson Bay lowlands. In: *Proceedings, Symposium on the Physical Environment of the Hudson Bay Lowlands*, University of Guelph, pp. 17–34.

Zoltai, S. C. (1983). Pingos on Bylot island, NWT, Canada. *Polarforschung*, **53**, 43–48.

Zoltai, S. C., Pettapiece, W. W. (1973). Terrain, vegetation and permafrost relationships in the northern part of the Mackenzie Valley and Northern Yukon. Environmental-Social Committee Northern Pipelines, Task Force on Northern Oil Development, Information Canada, Ottawa, report 73-4, 105 pp.

Zoltai, S. C., Tarnocai, C. (1974). Soils and vegetation of hummocky terrain. Environmental-Social Committee Northern Pipelines, Task Force on Northern Oil Development, Information Canada, Ottawa, report 74-5, 86 pp.

Zoltai, S. C., Tarnocai, C. (1975). Perennially frozen peatlands in the western Arctic and Subarctic of Canada. *Canadian Journal of Earth Sciences*, **12**, 28–43.

Zoltai, S. C., Tarnocai, C., Pettapiece, W. W. (1978). Age of cryoturbated organic materials in earth hummocks from the Canadian Arctic. In *Permafrost, Proceedings of the Third International Conference on Permafrost*, 10–13 July, Edmonton, Alberta, Canada. National Research Council of Canada, Ottawa, vol. 1, pp. 325–331.

Zuidhoff, F. S., Kolstrup, E. (2000). Changes in palsa distribution in relation to climate change in Laivadalen, Northern Sweden. *Permafrost and Periglacial Processes*, **11**, 55–69.

Index

Note: because of frequent repetition, Arctic, sub-arctic, northern Canada, western Canadian Arctic, Europe, and North America are not indexed.

The Periglacial Environment, Third Edition   Hugh M French
Copyright © 2007 John Wiley & Sons Ltd

Lightning Source UK Ltd.
Milton Keynes UK
UKOW022308190413

209501UK00004B/73/P